ARENA2036

Series Editor

ARENA2036 e.V., ARENA2036 e.V., Stuttgart, Germany

Die Buchreihe dokumentiert die Ergebnisse eines ambitionierten Forschungsprojektes im Automobilbau. Ziel des Projekts ist die Entwicklung einer nachhaltigen Industrie 4.0 und die Realisierung eines Technologiewandels, der individuelle Mobilität mit niedrigem Energieverbrauch basierend auf neuartigen Produktionskonzepten realisiert. Den Schlüssel liefern wandlungsfähige Produktionsformen für den intelligenten, funktionsintegrierten, multimateriellen Leichtbau. Nachhaltigkeit, Sicherheit, Komfort, Individualität und Innovation werden als Einheit gedacht.

Wissenschaftler verschiedener Disziplinen arbeiten mit Experten und Entscheidungsträgern aus der Wirtschaft auf Augenhöhe zusammen. Gemeinsam arbeiten sie unter einem Dach und entwickeln das Automobil der Zukunft in der Industrie 4.0.

More information about this series at http://www.springer.com/series/16199

Philipp Weißgraeber · Frieder Heieck ·
Clemens Ackermann
Editors

Advances in Automotive Production Technology – Theory and Application

Stuttgart Conference on Automotive Production (SCAP2020)

Springer Vieweg

Editors
Philipp Weißgraeber
ARENA2036 e.V.
Stuttgart, Germany

Frieder Heieck
ARENA2036 e.V.
Stuttgart, Germany

Clemens Ackermann
ARENA2036 e.V.
Stuttgart, Germany

ISSN 2524-7247 ISSN 2524-7255 (electronic)
ARENA2036
ISBN 978-3-662-62961-1 ISBN 978-3-662-62962-8 (eBook)
https://doi.org/10.1007/978-3-662-62962-8

Responsible Editor: Dr. Alexander Grün
This Springer Vieweg imprint is published by the registered company Springer-Verlag GmbH, DE part of Springer Nature.
The registered company address is: Heidelberger Platz 3, 14197 Berlin, Germany

Editorial

Clemens Ackermann, Philipp Weißgraeber, and Frieder Heieck
Reseach Campus ARENA2036, Stuttgart, Germany
clemens.ackermann@arena2036.de

Stuttgart Conference on Automotive Production: Advances in Automotive Production Technology – Theory and Application

Mobility as well as the production of its means currently undergoes the vastest changes since Henry Ford introduced the moving assembly line for its Model T in 1908. Today, the very industry that produces interconnected automobiles sees itself constantly confronted with questions regarding interconnected and smart production systems, with the necessity of an increasingly rapid incorporation of various enabling technologies, and issues of data management & interoperability. It does not come as a surprise then that there is a promising intersection of product and production technologies, at which the intelligent product becomes part of the production process already. Vice versa, an intelligent product has all the technical requirements to inform production over the course of its entire life-cycle whilst simultaneously benefiting from the data produced by every single comparable vehicle; i.e. the "fleet-intelligence" informs both product and production.

Now, the practical questions that arise from the above stated hypotheses are obviously manifold. And, more importantly, not to be answered or solved by any single researcher, developer, or disruptive inventor. What they actually require is the exchange of solution approaches and expert knowledge as well as a practical take on collaboratively answering some of the more pressing issues.

The successor to last year's "Stuttgarter Tagung zur Zukunft der Automobilproduktion"[1], namely, the *Stuttgart Conference on Automotive Production (SCAP2020)* set out to be a forum that would not only allow for the exchange of concepts and ideas but also for very specific answers within precisely

[1]Stuttgart Congress on the Future of Automobile Production.

defined solution spaces. The framework in which all contributions of the conference would operate was defined by the questions mentioned at the beginning and given the following headline: *Advances in Automotive Production Technology – Theory and Application.*

The SCAP2020, organized by ARENA2036 in collaboration with Fraunhofer IPA, University of Stuttgart, Startup Autobahn powered by Plug and Play and IEEE TEMS, has proven to be a stimulating forum for researches from the sciences, the industry, and startups allowing every participant to learn about important current trends, gain insights regarding the overall research landscape, and to find ways in which a transfer from theoretical approaches to practical applications becomes feasible.

Every single contribution up for discussion was peer-reviewed by either members of the scientific committee comprised of 19 international experts or by individual domain experts for specific subject matters. Accordingly, and in order to ensure the scientific quality of the conference in general and of this volume in particular, the organizing committee of the SCAP2020 was in the position to choose the contributions to the conference from a far larger number of submissions.[2]

The contributions in this volume are arranged thematically in four parts, allowing the readers to choose their fields of interest from a broad range of automotive production technologies. Part A focusses on *Novel Approaches for Efficient Production and Assembly Planning*, Part B on *Smart Production Systems and Data Services*, Part C discusses *Advances in Manufacturing Processes and Materials*, and Part D presents *New Concepts for Autonomous, Collaborative Intralogistics.*

Now, we would also like to thank everyone involved in planning and running the conference, as well as all the contributors to and attendees of the conference – especially Dr. Jörg Burzer, Rainer Brehm, Prof. Dr. Thomas Bauernhansl, and Prof. Dr. Soumaya Yacout for their inspiring and insightful keynotes.

Finally, we would like to invite you to stay in touch with ARENA2036, to stay tuned for SCAP2022, and to enjoy the following papers.

Stuttgart
11/30/2020

Philipp Weißgraeber
Frieder Heieck
Clemens Ackermann

[2]This book includes contributions submitted directly by the respective authors. The editors cannot assume responsibility for any inaccuracies, comments, and opinions.

Contents

Part C Advances in Manufacturing Processes and Materials

Part A New Approaches for Efficient Production and Assembly Planning

Agile Hybrid Assembly Systems: Bridging the Gap Between Line and Matrix Configurations

Amon Göppert[1(✉)], Esben Schukat[1], Peter Burggräf[1,2], and Robert H. Schmitt[1]

[1] Laboratory for Machine Tools and Production Engineering (WZL), RWTH Aachen University, Campus Boulevard 30, 52074 Aachen, Germany
a.goeppert@wzl.rwth-aachen.de

[2] Chair of International Production Engineering and Management (IPEM), University of Siegen, Paul-Bonatz-Straße 9-11, 57076 Siegen, Germany

Abstract. The ongoing transition towards electro-mobility requires an increased reactivity and reconfigurability in automotive assembly. However, the traditional line assembly, which is characterized by rigid cycle times and linear product flow, has already been pushed to its flexibility limits. Drivers are the increase of product changes, variants and derivatives within assembly lines. To further increase reactivity and reconfigurability, matrix structured assembly configurations are a possible solution. Several studies highlight the theoretical advantages, but it has not been applied and validated in industrial use-cases, due to the high transformational gap between line and matrix configurations. In contrast, segment-wise line-less structures show a high potential for this.

A use-case oriented approach improves reactivity and reconfigurability by implementing an agile hybrid assembly system that combines the advantages of line and matrix structured (also referred to as line-less) assembly systems and offers a lower investment threshold. Three fields of action are presented: The first consists of flexible planning and control software modules. Within the planning phase, an automated scenario analysis is performed for optimization by applying simulations. During the production phase, the simulated model is re-used for the operation of a dynamical multi-agent manufacturing execution system with online scheduling algorithms. The second field of action deals with reconfigurable infrastructures, which comprises short-term dispatching intralogistics and a flexible layout, facilitated by AGV transport routes and reconfigurable self-adaptive workstations. The third field of action comprises a system model that is an underlying fully integrated digital twin. Control interfaces integrate the infrastructure into the manufacturing execution system to enable rapid system changes.

The presented hybrid system contributes to the design of future assembly systems by showing which aspects of line and matrix configuration can be combined to have a beneficial impact on a broad spectrum of production scenarios. By considering all relevant fields of action in a holistic way and by analyzing a hybrid configuration, the arising challenges for producing companies are addressed in a practical and functional manner.

P. Weißgraeber et al. (Eds.): *Advances in Automotive Production Technology – Theory and Application,* ARENA2036, pp. 3–11, 2021. https://doi.org/10.1007/978-3-662-62962-8_1

1 Introduction

The transition towards electro-mobility has a profound impact on the development of Original Equipment Manufacturers (OEMs) and the entire value chain of the automotive industry [1, 2]. German, American and Japanese OEMs are announcing over 80 new electric models for 2019/20 alone [3]. The parallel production of conventional, hybrid and purely electrically powered vehicles confronts OEMs with major challenges and the growing product variance on integrated assembly lines is leading to far-reaching efficiency losses [4]. In addition to the high variety of products OEMs are facing, product lifecycles are being shortened, making even more reconfigurations of the production line necessary [5]. As today's globalized society opens new markets for manufacturers, competition is increasing accordingly. A customizable product and efficient, cost saving manufacturing remains the best way to gain an edge over competitors and increase product value [6].

The stated trends are particularly evident for automotive assembly. Assembly has a significant impact on the value chain, accounting for 50% of production time and up to 20% of total costs [7, 8]. Since the final assembly will remain a core competence of OEMs in the future [9], novel strategies for the successful transformation of the industrial value-chain towards electro-mobility must take the design of assembly systems into account.

Currently, assembly systems for automotive production are designed for stable market environments and only a few changes at a time [10]. They are limited by fixed transfer systems (e.g. roll conveyors) and only very few buffers. To further increase reactivity and reconfigurability and thus meet future requirements, matrix structured assembly configurations (also referred to as line-less) present a promising solution [11, 12].

The basis for matrix structured assembly systems is the removal of the restrictions imposed by fixed transfer systems, enabling movements between different assembly stations [5]. However, due to the high transformation gap between line and matrix structured assembly systems, industrial applications have not yet reached a practical level [11]. Further, the full potential of matrix structured assembly systems can only be explored when the product´s precedence graphs contain a certain level of flexibility.

Accordingly it can be assumed, that an assembly system should contain both, elements from matrix and line configurations, creating a hybrid form. Thus, this paper presents a use case based design approach for hybrid assembly systems, which incorporate the advantages of both matrix and line structured assembly systems.

2 Theoretical Background

Matrix-structured assembly systems have been well studied and explored over the past years. However, there exists no uniform terminology and classification for the description of matrix-structured assembly systems yet. Thus, the following explanations are intended to highlight the most important characteristics in a cross-section manner.

The aim of matrix-structured assembly systems is to design a more flexible assembly system in comparison to line assembly, while maintaining the same efficiency and profitability [12]. Flexibility is achieved by decoupled assembly stations and assembly stations arranged in a matrix structure. This allows for a dynamical adjustment of assembly process sequences within the restrictions of the assembly precedence graph as required during operation. [13, 14]. The assembly sequence as well as the route of each job is not proactively planned and determined, but defined according to the availability of resources and other situational circumstances such as the availability of workers, station efficiency, transport times or even malfunctions at stations [13, 14]. The absence of a higher-level cycle time eliminates the need for assembly scheduling or line balancing [12, 15]. Sequence flexible assembly thus enables the realization of flow assembly with different cycle times or cycle-independent assembly stations, as well as the production of highly individualized products within the same assembly system [12]. A requirement for the operational feasibility is the existence of a real-time control system, e.g. based on multi-agent system [16].

Further advantages of the matrix structured assembly system are the scalability and reconfigurability. Scalability can be achieved by duplicating bottleneck resources at station or equipment level. Reconfigurability is realized by the modular design of the assembly stations as well as associated resources [17, 18]. When reaching a situational and near-real time adaptation of the assembly system, the term "agile assembly system" is used. The planning process is characterized by a comparatively later as well as smaller reduction in systemic degrees of freedom compared to line assembly [19].

All outlined aspects show that the tasks of planning and controlling matrix structured assembly systems are increasingly merging [20]. In case of strong restrictions such as limited flexibility of the precedence graph or space availability, it is sensible to transfer only specific manufacturing segments into a matrix structure. This will reduce complexity as well as the transformation gap and costs. For these reasons, a framework for agile hybrid assembly systems is presented below, which addresses the segment-by-segment break-up of line structures both in terms of the relevant fields of action and the selection of potential production segments.

3 A Framework for Agile Hybrid Assembly Systems

The framework for an agile hybrid assembly system combines the advantages of line and matrix assembly systems. This way, the high efficiency and output of line configured assembly systems are expanded by the adaptability and flexibility of matrix structured system. Therefore, elements and principles from both configurations are considered for the design of an agile assembly system (see Fig. 1).

A boundary condition for efficiency is a production scenario that clearly shows potential caveats regarding key performance indicators e.g. adherence to due dates, utilization and reconfiguration cost. Such a production scenario could be the described parallel production of vehicles with various powertrain systems, which would result in an increasing complexity of tasks and planning efforts. In addition to the efficiency the profitability can be maintained. Operational costs, as one measure of the profitability, correlates with the system's efficiency. In addition, profitability includes investment costs, which need to be taken into account for a transition towards a matrix system. Thus, possible circumstances to maintain profitability are savings in operational costs, due to a higher system's efficiency in production scenario demanding for a flexible system.

One key enabler of a hybrid assembly system is the one directional flow used in line production. To dissolve bottlenecks, multifunctional assembly stations (i.e. stations capable of performing two or more assembly processes) can be duplicated und operated in parallel, a concept taken out of matrix structured assembly systems. Based on a simulation-based analysis of the required level of agility, it is determined which assembly stations should be duplicated, since highly efficient production segments can remain in the line configuration.

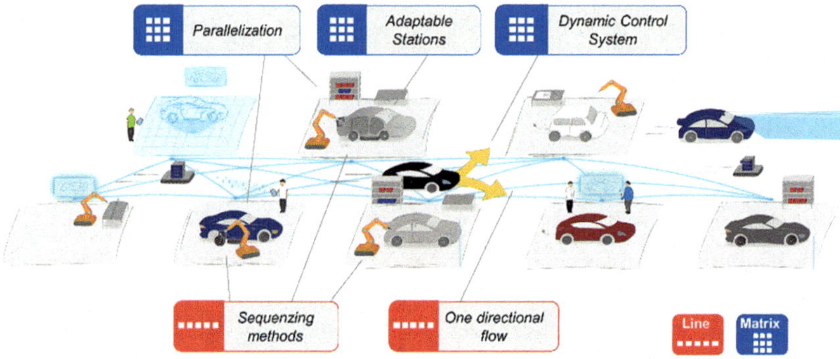

Fig. 1. Benefits of matrix and line configuration combined in hybrid assembly

To easily dissolve bottlenecks and allow for high utilization, stations must be highly adaptable. This includes the capability of stations to process multiple products and their variants. The utilization of the described flexibility requires the implementation of a control system. Various control architectures exist. A fully decentralized, autonomous system without a central control unit would be one implementation of a heterarchical architecture. Another approach would be a hierarchical control architecture, which is chosen when a set of tasks is required to be centralized. For an agile hybrid assembly system such a set of tasks demands for a hierarchical control architecture. The tasks are described in the following. The control system is responsible for the assignment of products to a specific work station. This is based on the product requirements and the work station abilities regarding the assembly operations. Also, it is responsible for the sequencing of assembly operations at the chosen work station. For these decisions, the control system may consider different factors such as the transport time, the redundancy of equipment at a work station or possible breakdowns at work stations. Since unforeseen changes on the shop floor can occur at any time, the control system needs to dynamically and frequently reassess decisions.

The framework for agile hybrid assembly systems adopts scheduling approaches for mix-model lines as they represented a validated method for optimizing the sequencing of orders. Since transport times are gaining considerable significance in matrix-structured assembly the scheduling approaches must be enhanced.

Operating a hybrid assembly system with maximum efficiency requires multiple components. These components can be grouped into three fields of action (see Fig. 2).

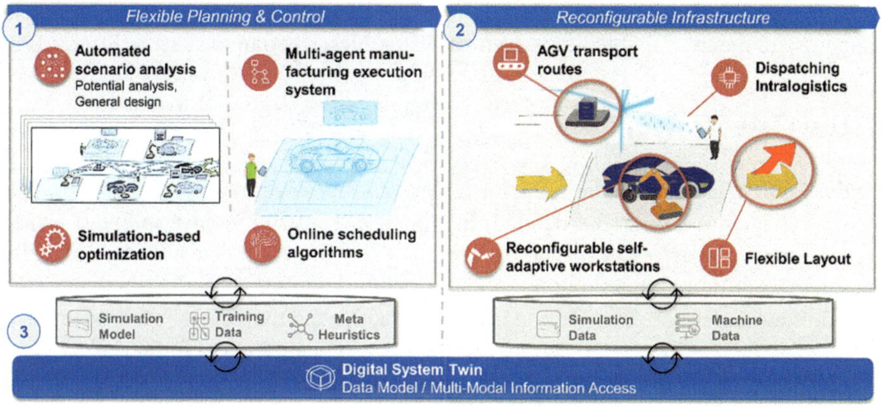

Fig. 2. System architecture and required technologies for an agile hybrid assembly system.

Flexible Planning and Control includes the before mentioned control system, also called the multi-agent manufacturing execution system, that coordinates every movement in the hybrid assembly system. The control system uses an online scheduling algorithm to assign each product its next process and the station that will carry out this process, planning an individual route for each job. For the planning phase of the hybrid assembly system, an automated scenario analysis is included. Its goal is the optimization of the production system by applying discrete event simulations (DES). Once production begins, the scenario analysis can be used to further improve production, analyzing data that was not available in the planning phase.

Reconfigurable Infrastructure enables the dynamic adjustment of production capacities. Autonomously reconfigurable workstations can adjust their capability profiles to handle an increasing and changing number of different processes. This makes a short-term dispatching intralogistics system crucial. The intralogistics system adapts to the flexible production layout and utilizes automated guided vehicles (AGVs) to ensure that all workstations receive necessary components and equipment for assembly. Although other transport vehicles can be used, AGVs are used as a representative vehicle form in this context. One feature of the system is the dynamic calculation of the AGV transport routes, reacting to sudden changes in production, like prioritization of certain jobs and breakdowns. Since the workstations are autonomously reconfigurable and AGVs can easily change routes, the infrastructure can be arranged in a flexible layout. This allows improvements if possible enhancements are uncovered during the simulation-based optimization process.

The underlying fully integrated **Digital System Twin** builds the connection between the first two fields of action as a structured and hierarchical data model. For the first field of action, the digital twin provides the data for training the online scheduling algorithms as well as the data for the simulation runs, done by the scenario analysis. To generate this information, the digital twin retrieves machine data from the reconfigurable infrastructure, e.g. movement information from the AGVs or processing times from the work stations. To solve optimization problems during the scenario analysis, meta-heuristics are made available to the control system described in the first field of action.

4 Use Case Development

For the implementation of the fields of action and their components described in the previous chapter, use cases are defined. In theory, a use case is a description of actions that a system can perform with the participation of actors. An actor can be any entity that interacts with a system: a user, another system, but also the physical environment of the system itself [21–23]. Thus, an actor can activate a use case of the system. This use case can then activate applications within the system or request further information from other actors. In this way, use cases enable the attainment of a defined goal for the respective actors by describing the functions of a system and the benefits for the actors involved [22].

In the context of this paper, examples of relevant actors are infrastructure, automated scenario analysis, a dynamic multi-agent manufacturing execution system, shop floor employees, orders and resources. Within the production structures, actors can, for example, have the option to evaluate the potential of a section-wise parallelization or initiate the corresponding restructuring. The use case oriented approach guarantees the practical feasibility and reduces the transformational gap of the agile hybrid assembly system. The use cases themselves are planned in brownfield and are thus aligned with the restrictions of existing production environments. They aim to solve the production challenges addressed by the components of each field of action (see Fig. 3).

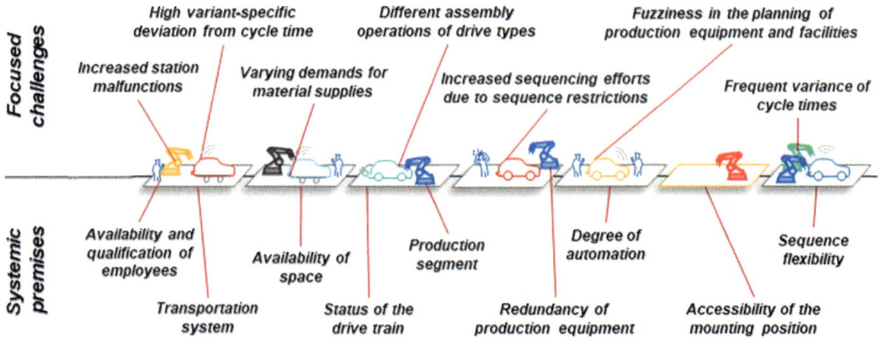

Fig. 3. Focused challenges and systemic premises for the use case development

A multi-stage procedure is applied for the collection and evaluation of the use cases. First, the three fields of action and each of their components are evaluated in an interdisciplinary project team regarding their possible integration into the current assembly environment. For this purpose, current structural improvement potentials of the assembly system, as well as assembly sections with restrictive and planning-intensive requirements are examined. The resulting integration concepts can then be consolidated in a list with specialist planners and evaluated using a qualitative criteria-based assessment of their potential. Selected integration concepts are then transferred into a detailed, standardized description, which include the basic functionalities, the systemic premises, the interrelationships of the systems and actors as well as description models for resources, processes and products. Those descriptions reflect the preliminary use cases and include alternative system configurations.

The preliminary use cases will then be transferred into a simulation model to further quantify their benefits. If a sufficient added value is proven, the relevant preliminary use cases need to be detailed with regard to their technological embedding and interaction in the existing system, e.g. the connection of the resources to the control system and concrete decision algorithms for decision making. The further development of the use cases is based on a hybrid planning strategy. This means that the development steps are divided into increments, which are further detailed either in a plan-driven way or developed in an agile way. By doing so, a late reduction of the degrees of freedom of the assembly system is guaranteed. This leads to a shorter development time by parallelizing work steps and also enables late modifications with little effort.

The necessary technological development requires cross-functional competencies and a close collaboration with the OEM companies. Parallel to this, the integration concept for the later system reconfiguration needs to be elaborated. This ensures that the necessary infrastructure and employee's competence are available in time for start of operation and that negative effects on the existing production system are minimized. By introducing use cases step-by-step the new components, e.g. the control system, can be tested and improved. Gaining experience with the concept will allow OEMs to apply the concept of matrix structured assembly on a bigger scale, integrating larger parts of the plant into the matrix, ultimately leading from hybrid manufacturing to a fully matrix configured assembly system, if reasonable. However, this is not always the ultimate goal. Some parts of production will always function best in line configuration, making hybrid assembly the most efficient manufacturing system in certain cases.

5 Conclusion

The presented hybrid system contributes to the design of future assembly systems by showing how aspects of line and matrix configurations can be combined to have a beneficial impact on a broad spectrum of production scenarios. By considering the relevant fields of action, i.e. flexible planning and control, reconfigurable infrastructure and digital system twin, in a holistic way and by analyzing a hybrid configuration, the arising challenges for producing companies are addressed in a practical and functional manner.

In addition to the presented fields of action an approach for the use-case development as a method for a practical implementation of an agile hybrid assembly system including the focused challenges and systemic premises was proposed. Further evaluation potentials would be the analysis of implemented use-cases regarding key performance indicators to achieve design guidelines for future implementations.

Acknowledgement. This work is part of the research project "AIMFREE" that is funded by the German Federal Ministry for Economic Affairs and Energy (BMWi) within the indirective on a joint funding initiative to fund research and development in the field of electromobility (funding number: 01MV19002A) and supported by the project management agency German Aerospace Center (DLR-PT). The authors are responsible for the content.

References

1. PWC: The turning of the tide. Impacts of the automotive transformation on the value chain. Study, PWC Autofacts (2018)
2. Seeberger, M.: Der Wandel in der Automobilindustrie hin zur Elektromobilität. Dissertation, Universität St. Gallen (2016)
3. Wirtschaftswoche: So viele E-Autos sind bis 2022 angekündigt. https://www.wiwo.de/unternehmen/auto/elektroautos-so-viele-e-autos-sind-bis-2022-angekuendigt/21262218.html. Accessed 27 Feb 2019
4. Kampker, A.: Elektromobilproduktion. Springer Vieweg, Berlin (2014)
5. Göppert, A., Hüttemann, G., Jung, S., Grunert, D., Schmitt, R.: Frei verkettete Montagesysteme. Ein Ausblick. ZWF **113**(3), 151–155 (2018)
6. Park, M., Yoo, J.: Benefits of mass customized products: moderating role of product involvement and fashion innovativeness. Heliyon **4**, 1–25 (2018)
7. Lotter, B.: Einführung. In: Lotter, B., Wiendahl, H.-P. (eds.) Montage in der industriellen Produktion. Ein Handbuch für die Praxis, 2nd edn., pp. 1–8. Springer Vieweg, Berlin (2012)
8. Hu, S.J., Ko, J., Weyand, L., ElMaraghy, H.A., Lien, T.K., Koren, Y., Bley, H., Chryssolouris, G., Nasr, N., Shpitalni, M.: Assembly system design and operations for product variety. In: CIRP Annals – Manufacturing Technology. 60th year, Nr. 2, pp. 715–733 (2011)
9. Deloitte: The Future of the Automotive Value Chain. 2025 and Beyond. Study, Deloitte (2017)
10. Hüttemann, G., Gaffry, C., Schmitt, R.: Adaptation of reconfigurable manufacturing systems for industrial assembly – review of flexibility paradigms, concepts, and outlook. Procedia CIRP **52**, 112–117 (2016)
11. Lettmann, P., Hüttemann, G., Schmitt, R.: Produktrouten in frei verketteten Montagesystemen. Ermittlung und Bewertung von Produktrouten mittels Merkmalsklassifizierung. ZWF **114**(9), 517–520 (2019)
12. Schönemann, M., Herrmann, C., Greschke, P., Thiede, S.: Simulation of matrix-structured manufacturing systems. J. Manuf. Syst. **421**, 1–25 (2015)
13. Hüttemann, G., Göppert, A., Lettmann, P., Schmitt, R.: Dynamically interconnected assembly systems – concept definition, requirements and applicability analysis. WGP-Jahreskongress **7**(1), 1–25 (2017)
14. Burggräf, P., Dannapfel, M., Adlon, T., Schukat, E., Kahmann, H., Holtwiesche, L.: Modeling and evaluating agile assembly systems using mixed-integer linear programming. In: 53rd CIRP Conference on Manufacturing Systems (2019)
15. Greschke, P.: Matrix-Produktion als Konzept einer taktunabhängigen Fließfertigung. Dissertation (2015)
16. Burggräf, P., Dannapfel, M., Adlon, T., Kahmann, H., Schukat, E., Holtwiesche, L.: Multi-agent systems in agile assembly – real-time scheduling of flexible operation sequences on multifunctional assembly stations. wt Werkstatt online **110**(4), 170–176 (2020)
17. Kampker, A., Bartl, M., Bertram, S., Burggräf, P., Dannapfel, M., Fischer, A., Grams, J., Knau, J., Kreisköther, K., Wagner, J.: Agile low-cost montage. In: Internet of Production für agile Unternehmen, pp. 231–259 (2017)
18. Kamper, A., Kreisköther, K., Wagner, J., Fluchs, S.: Mobile assembly of electric vehicles: decentralized, low-invest and flexible. In: 18th International Conference on Automotive and Mechanical Engineering. ICAME 2016, Sydney, Australia (2016)

19. Burggräf, P., Dannapfel, M., Adlon, T., Riegauf, A., Müller, K., Fölling, C.: Agile Montage - Montageplanung und -system als integrale Bestandteile der Fabrikplanung. **109**, 622–627 (2019)
20. Burggräf, P., Dannapfel, M., Adlon, T., Riegauf, A., Schukat, E., Schuster, F.: Optimization approach for the combined planning and control of an agile assembly system for electric vehicles. In: Nyhuis, P., Herberger, D., Hübner, M. (eds.) Proceedings of the 1st Conference on Production Systems and Logistics (CPSL 2020), pp. 137–146 (2020)
21. Jacobson, I., Ericsson, M., Jacobson, A.: The object advantage. Business process reengineering with object technology. Series: ACM Press books, Wokingham, England: Addison-Wesley (1995)
22. Bruegge, B., Dutoit, A.H.: Object-oriented software engineering. Using UML, patterns, and Java, 3rd edn. Prentice Hall, Boston (2010)
23. Jacobson, I.: Object-oriented software engineering. A use case driven approach. Repr. Addison-Wesley, Harlow (1992)

Economic Feasibility of Highly Adaptable Production Systems

Urs Leberle(✉) and Yannick-Léon Weigelt

Robert Bosch GmbH, Robert-Bosch-Campus 1, 71272 Renningen, Germany
Urs.Leberle@de.bosch.com

Abstract. An increasingly uncertain market environment, high product variety and shortened product life cycles lead to an increased demand for adaptable production systems. Due to higher initial investment costs, it becomes more difficult to assess the profitability of such production systems with conventional methods, since the advantages of adaptable production systems are not considered sufficiently. This article presents an approach allowing to determine the economic feasibility of highly adaptable production systems which are repeatedly undergoing reconfiguration processes to adapt to products, processes and technologies that are unknown during planning and launch. In contrast to others, this approach considers a preferably high level of adaptability enabling the production system to change extensively and quickly. To test the method a scenario from the publicly funded project Fluid Production is used.

1 Introduction and Motivation

An increasingly uncertain market environment, high product variety and shortened product life cycles lead to an increased demand for adaptable production systems. In highly adaptable production systems, production resources are no longer used exclusively for one product family or production process, but instead are reconfigured repeatedly adapting to products, processes and technologies that are unknown during planning and launch. Due to higher initial investment costs, it becomes more difficult to assess the profitability of such production systems with conventional methods, since the advantages of adaptable production systems are not considered sufficiently.

2 State of the Art

To evaluate long-term investment projects dynamic investment calculation methods such as the internal rate of return (IRR) and the net present value (NPV) are frequently used in the industry [1]. In contrast to static methods these approaches consider the time value of money by taking into account the time payments are made. In addition, the life-cycle costing (LCC) and total cost of ownership (TCO) make it possible to consider costs and revenues over all life phases of an investment. Unfortunately, the application of these presented methods lacks the possibility to consider the flexibility and adaptability of production systems [1].

P. Weißgraeber et al. (Eds.): *Advances in Automotive Production Technology – Theory and Application*, ARENA2036, pp. 12–19, 2021. https://doi.org/10.1007/978-3-662-62962-8_2

Previous research to determine and evaluate the economic efficiency of adaptable production systems focuses on determining the optimal reconfiguration potential [cf. 2–7]. This approach assumes that the principle of diminishing marginal utility also applies to the ability to change, hence cost and benefit are not linearly related [5]. The procedure is not suitable for highly adaptable production systems, since maximizing reconfigurability is a fundamental component of this production concept. It is considered necessary to operate sustainably in a highly volatile and uncertain production environment. The close involvement of humans as well as the dynamic and needs-based configuration are intended to reduce the cost of versatility in production [8].

Life-cycle-oriented assessments based on the VDMA34160 [9], which include not only procurement costs but also operating and disposal costs, are presented by Schweiger and Pachow-Frauenhofer [10, 11]. The resources of highly adaptable production systems are composed of individual modules that are solution-neutral and not linked to a specific product in order to minimize pre-determinations and complexity costs. Additionally, the individual modules and their respective composition is changed continuously which leads to difficulties calculating the system's service life, since each module has its own useful life. Therefore, neither the product life cycle nor the service life of the system can be used as a basis for cost considerations [12].

The uncertainty of future developments represents another challenge during the evaluation of adaptable production systems. Möller [4] applies the approach of the real option theory known from financial mathematics to the problem of determining the economic feasibility of reconfigurable production systems under uncertainty. This enables the time-dependent consideration of uncertainty, but due to the calculation effort only a few parameters can be considered. Since highly adaptable production systems intend to improve the ability to act in a particularly volatile market environment, many different parameters must be analyzed.

In summary, it can be stated that the determination of the optimal reconfiguration potential is not feasible for highly adaptable production systems, since the planning framework is too uncertain and the necessary adaptability depends strongly on the respective operating phases. Furthermore, the requirements for the assessment of a variable evaluation period and the consideration of short-term and dynamic changes of resources in production have so far hardly been taken into account. Ultimately, an appropriate approach needs to be developed to allow the monetary measurement over a variable observation period and consider uncertainty in the production environment.

3 Approach

To periodically allocate occurring costs during the use of a configuration the model shown in Fig. 1 was developed. The observation period can be freely selected. The incurring costs are determined based on a component-wise evaluation of residual values at the end of each period. The occurrence of an adaption leads to a reduction of the residual value if components of the system are no longer required. This procedure was chosen because within highly adaptable production system it is very likely that most components can be reused, thus minimizing the number of obsolete components. As an outcome of the economic evaluation and foundation for an investment decision the NPV was chosen. It is determined in six steps.

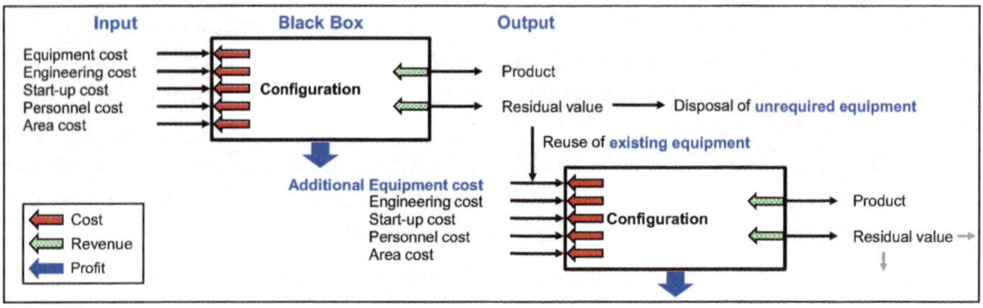

Fig. 1. Description model for highly adaptable production systems

In the first step, an analysis of possible production scenarios as well as the definition of general production conditions must be executed. The goal of the analysis is to determine key data like the annual quantity of units or product variants that are expected within the observation period. To set up a scenario funnel and to be able to consider possible future developments, worst- and best-case scenarios must be determined in addition to the forecast scenario [13, 14]. The production conditions include general production data like the shift model, the working days per year or the payment rate of workers.

The subject of the second step is the planning of the production system. This includes capacity planning by determining the production resources, such as type and number of machines and workstations, the linking in-between and the number of employees required for the production system in each period. The planning is based on the scenarios developed before, the required process technology and the assembly sequence.

In the next step the reconfiguration potential of the production system is determined according to Heger [2], allowing to estimate the share of components of a production system that can be adapted to new products, processes or technologies regarding certain conditions such as robot payload or dimensions of the assembly cell. However, Heger's method was reduced to essential aspects to evaluate the resources of a production system. The value of a plant object, such as a production resource or an entire production system, results from the sum of the individual normalized and weighted reconfiguration potential values of the system components of the object under consideration.

The fourth step involves the periodic compilation of costs arising in each period. This is done according to the LCC method presented within the VDMA 36160 guideline using the description model presented above. The costs of a period A_t consist of acquisition costs EK_t, operating costs BK_t and liquidation costs VK_t (see Eq. 1).

$$A_t = EK_t + BK_t + VK_t \tag{1}$$

The acquisition costs EK_t include investment costs for machinery equipment and tools as well as engineering and start-up costs. The operation costs BK_t comprise for example worker, area and energy costs. The liquidation costs VK_t consist of the disposal costs, the residual value of the production resources and other possible liquidation costs. Depending on the availability of data as well as the analyzed object and the degree of abstraction, the scope of considered costs can be adjusted as required. In the case of an adaption between two consecutive periods according to Stähr [6], the residual value RW_t results from the sum of the products of the reconfiguration potential values WP_y of the plant objects y and their present value $BW_{y,t}$ at the end of the corresponding period (see Eq. 2). If no adaption takes place between the individual periods, the residual value of the production resources RW_t is the sum of the book values of all plant objects used at the end of the period under consideration.

$$RW_t = \sum_{y=1}^{z} WP_y * BW_{y,t} \qquad (2)$$

In the fifth step, the NPV of the production system is calculated based on the costs occurring in each period. It results from the sum of all incoming and outgoing payments per period within the observation period, discounted to the time of consideration.

In the last step, the results of the evaluation method are to be checked for accuracy and stability by means of a local sensitivity analysis [15]. By examining the dependence of the planning variants on changes in the production environment or on assumptions made initially, the resulting investment decision can be secured.

4 Example of Application

The application of the developed method is demonstrated by comparing a fluid manufacturing system (FLMS) with a designated manufacturing line (DML) using an exemplary product and quantity scenario. FLMS can be specified as highly adaptable production systems characterized by the ability to adapt and change dynamically to cope with challenges from increasingly volatile markets.

The comparison is based on a simple demo product. The product is composed of a housing with cover, a printed circuit board (PCB) and a battery holder which is mounted in the housing. While the mounting of the PCB and the battery holder are automated, the remaining processes are carried out at a manual workstation. These steps include inserting the batteries into the holder, connecting the wire to the PCB, flashing the software, adjusting the integrated potentiometer and final testing and mounting the housing with the customer label attached. It is assumed that the product will be available in three variants within a period of ten years. For product variant A all components are fixed by screws in the housing. A second product variant B is launched replacing parts of the screwing process with a bonding process and with a faster flashing of the software to achieve shorter cycle times and cheaper process costs for high quantities. Later, enabled by a technological innovation, variant C is launched including a friction welding process to further enhance the mounting and an automatic adjustment of the integrated potentiometer. However, the production of previous variants must be continued for a certain time. The assumed scenario (see Fig. 3) results from the four periods of the economic cycle (expansion, boom, recession, depression) and other expected fluctuations. As general conditions for the production in Germany 17 shifts per week with 7 working hours each shift and 272 working days per year were assumed. The respective batch numbers to be produced were set to be constant at 3000 units for product variant A, 1000 units for variant B and 2500 units for variant C.

An automatic assembly cell in the form of the highly adaptable CESA3R system [16] as well as the modular manual working station Active-Assist of Bosch Rexroth and a flexible linking with intermediate buffers are used in the FLMS. The modular concept of the CESA3R system, consisting of mechatronic objects with standardized hard- and software interfaces, allows the removal, replacement or supplementation of individual processes or technologies in the assembly cell or of whole assembly cells in the production system. Only the characteristics of the base cell, like the dimensions or the robot payload limit the reconfiguration potential and the productivity of the CESA3R system. For operation and setup of these systems, a continuous adaption of qualification requirements is necessary [17]. The concept allows a simple and accelerated start-up, e.g. in the form of a software-assisted safety assessment concept [18], which requires no additional specialists. A gradual increase in output is achieved by the parallel linking of cells. Thereby the assembly is carried out on supplementary cells in the same steps.

For the DML an automatic station and two or three manual workstations (depending on the product variant) were planned. The assembly takes place in a sequential process with a serial linking by a belt driven transfer system. The product specific process equipment and the sequential assembly allow a high productivity resulting from minimal cycle and setup times as well as a simple operation by auxiliary stuff. However, the production of new variants of an existing product or new products requires complex reconfiguration or even new construction of mechanic and electric components as well as software. In addition, changes to individual stations or the assembly line require extensive re-commissioning and process approval by specialists. An increase in output beyond the maximal capacity can only be achieved by a second assembly line.

The assumed investment and operating costs, cycle and setup times are compared in Fig. 2. The capacity planning is based on the assumption that a new configuration of the production is always necessary when either a new product variant is launched or the maximum possible capacity utilization of the current configuration is exceeded.

	Configuration	Investment cost			Operating cost		Cycle time [s]			Setup time [min]		
		Equipment [€]	Engineering [€]	Start-up [€]	Personnel [€/P*a]	Area [€/m²*a]	Product			Product		
							A	B	C	A	B	C
FLMS	CESA^{3}R Screwing	100.000 €					96			30		
	CESA^{3}R Screwing + Bonding	120.000 €	21.385 €		96.300 €	540 €	96	76		30	60	
	CESA^{3}R Screwing + Friction welding	135.000 €					96		59	30		90
	Manual Workplace	25.000 €				291 €	83	78	68	0	0	0
DML	Screwing	225.000 €					48			30		
	Screwing + Bonding	260.000 €	30.662 €	10.621 €	70.600 €	1.328 €	48	31		30	60	
	Screwing + Friction welding	275.000 €					48		29	30		60

Fig. 2. Investment- and operating cost & cycle and setup time for DML and FLMS

Using the evaluation approach by Heger [2] the reconfiguration potential of the DML was rated with 30%. Due to reduced product commitment the FLMS configuration is highly adaptable but product specific requirements like the fixing equipment are limiting the reconfiguration potential at 90%.

Figure 3 shows the total costs of both production concepts and the quantity of the three product variants produced in each period of the example scenario. The initially lower costs of the FLMS result from the significantly smaller scaling steps per module. Combined with the faster start-up time this increases the degree of utilization of the FLMS for small piece numbers. On the other hand, at high production numbers many modules must be purchased and operated because of the required capacity. This reduces the economic efficiency of the FLMS with increasing quantities and explains its higher cost in the seventh period compared to the DML. The high costs of the DML in the third period result from an extraordinary depreciation that is incurred in this period because of the change of technology by the introduction of the friction welding process at the transition from the third to the fourth period. The FLMS can reuse most of the existing components which leads to reduced acquisition costs. Whereas the DML reaches its full potential in the seventh period due to optimal utilization it lacks the ability to adapt to the decrease in quantity in period eight. The FLMS concept can handle the changes in a more sufficient way and enables the production to operate sustainable even when the number of produced units is declining.

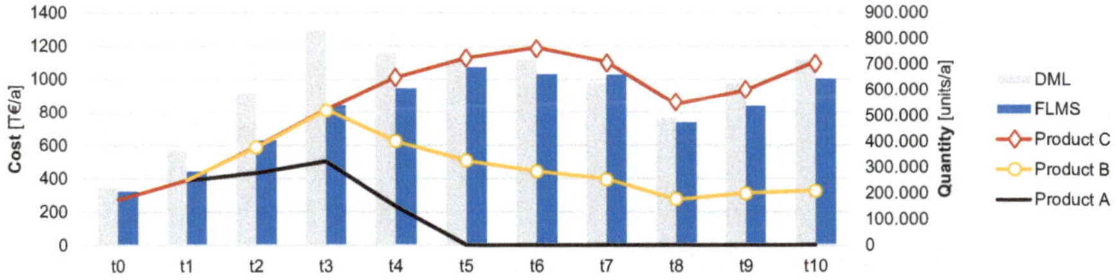

Fig. 3. Periodic cost analysis and accumulated quantity for DMS and FLMS

According to the periodic costs of this exemplary scenario, FLMS may be a more suitable alternative than DML. The accumulated NPV difference regarding an internal rate on return of 9% is 1.075.473 €. The difference in economic sustainability mainly results from the ability of FLMS to react more cost-efficient to fluctuations in quantity or the introduction of new products and technologies.

In the sensitivity analysis the reconfiguration potential value as well as the planning and start-up times were varied exemplarily, and a more intense development of the extreme scenarios was analyzed. It could be shown that the stability of the output variable is guaranteed in relation to the considered input variables. Nevertheless, the variation resulted in changes, which prove the influence of the selected input variables.

5 Discussion and Evaluation

The presented method allows to compare the economic feasibility of FLMS and DML. Due to the reconfiguration potential value determined according to Heger and the reconfiguration costs calculated therefrom according to Stähr, the adaption capability has a direct influence on the overall evaluation result. The developed description model allows the application of the LCC method for each individual period by describing it as a closed operating state with cost of acquisition, operation and liquidation. As a result, it is possible to consider the short-term and dynamic combination of resources in production systems and to analyze the profitability in an uncertain production environment over a variable observation period. Depending on requirements, the method can also be used to develop several scenarios with deviating forecasts, which can then be examined with the sensitivity analysis for their stability regarding varied input parameters. The determined NPVs can be compared in a results matrix to describe the situation under uncertainty. Depending on the risk tolerance of the management a suitable option can be chosen [19].

The application of the method to the exemplary product and quantity scenario resulted in the following findings for the comparison of FLMS and DML. The fast start-up time and the possibility of scaling in small steps increase the efficiency of the FLMS compared to the DML for low volumes significantly. This makes it possible to reduce the required capacities resulting in a reduction of the necessary acquisition costs. Even in the case of a technology or a product change, the individual acquisition costs are significantly lower for the FMLS than for the DML. On the other hand, the low scaling effect reduces the cost-effectiveness of FLMS at high volumes.

6 Summary and Outlook

The presented method allows to compare DML and FLMS during a selected observation period using the determined NPVs. The consideration of uncertainty in the production environment within the method is based on three scenarios determined in a scenario analysis (forecast, best- and worst-case) and a sensitivity analysis. Implementing methods for uncertainty evaluation like the real option theory, could lead to a more specific consideration of the aspect of uncertainty in the investment decision. But due to the complexity of the decision, the NPVs should not be the only valuated dimension of the investment decision. It is advisable to consider other factors that have a direct or indirect influence on the result. Possible factors are quality, working conditions or environmental impact.

The FLMS creates new degrees of freedom in the planning and operation of production. The so far only discrete adaption becomes a steady adaption and the solution space for possible adaptations is considerably larger due to short-term and dynamic combination of resources in production systems. Previously strategic decisions may become operational decisions. However, the additional degrees of freedom also go hand in hand with a much greater complexity of production. Digital planning tools could, for example, help to control the degrees of freedom and efficiently use the possibilities of FLMS.

Acknowledgements. The research presented in this paper has received partial funding under administration of the Project Management Agency (PTKA) inside the research campus ARENA2036. Our sincere thanks go to the Federal Ministry for Education and Research (BMBF) for supporting this research project by the grant agreement 02P18Q625.

References

1. Fechter, M., Dietz, T., Bauernhansl, T.: Cost calculation model for reconfigurable, hybrid assembly systems. In: 2019 IEEE 15th International Conference on Automation Science and Engineering (CASE), pp. 836–841
2. Heger, C.L.: Bewertung der Wandlungsfähigkeit von Fabrikobjekten. Dissertation, Univ., Hannover, 2006. PZH, Garbsen (2007) (Berichte aus dem IFA 2007, 1)

3. Wiendahl, H.-P., ElMaraghy, H.A., Nyhuis, P., Zäh, M.F., Duffie, N., Brieke, M.: Changeable manufacturing – classification, design and operation. CIRP Ann. **56**(2), 783–809 (2007)
4. Möller, N.: Bestimmung der Wirtschaftlichkeit wandlungsfähiger Produktionssysteme. Dissertation, Techn. Univ. Utz, München (2008) (Forschungsberichte IWB, 212)
5. Schuh, G., Harre, J., Gottschalk, S., Kampker, A.: Design for changeability (DFC): Das richtige Maß an Wandlungsfähigkeit finden. wt Werkstatttechnik online **94**(4), 100–106 (2004)
6. Stähr, T.: Methodik zur Planung und Konfigurationsauswahl skalierbarer Montagesysteme Ein Beitrag zur skalierbaren Automatisierung. Dissertation, Karlsruher Institut für Technologie, Karlsruhe (2020) (Shaker: wbk Institut für Produktionstechnik)
7. Eilers, J.: Methodik zur Planung skalierbarer und konfigurierbarer Montagesysteme. Dissertation, Techn. Univ., München. Utz, München (2008) (Forschungsberichte IWB 212)
8. Bauernhansl, T., Fechter, M., Dietz, T. (eds.): Entwicklung, Aufbau und Demonstration einer wandlungsfähigen (Fahrzeug-) Forschungsproduktion, 1st edn., pp. 1–4. Springer, Berlin (2020)
9. VDMA-Einheitsblatt VDMA34160: Prognosemodell für die Lebenszykluskosten von Maschinen und Anlagen
10. Schweiger, S. (ed.): Lebenszykluskosten optimieren: Paradigmenwechsel für Anbieter und Nutzer von Investitionsgütern, 1st edn. Gabler, Wiesbaden (2009)
11. Pachow-Frauenhofer, J.: Planung veränderungsfähiger Montagesysteme. Dissertation, Univ. PZH, Garbsen (2012) (Berichte aus dem IFA 2012, 1)
12. Dietz, T., Fechter, M.: Einleitung. In: Bauernhansl, T., Fechter, M., Dietz, T. (eds.) Entwicklung, Aufbau und Demonstration einer wandlungsfähigen (Fahrzeug-) Forschungsproduktion, 1st edn., pp. 5–10. Springer, Berlin (2020) (ARENA2036)
13. Mietzner, D.: Strategische Vorausschau und Szenarioanalysen: Methodenevaluation und neue Ansätze, p. 117 ff. Gabler, Wiesbaden (2009)
14. Reibnitz, U.: Szenario-Technik: Instrumente für die unternehmerische und persönliche Erfolgsplanung, 2nd edn., p. 23 ff. Gabler, Wiesbaden (1992)
15. Siebertz, K., Van Bebber, D., Hochkirchen, T.: Statistische Versuchsplanung: Design of Experiments (DoE), p. 247 ff. Springer, Heidelberg (2010) (VDI-Buch)
16. Vorderer, M., Junker, S., Lechler, A., Verl, A.: CESA^{3}R: highly versatile plug-and-produce assembly system. In: 2016 IEEE International Conference on Automation Science and Engineering (CASE), pp. 745–750
17. Bauernhansl, T., Fechter, M., Dietz, T. (eds.): Entwicklung, Aufbau und Demonstration einer wandlungsfähigen (Fahrzeug-) Forschungsproduktion, 1st edn., pp. 145–157. Springer, Berlin (2020) (ARENA 2036)
18. Koo, C.H., Schröck, S., Vorderer, M., Richter J., Verl, A.: A model-based and software-assisted safety assessment concept for reconfigurable PnP-systems. In: 53rd CIRP Conference on Manufacturing System (2020)
19. Wöhe, G., Döring, U.: Einführung in die allgemeine Betriebswirtschaftslehre, 25., überarbeitete und aktualisierte edn., pp. 88–96. Vahlen, München (2013)

Reconfiguration of Production Equipment of Matrix Manufacturing Systems

Michael Trierweiler[1,2(✉)] and Thomas Bauernhansl[1,2]

[1] Fraunhofer Institute for Manufacturing Engineering and Automation IPA, Stuttgart, Germany
{Michael.Trierweiler,miht}@ipa.fraunhofer.de
[2] Institute of Industrial Manufacturing and Management IFF, University of Stuttgart, Stuttgart, Germany

Abstract. Since the introduction of the assembly line in production around 100 years ago, the principal of mass and series production has not changed much. However, in the last decades, more individualized products lead to higher product variants, which challenge rigidly linked assembly lines. To provide higher adaptability to changing product variants and volumes, in manufacturing as well as in assembly, the concept of a production system structured as a matrix is developed (abbreviated as MMS). Here, the equipment of the production system is composed of various process modules providing the needed functions. Depending on the needed functions, the work pieces literally search by and by their way through production. The process modules themselves consist of one or more stations providing process functionalities. Assuming that these stations can be distributed to the various process modules in short time, this production structure offers a high changeability during operation. It can be used to reconfigure the system continuously to changing production programs. Through the high degrees of freedom of a matrix production system, finding this optimal configuration of the equipment can be seen as a complex task. For the initial planning of the system, several approaches exist. However, so far, there is no method for reconfiguring the system to changed requirements, mainly to changes of the composition of the production programs, during the operation of the production system. This paper gives an overview of the task and sketches an approach of how to indicate that a reconfiguration is beneficial and subsequently to find and realize an optimized one. Therefore, the feedback control technique is introduced and it is shown, how it can be applied in continuous change processes of production entities. Then, the technique is adapted to apply it to the reconfiguration problem of MMS. Finally, the needed research to realize that approach is outlined.

1 Introduction and Motivation

More than 100 years after introducing the assembly line by Ford, most production systems still follow the principles of line and tact dependency. However, for years now, an increase of product variants can be observed [1], which challenge rigidly-linked assembly

P. Weißgraeber et al. (Eds.): *Advances in Automotive Production Technology – Theory and Application*, ARENA2036, pp. 20–27, 2021. https://doi.org/10.1007/978-3-662-62962-8_3

lines [2]. Accordingly, there is research on new production systems, based on cyber-physical systems dissolving the line characteristic with tact time dependency [3–5]. In manufacturing as well as in assembly, a concept following this approach is the so-called matrix manufacturing system (MMS). It consists of flexibly linked process modules providing the functionalities [6]. That allows an individual flow of each product through the system and thereby provides process and product flexibility [7]. So far, several initial planning approaches for matrix production systems exist [6, 8, 9]. As in all manufacturing and assembly planning processes, a prediction of the production program is an important input value. The capacity planning and alignment of the production resources is based on that. The in that way designed system is optimized to the forecasted production program. However, it is very likely that after implementing and operating the system the composition of the production program will be quite different. Especially, in production programs with many variants it is very likely that the composition changes. Accordingly, the requirements on the system will change. This affects the time required for production resources as well as flows of process sequences. Actually, due to its flexibility, the MMS could still manufacture the products but with lower efficiency.

The changeability enablers modularity, scalability and mobility [10] can be seen as immanent properties of a MMS. Accordingly, it is seen as highly changeable [2, 4]. The present paper describes the options for changing MMS while focusing mainly on reconfiguration. Furthermore, it shows how reconfiguration can be used to adapt a MMS to changed requirements caused by varying production programs to maintain and increase its efficiency. Therefore, a method based on the feedback control technique is outlined.

2 Changeability of Matrix Manufacturing Systems

According to Westkämper and Zahn [11], changeability aims to continuously adapt a company to changing requirements to reach and maintain high efficiency and stay competitive. Therefore, they define a system as changeable, when it provides variability in process, structural and behavioural aspects. Furthermore, they claim the need, that these variabilities can be activated in a short time with minimal effort [11].

ElMaraghy and Wiendahl [12] give a good overview of changeability classes on the different company levels. Figure 1 shows an extract, focusing on production, with the levels station, cell and system. Here, the changeability classes change over ability, flexibility, and reconfigurability are assigned. The authors define change over ability as the ability to adapt a production resource to certain known operations with a minimal effort. Flexibility relates to the operative ability on cell, station and system level to reprogram, reroute and reschedule them to a known family of work pieces. Reconfigurability is defined as the tactical ability to adapt cells and systems to new part groups. It includes a physical change of the structure of processes and material flows (recombination) as well as the adding or removing of production components [12].

To identify options of changes on the three production levels of a MMS in a systematic way, the MMS can be described with the help of the structural concept of the system theory [13]. Its objective is to understand a system by dividing it into its elements [14]. Then options of changes can be derived. Tables 1 and 2 summarize them structured

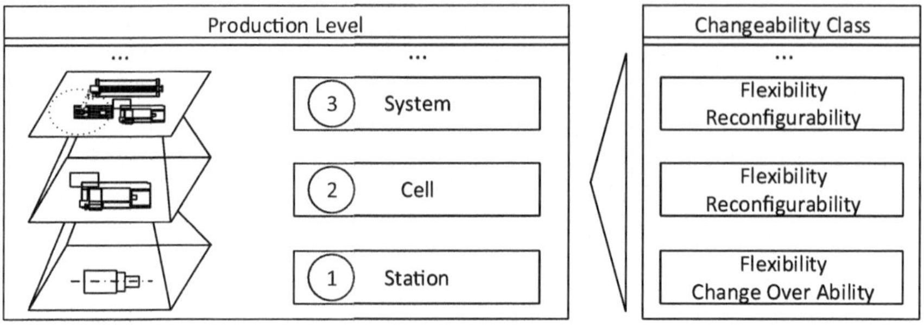

Fig. 1. Hierarchy of production with corresponding changeability classes, extract, based on [12]

column-wise by the need for adaptions which are ability, sequence and capacity. The lines structure them regarding the changeability classes flexibility and reconfigurability. Table 1 focuses on system level, Table 2 on cell level. Station level is not further considered since there are no MMS-specific options of change. More detailed information can be found in Trierweiler et al. [13].

Table 1. Options of changes on system level of a MMS, based on [13]

	Need for adaptions		
	Ability	Sequence	Capacity
Flexibility	–	Change of routing	Variation of shift model
Reconfigurability	Add/remove modules with abilities	Change position of modules	In-/decrease number of modules

Table 2. Options of changes on cell level of a MMS, based on [13]

	Need for adaptions		
	Ability	Sequence	Capacity
Flexibility	–	–	Variation of shift model
Reconfigurability	Add/remove stations with certain abilities	Change position of stations inside module	In-/decrease number of stations

As shown, a MMS provides many options for changes, so it is accurate to see it as a highly changeable system. The system is designed to use the options of flexibility during its daily operation to provide the full set of functionalities to produce each product and variant of a diverse production program. The options of reconfiguration can be used occasionally to adapt the system to changes in the production program to increase its efficiency. Here, the following questions evolve: How can the advantageousness of reconfiguring the system be detected? What does indicate that reconfiguration can increase or maintain the efficiency of the system? How can a better configuration of the system be derived and implemented? In fact, these questions can be summarized in the demand for a method detecting inefficiencies caused by the configuration of the system and designs a reconfiguration to maintain and increase efficiency. The following chapter introduces the feedback control technique as a meta model to solve this question. In addition, it shows how it can be applied to adapt production entities.

3 Continuous Adaption of Production Entities

In general, industrial production systems are seen as complex [15]. The complexity is caused through a high number of different elements and their various relations [14]. Since a change of one factor can have many influences on others [16], the task of adapting a system to changing requirements can be difficult. Accordingly, an effective adjustment needs to be planned properly, by considering all possible influences. To handle this kind of complex tasks, the systems theory has developed several approaches. The black-box and hierarchy concept as well as the control technology can be mentioned among others [14]. For adaption processes, the feedback control technique is often used as a meta model, since it can be applied to control technical as well as sociotechnical systems as production systems [17]. In general, to control a system, the system's status is monitored by the so-called feedback variables. Their values are then compared to reference values to determine the deviations. Based on that, a control unit calculates a value to control an actuator which influences the system to align the feedback to the reference variables [18].

Nofen has applied this technique to model the adaption of a production facility to changeability drivers [19]. This task shows similarities to the question assessed in this paper – the continuously reconfiguration of a MMS. Accordingly, it is described more in detail in the following.

In the mentioned approach, the control loop starts with monitoring the operation of the factory by indicators. They can be seen as feedback variables and conclude the need for changes. In the controller, these variables are compared with the reference variables. These are guidelines in form of key performance indicators (KPI) given by the management. The controller determines the need for adaption and derives the kind of adjustment process. Regarding the reconfiguration, this can be a structural change or the adjustment of elements. Here, all possible interdependencies need to be considered to take into account all influences of changes. Nofen therefore uses effect chains. A following actuator is responsible to execute the adaption of the factory. Finally, the operation of the changed factory is again monitored by the indicators. This is where the control loop is closed [19].

The described work uses the feedback control technique for continuous change processes of complete factories. Thereby it is shown, that this technique works for adaption processes of production entities. However, the approach is designed for complete factories. Azab applies the feedback control technique to the adaption of production facilities. He assesses the reconfiguration of manufacturing systems more closely [20]. However, the given approach is not directly transformable to the reconfiguration process of MMS. Therefore, the following chapter discusses a possible way to apply the feedback control technique to reconfigure a MMS.

4 Approach to Reconfigure MMS

Section 2 shows the specific options to reconfigure the production equipment of a MMS. Section 3 introduces the feedback control technique as a way to adapt production entities continuously. The following gives an overview of using this approach to reconfigure the production equipment of MMS to increase and maintain efficiency. The approach is limited to adapt to changes caused by the variation of the composition of production programs, since this is expected to be the main reason to reconfigure a MMS during operation. Accordingly, needed reconfiguration due to the introduction of complete new products with different production processes into the MMS is out of scope. In that case, a more profound approach with one of the existing initial planning methods mentioned in Sect. 1 seems favourable.

The approach is divided into 1) indicators and measuring, 2) determination of required reconfiguration and 3) the designing of reconfiguration (Fig. 2).

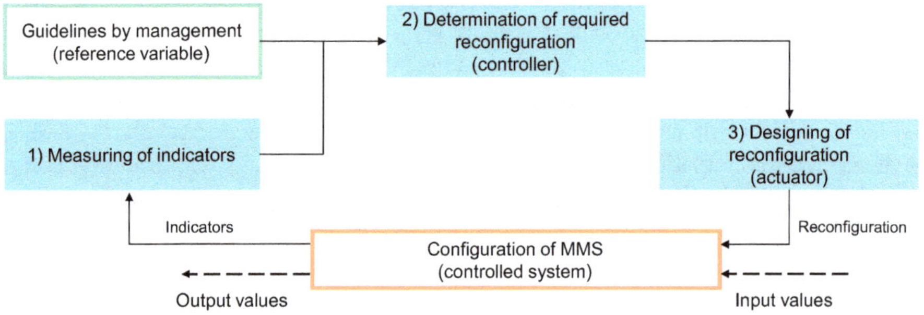

Fig. 2. Feedback control technique applied to reconfigure the production equipment of MMS

1) Definition of Indicators and Measurement of Those

First, losses of productivity, which are caused by the configuration of MMS need to be identified. In general, losses of productivity can be detected by waste in the several processes. Therefore, Ohno defined the seven waste classes which are overproduction, storing, wrong processes, defects, movements, transportations, and waiting [21]. It can be stated that the three first mentioned classes are not specific to the MMS, since these classes are mainly influenced by the production planning and control as well as in manufacturing and assembly planning. The other four can be seen as linked with the configuration.

Defects, related to the system's configuration, can be caused, among others, by the distance between processes. For example, when in a gluing process the distance between the dispensing of the glue bead and the final mounting of two parts is too far, the glue can be already too dry which can influence the quality of the glued joint. Thus, a reconfiguration through repositioning of the dispensing and mounting could avoid defects. Another configuration-specific category is the movements of workers and parts. A repositioning of stations inside a module, for example, can reduce waste. To reduce the effort for transportation, depending on the product-specific assembly sequences, a repositioning of modules can be useful. The waste class waiting indicates needed capacity adaptions. It can be divided into waiting time related to orders and related to resources. On the one hand, order-related waiting time can indicate a too low capacity of a certain process inside the MMS. Accordingly, adding an additional resource with that functionality into the system can decrease that waste. On the other hand, downtimes of production resources due to missing orders can indicate an overcapacity of a certain process. In case this process is represented by more than one resource, taking out one of the resources could decrease the overcapacity. Furthermore, KPI, which are given by the management, need to be captured.

Secondly, after defining the certain indicators and KPI the recording of those needs to be specified. Therefore, order and production resource-related data need to be acquired. Data sources can be ERP and MES as well as additional sensor systems, specialized in monitoring assembly systems as described by Kärcher [22]. Thereby, a system-specific so-called digital shadow can be implemented, providing the needed data for optimizing the system via reconfiguration.

2) Determination of Required Reconfiguration
During this step, the indicators are analysed and the deviations of the measured to the reference KPI need to be determined. Here, a very crucial point is to investigate the extent of the wastes and deviations to the reference KPI to determine their importance. Regarding the wastes, it needs to be decided between rarely and regularly occurring phenomena. Subsequently, the option of reconfiguration can be derived which can consists of one or a combination of the options summarized in Tables 1 and 2.

3) Designing of Reconfiguration
After the decision about the reconfiguration, it needs to be concretely designed. For example, when elements are to be repositioned and additional elements should get implemented, it has to be decided about the new locations. Since this should be done by ensuring a production with minimum waste and a maximum adherence to the reference KPI as well as keeping the effort for reconfiguration lower as the potential benefits, this question can be seen as a multi-criterial optimization task. Optimization approaches applied to production-related questions can be found in [23–25]. In addition, applying machine learning techniques seem promising. After designing a new reconfiguration it should be modelled in a material flow simulation to validate the effectiveness. When the results of the simulation prove the advantageousness of the reconfiguration the responsible workshop managers should finally assess it and create a reconfiguration schedule. Finally, they can instruct the set-up staff to implement the changes physically in the production system.

Here, the control as well as the optimization loop close and the operating system can be monitored and optimized all over again in the following production period.

5 Conclusion and Outlook

This paper gives an overview of how to adapt a MMS to changing production programs. Effective ways of reconfiguration are shown and the feedback control technique is introduced as a meta model for adaption processes. Section 4 discusses the application of the feedback control to detect the advantageousness, determine and design a reconfiguration. However, to apply that approach to a MMS in reality it needs to be worked out in detail. First of all, the indicators and the corresponding recording need to be developed. Then, the controller part of the control loop needs to be designed. Furthermore, the designing process of the concrete reconfiguration needs to be investigated. Here, a main challenge is to choose the suitable optimization technique. Promising ones are found in the field of operations research and in machine learning approaches. However, this paper has structured the problem clearly. It can be seen as the basis to elaborate the problem sequentially in further research with the objective to allow a continuous reconfiguration of MMS to adapt to changing, diverse production programs.

References

1. Schuh, G. (ed.): Produktkomplexität managen, 3rd edn. Hanser, München (2017)
2. Bauernhansl, T.: 23. Deutscher Materialfluss-Kongress. Mit Fachkonferenz Automobillogistik; TU München, Garching, 20. und 21. März 2014. In: 23. Deutscher Materialfluss-Kongress, vol. 2232, pp. 269–276
3. Bauernhansl, T.: Wandlungsfähige Automobilproduktion der Zukunft. https://publica.fraunhofer.de/eprints/urn_nbn_de_0011-n-5590909.pdf. Accessed 7 July 2020
4. Kern, W., Rusitschka, F., Kopytynski, W., Keckl, S., Bauernhansl, T.: Alternatives to assembly line production in the automotive industry. In: 23rd International Conference for Production Research, ICPR (2015)
5. Foith-Förster, P., Bauernhansl, T.: Changeable assembly systems through flexibly linked process modules. In: Roberto TETI (ed.) Procedia CIRP. Research and Innovation in Manufacturing: Key Enabling Technologies for the Factories of the Future, pp. 230–235 (2016). https://doi.org/10.1016/j.procir.2015.12.124
6. Greschke, P.: Matrix-Produktion als Konzept einer taktunabhängigen Fließfertigung. Dissertation, Technische Universität Braunschweig (2016)
7. Greschke, P., Schönemann, M., Thiede, S., Herrmann, C.: Matrix structures for high volumes and flexibility in production systems. In: ElMaraghy, H. (ed.) Procedia CIRP. Variety Management in Manufacturing, pp. 160–165 (2014). https://doi.org/10.1016/j.procir.2014.02.040
8. Foith-Förster, P., Bauernhansl, T.: Changeable and reconfigurable assembly systems – a structure planning approach in automotive manufacturing. In: Bargende, M., Reuss, H.-C., Wiedemann, J. (eds.) 15. Internationales Stuttgarter Symposium, pp. 1173–1192. Springer Fachmedien Wiesbaden, Wiesbaden (2015). https://doi.org/10.1007/978-3-658-08844-6_81
9. Kern, W., Rusitschka, F., Bauernhansl, T.: Planning of workstations in a modular automotive assembly system. In: Westkämper, E., Bauernhansl, T. (eds.) Editorial 49th CIRP International Conference on Manufacturing Systems (CIRP CMS), pp. 327–332 (2016). https://doi.org/10.1016/j.procir.2016.11.057
10. Wiendahl, H.-P., Wiendahl, H.-H.: Betriebsorganisation für Ingenieure, 9th edn. Hanser, München (2020)
11. Westkämper, E., Zahn, E.: Wandlungsfähige Produktionsunternehmen. Springer, Berlin (2009)
12. ElMaraghy, H.A. (ed.): Changeable and Reconfigurable Manufacturing Systems. Springer, London (2009)
13. Trierweiler, M., Foith-Förster, P., Bauernhansl, T.: Changeability of matrix assembly systems. In: CIRP Proceedings. Conference on Manufacturing Systems (2020)

14. Schiemenz, B.: Komplexität von Produktionssystemen. In: Kern, W. (ed.) Enzyklopädie der Betriebswirtschaftslehre. Handwörterbuch der Produktionswirtschaft, 2nd edn., pp. 899–900. Schäffer-Poeschel, Stuttgart (1996)
15. REFA: Planung und Gestaltung komplexer Produktionssysteme, 2nd edn. Methodenlehre der Betriebsorganisation. Hanser, München (1990)
16. Ashby, W.R.: Introduction to Cybernetics. Chapman & Hal, London and Becclesl (1961)
17. Ropohl, G.: Allgemeine Technologie. Eine Systemtheorie der Technik. KIT Scientific Publishing, Karlsruhe (2009)
18. DIN Deutsches Institut für Normung e. V.: International Electrotechnical Vocabulary. Part 351: Control Technology. Beuth Verlag GmbH, Berlin 01.040.35; 01.040.29; 35.240.50; 29.020 (DIN IEC 60050–351) (2013)
19. Nofen, D.: Regelkreisbasierte Wandlungsprozesse der modularen Fabrik. Berichte aus dem IFA, 2006, vol. 1. PZH, Produktionstechn. Zentrum, Garbsen (2006)
20. Azab, A., ElMaraghy, H., Nyhuis, P., Pachow-Frauenhofer, J., Schmidt, M.: Mechanics of change: a framework to reconfigure manufacturing systems. CIRP J. Manuf. Sci. Technol. (2013). https://doi.org/10.1016/j.cirpj.2012.12.002
21. Ohno, T.: Das Toyota-Produktionssystem. Campus, New York (2005)
22. Kärcher, S., Bauernhansl, T. (eds.): Approach to generate optimized assembly sequences from sensor data. In: 52nd CIRP Conference on Manufacturing Systems (2019)
23. Bogatzki, A.: Fabrikplanung. Verfahren zur Optimierung der Maschinenaufstellung. Dissertation, Univ., Wuppertal. Theorie und Forschung, vol. 534. Roderer, Regensburg (1998)
24. Kettner, H., Schmidt, J., Greim, H.-R. (eds.): Leitfaden der systematischen Fabrikplanung. Mit zahlreichen Checklisten. Hanser, München (2010)
25. Krüger, T.: Entwicklung einer Gesamtmethodik zur Kombination von mathematischer Anordnungsoptimierung und Materialflusssimulation für die Produktionslayoutplanung, Universitätsbibliothek Der TU Clausthal (2019)

A User-friendly Planning Tool for Assembly Sequence Optimization

Dominik Schopper(✉) and Claudia Tonhäuser

University of Stuttgart, Stuttgart, Germany
schopper@ifb.uni-stuttgart.de
https://www.ifb.uni-stuttgart.de/en/research/pi-group/

Abstract. Digitalization offers new opportunities to improve the quality of planning, adaptation and optimization of assembly processes. The approach presented in this paper allows for the (re-)planning of the assembly process with alternative assembly paths. Since usually multiple valid assembly sequences exist for one and the same product, an automated assembly path analysis is realized for the identification of a time-optimal sequence. The implementation is realized using a combination of Petri nets and graph-based design languages. This framework allows an easy integration of the assembly planning process into the digital design process and the automatic evaluation of possible assembly sequences.

1 Introduction and Fundamentals

Today, companies face the challenge of reducing assembly costs, which account for 20–70% of the total manufacturing costs depending on the domain [1]. In the manufacturing sector automation is demanding, since assembling geometrically complex components is difficult. This paper describes a framework for the individual planning and design of assembly processes in the form of a graph-based design language. All possible assembly options are compared, evaluated and an optimal solution is derived *automatically*. To provide the required input in the form of a Petri net, a user-friendly GUI was implemented.

1.1 Graph-based Design Languages

Inspired by natural languages, *graph-based design languages* (GBDLs) make it possible to digitally represent the whole design process in engineering in a new way. GBDLs are essentially based on a set of *vocabulary* and *rules* that are applied in a defined sequence [2,3]. The philosophical justification for the connection between languages and engineering is derived in [4], stating that "*a new*

P. Weißgraeber et al. (Eds.): *Advances in Automotive Production Technology – Theory and Application,* ARENA2036, pp. 28–36, 2021. https://doi.org/10.1007/978-3-662-62962-8_4

device or method is composed of the available components – the vocabulary – of a domain. In this sense, a domain forms a language; and a new technological artifact constructed from components of the domain is an utterance in the language of the domain" [4]. The representation of the design in a GBDL is a graph that expresses the topological arrangement of the *vocabulary*. The *vocabulary* itself is a formal model (i.e. an ontology) of the domain in the form of a class diagram. The available *vocabulary* is combined in *rules*. The *rules* can then be executed in a defined order (in a *production system*) by the design compiler and is translated into the *design graph*. From this central model, the domain-specific models can automatically be transformed (i.e. *design graph* to CAD). Figure 1 illustrates the information architecture and the role of the individual components.

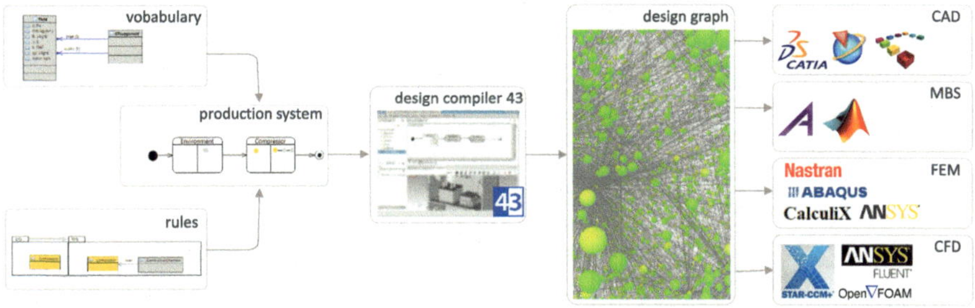

Fig. 1. Information architecture of graph-based design languages [5]

The compilation of the vocabulary with an individual rule sequence to a complete system model (i.e. a digital model of the assembly as will be shown later in this paper) leads to a digital design method for a specific problem. For a detailed description of the functionality of GBDLs, the reader is referred to [5].

1.2 Representation of Assembly Systems

According to the VDI Guideline 2860 "*Assembly and Handling Technology*" [6], assembly is defined as the sum of all processes that serve to assemble geometrically defined bodies. These include handling, controlling, joining, adjusting and special operations. Figure 2 shows the classification scheme of these processes.

Each of these five main groups can be subdivided into sub-processes which in turn split in subgroups (neglected for clarity). For the digital representation of the assembly process in the presented framework, all aforementioned assembling methods are modeled as vocabulary and serve as an ontology for the GBDL.

1.3 Petri Nets

A Petri net can be represented as a directed graph in which nodes called *places* and *transitions* are connected by edges [10]. *Places* are drawn as circles and represent passive elements (e.g. presence of a component), while *transitions* are drawn

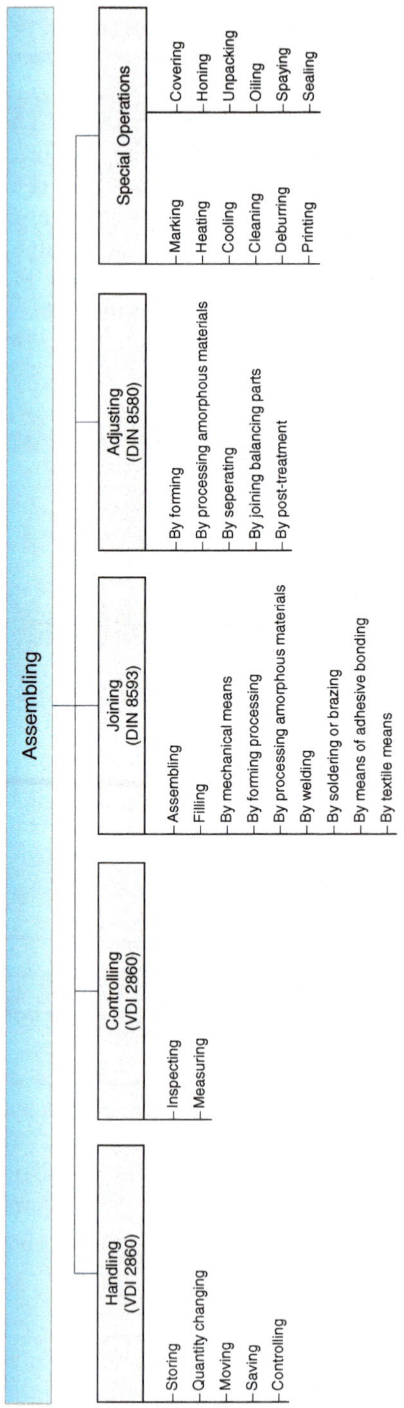

Fig. 2. Hierarchy of assembling [6]

as rectangles to model activities (e.g. manufacturing steps). The dynamic behavior of the modeled system is represented by a transition *firing* which means that an activity is executed. This *firing* can be triggered if the necessary preconditions marked by so-called *labels* are met. In this manner Petri nets ensure a correct dynamic behavior of the modeled system, since the existence or non-existence of relevant *labels* controls the workflow. The modelling of manufacturing systems is one of the oldest applications [7–9].

2 Implementation

For the implementation of the user-friendly planning tool for assembly sequence optimization two elementary building blocks are combined in a framework: A graphical user interface (GUI) for Petri net input definition and a core GBDL for model execution and optimization. The core GBDL consists of a Unified Modeling Language (UML) based representation of Petri nets and assembly systems (see Sect. 1.2), an executable rule-based activity diagram for model-to-model transformation into a design graph and lastly the optimization code.

2.1 Graphical User Interface

For the assembly sequence optimization within the presented framework, an initial input of the system in form of a Petri net has to be given. In GBDLs this would normally require the formulation of graphical rules for model-to-model transformation which in turn would be executed and generate the Petri net model. Since the formulation of the rules requires a certain familiarity with GBDLs, an easy-to-use GUI was implemented for this initial manual task. The user is kept away from rule modeling and does not need to have any prior knowledge about UML and GBDLs. The GUI is therefore not an essential part of the framework, but is indeed crucial for the aforementioned user-friendliness.

The first step in the GUI is the specification of all given assembly processes, resources and product parts. The assembly process definition is based on the given assembly systems ontology (see Sect. 1.2) whereas the resources and product parts can be entered in any specification standard. Company specific information such as part number, operating costs per time, acquisition costs, resource availability and further criteria can be defined and stored as well. If already done, this step can be skipped and the specifications can be loaded into the model instead. Based on this input in a next step user specific Petri nets can be modeled including all dependencies in a specially provided modeling field. This task is simply done by drag and drop activities. The defined Petri net is finally saved in a separate file and can be loaded anytime for reuse or adaption. After these steps have been performed the input is ready for use.

In the future, the GUI will be extended by an advanced selection menu to define the optimization goal and the optimization method[1]. After the execution of the framework the found solution is given as output via the GUI.

[1] In our example the assembly time is minimized and a simple depth first search is implemented.

2.2 Graph-based Design Language for Assembly Planning

As described in Sect. 1.1, the representation of a domain is realized via a class diagram. The presented GBDL for assembly sequence optimization contains two of them: One for the representation of the Petri net structure, the other for the representation of the assembly system structure (see Sect. 1.2). Since Petri nets can be formulated on a very high level of abstraction, only the two classes `Place` and `Transition` are needed. The generation and processing of data is modeled by an activity diagram. Figure 3 shows an overview of the framework.

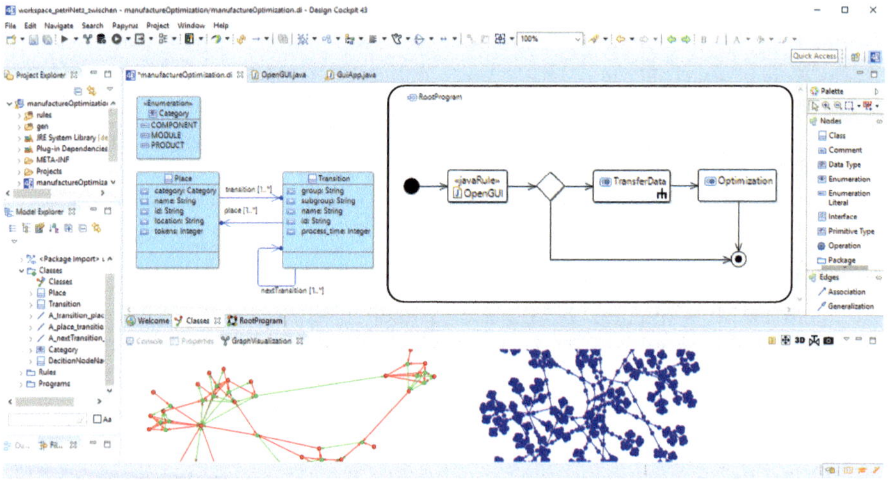

Fig. 3. Framework for assembly sequence optimization

On the upper left side you can see the Petri net class diagram. It becomes apparent that a `Place` is always a representation of a product entity which can be a single component, a module (assembly of two or more components) or the final product. For this purpose the attribute *category* can be set to COMPONENT, MODULE or PRODUCT. The class `Transition` represents any assembly process. The dependency between a transition and the given assembly systems class diagram is enforced using predefined *group* and *subgroup* attributes. The values they can take are directly linked to the assembly systems ontology. The `Transition` class also has a *nextTransition* association, which is used to build up the search tree when using the depth first search.

On the right hand side of Fig. 3 the implemented activity diagram containing the GUI call, the model transformation into a design graph and the optimization algorithm is shown. The diamond-shaped symbol stands for a decision node that ensures that the program will run without executing the following rules if the GUI is closed without further instructions. The `TransferData` subprogram reads all data from the GUI and performs the transformation. This is done by multiple instantiations of the aforementioned classes and associations.

The result is a so called *design graph*, where each node represents an object and each edge represents a link between two objects. The design graph is shown in Fig. 3 on the left hand side of the bottom window. The `Optimization` subprogram contains a simple depth first search optimization and creates a search tree as can be seen on the right hand side of the lower window. This tree represents all successful[2] assembly sequences implied in the Petri net. At the same time the optimization is done by evaluating and comparing measures of interest for all sequences. The `Optimization` sub program can be extended easily.

In the future more efficient search methods can be integrated here. However the possibilities go far beyond a simple search of the Petri net. The rule-based structure makes it possible to delete, add or replace entire parts of the design graph. Since the rules are able to automatically identify certain parts of the graph, further heuristics can be implemented to restructure the graph and optimize it in this way. The presented framework combines the following advantages:

- The framework is based on UML: The UML is the dominant language for modeling software systems. It ensures a manageable and clear representation of data through the integration of standardized hierarchical structures.
- The execution of the framework is rule based: The rule-based structure enables a completely new type of optimization, since the given Petri net can be completely restructured within an optimization heuristic.
- The model is represented as a graph: The graph representation allows the application of all graph algorithms already existing in literature.
- The framework is implemented as GBDL: GBDLs can be formulated for any domain. They are equally suited for the representation of the product to be produced or the manufacturing resources. If further GBDLs are available for these domains, the assembly planning can easily be linked to them.
- The framework enables the integration of a GUI: Users of the framework not necessarily need to have knowledge about GBDLs. This is only required if the framework shall be extended.

2.3 Workflow

Figure 4 provides an overview of the workflow of the whole framework. In the first step, the assembly specific elements are defined as input by a user using the GUI. These elements are then arranged into a Petri net (step 2). From here the design language execution begins. The user-specific Petri net model is automatically transformed into a design graph using an underlying GBDL (step 3). In step 4 the design graph is analyzed by a depth first search for optimization. Synchronously to the generation of the search tree the search for the critical path is executed. Note that any optimization algorithm could be used here instead.

[2] In this context successful means that a product could be produced under the given conditions.

After the optimization the output is visualized in the GUI (step 5). In step 6 the most suitable assembly process found is given as solution of the manufacturing planning phase within the product life cycle. In future, the same mechanism could also be used to give instructions to workers in real time for differing basic situations in highly individualized assembly processes.

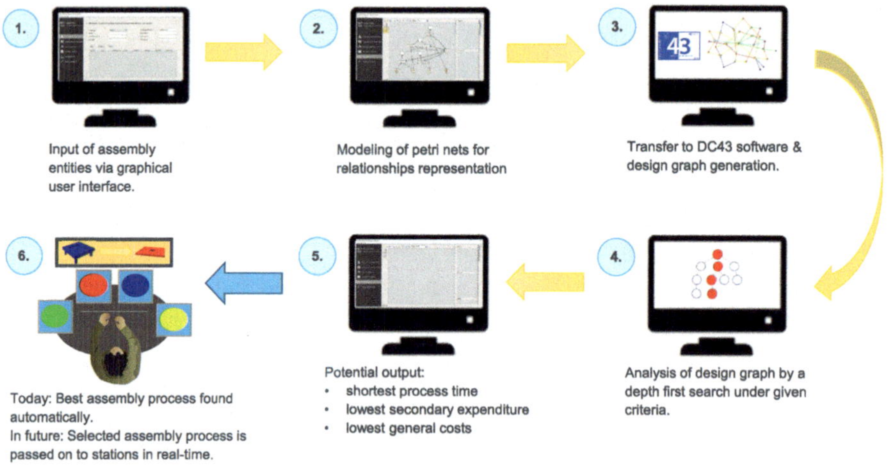

Fig. 4. Workflow steps when using the presented user-friendly assembly planning tool

3 Application Example

For a better understanding of the presented framework we use the example of a ballpoint pen. Figure 5 shows the Petri net that was modeled by a user via the GUI. Three different assembly sequences are included. The given Petri net is transformed into a design graph when executing the GBDL. In a next step the implemented depth first search is performed on the Petri net graph. Figure 6 shows the resulting search tree in theory on the left hand side. For better clarity, only those assembly sequences are shown which successfully lead to the given product, taking label availability into account. The best sequence found by the algorithm is highlighted by a circle. On the right side you can see the whole search tree as it appears within the framework after execution.

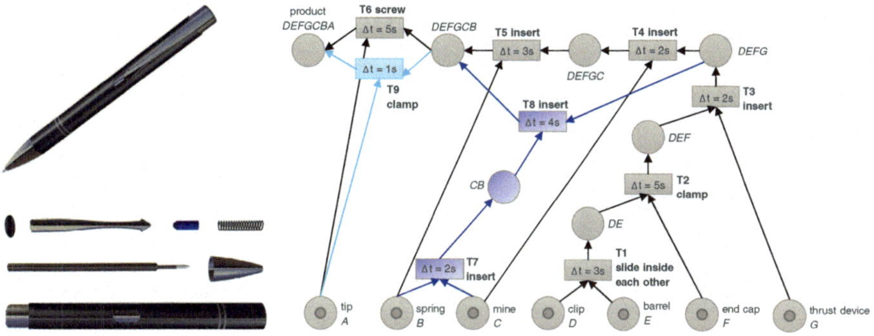

Fig. 5. Ballpoint pen (left) and Petri net with various assembly sequences (right)

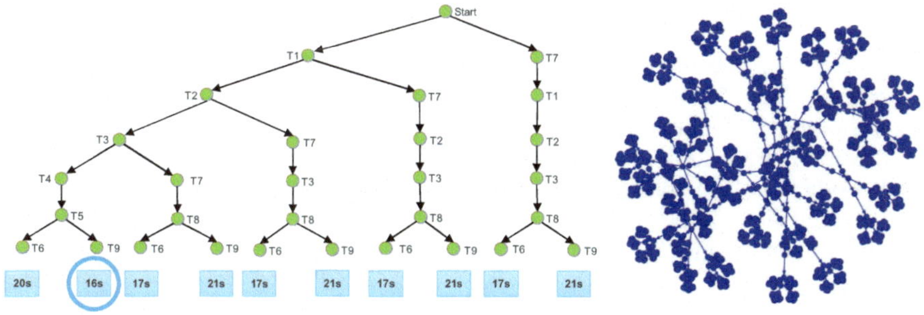

Fig. 6. Search tree for ballpoint pen example above. Left: Only successful sequences are shown. Right: Search tree with all sequences as design graph.

4 Summary and Discussion

In this paper we have shown the framework for a user-friendly planning tool for assembly sequences and outlined how graph-based design languages and Petri nets can be used to derive the time-optimal sequence of an assembly. Although the shown example is kept simple, the power of the presented framework is emphasized in the presented paper. It forms the basis for the following potential steps in future: Other optimization criteria and more efficient search algorithms can be taken into account. Since it is now possible to create rule-based patterns for the automated restructuring of the designed Petri net graph a more extensive optimization can be implemented. Besides a separate design language for an automated Petri net generation can replace the graphical user interface. In this way, the process can be further automated and ultimately run fully automatically. For further digitalization and automation of the production planning, further design languages for the product structure or manufacturing machinery can be created and connected to the presented design language.

Acknowledgments. Part of this work is supported by a grant of the Ministry of Science, Research and the Arts of the State of Baden-Württemberg and the European Regional Development Fund (EFRE), by name the ZaFH-Project "Digital Product Lifecycle (DiP)" (grant no. 43031423). Further information at www.dip.rwu.de and www.rwb-efre.baden-wuerttemberg.de.

EUROPÄISCHE UNION
Europäischer Fonds für regionale Entwicklung

EFRE
Investition in Ihre Zukunft.

Baden-Württemberg

References

1. VDI-Verlag: Rechnerintegrierte Konstruktion und Produktion. Düsseldorf (1990). http://d-nb.info/901184942/04
2. Rudolph, S.: Übertragung von Ähnlichkeitsbegriffen. Habilitation. Institute of Statics and Dynamics of Aerospace Structures – University of Stuttgart (2002)
3. Kröplin, B., Rudolph, S.: Entwurfsgrammatiken – Ein Paradigmenwechsel? Der Prüfingenieur (2005)
4. Arthur, W.B.: The Nature of Technology – What it is and How it Evolves. Free Press, New York (2009)
5. Schmidt, J.: Total Engineering Automation. Vision and Realization with Graph-based Design Languages and the Design Cockpit 43 ® software suite. https://iils.de/downloads/IILS-WhitePaper-TotalEngineeringAutomation.pdf
6. Lotter, B.: Einführung. In: Lotter, B., Wiendahl, H.P. (eds.) Montage in der industriellen Produktion. VDI-Buch. Springer, Berlin (2012). https://link.springer.com/chapter/10.1007/978-3-642-29061-91
7. Seatzu, C.: Modeling, analysis, and control of automated manufacturing systems using Petri nets. In: 24th IEEE International Conference on Emerging Technologies and Factory Automation (ETFA), Zaragoza, pp. 27–30 (2019). https://doi.org/10.1109/ETFA.2019.8869012
8. Recalde, L., Silva, M., Ezpeleta, J., Teruel, E.: Petri nets and manufacturing systems: an examples-driven tour. In: Desel, J., Reisig, W., Rozenberg, G. (eds.) Lectures on Concurrency and Petri Nets. ACPN 2003. Lecture Notes in Computer Science, vol. 3098. Springer, Berlin (2004)
9. Moore, K.E., Gupta, S.M.: Manufacturing systems modeling using Petri nets. In: Swamidass, P.M. (ed.) Encyclopedia of Production and Manufacturing Management. Springer, Boston (2000)
10. Petri, C.: Kommunikation mit Automaten. Dissertation. Schriften des Institutes für Instrumentelle Mathematik, Bonn (1962)
11. Murata, T.: Petri nets properties analysis and applications. Proc. IEEE **77**(4), 541–580 (1989)
12. Zimmermann, A.: Modellierung und Bewertung von Fertigungssystemen mit Petri-Netzen. Dissertation. Berlin (1997)

Fluid Manufacturing Systems (FLMS)
A Novel Approach for Versatility in Production

Christian Fries[1,2](✉), Manuel Fechter[1], Daniel Ranke[1,2], Michael Trierweiler[1,2], Anwar Al Assadi[1], Petra Foith-Förster[1], Hans-Hermann Wiendahl[1,2], and Thomas Bauernhansl[1,2]

[1] Fraunhofer Institute for Manufacturing Engineering and Automation IPA, 70569 Stuttgart, Germany
christian.fries@ipa.fraunhofer.de

[2] Institute of Industrial Manufacturing and Management IFF, University of Stuttgart, 70569 Stuttgart, Germany

Abstract. Volatile market demands, stronger regionalization of markets, ever-shortening product and innovation cycles as well as an ongoing demand for individualized products increase the need for adaptable production systems. More than a century after the start of mass production, alternative production systems are required to go beyond the current state of the art concerning adaptability, flexibility and reconfigurability to market requirements and demands. Fluid Manufacturing Systems (FLMS) describe such new production system concepts. The basic idea is to dynamically adapt and change all logistics and production processes, based on the comprehensive application of cyber-physical production systems (CPPS), thus enabling ongoing change in setup, configuration and product scope. CPPS provide a high degree of changeability, thus allowing for fast adaptions of the system to the changing requirements. Therefore the processes are continuously assessed, benchmarked and reconfigured to match the functional capabilities of production and logistic resources to the actual requirements originating from products and external influencing factors. Within this paper, conventional production systems such as Dedicated Manufacturing Lines (DML), Matrix Manufacturing System (MMS) and Flexible Manufacturing Systems (FMS) are described and characterized using defined criteria. The paper closes with a description of the Fluid Manufacturing System (FLMS), the core hypotheses and the advantages of the presented concept compared to conventional productoion systems.

1 Introduction and Motivation

The growing world population, ageing workforce, ongoing urbanization and the need for sustainable economic actions are only a few challenges to be named affecting future manufacturing [1,2]. The continuous trend towards personalized

P. Weißgraeber et al. (Eds.): *Advances in Automotive Production Technology – Theory and Application,* ARENA2036, pp. 37–44, 2021. https://doi.org/10.1007/978-3-662-62962-8_5

products [3] leads to volatile and fluctuating market demands and reinforces the need for lower prices in shorter lead times in production engineering [4]. A fixed scope of production output and technological capability per production system does not seem to be adequate anymore. New technologies with disruptive character and the advancing speed of technological development further increase the need for adaptable and fast reconfigurable production systems allowing mass production for individualized products at a competitive price [1,3–6].

Manufacturing companies have to consider these challenges in order to meet the customer demands, choose the right production concept and maintain a competitive business [7]. High price pressure combined with high labor costs have forced manufacturing companies especially from high-wage countries to gradually globalize operations. The resulting globally connected manufacturing networks have had a positive impact regarding higher sales volumes, increased turnover and lower production costs. On the downside, the most recent economic world crisis, following the pandemic outbreak, brings to bear the downsides of globally connected manufacturing networks, which were not able to respond in time to required changes in product portfolio and market demands.

Due to the systematic restrictions of conventional production systems, such as Dedicated Manufacturing Lines (DML) or Flexible Manufacturing Systems (FMS) with limited changeability and long transition times, alternative production systems receive increasing attention over the past years [3,8–11].

The functionality-based, system-driven [12] continuous reconfiguration of flexibly linked process modules used in Matrix Manufacturing Systems (MMS) [8] and Reconfigurable Manufacturing Systems (RMS) [3] enables current trends in personalized production. Simulations showed that MMS can help to improve production performance when the product variance is high and the production volumes per product are low [9]. However, there is further potential for improvement especially in terms of delay of transformation and scope of adaptions to the production demands.

This paper outlines the concept of Fluid Manufacturing Systems (FLMS) combining the conceptual ideas of flexible linkage of production cells as well as continuous adaption and reconfiguration of process modules. FLMS further refines the granularity of process module design making use of the benefits in connectivity and data transparency provided by the comprehensive use of cyber-physical production systems (CPPS). In doing so, the production system is empowered to smoothly trace the required market demands and continuously adapts the system functional capabilities.

2 State of the Art

As stated in Sect. 1, companies face the need for changeable production systems. Different types of production systems have been developed over time with each system having its benefits and drawbacks. Table 1 evaluates (based on an expert survey) the introduced production systems regarding different criteria. The criteria are chosen to define a qualitative base to compare common production systems regarding their suitability for mass production of individualized products. This includes criteria indicating the ability to produce high quantities to competitive prices. Additional indicators such as the flexibility to variance-mix and the ability for reconfiguration are assessed considering the adaptability of the system design and the required efforts for adaptions.

Table 1. Comparison of DML, FMS, RMS and MMS

	DML	FMS	RMS	MMS
High capacity	●	◔	◕	◕
Permanent availability of full set of functionalities	◔	●	◑	◑
Ability to integrate multiple products and variants	○	●	◕	●
Flexibility to volatile market demand	◔	◑	◕	●
Flexibility to variance-mix	○	●	◕	◕
Integration of personalized products	○	●	◑	◕
Ability for reconfiguration	○	◔	◕	●
Integration of new technologies	◔	◑	●	●
Low operative complexity	●	◕	◑	◑
Granularity of adaption/reconfiguration	◔	◑	◕	◕
Low transition time for adaption	○	◑	◑	◕
Responsiveness to turbulence	○	◑	◑	◕
Low operation costs	●	○	◑	◑
Short lead times	●	◕	◕	◑
Suitability to produce in lot-size one	◔	◕	◑	◕

Legend: ●: well suited, ◕: mainly suited, ◑: partly suited, ◔: fractionally suited, ○: not suited

Dedicated Manufacturing Lines (DML)
Dedicated manufacturing lines are designed to produce one product at a defined volume. The system is characterized by the sequential organization of dedicated process modules and machines in one, unidirectional flow [13]. Through focusing on one product and only few product variants, the investment into hardware is low. The uniform product flow provides high productivity. This allows low operation costs, short lead times and minimal efforts for production control. On the other side, DML are limited in terms of flexibility. Therefore, fluctuations to product variety or production volume lead to system inefficiencies.

Flexible Manufacturing Systems (FMS)
To overcome the restrictions of DML regarding the aspect of flexibility, the concept of FMS was developed. Its approach aims at providing a permanently linked production system with various flexible process modules, offering a high degree of functionality [3]. Accordingly, a FMS is suitable to a production program

consisting of multiple product-variants at small quantities [14], as it can flexibly change-over its functionality to the required product variety. FMS fulfill the criteria of producing multiple product families and variants at a time. However, despite their flexibility, FMS can hardly be reconfigured. Therefore, it is impossible to integrate additional products requiring further functionalities, unknown at the time of system design. Additional drawbacks are high initial investments and a lower outputs which challenge economic operations [3,14].

Reconfigurable Manufacturing Systems (RMS)
RMS again, were designed to overcome the limitations of FMS. They aim to achieve fast system reconfiguration by low transformation efforts based on a modular and adaptable production setup. Through modular design in electronics and mechanics, separate parts of the system can be easily exchanged to encounter volume change or new product variants [14]. Thereby, the advantages of DML (e.g. high throughput) are combined with the flexibility of FMS [3]. Accordingly, the assessment of RMS characteristics in Table 1 show a high fulfillment of all criteria. RMS are particularly beneficial regarding the integration of new technologies as well as all criteria related to the scope of flexibility. However, RMS lack in responsiveness to turbulences and system adaption times. This makes it a potentially unsuitable system for high market volatilities.

Matrix Manufacturing Systems (MMS)
The MMS consists of flexibly linked, usually dedicated, process modules. Each process module offers predefined sets of technological functionalities necessary for production. MMS enable new features of production control as each product is capable of defining an individual production path by choosing its process modules, depending on the available process module functionalities, assembly priority graph and current state of production resources. Cycle times of the process modules are no longer uniform and functionalities can be reconfigured considering a mid-term perspective [6,8]. Due to its structure and the optional flexible linkage of process modules, MMS show the highest capability for reconfiguration (see Table 1). Furthermore, the system structure shows advantages considering the integration of multiple products and variants at the same time.

It can be concluded, that the MMS design already combines many advantages of DML, FMS and RMS. However, the approach still lacks in terms of operation costs and the granularity of adaption. At the same time, the requirements for high production capacity and short changeover times are not completely fulfilled. To combine an extra granularity of adaption and shorter changeover times (low latency) at high production capacity, a new approach is required. This approach is represented by the proposed Fluid Manufacturing System (FLMS).

3 Definition of Fluid Manufacturing Systems (FLMS)

FLMS are an evolution of the Matrix Manufacturing System (MMS). FLMS concepts bases on the principle of ad-hoc resource allocation and reconfiguration to individual process modules for optimal manufacturing performance results. Comprehensivly using the benefits of cyber-physical production systems (CPPS) and their capability to self-integrate and -parametrize, process modules can be easily aggregated from single resources to advanced manufacturing systems. To fully leverage the potential of FLMS, all process modules are intended to be modular and mobile. The requirements into mobility and modularity allow for on-demand adjustments of capabilities and functionalities as well as adaptable production layouts. Thus, the manufacturing system is capable to iteratively reconfigure in variable steps to the currently required product configuration. This reconfigurability requires complex production planning and control logic considering previously unknown degrees of freedom in production system design.

The specific degrees of freedom for a MMS, such as **Operation Sequence** (specifies the sequence of work operations) and the **Work Distribution** (assigns the process modules to the production order) need to be expanded. FLMS open up the **Work Content** (defines the competencies of a specific process module) and the **Layout Position** (defines the position of production equipment within the shop floor).

The described bifurcation in process planning, making use of the additional degrees of freedom, have to be controlled efficiently in order to fully exploit the potential of FLMS. Either the product can be further manufactured at the next available process module, being capable to perform the required tasks or at the next idle process module, which can be reconfigured to the desired functional range in a feasible amount of time. So far, available control procedures do not cover these extended degrees of freedom, which reinforces the need for new procedures to be developed [15].

To understand the functional capabilities of every mentioned production system, it is important to distinguish three major process module types, see Fig. 1.

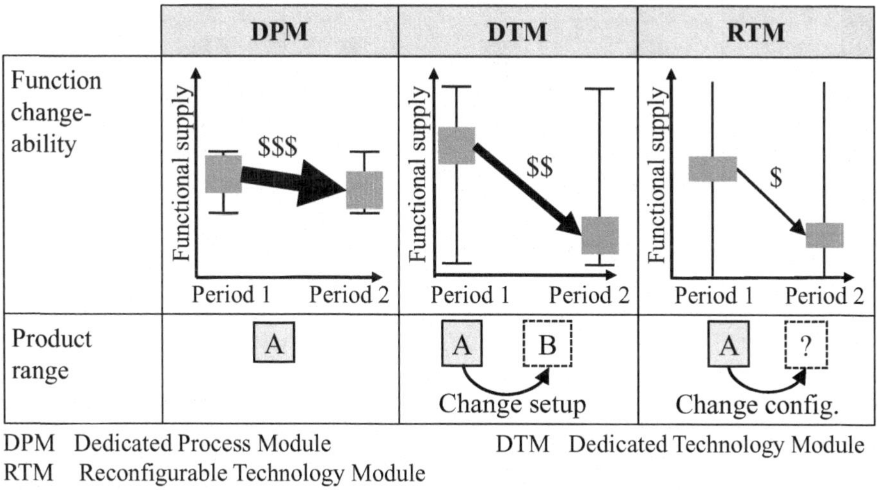

Fig. 1. Comparison of different process module types

- **Dedicated Process Modules (DPM)** are performing one single process task at a time. Functional changes are time-consuming and of high effort.
- **Dedicated Technology Modules (DTM)** represent the functional base of FMS. DTM are capable to deliver a wide range of built-in functionalities, meanwhile requiring high efforts in resources and time during design and implementation.
- **Reconfigurable Technology Modules (RTM)** are defined with narrow functional scope but wide modularity concerning electric and mechanical interfaces. The efforts for change are much lower due to the comprehensive use of CPPS (e.g. self-description of resources) in process module design and implementation.

Figure 2 depicts the concept and classification of FLMS in comparison to common production systems described in Sect. 2.

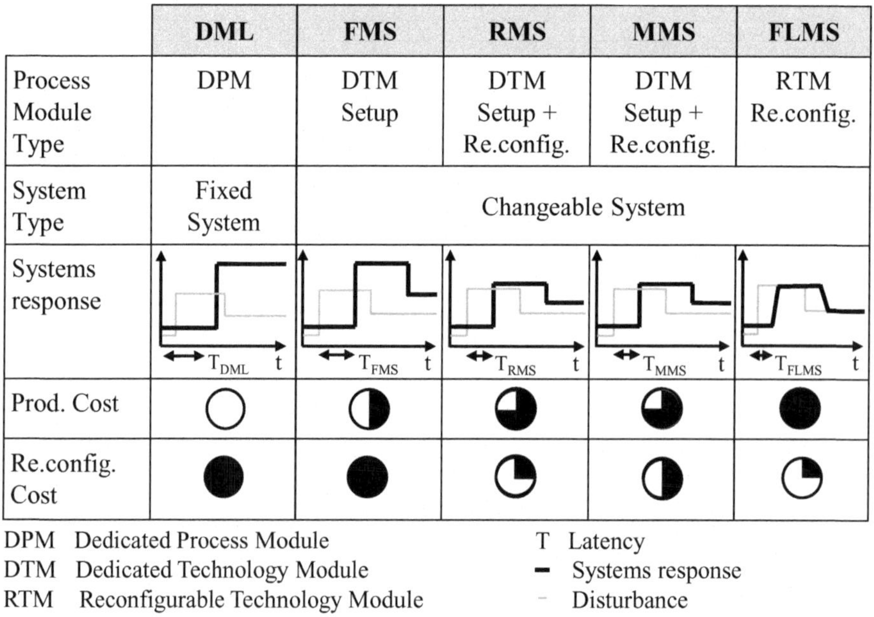

Fig. 2. Comparison of manufacturing systems

The system response of FLMS runs smoother and is not represented by box-shaped behavior, following the external triggers initiated by e.g. fluctuating demands or technical progress. Accordingly, the gap between functional requirements (gray line) and functional scope (black line) gets smaller and over-engineering in system design is avoided. According to Eq. (1), the system defined non-productive downtime due to adaption and reconfiguration (T_{FLMS}) for

FLMS is shorter than in any other production system. The non-productive downtime, a major cost driver in production systems, is considered to be the time between receiving a change order and the start of production.

$$T_{DML} > T_{FMS} > T_{RMS}, T_{MMS} > T_{FLMS} \tag{1}$$

The main reason lies within the FLMS architecture which fully consists of CPPS as the basic process module. This architecture implies the use of RTM instead of DPM or DTM and less engineering efforts during reconfiguration to guruantee a smooth approximation of the functional scope to external demands. A comparison of economic indicators reveals:

- DML and FMS are represented by low variable production costs, but higher costs during transformation.
- MMS, RMS and FLMS incorporate built-in flexibility due to versatile process modules for continuous integration and lower transformation costs.

4 Conclusion

The presented paper delivers fundamental definitions of the differences and limitations of common production systems used in today's manufacturing companies. In order to use the improvement potentials, highlighted in Sect. 2, the Fluid Manufacturing System (FLMS) is introduced and defined as a combined evolution of the MMS and RMS approach. FLMS are capable to ideally trace the systems' demand curve defined by production constraints and volatile market environments. Compared to other production systems, the system downtime caused by reconfiguration due to external trigger events is the lowest. Furthermore, the comprehensive use of CPPS in production system design might lead to better fitting functional scopes, the avoidance of over-engineering in system design, better adaptability to market demands, faster production ramp-ups and prevention of potential inefficiencies.

Further research shall investigate the production planning and control of FLMS, where additional degrees of freedom have to be controlled in order to fully exploit the potentials of FLMS. Furthermore, the implications of FLMS on the logistical processes and material supplies need to be considered.

Acknowledgments. The research presented in this paper has received partial funding under administration of the Project Management Agency (PTKA) inside the research campus ARENA2036. Our sincere thanks go to the Federal Ministry for Education and Research (BMBF) for supporting this research project by the grant agreement 02P18Q620 and 02P18Q626.

References

1. Westkämper, E., Löffler, C.: Strategien der Produktion. Springer, Berlin (2016). https://doi.org/10.1007/978-3-662-48914-7
2. Abele, E., Reinhart, G.: Zukunft der Produktion: Herausforderungen, Forschungsfelder, Chancen. s.l.: Hanser (2011). https://doi.org/10.3139/9783446428058
3. Koren, Y.: The global manufacturing revolution: product-process-business integration and reconfigurable systems. In: Wiley Series in Systems Engineering and Management. Wiley, Hoboken (2010). https://doi.org/10.1002/9780470618813
4. Booth, R.: Agile manufacturing. Eng. Manag. J. **6**(2), 105 (1996). https://doi.org/10.1049/em:19960206
5. Wiendahl, H.-P., ElMaraghy, H.A., Nyhuis, P., Zäh, M.F., Wiendahl, H.-H., Duffie, N., Brieke, M.: Changeable manufacturing – classification, design and operation. CIRP Ann. Manuf. Technol. **56**(2), 783–809 (2007). https://doi.org/10.1016/j.cirp.2007.10.003
6. Foith-Förster, P., Bauernhansl, T.: Changeable assembly systems through flexibly linked process modules. Procedia CIRP **41**, 230–235 (2016). https://doi.org/10.1016/j.procir.2015.12.124
7. Nyhuis, P. (ed.): Wandlungsfähige Produktionssysteme. Schriftenreihe der Hochschulgruppe für Arbeits- und Betriebsorganisation e. V. (HAB). GITO, Berlin (2010)
8. Greschke, P.: Matrix-Produktion als Konzept einer taktunabhängigen Fließfertigung. Dissertation. Technische Universität Braunschweig (2016)
9. Foith-Förster, P., Eising, J.-H., Bauernhansl, T.: Effiziente Montagesysteme ohne Band und Takt. wt Werkstattstech. Online **107**(3), 169–175 (2017)
10. Bauernhansl, T., ten Hompel, M., Vogel-Heuser, B.: Handbuch Industrie 4.0, Band 1, Produktion, 2., erweiterte und bearbeitete edn. Springer Reference Technik. Springer Vieweg, Berlin (2017)
11. Hofmann, C., Brakemeier, N., Krahe, C., Stricker, N., Lanza, G.: The Impact of Routing and Operation Flexibility on the Performance of Matrix Production Compared to a Production Line, pp. 155–165. Springer International Publishing, Cham (2019). https://doi.org/10.1007/978-3-030-03451-116.
12. Cuiper, R.: Durchgängige rechnergestützte Planung und Steuerung von automatisierten Montagevorgängen. Dissertation. Techn. Univ., München, vol. 143, p. 2000. Utz, Forschungsberichte/IWB, München (2000)
13. Ford, H., Crowther, S.: My Life & Work. Doubleday, Page & Company, New York (1924)
14. ElMaraghy, H.A.: Changeable and Reconfigurable Manufacturing Systems. Springer, London (2009). https://doi.org/10.1007/978-1-84882-067-8
15. Fries, C., Wiendahl, H.-H., Foith-Förster, P.: Planung zukünftiger Automobilproduktionen. In: Bauernhansl, T., Fechter, M., Dietz, T. (eds.) Entwicklung, Aufbau und Demonstration einer wandlungsfähigen (Fahrzeug-) Forschungsproduktion. ARENA2036, pp. 19–43. Springer Vieweg, Berlin (2020). https://doi.org/10.1007/978-3-662-60491-5_4

Automated Environmental Impact Assessment (EIA) via Asset Administration Shell

Anwar Al Assadi[1(✉)], Lara Waltersmann[1], Robert Miehe[1], Manuel Fechter[1], and Alexander Sauer[1,2]

[1] Fraunhofer Institute for Manufacturing Engineering and Automation IPA, Nobelst. 12, 70569 Stuttgart, Germany
Anwar.Alassadi@ipa.fraunhofer.de

[2] Institute for Energy Efficiency in Production, EEP, University of Stuttgart, Nobelst. 12, 70569 Stuttgart, Germany

Abstract. Due to growing public awareness and rising requirements of legislation and customers' expectations in the field of sustainability, it is increasingly important for enterprises to assess and subsequently reduce their environmental impact. However, the acquisition of environmental data in enterprises still causes considerable effort, due to the necessary manual acquisition.

A unified asset administration shell (AAS) potentially provides data transparency and environmental data interoperability along the value chain and thus a more detailed (real-time capable) accounting of the environmental impact of products and services. Hence, this paper presents an approach for an automated environmental impact assessments (EIA) of products and production sites.Thereby, a first application of the AAS in the context of automated EIAs was implemented at the ARENA2036 research factory. The AAS automatically collects energy and emission data throughout a production process and thus allows the allocation of actual emissions to product and equipment (environmental wallet). The results reveal a first starting point for automated EIA, facilitating individual EIAs to address increasing product variety.

1 Introduction

Increasing requirements of legislation and public awarenessdrive the sustainability efforts of enterprises. Various enterprises react to these requirements and set their own goals for decabonization [1, 2]. Therefore climate neutrality along the value chain, will become increasingly a competitive factor. Environmental impact assessments (EIAs), based on the determination of the current environmental impact caused by a company's activities, are thus becoming indispensable. However, the acquisition of environmental data in enterprises still causes considerable effort, due to the necessary manual acquisition procedures.

The efforts towards unified Asset Administration Shells (AAS) represent a promising approach for automated EIAs of products and production sites. An AAS potentially provides data transparency and environmental data interoperability along the value-added

P. Weißgraeber et al. (Eds.): *Advances in Automotive Production Technology – Theory and Application,* ARENA2036, pp. 45–52, 2021. https://doi.org/10.1007/978-3-662-62962-8_6

chain and thus a more detailed (real-time capable) accounting of the environmental impact of products and services. The general potential is indicated in the German Standardization Roadmap on Industry 4.0, there is no detailed discussion about the potential for automated EIA. Therefore, this paper addresses the following research question: How can the AAS be used for EIAs?

To answer this research question, this paper is divided into five sections. Following the introduction, the state of the art is presented divided into EIA within enterprises, usage of AAS and the comprehensive use of AAS for EIA. Sections 3 and 4 substantiate the application of the AAS for EIA in context of energy consumption and CO_2-emissions. The results show the specific energy consumption and CO_2-emissions of the assembly process of two products. Section 4 describes the limitations of the conducted research and open questions. Section 5 concludes with a summary and an outlook.

2 State of the Art

2.1 Environmental Impact Assessment (EIA) Within Enterprises

According to [3], EIA is defined as "an assessment of the impact of a planned activity on the environment". There are numerous approaches to EIA in literature and industrial practice [3]. On the one hand, companies need EIAs for their internal decision-making processes concerning product and production decisions. On the other hand, an increasing number of companies has to publish information to stakeholders externally, e.g. in the form of sustainability reports. In the future, CO_2 labelling of products could become mandatory, in analogy to already existing energy labels.

The most widely used method for EIA is Life Cycle Assessment (LCA) [4]. This standardized method enables the determination of environmental impacts of a product or entire company. However, when applying the method, assumptions and simplifications have to be made, so that the results cannot be easily compared. Due to the detailed consideration of all life cycle phases, the data acquisition causes high efforts and requires huge databases. Here, only average values, e.g. for raw material mining or production of exemplary products, are available [5, 6]. Although these average values considerably reduce the effort for data collection and make a complete LCA of a product life cycle possible in the first place, it nonetheless inhibits an accounting based on the actual environmental impacts [6]. Further methods for EIA are e.g. Material Flow Cost Accounting or specific environmental performance indicators [7]. The data collection shapes the basis for each EIA and is a prerequisite for a valid and meaningful assessment [8]. In a best case, the environmental impacts can be directly assigned to products or the processes that cause them. For decision support, it should also be possible to analyze the data in close-to-real time. However, there are currently hardly any approaches to structurally organize the acquisition and transfer of environmental data in companies [4].

Without the support of digitization or software, the acquisition and structuring of sustainability data in companies is hardly possible [9]. Current existing software soultions do mostly not allow real-time monitoringor are very specifically designed for individual use cases, such as energy monitoring. An extension of systems already in use, like ERP systems, have been suggested. Mostly, they can only provide a small fraction of the data required for an EIA, e.g. material data [10,11, 12]. Thus, they do not fulfill the requirements for a complete and efficient EIA within manufacturing companies.

2.2 Usage of AAS

The approach of an AAS can by derived from the approach of an industry 4.0 component (I4.0 component). The I4.0 component combines an asset (e.g. a machine or a machine sensor; the granularity is application-dependent) and a virtual interface, which is referred to as AAS [13]. The AAS bundles functions and data via sub-models, which can be delivered as active or passive implementation [14].

The reference architecture model for the I4.0 component referes the entire life cycle value stream, which includes the product development, production, operation and recycling process [15]. The requirements for an AAS can be found in [16]. Several works [17, 18] have demonstrated the potentials of application of the AAS in various manufacturing areas e.g. for Wenger et al. demonstrated the PLC connection by using an asset administration shell for configuration purposes [19]. Lang et al. propagated the introduction of sub-models within the AAS for blockchain technology, which serves as security component during the plug and produce approach [20]. Additional work has been carried out by Lang et al., where the AAS has been used for maintenance procedures [21]. Al Assadi et al. presented an AAS for human workers in production, which included worker preferences to improve the human-machine-interface [22].

The AAS has not been used as an enabler for automated EIA. However, an energy efficiency sub-model has been mentioned in [16]. Further explanation or detailed discussion is missing. According to [16] a potential source of the sub-model can be the ISO 20140-5.

2.3 AAS in the Context of EIA

In order to apply AAS for EIA, it is first necessary to develop a general picture on how an AAS may be used for EIA. Thereby, intra and intercompany aspects have to be differed in relation to the features of AAS. The interrelation between features, advantages and the resulting intra- and inter-company applications can be seen in Fig. 1.

Sub-models contain structured properties, events and operation which include a self-description for automation purposes [23, 24]. This includes the standardization of interfaces and semantic knowledge representation formalism, which enable interoperability across components. The automated data acquisition reduces manual efforts [25]. A standardized data structure of assets facilitates data processing and aggregation and thus, supports decisions by improving the data basis. The data content of AAS is enhanced successively during the planning, runtime and recycling period [26].

Within the company real-time capability enables control, decision-making and modifications while running production and a faster detection of anomalies or errors which can lead to an environment-oriented production planning and control. Higher data availability within production and, if applicable, of additional product life cycle phases, like usage or end of life, can enable an environment-oriented product development. Due to the detailed and structured data acquisition, resource consumption and monetary as well

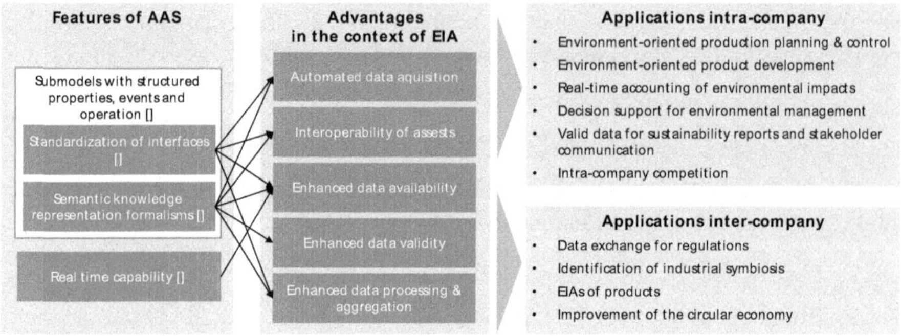

Fig. 1. Features, resulting advantages in the context of EIA and possible intra-/intercompany applications of AAS

as environmental costs can be assigned to the corresponding cost center. Further applications are sustainability reports and intra-company competition, e.g. for a minimization of CO_2.

Intercompany applications in the context of EIA can be data exchange for regulations (e.g. REACh, RoHS, WEEE). An EIAs of products can be enabled and the manual effort drastically reduced so that a LCA for products can be automated and based on real data, which could be used for product labels. By providing information on product composition, use circular economy can be accelerated.

3 Implementation of the AAS for EIA

In order to enable an automated product related allocation of CO_2 emissions and energy consumption, the implementation of the presented approach requires three AAS. Namely one AAS for the robot and one for the energy logger, compare Fig. 2. A third AAS serves as EIA service in which the power and energy consumption as well as CO_2 footprint is calculated.

The AAS of the robot contains the robot process related properties (e.g. joint positions, torques and forces) and the lable of the product programm. This additional information serves as allocation point for the product related energy consumption and CO_2 emission. Technically, the AAS approach of Ewert et al [17] is applied. The Message Queuing Telemetry Transport (MQTT) serves as communication protocol for the demonstrated use case. The different assets are publishing, respectively subscribing MQTT topics, compare Fig. 2.

4 Result

Figure 3 illustrates the graph of electric power over time (dark green) and cumulated CO_2 emissions (red) during an experiment of 7 min, where two different products (A & B) are assembled. The tact time of an assembly can be considered to be 3 min and 20 s, 1 min and 10 s respectively. Figure 3 shows that during assembly, the power ranges

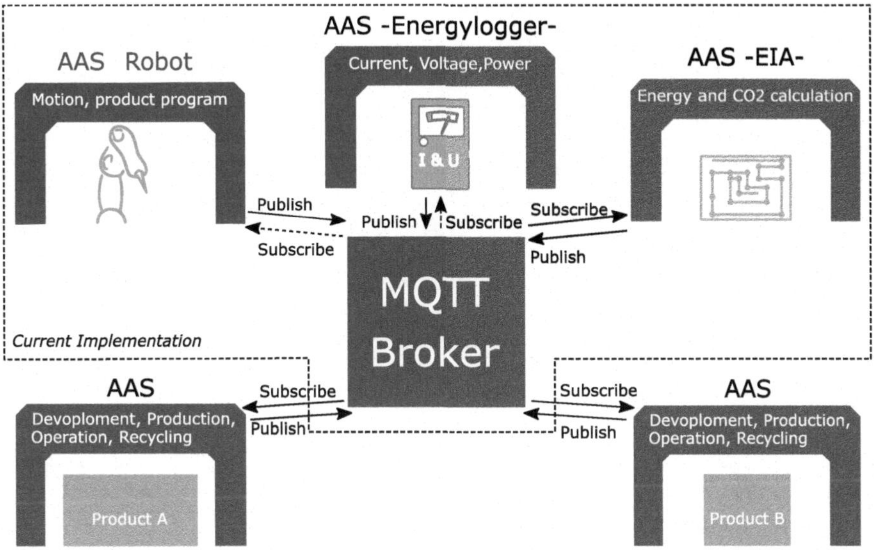

Fig. 2. The implementation setup of an automated energy/EIA via AAS

between 210 and 243 W. The different path velocities of the robot program causes the fluctuation in power consumption. The steady state is justifiable by the constant power consumption of the robot controller in idle mode.The linear approximation of the cumulative CO_2 emission shows the different slope/gradient of the increasing CO_2 emssion during assembly. The cycle time of the assembly takes 3 min and 20 s for A, 1 min and 10 s for B, respectively. The resulting emission of carbon dioxide during the assembly of A is 5 g, whereas the assembly of B is 2 g.

5 Discussion

This paper showed the potential of AAS in context of energy and CO_2 emission tracking. The further allocation of emissions to specific products opens up the window to foot print calculations. It seems that there is a linear relationship between cycle time and CO_2 emissions, however we conducted a short and simple assembly process. We expect a non-linear relationship between cycle time and CO_2 emissions in case of an entire production job. For reasons of poor data availability of the specific CO_2 emissions due to different energy sources, we have used the average value of 401 g CO_2 emissions per kilowatt hour of the specific electricity mix in Germany [27]. Further work will take into account the individual power supply and higher data resolution, which can be integrated in the service-oriented approach of the AAS. The adjustments will result in improved accuracy of the caused CO_2 emissions estimations[1].

The described approach demonstrates the ability of product-related sustainability footprint tracking in the context of energy consumption and CO_2 emissions. The general

[1] The specific CO_2 emissions of purchased electricity are available as 15 min average calculated by energy supply companies.

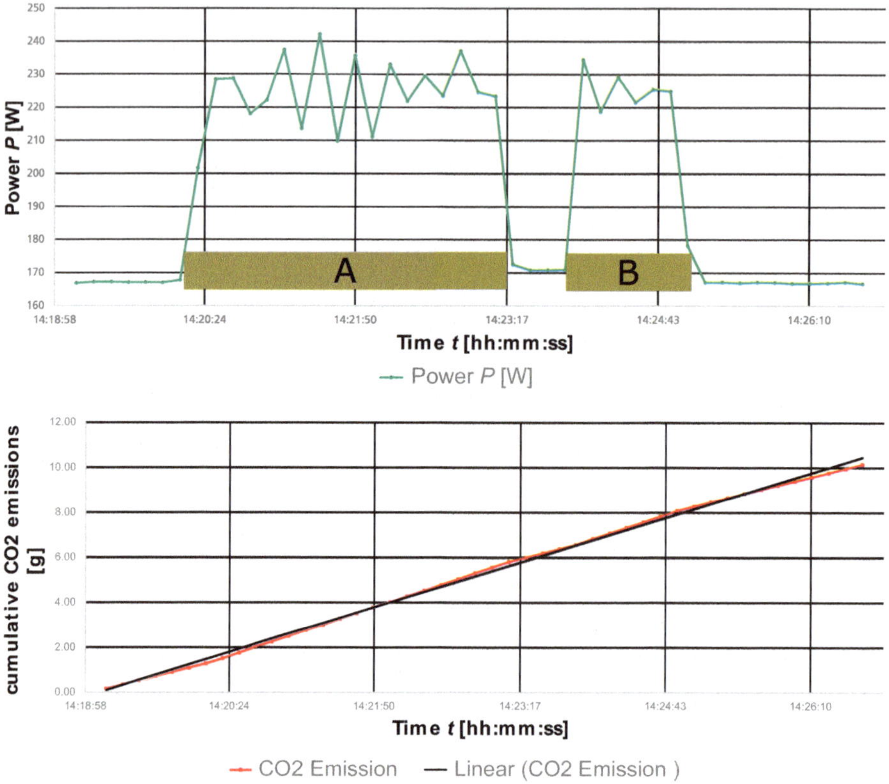

Fig. 3. Electric power and culmative CO_2 emisions during the assembly of product A & B

potential of an AAS in the context of EIA and implemented an AAS for the monitoring of CO_2 emissions of an assembly processs has been shown, the following research questions remain unanswered. They need to be addressed in future research in order to develop and evolve thepotential of AAS for EIA:

- Which data for EIA acquired by AAS could create added value for companies, societies, authorities and customers?
- How could data for EIA be semantically described and structured in order to create a semantic sub-model for EIA?
- Which new services or business models are enabled via AAS in the context of EIA?
- How can a production process be environmentally optimized by applying AAS?

6 Summary and Outlook

This paper presented the potential of the AAS regarding an automated EIA. Besides the presentation of the state of the art, features and advantages of AAS in the context of EIA were adresssed. A first demonstration and calculation of the product- specific carbon dioxide emissions of an assembly process have been performed. Results show that the real electric energy consumption and CO_2 emissions can be assigned to the product using AAS. This highlights the usability and advantages of using AAS for EIA. However, there are some limitations of the preliminary implementation which are emphasized within the discussion. Furthermore, still unanswered research questions according to AAS and EIA are highlighted.

It is intended to pick up these unanswered research questions and eliminate some of the limitations of the research described. In analogy to the described approach in production tracking, the usability of AAS for EIA of products should be further demonstrated by extending the tracking of CO_2 emissions to the entire production.

Acknowledgements. The research presented in this paper has received partial funding under administration of the Project Management Agency (PTKA) inside the research campus ARENA2036. Our sincere thanks go to the Federal Ministry for Education and Research (BMBF) for supporting this research project by the grant agreement 02P18Q620. The authors would like to thank Mr. Matthias Paukner from KUKA Systems GmbH for equipment support.

References

1. Hogh-Binder, D.: Klimaschutz: Bosch ab 2020 weltweit CO_2-neutral. https://www.bosch-presse.de/pressportal/de/de/klimaschutz-bosch-ab-2020-weltweit-co2-neutral-188800.html. Accessed 30 June 2020
2. UNFCCC: I am a company or an organization. https://unfccc.int/climate-action/climate-neutral-now/i-am-a-company-or-an-organization. Accessed 30 June 2020
3. Büyüközkan, G., Karabulut, Y.: Sustainability performance evaluation: literature review and future directions. J. Environ. Manag. **217**, 253–267 (2018)
4. Schaltegger, S., Burritt, R.: Contemporary Environmental Accounting. Issues, Concepts and Practice. Taylor and Francis, London (2017)
5. Miah, J.H., Griffiths, A., McNeill, R., Halvorson, S., Schenker, U., Espinoza-Orias, N., Morse, S., Yang, A., Sadhukhan, J.: A framework for increasing the availability of life cycle inventory data based on the role of multinational companies. Int. J. Life Cycle Assess. **23**(9), 1744–1760 (2018)
6. Teh, D., Khan, T., Corbitt, B., Ong, C.E.: Sustainability strategy and blockchain-enabled life cycle assessment: a focus on materials industry. Environ. Syst. Decis. **40**(4), 605–622 (2020)
7. Jasch, C., Tukker, A.: Environmental and Material Flow Cost Accounting. Principles and Procedures. Eco-efficiency in Industry and Science, 1st edn., vol. 25. Springer, Netherlands (2009)
8. Schebek, L., Kannengießer, J., Campitelli, A.: Ressourceneffizienz durch Industrie 4.0: Potenziale für KMU des verarbeitenden Gewerbes. VDI Zentrum Ressourceneffizienz GmbH (VDI ZRE) (2017)
9. Beier, G., Niehoff, S., Xue, B.: More sustainability in industry through industrial internet of things? Appl. Sci. **8**(2), 219 (2018)
10. Jacob, M.: Digitalisierung & Nachhaltigkeit. Eine unternehmerische Perspektive, 1st edn. (2019)
11. Lang-Koetz, C.: Ein Vorgehensmodell zur Einführung eines integrativen Umweltcontrollings auf Basis eines ERP-Systems. Dissertation, Univ., Stuttgart. [IPA-IAO-Forschung und -Praxis], vol. 440. Univ; Jost-Jetter, Stuttgart (2006)
12. Scholtz, B., Calitz, A., Haupt, R.: A business intelligence framework for sustainability information management in higher education. Int. J. Sustain. High. Educ. (2018)

13. Adolphs, P., Bedenbender, H., Dirzus, D., Ehlich, M., Epple, U., Hankel, M., Heidel, R., Hoffmeister, M., Huhle, H., Kärcher, B.: Reference architecture model industrie 4.0 (rami4. 0). ZVEI and VDI, Status report (2015)
14. Fuchs, J., Schmidt, J., Franke, J., Rehman, K., Sauer, M., Karnouskos, S.: I4.0-compliant integration of assets utilizing the Asset Administration Shell. In: Proceedings, 2019 24th IEEE International Conference on Emerging Technologies and Factory Automation (ETFA). Paraninfo Building, University of Zaragoza, Zaragoza, Spain, 10–13 September, 2019. 2019 24th IEEE International Conference on Emerging Technologies and Factory Automation (ETFA), Zaragoza, Spain, 9/10/2019–9/13/2019, pp. 1243–1247. IEEE, Piscataway (2019)
15. Hankel, M., Rexroth, B.: The reference architectural model industrie 4.0 (rami 4.0). ZVEI **410** (2015)
16. BMWI: The structure of the administration shell: TRILATERAL PERSPECTIVES from France, Italy and Germany. https://www.plattform-i40.de/PI40/Redaktion/DE/Downloads/Publikation/hm-2018-trilaterale-coop.html. Accessed 19 June 2020
17. Daniel, E., Stiedl, T., Jung, T., Tasci, T.: Assets2036 – Eine leichtgewichtige Implementierung der Verwaltungsschale für einfache Adaption (2020)
18. Palm, F.: openAAS development repository for open asset administration shell. https://acplt.github.io/openAAS/. Accessed 19 June 2020
19. Wenger, M., Zoitl, A., Müller, T.: Connecting PLCs with their asset administration shell for automatic device configuration. In: 2018 IEEE 16th International Conference on Industrial Informatics (INDIN), pp. 74–79 (2018)
20. Lang, D., Friesen, M., Ehrlich, M., Wisniewski, L., Jasperneite, J.: Pursuing the vision of Industrie 4.0: secure plug-and-produce by means of the asset administration shell and blockchain technology. In: 2018 IEEE 16th International Conference on Industrial Informatics (INDIN). 2018 IEEE 16th International Conference on Industrial Informatics (INDIN), Porto, 18.07.2018–20.07.2018, pp. 1092–1097. IEEE (2018)
21. Lang, D., Grunau, S., Wisniewski, L., Jasperneite, J.: Utilization of the asset administration shell to support humans during the maintenance process. In: 2019 IEEE 17th International Conference on Industrial Informatics (INDIN), pp. 768–773 (2019)
22. Al Assadi, A., Fries, C., Manuel, F., Maschler, B., Ewert, D.: User-friendly, requirement based assistance for production workforce using an asset administration shell design. Procedia CIRP (2020) (in Press)
23. Ye, X., Hong, S.H.: Toward Industry 4.0 components: insights into and implementation of asset administration shells. EEE Ind. Electron. Mag. **13**(1), 13–25 (2019)
24. Marcon, P., Diedrich, C., Zezulka, F., Schröder, T., Belyaev, A., Arm, J., Benesl, T., Bradac, Z., Vesely, I.: The asset administration shell of operator in the platform of Industry 4.0. In: 2018 18th International Conference on Mechatronics – Mechatronika (ME), pp. 1–5 (2018)
25. Uhlemann, T.H.-J., Schock, C., Lehmann, C., Freiberger, S., Steinhilper, R.: The digital twin: demonstrating the potential of real time data acquisition in production systems. Procedia Manuf. (2017)
26. Wagner, C., Grothoff, J., Epple, U., Drath, R., Malakuti, S., Gruner, S., Hoffmeister, M., Zimermann, P.: The role of the Industry 4.0 asset administration shell and the digital twin during the life cycle of a plant. In: 2017 22nd IEEE International Conference on Emerging Technologies and Factory Automation. September 12–15, 2017, Limassol, Cyprus. 2017 22nd IEEE International Conference on Emerging Technologies and Factory Automation (ETFA), Limassol, 9/12/2017–9/15/2017, pp. 1–8. IEEE, Piscataway (2017)
27. Umweltbundesamt: Strom- und Wärmeversorgung in Zahlen. https://www.umweltbundesamt.de/themen/klima-energie/energieversorgung/strom-waermeversorgung-in-zahlen. Accessed 14 July 2020

Business Model Innovation in Manufacturing Equipment Companies
Joint Project Fluid Production, ARENA2036

Alberto Mesa Cano, Tobias Stahl(✉), and Thomas Bauernhansl

Fraunhofer IPA, Nobelstr. 12, 70569 Stuttgart, Germany
tobias.stahl@ipa.fraunhofer.de

Abstract. Higher individualization levels, shorter product life cycles and fluctuating market demand influence manufacturing environments considerably, leading for example to a greater planning complexity and more frequent reconfiguration of production equipment and systems. These factors generate high non-value-added production costs that reduce factory adaptability.

In this context, automobile manufacturers demand a reduction in installation and integration activities that allow for an increase in the versatility of their production facilities. Business model innovations are hence required for manufacturing equipment suppliers to meet their customers' demands and stay competitive in the long term.

This paper identifies promising approaches for manufacturing business model innovation, including preliminary assessment within the research campus ARENA2036. The approaches lead to alternative business models that represent a possible way out of the conventional business dynamics between automobile manufacturers and their equipment suppliers, currently inhibiting innovation towards more flexible production systems.

1 Introduction

Increasing uncertainty resulting from higher individualization levels, shorter product life cycles and fluctuating market demand leads to greater product and market complexity, to which companies must constantly adapt in order to survive. High dynamics and intensity of changes call for a short-term adaptability which in turn requires a holistic interoperability of production equipment, products, value chains and business processes [1–3]. Existing production concepts in industrial manufacturing have not yet been able to satisfy these requirements [2, 4, 5].

Reconfigurable production systems promise automobile manufacturers a quick and cost-effective adjustment of production facilities in terms of product variability and production capacity [6]. The joint project Fluid Production within the research campus ARENA2036 aims at a smart factory concept that meets the versatility requirements in the automotive industry [7]. However, the transformation of business processes is

P. Weißgraeber et al. (Eds.): *Advances in Automotive Production Technology – Theory and Application,* ARENA2036, pp. 53–62, 2021. https://doi.org/10.1007/978-3-662-62962-8_7

only possible if sustainable business models are in place [8, 9]. Business models are described as "the rationale of how an organization creates, delivers, and captures value" [10], finding an intense debate around the term [11]. A business model innovation takes place when several of its elements, such as value offer or target customer, are changed to achieve better outcomes [12]. Olthough our research is focused on Gassmann et al., Sustainable Business Model Innovation receives much attention of other academicians too [13–15].

The purpose of this paper is to identify promising approaches to business model innovation for manufacturing equipment companies in the context of reconfigurable production systems, taking into account the research activities within ARENA2036.

Our starting point were the 55 business model patterns provided by the BMI-Lab [12]. We began by defining several selection criteria to screen the most relevant patterns according to the requirements in the joint project Fluid Production. We then combined the selected patterns into eight approaches to business model innovation, and used the Value Proposition Canvas methodology [16] to assess their suitability to address automobile manufacturers needs in the context of reconfigurable production systems. Further literature research complemented the process, paying special attention to service economy in the manufacturing field. Later, we evaluated the approaches potential through a series of interviews within the research campus ARENA2036, a first step to prioritize the most promising concepts and discard those that should not be further developed into detailed business models.

2 Approaches to Business Model Innovation

2.1 Need for Business Model Innovation

Fluid Manufacturing Systems (FLMS) within the ARENA2036 research campus have ambitious goals with regard to production reconfiguration: Reduction of 50% of the costs generated by indirect areas, 60% reduction in integration costs, new product introduction with 80% re-use of equipment without re-programming [7].

35% of the purchase price of a spot-welding robot in the American automotive industry corresponds to systems engineering tasks, which include programming and installation [17]. This value was used to estimate the impact of FLMS in the revenue of equipment suppliers, relying on the description of the life cycle costs (LCC) provided by VDMA [18]. The 60% reduction in integration costs (installation, start-up and training costs) lowers the procurement value of an industrial robot by around 21%. Table 1 shows that equipment suppliers face further reduction in their turnover, due to the fact that the reconfigurable production system aims at various LCC elements that represent important sources of revenue. New business models are therefore required for these companies to compensate for these declines.

Table 1. Relationship between LCC elements and FLMS goals within ARENA2036 [7, 18]

LCC – Elements		Reduction
E1 Acquisition	E1.5 Installation	−60%
	E1.6 Start-up costs	−60%
	E1.9 Training costs	−60%
E2 Infrastructure costs		−60%
IH1 Maintenance and inspection		−50%
IH2 Repairs		−50%
IH3 Unscheduled repairs		−50%
V1 Dismantling	V1.1 Dismantling and decommissioning	−60%
	V1.5 Renovation	−60%

2.2 Screening of Business Model Patterns

In our study we screened 55 business model patterns from the BMI-Lab [12] to identify promising approaches to business model innovation in the context of reconfigurable production systems. Although the need to adapt traditional business models has been clearly highlighted in the project Fluid Production [7], the financial issues that manufacturing equipment suppliers may have to confront in the new production environment are rather openly described, primarily speaking about decreasing turnover.

We favoured the explorative approach from Gassman et al. [12] and defined two criteria for the pattern selection. At first, we only considered patterns that benefit manufacturing equipment suppliers as well as their customers, since a supplier is unlikely to modify its current business model if the change only creates additional value for the companies it serves [19]. In a second step, we selected business model patterns that offer possible mechanisms to mitigate the potential negative effects of reconfigurable production systems on the suppliers' financial results. We screened the patterns according to these selection criteria and on the basis of the detailed patterns description provided by the BMI-Lab [12], further literature (Table 2, Pertinence) and the insights gained during research activities within ARENA2036. This process resulted in eleven concepts we then combined into eight (Table 2). The combination of business model patterns is one of the basic strategies to generate new business ideas, using reinforcing effects to make imitation by competitors more difficult [12].

Given the exploratory nature of our study, this selection of promising approaches to business model innovation represents only a few of the possible ways to encourage greater versatility of production equipment in the mechanical engineering sector.

Next, we used the Value Proposition Canvas [16] to assess the extent to which these approaches match the goals and challenges of reconfigurable production systems. In our study the customer profile was defined according to the requirements of reconfigurable production systems and the value proposition was represented by the approaches to business model innovation (Table 2).

Table 2. Approaches to business model innovation [12]

Approach	Description	Pertinence
Cross Selling & Solution Provider	The value offer is extended with complementary products and services up to a complete integrated solution. The customer reduces procurement costs and the risk associated with new suppliers. He can also fully concentrate on his core business, which usually leads to a performance increase. The supplier aims at a higher revenue, without investing in the acquisition of new customers, strengthened customer relationships and differentiation advantages, since he often becomes the "single point of contact"	[20–23]
Mass Customization & Add On	Customization of production equipment is made possible based on the combination of standardized modules. Extras, like additional attributes or bundled services, are charged separately. The customer benefits from tailor-made products without having to pay for functionalities he does not value. At the same time, the customer is likely to pay more for the final customized product than a similar offer would cost in the market	[20, 22]
Guaranteed Availability	The equipment is offered along with an availability guarantee, usually in form of a contract agreement. The customer benefits from guaranteed equipment functionality and possible savings through maintenance outsourcing. The supplier generates recurring revenues while strengthening his relationship with the customer [24]	[21, 23]
Fractionalized Ownership	This approach involves the purchase of production equipment by several companies that are willing to share it. Shares are usually offered in different amounts, entitling the owner to a certain quota of use hours or produced units. The customer can use equipment that would otherwise not be affordable while the supplier access new customer segments. The supplier can also manage the arrangement and generate new income through administration fees	[25]

(continued)

Table 2. (*continued*)

Approach	Description	Pertinence
Digitalization & Leverage Customer Data	Production facilities are equipped with sensors and connected to the internet, generating customer value through data streams analytical evaluation and internet-based services. The customer benefits from a wide range of new and improved services which in turn enable the supplier to differentiate himself. The potential of data-related business models for all involved stakeholders is also based on high measurability, reach and scalability [26–28]	[14, 20–22, 29, 30]
Rent Instead of Buy	The customer acquires the right to use production equipment for a certain period of time, thereby avoiding investment costs and gaining access to certain resources that might otherwise not be affordable. The supplier generates recurring income, allowing for a possible increase in turnover per machine. He can also increase its customer base targeting smaller companies	[23, 25]
Pay-per-Use	The customer pays based on his actual use of production equipment, often committing to a minimum figure. His advantages are related with increased financial flexibility and cost transparency. As with the previous approach, the supplier benefits from recurring revenues and the potential for a higher turnover	[20, 21, 23, 25]
Performance-based Contracting	Customer pays according to the obtained outcomes. Other than the Pay-per-Use approach it makes no difference how often or how intensively the equipment is used. The services often include the entire product life cycle [27]	[20, 21, 23, 25]

2.3 Requirements of Reconfigurable Production Systems

In line with the Value Proposition Canvas, we expressed the goals and challenges of reconfigurable production systems as customer jobs, pains and gains, thus facilitating the subsequent analysis of the value proposition suitability to address them [16]. These requirements were primarily extracted from the Fluid Production framework (Table 3). [7].

Table 3. Value Proposition Canvas – Customer Jobs, Pains and Gains of the FLMS

Customer Jobs (tasks, problems and needs)	a. To plan production to meet growing product customization and shorter product life cycles, aiming at the highest possible profitability [31–33] b. To adapt production capacity to the constantly fluctuating market demand as fast and economically as possible [6]
Customer Pains (annoyants, obstacles and risks)	c. Production facilities are constrained by monolithic structures that obstruct their adaptability to changing conditions [33] d. Decreasing responsiveness due to frequent, complex and time-consuming setup, start-up and re-parameterization of production equipment following adjustments. This results in greater complexity in production planning and control [33] e. Fixed sequence in production line [34]
Customer Gains (benefits and desired outcomes)	f. Production reacts quickly and cost-effectively to changes and deviations [2, 34] g. Easier remodelling of factory equipment [32, 34] h. Reduction of integration efforts via software tools for automatic system calibration, enabling "Plug and Produce" capability of production equipment [34–36] i. Easier ramp-up and reduction of indirect processes to control automated production processes [34, 37] j. Learning processes through feedback and complete data transparency [8, 31, 38] k. Breakdown of monolithic systems into smart modules in the form of cyber-physical systems [34, 36, 38]

2.4 Problem-Solution Fit

Once we had defined the customer profile, i.e. customer jobs, pains and gains, and the value proposition, i.e. approaches to business model innovation, we analyzed the fit between them. In Table 4 they are matched directly, without subdividing the approaches into the products and services, pain relievers and gain creators they entail, as Osterwalder et al. propose [16]. This fit analysis was a process based on both literature research and expert interviews within ARENA2036.

The potential of the business model innovation approaches to alleviate eventual turnover losses at manufacturing equipment suppliers, on which one criterion for selecting business model patterns was based, does not address any customer job, pain or gain, since improving the financial performance of suppliers is obviously not an objective of a reconfigurable production. It is, however, a basic requirement for achieving suppliers engagement in the development and implementation of the new system [7]. Thus the approaches help to adapt the business to the new production context while facilitating its realization.

Table 4. Fit between approaches to business model innovation and reconfigurable production systems [12, 16]

Value Proposition	Jobs		Pains			Gains					
	a	b	c	d	e	f	g	h	i	j	k
Cross Selling & Solution Provider[1]	■	■				■					
Mass Customization & Add On[2]		■				■	■				
Guaranteed Availability[3]									■		
Fractionalized Ownership[4]		■									
Digitalization & Leverage Customer Data[5]	■	■	■	■	■	■	■	■	■	■	■
Rent Instead of Buy[6]	■	■				■					
Pay-per-Use[6]	■	■				■					
Performance-based Contracting[6]	■	■				■					

(1) The approach Solution Provider covers several concepts, including "Factory within Factory", "Fence-to-Fence" and "Factory-as-a-Service" [39], which offer different options to handle increasing customization demands, shorter product life cycles and fluctuating market demand. (2) The modularization of production equipment increases its flexibility as well as the overall scalability of production capacity [20, 22, 34]. (3) A greater know-how from supplier leads to an increase of maintenance efficiency [23, 34]. (4) Sharing concepts are an option to maximise utilisation or ensure back-up production capacity [25]. (5) Increased data transparency and interoperability offers a wide range of new possibilities for process optimization [20, 29, 34]. (6) Alternatives to capital expenditures reduce investment risks, addressing shorter product life cycles as well as changes in required production capacity [19, 23]

3 Feedback Within ARENA2036 and Final Discussion

The approaches to business model innovation were reviewed through a series of interviews within the ARENA2036 research campus, as part of the FLMS project. In order for the industry experts to easily assess the potential perceived in every approach, we defined an evaluation template based on business model elements [10], business hypothesis [16, 40] and assessment criteria used previously in the field: novelty, context-related relevance, practicability and intrinsic value [41]. This procedure does not claim to be complete. The results can also be transferred to other areas of manufacturing engineering, although in a different context a new screening and evaluation of the business models patterns is advised.

Based on the evaluation results we formulated several theses which should be further studied in future research work to increase the so-called "certainty of evaluation" [10], for example through the calculation of business cases and subsequent implementation of pilot projects with target customers:

- Technology-oriented approaches show the highest potential. These are Digitalization & Leverage Customer Data and Pay-per-Use [30, 42].
- The approach Digitalization & Leverage Customer Data offers its greatest potential in terms of key resources and key partners. However, it has a limited impact on the cost structure [43].
- External financing plays a key role in the viability of approaches based on a shift away from capital costs. These are Rent Instead of Buy, Pay-per-Use and Performance-based Contracting [23].
- One of the biggest challenges for manufacturing equipment suppliers regarding the implementation of the reviewed concepts is to determine a price policy that keeps revenue streams above costs while being proportional to the value perceived by the customer, thereby encouraging his acceptance of the new framework [26, 43].

4 Summary

Achieving the flexibility levels targeted in the Fluid Production without adjusting the business models involved poses a major risk for equipment suppliers. An explorative analysis of business model patterns in a context of reconfigurable production systems resulted in eight approaches to business model innovation. Digitalization of factory equipment, process data analysis and alternative pricing models such as Pay-per-Use stood out as the most promising concepts to adapt traditional business models in Fluid Manufacturing Systems.

References

1. Bauernhansl, T., Tzempetonidou, M., Rossmeissl, T., et al.: Requirements for designing a cyber-physical system for competence development. Procedia Manuf. **23**, 201–206 (2018). https://doi.org/10.1016/j.promfg.2018.04.017
2. Delbrügger, T., Döbbeler, F., Graefenstein, J., et al.: Anpassungsintelligenz von Fabriken im dynamischen und komplexen Umfeld (2017)
3. Steegmüller, D., Zürn, M.: Wandlungsfähige Produktionssysteme für den Automobilbau der Zukunft. In: Vogel-Heuser, B., Bauernhansl, T., Hompel, M. ten (eds.) Handbuch Industrie 4.0: vol. 1: Produktion, 2., erweiterte und bearbeitete edn., pp. 27–44. Springer Vieweg, Berlin (2017)
4. Bauernhansl, T.: Die Vierte Industrielle Revolution – Der Weg in ein wertschaffendes Produktionsparadigma. In: Bauernhansl, T., Hompel, M. ten, Vogel-Heuser, B. (eds.) Industrie 4.0 in Produktion, Automatisierung und Logistik, pp 5–35. Springer Fachmedien Wiesbaden, Wiesbaden (2014)
5. Pallasch, C., Wein, S., Hoffmann, N., et al.: Edge powered industrial control: concept for combining cloud and automation technologies. In: 2018 IEEE International Conference on Edge Computing, pp. 130–134. IEEE, Piscataway (2018)
6. Bauernhansl, T.: Die Generation Smart Factory: Fraktal, wandlungsfähig, nachhaltig: Konzept einer zukunftsfähigen Fabrik. Fraunhofer IPA Interak. **2014**, 26–29 (2014)
7. Forschungscampus ARENA2036 e. V.: Fluide Fahrzeugproduktion für die Mobilität der Zukunft: Rahmenplan (2018)

8. Jerman, A., Pejić Bach, M., Aleksić, A.: Transformation towards smart factory system: examining new job profiles and competencies. Syst. Res. Behav. Sci. **37**, 388–402 (2020). https://doi.org/10.1002/sres.2657
9. Klostermeier, R., Haag, S., Benlian, A.: Geschäftsmodelle digitaler Zwillinge. Springer Fachmedien Wiesbaden, Wiesbaden (2020)
10. Osterwalder, A., Pigneur, Y.: Business Model Generation: A Handbook for Visionaries, Game Changers, and Challengers. Wiley, New York (2013)
11. Wirtz, B.W., Pistoia, A., Ullrich, S., et al.: Business models: origin, development and future research perspectives. Long Range Plan. **49**, 36–54 (2016). https://doi.org/10.1016/j.lrp.2015.04.001
12. Gassmann, O., Frankenberger, K., Csik, M.: Geschäftsmodelle entwickeln: 55 innovative Konzepte mit dem St. Galler Business Model Navigator (2017)
13. Shakeel, J., Mardani, A., Chofreh, A.G., et al.: Anatomy of sustainable business model innovation. J. Clean. Prod. **261**, 121201 (2020). https://doi.org/10.1016/j.jclepro.2020.121201
14. Hoerlsberger, M.: Innovation management in a digital world. J. Manuf. Technol. Manag. **30**, 1117–1126 (2019). https://doi.org/10.1108/JMTM-12-2019-461
15. Improving your business posture. Strateg. Dir. **34**, 31–33 (2018). https://doi.org/10.1108/SD-02-2018-0025
16. Osterwalder, A., Pigneur, Y., Bernarda, G., et al.: Value Proposition Design: How to Create Products and Services Customers Want. Strategyzer Series, Wiley, Hoboken (2014)
17. Statista: Prognose zur Preisentwicklung eines Industrieroboters in den USA nach Posten bis zum Jahr 2025 (2015). https://de.statista.com/statistik/daten/studie/416525/umfrage/preisentwicklung-eines-industrieroboters/. Accessed 12 Feb 20
18. VDMA: VDMA 34160: 2006-06: Prognosemodell für die Lebenszykluskosten von Maschinen und Anlagen, Frankfurt (2006)
19. Lay, G., Meier, H., Schramm, J., et al.: Betreiben statt Verkaufen: Stand und Perspektiven neuer Geschäftsmodelle für den Maschinen- und Anlagenbau. Industrie-Management **19**, 9–14 (2003)
20. Demont, A., Paulus-Rohmer, D.: Industrie 4.0-Geschäftsmodelle systematisch entwickeln. In: Schallmo, D., Rusnjak, A., Anzengruber, J., et al. (eds.) Digitale Transformation von Geschäftsmodellen: Grundlagen, Instrumente und Best Practices, vol. 3, pp. 97–125. Springer Gabler, Wiesbaden (2017)
21. Bollhöfer, E., Buschak, D., Moll, C.: Dienstleistungsbasierte Geschäftsmodelle für Industrie 4.0 – aktueller Stand und Potenziale für KMU. In: Nissen, V., Stelzer, D., Straßburger, S., et al. (eds.) Multikonferenz Wirtschaftsinformatik (MKWI) 2016: Technische Universität Ilmenau, 09.–11. März 2016, vol. 3, pp. 1287–1298. Universitätsverlag Ilmenau, Ilmenau (2016)
22. Born, H.-J.: Geschäftsmodell-Innovation im Zeitalter der vierten industriellen Revolution: Strategisches Management im Maschinenbau. Springer Vieweg, Wiesbaden (2018)
23. Schuh, G., Gudergan, G., Grefrath, C.: Geschäftsmodelle für industrielle Dienstleistungen. In: Schuh, G., Gudergan, G., Kampker, A. (eds.) Management industrieller Dienstleistungen: Handbuch Produktion und Management 8, 2nd edn., pp. 65–104. Springer, Heidelberg (2016)
24. Granig, P., Hartlieb, E., Lingenhel, D.: Geschäftsmodellinnovationen. Springer Fachmedien Wiesbaden, Wiesbaden (2016)
25. Scholl, G., Gossen, M., Grubbe, M., et al.: Alternative Nutzungskonzepte – Sharing, Leasing und Wiederverwendung: PolRess AP2 – Politikansätze und -instrumente, Berlin (2013)
26. Acatech: Smart Service Welt 2018: Wo stehen wir? Wohin gehen wir? München (2018)
27. Oberländer, A.M., Übelhör, J., Häckel, B.: IIoT-basierte Geschäftsmodellinnovation im Industrie-Kontext: Archetypen und praktische Einblicke. HMD **56**, 1113–1125 (2019). https://doi.org/10.1365/s40702-019-00570-1

28. Cagle, M.N., Yılmaz, K., Doğru, H.: Digitalization of business functions under Industry 4.0. In: Hacioglu, U. (ed.) Digital Business Strategies in Blockchain Ecosystems, pp. 105–132. Springer International Publishing, Cham (2020)
29. Acatech: Smart Service Welt: Umsetzungsempfehlungen für das Zukunftsprojekt Internet-basierte Dienste für die Wirtschaft. Abschlussbericht, Berlin (2015)
30. Dremel, C., Herterich, M.: Digitale Cloud-Plattformen als Enabler zur analytischen Nutzung von operativen Produktdaten im Maschinen- und Anlagenbau. HMD **53**, 646–661 (2016). https://doi.org/10.1365/s40702-016-0250-9
31. 2018 IEEE International Conference on Edge Computing. IEEE, Piscataway (2018)
32. Möller, N.: Bestimmung der Wirtschaftlichkeit wandlungsfähiger Produktionssysteme. Dissertation, Techn. Univ., München. Forschungsberichte IWB, vol. 212. Utz, München (2008)
33. Huber, W.: Konzeption und Realisierung eines Produktionssystems für die modulare Fertigung in der Automobilindustrie. HMD **56**, 1250–1267 (2019). https://doi.org/10.1365/s40702-019-00572-z
34. Bauernhansl, T., Hompel, M. ten, Vogel-Heuser, B. (eds.): Industrie 4.0 in Produktion, Automatisierung und Logistik. Springer Fachmedien Wiesbaden, Wiesbaden (2014)
35. Kobzan, T., Boschmann, A., Althoff, S., et al.: Plug & Produce durch Software-defined Networking Thomas (2018). https://publica.fraunhofer.de/starweb/pub09/servlet.starweb
36. Abele, E., Berkholz, D., Bredow, M. von, et al.: Wandlungsfähige Produktionssysteme: Heute die Industrie von morgen gestalten (2008)
37. Hamm, J.: Vom Shopfloor bis ins ERP mit einer Sprache sprechen. MM – Maschinenmarkt. Das IndustrieMagazin 60–62 (2019)
38. Boss, B., Bader, S., Orzelski, A., et al.: Verwaltungsschale. In: Hompel, M. ten, Vogel-Heuser, B., Bauernhansl, T. (eds.) Handbuch Industrie 4.0, pp. 1–27. Springer, Heidelberg (2019)
39. Lay, G., Horst Meier, J.: Betreiben statt Verkaufen: Stand und Perspektiven neuer Geschäftsmodelle für den Maschinen- und Anlagenbau (2003)
40. BMI-Lab: The big challenge in business model development: how to systematically test your business model (2019). https://bmilab.com/blog/2019/4/18/the-big-challenge-in-business-model-development-how-to-systematically-test-your-business-model. Accessed 21 Mar 20
41. Csik, M.: Muster und das Generieren von Ideen für Geschäftsmodellinnovationen, Universität St. Gallen, Hochschule für Wirtschafts-, Rechts- und Sozialwissenschaften sowie Internationale Beziehungen (2014)
42. Lachenmaier, J.F., Haußmann, C., Kemper, H.-G., et al.: Entwicklung, Produktion und Einsatz von Industrie 4.0-Komponenten – Betriebswirtschaftliche Potenziale des digitalen Zwillings in der Produktion. In: Obermaier, R. (ed.) Handbuch Industrie 4.0 und Digitale Transformation, vol. 28, pp. 317–337. Springer Fachmedien Wiesbaden, Wiesbaden (2019)
43. Björkdahl, J.: Strategies for for digitalization in manufacturing firms. Calif. Manag. Rev. **62**, 17–36 (2020). https://doi.org/10.1177/0008125620920349

Identification of Reconfiguration Demand and Generation of Alternative Configurations for Cyber-Physical Production Systems

Timo Müller(✉), Simon Walth, Nasser Jazdi, and Michael Weyrich

Institute of Industrial Automation and Software Engineering, University of Stuttgart, 70550 Stuttgart, Germany
timo.mueller@ias.uni-stuttgart.de

Abstract. Ensuring high availability despite the growing frequency of changes in production requirements leads to an increased reconfiguration demand in the domain of industrial automation systems, which will be dominated by Cyber-Physical Production Systems (CPPS) in the future. Therefore, a concept, covering a methodology containing four steps, is utilized to answer the research question how to enhance CPPS with a self-organized reconfiguration management. This contribution focuses on the first two steps of this concept: the *identification of reconfiguration demand* and the *generation of alternative configurations.* A condensed presentation of the related work reveals that an interface-oriented formalized process description depicts an appropriate conceptual basis. Building upon this, the presented concept contains a capability model of the CPPS and the reasoning to determine a potential reconfiguration demand, as well as a procedure for the generation of alternative configurations. To evaluate the concept, an agent-based implementation is given, which uses an OPC-UA controlled simulated modular production system as a substitute CPPS in discrete manufacturing.

1 Introduction

The upcoming trends of shorter innovation and product life cycles [1], as well as the development towards mass-individualization, result in a frequent change of production requirements. These frequent changes cannot be fully foreseen as they occur during operation. Thus, no matter how big the flexibility corridor of a production systems is laid out in its design phase it will eventually fail to contain the appropriate functionality. Therefore, system changes during operation will become the rule instead of the exception [2, 3]. I.e. Adaptions by the means of reconfigurations are gaining more and more importance to ensure a high availability of production systems. Furthermore, Cyber-Physical Production Systems (CPPS), i.e. production systems consisting of Cyber-Physical Systems (CPS), form the baseline of future industrial automation [4–6]. Due to their high degree of connectivity, CPPS represent the vision of adaptive, self-configuring and partially self-organizing, flexible production systems [5]. Thus, they can lead to reduced set-up times and optimized use of energy and resources [5] amongst many other benefits. CPPS

P. Weißgraeber et al. (Eds.): *Advances in Automotive Production Technology – Theory and Application,* ARENA2036, pp. 63–70, 2021. https://doi.org/10.1007/978-3-662-62962-8_8

promise a broad spectrum of theoretical potentials, some of which are highlighted in [7]. However, corresponding concepts are required to utilize these theoretical potentials.

Therefore, the research question: *How to enhance CPPS with a self-organized reconfiguration management?* arises.

1.1 Requirements for the Self-organized Reconfiguration Management

In Order to determine the requirements for a self-organized reconfiguration management for CPPS, basics of reconfiguration have to be considered and current weaknesses have to be identified. This was accomplished through intensive literature research.

According to [8] RMS are one of the roots of CPPS. Additionally, the success of an RMS is primarily measured by the effort required for a reconfiguration process and the resulting benefits [9, 10]. Furthermore, to clarify that the topic reconfiguration spans more than only the conduction of reconfiguration measures, the term *reconfiguration management*, comprising the *identification of reconfiguration demand*, the *reconfiguration planning* and as an optional extend the *execution of reconfiguration measures* was introduced in [7]. Whereby, the reconfiguration planning can be further subdivided into the steps *generation of alternative configurations, evaluation of configurations* and *selection of a new configuration.*

So far, reconfigurations of production systems are rarely carried out [11] despite the fact that their necessity is undeniable [9]. The reasons for this are weaknesses concerning the current state of the art of reconfigurations, which are derived from [12, 13] and concluded as the following within [7]: *Time-consuming, error-prone, not optimum near* and *not based on objective criteria.*

Therefore, the requirements (RQ's) for the self-organized reconfiguration management for CPPS were derived as summarized in the following Table 1.

Table 1. Derived requirements for the self-organized reconfiguration management.

No. of RQ	Requirement
1	Appropriate representation of the reconfiguration management steps within the Cyber-Physical Production System
1.1	Identification of reconfiguration demand
1.2	Generation of alternative configurations
1.3	Evaluation of configurations
1.4	Selection of a new configuration
2	Automated execution of the reconfiguration management
3	Exploiting the potentials of Cyber-Physical Production Systems

A study carried out on the state of research has revealed that the formulated research question is not sufficiently answered with respect to the formulated requirements. Thus, [7] presents a basic concept in order to enhance CPPS with a self-organized reconfiguration management.

1.2 Basic Concept and Focus of this Contribution

This basic concept comprises the four steps *identification of reconfiguration demand*, *generation of alternative configurations*, *evaluation of configurations* and *selection of a new configuration*, covering the derived requirements. It was firstly presented in [7] and described on a rather abstract level. This contribution focuses on the first two steps *identification of reconfiguration demand* and *generation of alternative configurations*. In addition, [14] provides a detailed examination of aspects to be covered by information modeling.

2 Related Work

This chapter gives a brief overview of the work related to the focus of this contribution (*identification of reconfiguration demand* and *generation of alternative configurations*) and ends with a conclusion regarding the applicability of the reviewed approaches.

Even though not all derived requirements are met by existing concepts, to identify a reconfiguration demand and also to generate alternative configurations it is first necessary to be able to compare the requirements of the products with the capabilities of the CPPS, for which existing approaches are investigated. Most of the approaches within the literature rely on the Product-Process-Resource (PPR) concept, however, there is a wide range of variations.

In [15] an approach resulting in a framework for feasibility feedback dedicated to early design phases is presented. The framework relies on three kinds of ontologies to provide resource descriptions and product specifications based on domain specific knowledge in a reusable manner.

The development of the OWL-based manufacturing resource capability ontology (MaRCO) is presented in [16]. The authors give a detailed description of a resource description which, in contrast to others, supports an automatic inference of combined capabilities. In [17] the authors describe how information from a product model and a resource model is used to perform the matchmaking between product requirements and resource capabilities by utilizing a process taxonomy model.

In order to map CPPS capabilities with production requirements a virtual CPPS representation, containing the CPPS structure in the AML format, is applied in [18]. They convert the AML file into a UML data model for the attribute mapping between the CPPS capabilities and the production requirements based on specific rules. However, the approach is described on a rather abstract level.

A promising approach for the modeling of manufacturing resource capabilities is presented in [19]. Here, the PPR concept is applied through the formalized process description (FPD) based on the VDI/VDE 3682 guideline [20]. The special feature is that the processes are described based on their input and output elements, rather than by defined process parameters which is the common state of the art.

The interface-oriented FPD is chosen, as it offers further degrees of freedom for an intelligent generation of configurations at machine level by the Cyber-Physical Production Modules (CPPMs) themselves, amongst other benefits.

3 Identification of Reconfiguration Demand and Generation of Alternative Configurations

3.1 Identification of Reconfiguration Demand

To enable the comparison of a target production with the current configuration of the CPPS, the interface-oriented FPD is utilized as a conceptual basis for the modeling. The capabilities of a production resource (each part of the respective CPPM) are described by a process operator that defines two lists of characteristics and their values. The first list describes the requirements of the production input e.g. the *width* and *length* on an input product to be processable. The second list describes the possible output products that the production resource can produce. Thus, it describes the possible transformations that the production resource can carry out on a given input product.

Figure 1 depicts how the modeling approach is utilized to realize the capability model that describes the current configuration of the CPPS. In the example, the resources of the CPPS offer three process operators (*stamp, drill* and *mill*) with their defined in- and output products. Input characteristics with *preValues* are employed to indicate that values are copied from a predecessor. For outputs, they indicate that the process operator does not change the corresponding characteristic.

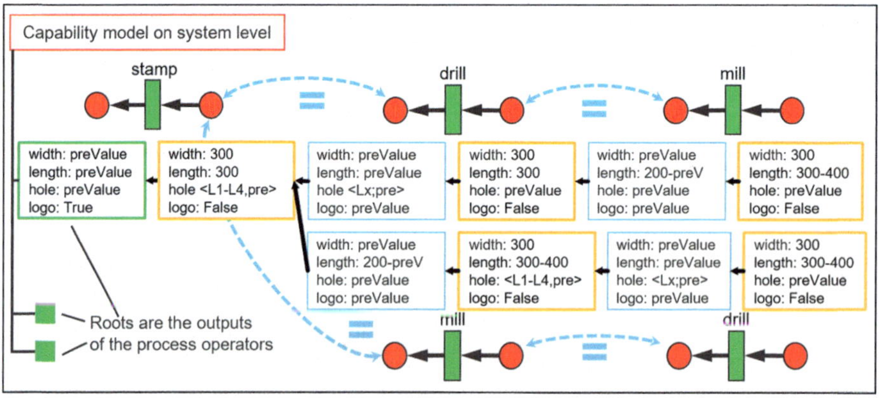

Fig. 1. Capability model of the CPPS.

Based on this description scheme the capability model is created by matching the output of a process operator with the input of another process operator yielding a tree structure with sequences of production processes. The roots of this tree structure are the outputs of the set of available process operators. They define all possible outputs of the current configuration of the CPPS in a generic way. Each possible output is connected to its corresponding process operator, with its respective input (orange) containing constraints with respect to following process operators. These inputs are then matched with the outputs of all of the next possible operators (blue). This procedure continues until no more process operators that provide a valid transformation can be found. Here it should be noted that process operator loops could occur which need to be addressed.

To perform the comparison of the target production and the current configuration, the target production is also described by the means of an input and an output product. Whenever its output matches one of the roots (green) of the tree, its input is compared with each input (orange) of the roots branches. Each match reveals a possible production sequence consisting of all process operators necessary to reach the respective root. If no branch of the capability model can produce the desired product, a reconfiguration demand is identified, and alternative configurations are generated in the second step.

3.2 Generation of Alternative Configurations

As soon as a reconfiguration demand is revealed, the first step of the *generation of alternative configuration*, the *generation of alternatives for production sequences* is performed, which is exemplified in Fig. 2.

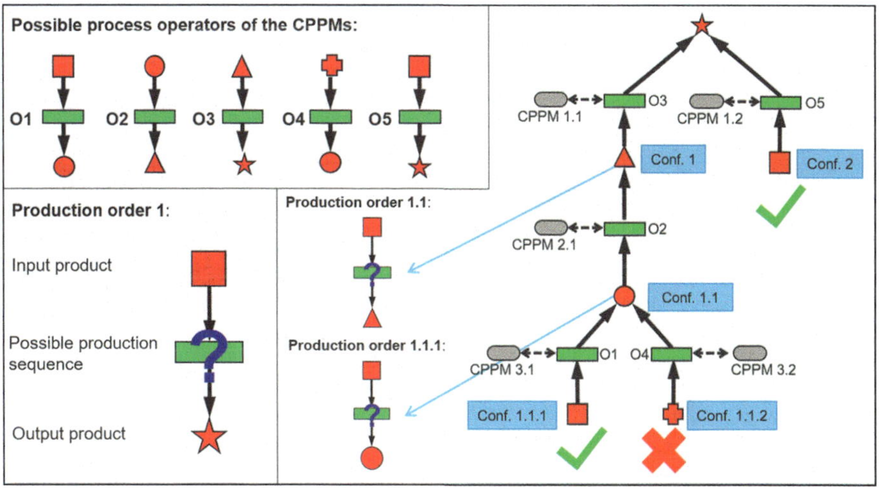

Fig. 2. Procedure for the generation of alternatives for production sequences (exemplified).

For the sake of simplicity, the FPD is enhanced by a generic in- and output product description, which is displayed in differently shaped red forms. Based on the production order, in form of an input and an output product description (see *production order 1* in Fig. 2), production sequences that are able to execute the transformation of the given input product into the required output product are determined.

At first, the description of the required output product is provided to all CPPMs. The CPPMs perform the *generation of alternatives at machine level* where they determine if they can offer a process operator (O), in their current or an alternative configuration, which can transform any kind of input product into the required output product (see O3 and O5 in Fig. 2). Subsequently, for each of the found process operators of the CPPMs a new *system configuration* is created and the process operator as well as its respective CPPM configuration are connected to the output product. After that, again a process operator is connected to the input product of each previously connected process operator until the input product of an added process operator matches the required input product, a loop occurs or no suitable process operator can be found.

To *determine layout variants of the alternatives,* the layout of the CPPS is modeled as a graph, in which the possible machine locations are numbered and transport connections are represented. All layout variants are determined by a simple brute force approach. Based on the layout information of the current configuration, the reconfiguration efforts of the CPPMs and the determined layout variant, the reconfiguration efforts for each alternative system configuration are calculated.

4 Proof of Concept

A modular production system, based on [21] with a matrix layout, simulated in Unity (Fig. 3, right) and controlled by a distributed service-oriented OPC UA control network providing discrete manufacturing services such as stamp, drill and mill is used to evaluate the presented concept.

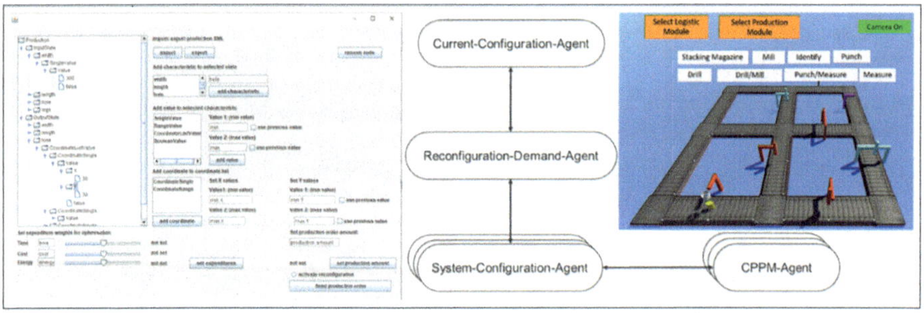

Fig. 3. Overview of the prototypical implementation.

The presented concept is realized using a multi-agent system (Fig. 3, mid), which is implemented with JADE as follows: A "Current-Configuration-Agent" provides the current configuration of the CPPS to a "Reconfiguration-Demand-Agent", which implements the step *identification of reconfiguration demand* described in Sect. 3.1. The step *generation of alternative configurations* is realized by "System-Configuration-Agents", each creating a new system configuration in cooperation with the "CPPM-Agents" and calculating its reconfiguration efforts. If a created system configuration is not able to produce the desired product, the corresponding "System-Configuration-Agent" terminates itself. Each "CPPM-Agent" represents a CPPM and checks, whether the output product requested by a "System-Configuration-Agent" can be reached by a process operator either in its current or a possible alternative resource configuration. Therefore, each "CPPM-Agent" contains its set of possible resource configurations and the reconfiguration efforts to conduct the corresponding reconfiguration. Furthermore, the GUI visualized in Fig. 3 (left) was implemented to enable the input of the target production.

5 Conclusion and Future Work

This contribution outlines the emergence of CPPS and the increasing demand for reconfigurations in future industrial automation. Consequently, the research question: *How to enhance CPPS with a self-organized reconfiguration management?* is postulated. Respective requirements to answer this research questions are derived and a basic concept that meets these requirements is presented. Subsequently, the scope of this paper is specified to the first two steps of this concept: The *identification of reconfiguration demand* and the *generation of alternative configurations.* The condensed presentation of the related work concerning the information modeling approaches reveals that an interface-oriented formalized process description depicts an appropriate conceptual basis to enable the functional comparison of the requirements of a product with the capabilities of a CPPS. A concept is presented which utilizes the interface-oriented formalized process description for:

- Creating a capability model of the CPPS and the reasoning with it to determine a potential reconfiguration demand.
- The decentralized generation of alternative configurations.

As a proof of concept an agent-based implementation with JADE is used and forms the substitute CPPS together with a service-oriented OPC UA control network and its modular production system simulated in Unity.

Regarding future work, the concept of the Intelligent Digital Twin [22] will be applied to perform a subsequent simulation-based optimization for each determined alternative configuration. The modeling of a CPPM already contains the required production efforts, which will be used for the optimization, evaluation and selection in accordance with the basic concept. However, this contribution did rather focus on the methodology and the applied modeling approach than on the concrete information modeling which will be presented in a future publication.

References

1. Järvenpää, E., Siltala, N., Lanz, M.: Formal resource and capability descriptions supporting rapid reconfiguration of assembly systems. In: 2016 IEEE International Symposium on Assembly and Manufacturing (ISAM), pp. 120–125 (2016)
2. Müller-Schloer, C., Schmeck, H., Ungerer, T.: Organic computing. Inform. Spektrum **35**, 71–73 (2012). https://doi.org/10.1007/s00287-012-0599-2
3. Vogel-Heuser, B., Fay, A., et al.: Evolution of software in automated production systems: challenges and research directions. J. Syst. Softw. **110**, 54–84 (2015)
4. Vogel-Heuser, B., Böhm, M., et al.: Interdisciplinary engineering of cyber-physical production systems: highlighting the benefits of a combined interdisciplinary modelling approach on the basis of an industrial case. Des. Sci. **6** (2020)
5. Bettenhausen, K.D., Kowalewski, S.: Cyber-physical systems: Chancen und Nutzen aus Sicht der Automation. VDI/VDE-Gesellschaft Mess-und Automatisierungstechnik 9–10 (2013)
6. Grochowski, M., Simon, H., et al.: Formale Methoden für rekonfigurierbare cyber-physische Systeme in der Produktion. at-Automatisierungstechnik **68**, 3–14 (2020)

7. Müller, T., Jazdi, N., et al.: Cyber-physical production systems: enhancement with a self-organized reconfiguration management. Procedia CIRP (2020)
8. Monostori, L.: Cyber-physical production systems: roots, expectations and R&D challenges. Procedia CIRP **17**, 9–13 (2014)
9. Stehle, T., Heisel, U.: Konfiguration und Rekonfiguration von Produktionssystemen. In: Neue Entwicklungen in der Unternehmensorganisation, pp. 333–367. Springer (2017)
10. Hees, A.F.: System zur Produktionsplanung für rekonfigurierbare Produktionssysteme, vol. 331. Herbert Utz (2017)
11. Järvenpää, E., Siltala, N., et al.: Capability matchmaking procedure to support rapid configuration and re-configuration of production systems. Procedia Manuf. **11**, 1053–1060 (2017)
12. Karl, F.: Bedarfsermittlung und Planung von Rekonfigurationen an Betriebsmitteln, vol. 298. Herbert Utz (2015)
13. Hoang, X.L., Fay, A., et al.: Systematization approach for the adaptation of manufacturing machines. In: 2016 IEEE 21st International Conference on Emerging Technologies and Factory Automation (ETFA), pp. 1–4 (2016)
14. Bauernhansl, T., Weyrich, M., et al.: Semantic structuring of elements and capabilities in ultra-flexible factories. Procedia CIRP (2020)
15. Ocker, F., Vogel-Heuser, B., Paredis, C.J.J.: Applying semantic web technologies to provide feasibility feedback in early design phases. J. Comput. Inf. Sci. Eng. **19** (2019)
16. Järvenpää, E., Siltala, N., et al.: The development of an ontology for describing the capabilities of manufacturing resources. J. Intell. Manuf. **30**, 959–978 (2019)
17. Järvenpää, E., Siltala, N., et al.: Product model ontology and its use in capability-based matchmaking. Procedia CIRP **72**, 1094–1099 (2018)
18. Hengstebeck, A., Barthelmey, A., Deuse, J.: Reconfiguration assistance for cyber-physical production systems. In: Tagungsband des 3. Kongresses Montage Handhabung Industrieroboter, pp. 177–186. Springer (2018)
19. Hoang, X.-L., Backhaus, S., et al.: An interface-oriented resource capability model to support reconfiguration of manufacturing systems. In: 2019 IEEE International Systems Conference (SysCon), pp. 1–8 (2019)
20. VDI/VDE Society for Measurement and Automatic Control: Formalised process descriptions (VDI/VDE 3682) (2015)
21. Schmidt, J.-P., Müller, T., Weyrich, M.: Methodology for the model driven development of service oriented plant controls. Procedia CIRP **67**, 173–178 (2018)
22. Talkhestani, B.A., Jung, T., et al.: An architecture of an intelligent digital twin in a cyber-physical production system. at-Automatisierungstechnik **67**, 762–782 (2019)

Method for Data-Driven Assembly Sequence Planning

Susann Kärcher[1(✉)] and Thomas Bauernhansl[1,2]

[1] Fraunhofer IPA, Nobelstraße 12, 70569 Stuttgart, Germany
susann.kaercher@ipa.fraunhofer.de
[2] IFF University of Stuttgart, Nobelstraße 12, 70569 Stuttgart, Germany

Abstract. In many manual assembly systems, there is great potential for optimization, especially when products in small quantities, high variants or with high complexity are produced. The more often the assembly is changed, the greater is the potential.

The main reason for the optimization potential is the still high effort required for an assembly planning. Especially in today's challenging and volatile environment, classic assembly planning often reaches its limits. As a result, assembly systems are often not planned in sufficient detail. The consequence is a lack of transparency: Workers in assembly do not get clear work instructions and planners do not get feedback from the assembly.

There are approaches to reduce the effort required for assembly planning meeting the challenge from two sides: On the one hand, there are approaches to further integrate assembly planning with previous processes, such as product development. On the other hand, there are approaches that optimize the processes from an assembly perspective.

This paper focuses on a method to optimize assembly sequence planning based on actual data. Data is collected, for example, via sensors in the assembly area. Afterwards, different runs of the assembly process are analyzed. Then, an algorithm derives the best practice to assemble the product. Best practice describes the assembly sequence that leads to the fastest assembly. The method fits into a methodology to transfer benchmarking to manual assembly and can be used for a one-time optimization project as well as for continuous optimization. The results generated in the algorithm are then made available to workers and planners.

1 Introduction

Today, companies face a volatile and challenging environment. Classical methods of assembly planning reach their limits especially when assembly systems are subject to frequent changes, the number of variants is high or quantities are small. Therefore, many manual assembly systems still have a high optimization potential. These systems lack in transparency for planners and workers. [1]

A methodology for the automated generation of an assembly sequence planning meets these challenges. It compares different solutions for assembling the product based on data from production, combines them and derives the best assembly sequence. [1]

P. Weißgraeber et al. (Eds.): *Advances in Automotive Production Technology – Theory and Application,* ARENA2036, pp. 71–79, 2021. https://doi.org/10.1007/978-3-662-62962-8_9

This paper focuses on the data analysis step. Section 2 presents other approaches in the field of automatic assembly planning. Section 3 provides the overall context (Sect. 3.1), then, the method is presented (Sect. 3.2) and illustrated with an example (Sect. 3.3). Section 4 concludes with a summary and an outlook.

2 Approaches in the Field of Automatic Assembly Planning

There are approaches to reduce the effort required for assembly planning meeting the challenge from two sides: On the one hand, there are approaches to further integrate assembly planning with previous processes, such as product development. On the other hand, there are approaches that optimize the processes from an assembly perspective (see Fig. 1). There are also combinations of the two perspectives.

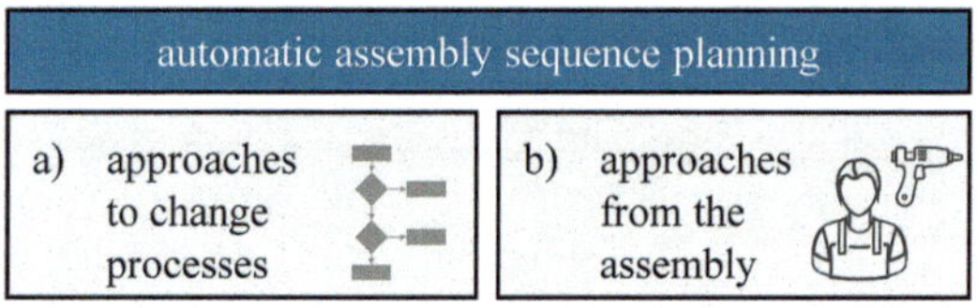

Fig. 1. Approaches in the field of automatic assembly planning

Approaches to process changes are, for example, the (feature-based) derivation of assembly planning from the product model [2, 3], the simultaneous planning of assembly during product development [4] and prototype phase [5, 6].

Approaches from the assembly are, for example, recording assembly activities, authoring systems or possibilities to improve the assembly instructions from the workers on the shop floor. In the following, a few approaches are presented:

[5] presents an approach to reduce the planning effort for an assembly, especially in the case of short product development cycles. Worker data is recorded during the assembly of the prototypes and the necessary information for series production is generated. [6] introduce an approach in which assembly activities are recorded via RFID during the construction of a prototype and bills of materials are created.

In both approaches [5, 6] the assembly planning is also integrated as a process in the product development process. Both approaches therefore belong to a) and b).

[7] describe an authoring system that enables workers in manual assembly to record their own assembly activities via pictures and videos. These are then made available to other workers through tutorials in a knowledge database.

[8] present a framework for the intelligent creation and management of assembly instructions. It enables workers on the shop floor to provide feedback on assembly instructions. They can add new or missing information, update poor quality instructions and identify unclear or irrelevant information.

In [9] the workers create new documentation or modify existing documentation after assembly.

The methodology presented in Sect. 3.1 [1, 10] is also an approach from the assembly.

3 Method to Generate Optimized Assembly Sequences from Actual Data

In the following, the context in which the method operates is described (Sect. 3.1). Then, the method will be introduced (Sect. 3.2) and applied to a simple example (Sect. 3.3).

3.1 Context

The method is part of an approach for optimizing manual assembly systems by deriving the best practice assembly sequence based on recorded data. The approach adapts the benchmark approach to manual assembly systems. [1] An overview is given in Fig. 2.

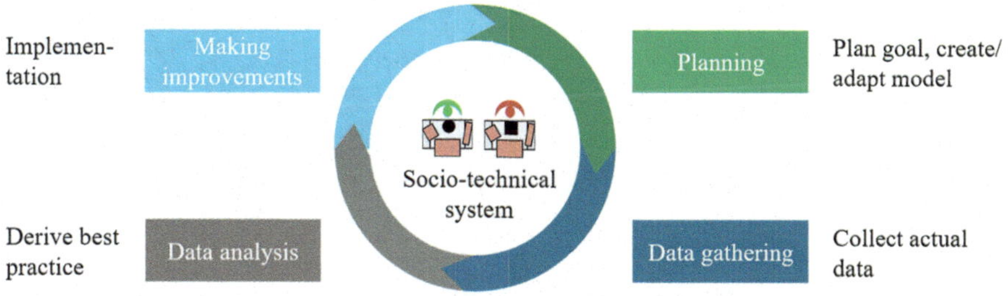

Fig. 2. Approach to apply benchmark in manual assembly [1, 10]

In the planning phase, the benchmark and its goal are planned. The goal of the benchmark is to record the assembly processes and times of different workers with different solution strategies to assemble a product and to derive the best solution strategy. The best solution strategy is defined as the assembly sequence that leads to the shortest assembly time. Moreover, the model is created or adapted. [1, 10]

In the data gathering phase, actual data from the shop floor is collected as basis for data analysis [1]. Data can be collected in many different ways, e.g. with a sensor system [11], via video recording or through a manual input via app.

In the next step, data analysis, the best practice to assemble the product is derived [1].

In the fourth step, the best practice needs to be adapted. Furthermore, the improvements are introduced. [1]

This paper focuses on the method during the data analysis phase.

3.2 Data Analysis

Overview The following Fig. 3 shows the steps of the data analysis (following the approach presented in [1]).

The input and basis of the method is the data collected during the data gathering phase. Then, the data has to be analyzed and the best practice among the recorded solutions to assemble the product needs to be derived. This is done in the following steps:

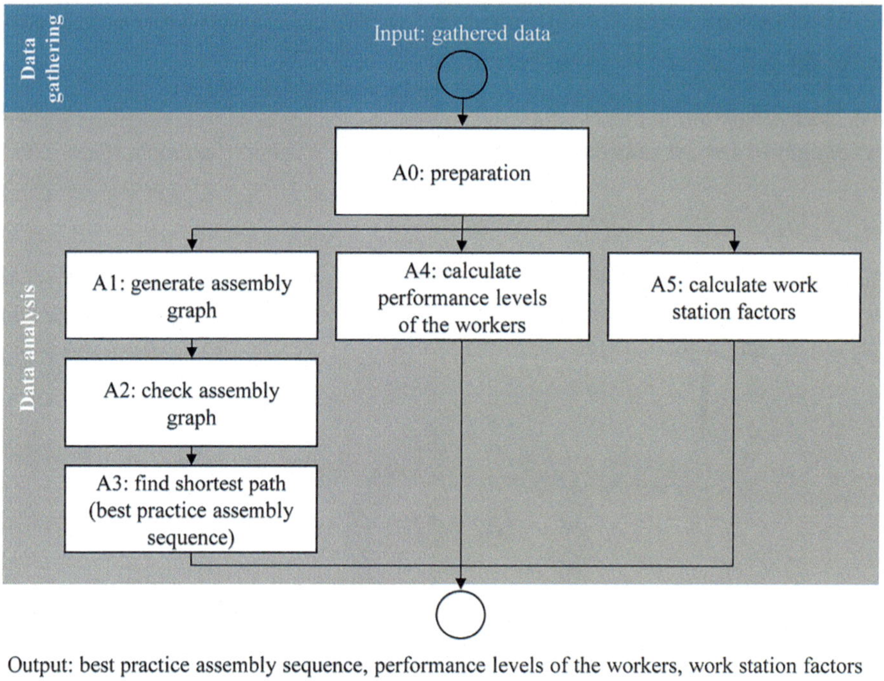

Fig. 3. Overview data analysis

Step A0 – Preparation In this step, the data is prepared. Therefore, the assembly sequences are first extracted from the collected data. One assembly sequence corresponds to one assembly run. Waste should not be included in the planning of an assembly sequence. Therefore, wasted processes are filtered from the data and reduced in the context of the optimization, e.g. through the optimization of processes and workplace design. Since waste can be significantly reduced most of the time during an effective assembly planning, it still cannot be completely eliminated. Therefore, it is possible to include a factor in the target time later. Second, processes that occur several times in one run, i.e. within an assembly sequence, are removed (e.g. for rework, corrections, interruptions). To find these processes, the method counts, how often each of the processes (uniquely determined by an id) occurs within each assembly sequence. If a process occurs more than once, the data of the assembly sequence will be adjusted as follows: The first part of the process step is retained, this is the process step where the activity was started (because the process may require for instance rework due to its position in the assembly sequence). The subsequent process steps that belong to this first process, such as rework, are removed and the time needed for their execution is added to the first part of the process step. The reason for this step is that a process may be fast in the first (incomplete) execution, but then requires rework.

Step A1 – Generate an Assembly Graph All solutions to assemble the product will be represented in an assembly graph [1].

At the beginning, a start and an end node are created. The first assembly sequence considered in the analysis is the first path in a directed graph $G = (V, R)$. The nodes $v \in V$ describe the process results. The directed edges (u, v) connect the single nodes (start node u, end node v) and thus represent the predecessor-successor relationships. The costs $c: R \to \mathbb{R}$ at the edges describe the assembly times per assembly process [1]. The assembly times are always related to the worker who carried out the process and the workplace where the assembly was carried out. The costs from the last process result to the end node are zero. If a process in a certain order occurs for the first time, the suffix "_1" is given. Then, subsequent runs or assembly sequences are added to the digraph as follows: At each process of the assembly sequence there is a comparison whether the (partial) assembly sequence of necessary processes is already illustrated in the graph or a new (partial) sequence has been executed (the suffix is not considered):

a) *The assembly order is the same as an order already illustrated in the graph:* If the assembly sequence is the same, the individual process durations must be compared. If the assembly duration of the process is shorter, the existing times (costs) in the directed graph are updated by the shorter time. [1]
b) *The assembly sequence is different:* If the assembly sequence is different, a new path is created in the directed graph. Due to the priority relationships and the assembly progress, new nodes must be generated. [1] This is done by appending a new suffix.

Paths in the assembly graph separate when the assembly sequence is different. They rejoin when the same processes are carried out till this node.

Step A2 – Check the Assembly Graph For each process step all executed times are stored and checked for outliers in this step. Outliers are possible in two directions: The time for a process step is significantly longer (outlier upwards) or significantly shorter (outlier downwards) than the average time for a process step. Since assembly times depend on the process sequence, due to which e.g. better or worse accessibility results, only times of processes in the same sequence are compared. The outliers upwards, i.e. process times that are significantly longer than the average, do not need to be corrected, as they will be overwritten by another run anyway, if the assembly sequence is the same. However, the outliers downwards are checked for plausibility at this point.

Step A3 – Find the Shortest Path (Best Practice Assembly Sequence) Now the shortest path, i.e. the assembly sequence that leads to the shortest assembly time, needs to be found in the assembly graph. The problem to be solved is an SSP-problem (single source shortest path problem). Starting from a directed, circle-free graph $G = (V, R)$ with assembly times as costs at the edges $c: R \to \mathbb{R}$ the shortest path through the graph from the start node $s \in V$ to the end node $e \in V$ is searched [1]. This is done by executing the Dijkstra algorithm. In this way, not only solutions executed by the workers are found, also combinations of different solutions.

Step A4 – Calculate Performance Levels of the Workers In this step, the performance levels of the workers are calculated, which are later used to adapt the target times. The average performance is used as reference performance, which is the average value of all runs for a product.

Step A5 – Calculate Work Station Factors The workplace with its layout and set-up also has an influence on assembly times, e.g. different material provisions at the work station lead to different assembly times. The method calculates how the assembly times at a workstation are in relation to the average. As a result, the influence of the workplace on the assembly times can be reduced. Moreover, an overview about workstations which are better or less optimal can initiate further optimization.

3.3 Example

Figure 4 shows an example product and the associated precedence graph. The components x and y have to be assembled on a base plate. Component z gets assembled on y. Finally, a housing is mounted.

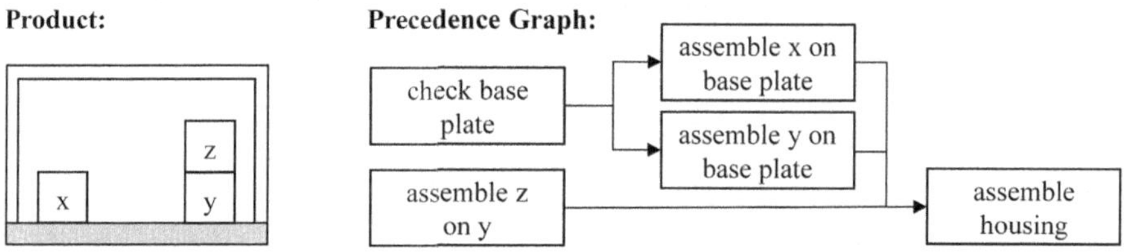

Fig. 4. Example product (left side) and associated precedence graph (right side)

The exemplary process data is visualized in Table 1. The product has been assembled twice (two runs). Worker 1 has performed the first run, worker 2 the second run.

Value-adding activities (V), supporting activities (S) and wasting operations (W) have been carried out. The id clearly identifies the process step.

Waste is filtered out and the data is adjusted so that a process step only occurs once within a run. Table 2 shows the adapted data (step A0).

Then, the solutions to assemble the product are transformed into a directed graph (Fig. 5). No outliers were found in the data (A2), also due to the small data base. There are two ways through the digraph, which represent two solution strategies to assemble the product. The shortest path in the graph, i.e. the best practice of assembling the product, is marked in blue (above) and has a length of 118 s. The other path (below) has a length of 135 s.

The performance level calculation results in a performance level of 106% for worker 1 and a performance level of 94% for worker 2 (A4). Since all assembly steps were carried out at the same workstation, the workstation factor is 1 for all processes carried out (A5).

In the next step, the derived best practice is adapted (e.g. inclusion of the performance level and the workplace factor) and then implemented (this is not considered in this paper).

Table 1. Process data

Worker	Run	No	Id	Name	Start [s]	End [s]	Duration [s]	Category	Necessary
1	1	0	9001	Search for cleaning fluid	0	25	25	W	0
1	1	1	1000	Check base plate	25	55	30	S	1
1	1	2	1001	Clean base plate	55	95	40	S	0
1	1	3	1002	Assemble x on base plate	95	117	22	V	1
1	1	4	1003	Assemble y on base plate	117	132	15	V	1
1	1	5	9002	Search for screws	132	203	71	W	0
1	1	6	1004	Assemble z on y	203	210	7	V	1
1	1	7	1005	Assemble housing	210	216	6	V	1
2	2	0	1000	Check base plate	0	28	28	S	1
2	2	1	1004	Assemble z on y	28	75	47	V	1
2	2	2	1003	Assemble y on base plate	75	89	14	V	1
2	2	3	1002	Assemble x on base plate	89	105	16	V	1
2	2	4	1003	Assemble y on base plate	105	130	25	V	1
2	2	5	9003	Search for housing	130	160	30	W	0
2	2	6	1005	Assemble housing	160	165	5	V	1

Table 2. Adapted data

Worker	Run	No	Id	Name	Duration [s]	Category	Necessary
1	1	1	1000	Check base plate	30	S	1
1	1	2	1001	Clean base plate	40	S	0
1	1	3	1002	Assemble x on base plate	22	V	1
1	1	4	1003	Assemble y on base plate	15	V	1
1	1	6	1004	Assemble z on y	7	V	1
1	1	7	1005	Assemble housing	6	V	1
2	2	0	1000	Check base plate	28	S	1
2	2	1	1004	Assemble z on y	47	V	1
2	2	2	1003	Assemble y on base plate	14 + 25	V	1
2	2	3	1002	Assemble x on base plate	16	V	1
2	2	6	1005	Assemble housing	5	V	1

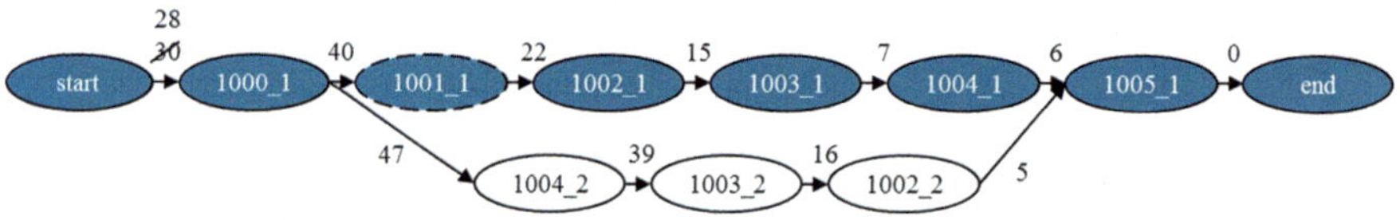

Fig. 5. Assembly graph

4 Conclusion and Outlook

In this paper, a method to automatically generate an assembly sequence planning based on actual data collected in an assembly has been presented. Afterwards, this method has been illustrated with an example.

The derived best practice assembly sequence consists of (partial) solutions provided by the workers. The method does not necessarily derive the optimum assembly sequence, as only sequences that have been carried out by the workers are included and combined. A major advantage of the solution is the process-parallel assembly sequence planning and the derivation of a best practice that is feasible. Furthermore, the performance levels of the workers and the equipment at the workplace are taken into account.

In the following work, the focus should be on providing the workers with the planned assembly sequences.

References

1. Kärcher, S., Bauernhansl, T.: Approach to generate optimized assembly sequences from sensor data. Procedia CIRP **81**, 276–281 (2019) (Ljubljana, Slovenia)
2. Leu, M.C., ElMaraghy, H.A., Nee, A.Y., Ong, S.K., Lanzetta, M., Putz, M., et al.: CAD model based virtual assembly simulation, planning and training. CIRP Ann. **62**(2), 799–822 (2013)
3. Neb, A.: Review on approaches to generate assembly sequences by extraction of assembly features from 3D models. Procedia CIRP **81**, 856–861 (2019) (Ljubljana, Slovenia)
4. Grunwald, S.: Methode zur Anwendung der flexiblen integrierten Produktentwicklung und Montageplanung. Forschungsberichte iwb, vol. 159. Herbert Utz, München (2002)
5. Molitor, M.: Generative Montageablaufplanung in der hochiterativen Produktentwicklung, 1st edn. Apprimus, Aachen (2019)
6. Schuh, G., Zeller, V., Prote, J.P., Molitor, M., Wenger, L.: Generative Stücklistenerstellung in der manuellen Montage. ZWF Z. wirtschaftlich. Fabr. **112**(6), 392–395 (2017)
7. Schuh, G., Prote, J. P., Gerschner, K., Molitor, M., Walendzik, P.: Technologiebasierte Externalisierung von Wissen. Aufbau einer Wissensbasis unter Verwendung von Autorensystemen in der manuellen Montage. wt Werkstattstech. Online **9**, 578–581 (2017)
8. Claeys, A., Hoedt, S., Schamp, M., Van De Ginste, L., Verpoorten, G., Aghezzaf, E.H., Cottyn, J.: Intelligent authoring and management system for assembly instructions. Procedia Manuf. **39**, 1921–1928 (2019)
9. Müller, M., Jäger, M., Ardelt, T., Metternich, J.: Facharbeitergestützte Dokumentation: Ansätze zur Digitalisierung in der komplexen Montage. ZWF Z. wirtschaftlich. Fabr. **115**(1–2), 91–93 (2020)
10. Kärcher, S., Görzig, D., Bauernhansl, T.: Modeling manual assembly system to derive best practice from actual data. In: IFIP International Conference on Advances in Production Management Systems, pp. 431–438. Springer, Texas (2019)
11. Kärcher S., Cuk, E., Denner, T., Görzig, D., Günther, L.C., Hansmersmann, A., Riexinger, G., Bauernhansl, T.: Sensor-driven analysis of manual assembly systems. Procedia CIRP **72**, 1142–1147 (2018) (Stockholm, Sweden)

Evaluation of Material Supply Strategies in Matrix Manufacturing Systems

Daniel Ranke[1,2](✉) and Thomas Bauernhansl[1,2]

[1] Fraunhofer Institute for Manufacturing Engineering and Automation IPA, Nobelstr. 12, 70569 Stuttgart, Germany
daniel.ranke@ipa.fraunhofer.de

[2] Institute of Industrial Manufacturing and Management IFF, Nobelstr. 12, 70569 Stuttgart, Germany

Abstract. Today's productions are driven by increasing variants, uncertainty of variants' distribution as well as volume and shorter innovation cycles.

A matrix manufacturing system aims to tackle these challenges. This new system concept consists of independent and flexibly linked process modules, which have no uniform cycle time and no fixed product order sequence in the system.

However, new challenges arise from this system setup. The changes in the structure have an impact on logistics. In research, there are only a few investigations regarding the consequences to logistics, especially to material supply. Common and new innovative supply strategies are used without knowing their suitability to the new system. The applicability of kanban, single product supply or kitting basket supply differs to the usage in a line assembly. A systematic derivation of suitability in the new context is missing.

The paper fills the gap of research in the outlined field. Firstly, changes and characteristics through the new structure which occur as challenges to the material supply are investigated. These are e.g. the flexibility of order and process sequences. In a second step, the material supply strategies are evaluated according to strategic requirements. As a result, each material supply strategy's suitability is evaluated for usage in matrix manufacturing systems.

The paper concludes with a derivation of guidelines for the planning and selection of a material supply strategy.

1 Introduction

Today's markets are driven by the individualization of products and a high dynamic of demand [1]. The number of variants is constantly increasing, coupled with a decrease of each variant's volume [1, 2]. Hence, market predictions are not easy. To meet the external market requirements, firms and production facilities need to provide internal changeability and scalability. The production system's changeability consists of flexibility, reconfigurability and changeover ability [3]. The "matrix manufacturing system" (MMS) provides the structure and logic to realize the requested changeability. The system, structured in the abstract form of a matrix, consists of flexibly linked process modules with independent cycle times. Product routes through the production are not

P. Weißgraeber et al. (Eds.): *Advances in Automotive Production Technology – Theory and Application*, ARENA2036, pp. 80–88, 2021. https://doi.org/10.1007/978-3-662-62962-8_10

determined a priori but short-term driven through real-time capacity demand and supply of multiple suitable process modules. Existing research focuses on designing and planning these systems [4, 5], whereas only few authors engage in the planning of material supply [6, 7]. Since a stable material supply is a main enabler of a properly working production system, changes influencing the material supply need to be investigated.

2 State of the Art

2.1 Material Supply

Material supply is defined by REFA [8] as the task of providing the required material, in the right kind and amount, to the right time and to the right point of use, to enable the further processing. It consists of several dimensions like storage and container policy, internal supply chain design, definition of material amount, transport and routing and further dimensions [9, 10]. A material supply strategy combines forms of multiple dimension. Common used strategies in practice are: combined order supply, total order supply, partial order supply, single product supply, periodic supplying, kanban, multiple container supply and handheld storage supply [9, 11]. Furthermore, in the automotive sector the concept of a kitting part system is used. In an upstream logistic area, a kitting part with parts demanded by multiple process modules is commissioned and added to the product on a certain process module. Each process takes its required parts of the kitting and uses them for its value creation. In the context of MMS, Popp [6] has introduced a concept of a new material supply strategy. It consists of a mobile storage box including a picking tool. The storage box gets filled with multiple parts which need to be supplied to a certain point of use with a high degree of certainty (but still some uncertainty). After moving to the point of use, the tool picks the needed part just in time and provides it to the worker. Not needed parts are restored in the warehouse. The strategy is called bar concept supply and offers a high degree of flexibility.

For choosing a feasible material supply strategy several approaches exist [9, 10, 12–14]. To select a strategy several analyses investigate e.g. the material demand, part-value, manufacturing structure and space. Still, in practice, choosing a material supply strategy is often done intuitively and without deeper analysis [15, 16].

2.2 Matrix Manufacturing System

A MMS consists of flexibly linked, but physically uncoupled process modules. It has characteristics of a workshop fabrication, but individual process modules are combined in a process flow structure which is required by the products' structure. Kern [17] outlines general characteristics which describe the system (Table 1, row "General"). E.g., the system offers a high flexibility of assembling similar but different products, it is easy to reconfigurate and offers a continuous scalability. These characteristics describe the high changeability of the system.

In addition to the general characteristics outlined by Kern, there are further characteristics of the system, which occur in operation. Firstly, there is no fixed relation between source and sink [18]. An ad-hoc production control assigns a product to its next

Table 1. General characteristics based on [14] and operational characteristics of a MMS

Characteristics		Forms
General	Movement (objects)	Products
	Process module (PM)	Modular/Matrix
	Coupling of PM	Uncoupled
	Distance of PM	Medium
	Order of processes	Variable
	Synchronous movement	No
	Batch size	One
	Division of labor	High
	Qualification level	Medium
	Lead time	Low
	Flexibility/changeability	High
	Robustness	Medium
	Volume of parts	High
	Variances of parts	High
Operational	Work distribution	Flexible
	Product assignment to PM	Short-term/ad-hoc
	Task-sequence	Flexible
	Order-sequence	Flexible
	Amount of product-/order-flows	Multiple
	Flow direction	For- and backward
	Process module's cycle time per product	Flexible
	Determination of system's configuration	Mid-term

point of operation. A deterministic prediction of the next sink is, thus, impossible. The assignment depends on the restrictions of the priority graph, the process modules' abilities, the current capacity utilization and the operation status of the system. In conclusion, the task sequence is flexible and not fixed, like in a dedicated manufacturing line [19]. Secondly, through the individual and ad-hoc decisions of choosing a new sink, there is no overall defined cycle time, which is similar to each process module [5]. Therefore, the sequence of different orders might change inside the system [20]. Furthermore, there are multiple product and order flows in one matrix system [20]. The flow's direction is mostly forward, but can be backward in a loop as well. Through reconfigurations of the system [19] the newly outlined characteristics, like a flexible product flow, change with system parameters. Also, when adding a new process module or changing a product, the system behaves differently.

In conclusion, a MMS offers new degrees of freedom during operations. This challenges the planning and control of the system and affects not only the manufacturing structure itself but also the material supply.

3 Material Supply in the Context of MMS

3.1 Challenges to Material Supply

Through the new degrees of freedom within the new system and differences to known manufacturing systems, new challenges to the material supply arise. Six main challenges and their consequences can be identified and are shown in Table 2. They arise through the characteristics of the new structure and organization [17, 18, 21].

Regarding the material supply, the overall challenge is uncertainty. The uncertainty refers to the location, volume and time. In addition, anchors like a defined product and order sequence do not exist, as e.g. in a dedicated manufacturing line. Further, the system is confronted with an increased number of articles in volume and variants, and with a continuously changing system. In addition, the MMS system itself and each system's element bases on a stochastic behavior, which effects depending processes like the material supply. The material supply has to readjust its strategies to this changing and stochastic system.

Table 2. Challenges and consequences to the material supply through the MMS

Challenges to mat. Supply	Consequences
Uncertainty of point of use	Risk of shortages
Uncertainty of demand quantity	Risk of shortages
Uncertainty of time of use	Just in time (JIT) delivery is impossible
No fixed product and order sequence	Combined order supply is not feasible & demand summarizing of products is uncertain
Increased quantity and variance	The existing supply area is fixed and stocking policy might be changed
Reconfiguration of the system	No long-term supply strategy is feasible, as system's and workstations' demand are changing, as well

3.2 Methodology of Evaluating the Suitability

According to the literature [11, 22] and the identified challenges, ten criteria are chosen to evaluate the suitability of material supply strategies in MMS (Fig. 1). Each of the strategies mentioned in chapter 2 perform on a scale between one and five in each criterion. Additionally, the MMS requires performance in each criterion. For each criterion of each strategy a delta between the required and achieved performance arises [11]. This approach is adopted from Richter [11]. The strategies' assessments and set requirements are done through a literature review [6, 7, 11, 19] and the authors' discussion.

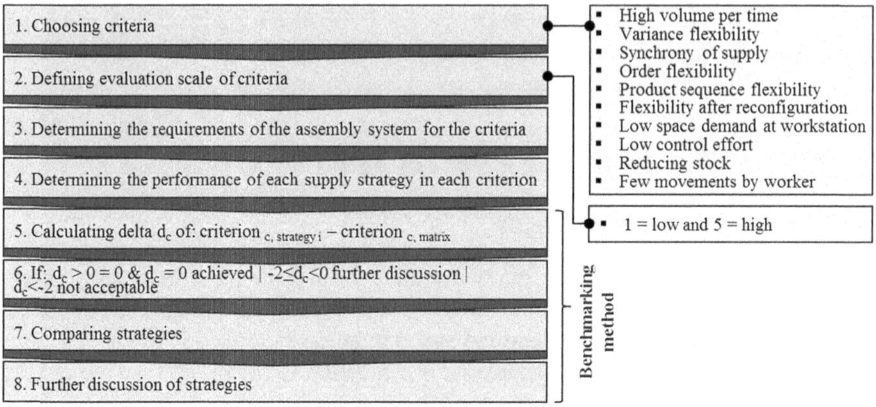

Fig. 1. Chosen method

Different methods exist to evaluate the results [23]. Methods adding multiple results to one number, like building an average, are not sufficient in this case. Well performing criteria cannot compensate a failing criterion. A multi-perspective evaluation is needed which takes all criteria and individual results under investigation. Still, an analytic and numerical evaluation is preferable, as it facilitates the objectivity and comparison.

In conclusion, the benchmark method is chosen, comparing each strategy's performance to the given requirements. The performance is the individual comparison between a strategy's assessment and the requirements. Therefore, the delta of assessment minus requirement is taken. An over performing delta greater than zero is equal to zero since there is no additional benefit in practice. A negative delta is no final excluding indicator in the evaluation. Thus, additional to the analytic calculation, the strategy is interpreted to close the gap between the numerical evaluation of a complex strategy and the authors' evaluation.

The requirements of the MMS are driven by its characteristics presented in Table 1.

3.3 Evaluations

Table 3 shows the result of the evaluation. The deltas are outlined in grey. No strategy achieves the desired requirements in all criteria. Only three strategies have no criterion's delta of less than 2. These strategies are: partial order supply, kitting part system supply and bar concept supply. Furthermore, only the named ones achieve a benchmark greater 80%.

The **combined order supply** is one of the less suitable strategies. Through summarizing different orders, it is inflexible to reconfigurations. In ad-hoc and short-term situations the strategy is less capable to respond in the required time. It needs a high degree of control effort to compromise the challenge of many variances and mid-term reconfigurations of the system. To the requirements of a MMS, it is not suitable. The **total and partial order supply** are characterized by focusing on a (partial) order. This does not suit the batch size of one in matrix systems and makes the strategy mostly unsuitable. The benefits of synchrony of supply, low control effort and few movements

Table 3. Requirements and performance of different strategies

	MMS: Requirements	Performance of each strategy									
		combined order supply	total order supply	partial order supply	single product supply	periodic supply	KANBAN	multiple container supply	handheld stortage supply	kitting part system supply	bar concept supply
High volume per time	4	3	4	5	2	3	4	5	5	2	3
Variance flexibility	4	2	3	4	5	2	2	2	4	5	5
Order flexibility	5	3	3	4	5	2	3	3	4	5	4
Product sequence flexibility	5	1	1	1	3	1	3	3	5	5	5
Synchrony of supply	4	3	5	5	5	4	3	3	2	5	4
Flexibility after reconfiguration	5	2	4	4	4	2	2	1	4	5	4
Low space demand	4	3	2	3	1	4	5	5	4	4	2
Low control effort	3	2	4	4	3	4	5	5	4	3	2
Reducing stock	4	4	4	4	4	3	2	1	2	3	3
Few movements by worker	4	4	4	4	4	4	4	4	1	5	4
Benchmark	100%	62%	74%	81%	79%	67%	71%	67%	79%	93%	81%

Legend: delta equal or greater 0 (no shading) · delta between 0 and -2 (light grey) · delta minor -2 (dark grey)

by the worker do not have an effect. Additionally, both differ in flexibility and low space demand to the desired value. The strategies are only suitable, if the defined batches are greater than one. The most suitable strategy is the **kitting part system supply**. It offers all desired values except the high volume per time. This is caused by the high effort in picking all parts. By using this strategy, all degrees of freedom of flexibility in variance, order, reconfiguration and assignment to a process module are given. Only the efforts and additional space demand in the previous order picking area are a disadvantage to the strategy. The newly introduced strategy of the **bar concept supply** achieves a benchmark of 83% by offering a synchronized supply of varying parts. Through the opportunity of restoring unneeded material it is quite flexible but needs control effort by predicting the needed material. Furthermore, through its design it requires space for commissioning at the process module and, finally, unneeded material requires space as well. The strategy is more suitable than an order supply, if unpredictability of multiple variants is high, ad-hoc assignments are certain and sufficient supply space at the process module is given. To the contrary, the **single product supply** is defined by a low supply volume, which can be only compromised by a higher space demand. The suitability of the strategy depends on the amount of processes that a process module can carry out. By many conducted processes and a feasible time to supply, the strategy can be suitable, as it supplies many parts at one time. It is like a less mobile and less flexible kitting part system supply. The **periodic supply** offers the benefit of low control effort and space demand. Normally, in the required time frame the supplies are synchronized as well. However, through its periodical perspective its flexibility to changes is low and thus less suitable to the character of the matrix system. The high volume strategies, **kanban** and **multiple container supply**, can supply a high volume per time, but do not offer the flexibility to changes in the reconfigurations, in orders or variants. Only by increasing the stocks of every variant changes can be ignored. A quite flexible strategy is the **hand-held storage supply**. By possessing stock apart from the individual process modules it can set up stock – and consequently flexibility – which can be used by multiple process modules. This increases the movements of the worker and leads to lower productivity, which makes the strategy less suitable for a matrix manufacturing system.

4 Findings

Across all investigated strategies, none of the strategies fully fulfills all desired requirements of the MMS. Only three strategies exist which perform within promising ranges. Still each strategy has advantages and disadvantages which go even beyond the assessment. The high volume strategies are less flexible to products with a high degree of variants, whereby flexible strategies are not applicable to all material, as they consume a lot of effort and resources. To manage the conflict, following dedicated guidelines arise for planning the material supply strategy: a. the selection of a supply strategy has to be done for each article (-group) individually; b. the selection depends on a multi-perspective consideration, including e.g. costs and space (of resources); c. the uncertainty of demand in MMS in terms of time, quantity and location must be taken into account; d. a planning approach should adopt to changes in the system's configuration.

5 Conclusions & Outlook

The paper investigates the suitability and performance of different material supply strategies in the context of MMS. Through the arising uncertainty within the system, the loss of a fixed order sequence and the batch size one, many strategies are affected and less suitable. Furthermore, the uncertainty and short reaction times are challenging existing approaches. As outlined in the findings, a new planning approach is needed. Furthermore, new supply strategies must be developed to cope with the matrix system's requirements. The strategies' assessment and evaluation is supported by a literature review and a discussion. Through rising importance and knowledge gain of MMS, new insights can be achieved and may extend the outlined approach. The assessment and evaluation offer first indications for further consideration.

References

1. Kampker, A. et al.: Agile low-cost montage. In: Schuh G (ed.) AWK Aachener Werkzeugmaschinen-Kolloquium 2017 Internet of Production für agile Unternehmen, 1st edn., pp. 231–259. Apprimus, Aachen (2017).
2. Koren, Y.: The global manufacturing revolution. Product-process-business integration and reconfigurable systems. Wiley series in systems engineering and management, v. 80. Wiley, Hoboken (2013)
3. Wiendahl, H.-P., et al.: Changeable manufacturing – classification. Design and operation. CIRP Annals **56**(2), 783–809 (2007)
4. Foith-Förster, P., et al.: Generic production system model of personalized production. MATEC Web Conf. **301**, 19 (2019)
5. Greschke, P., et al.: Matrix structures for high volumes and flexibility in production systems. Procedia CIRP **17**, 160–165 (2014)
6. Popp, J.: Neuartige Logistikkonzepte für eine flexible Automobilproduktion ohne Band. Dissertation, Wehking, Karl-Heinz Universität Stuttgart (2018)
7. Kern, W., Lämmermann, H., Bauernhansl, T. (eds.): An integrated logistics concept for a modular assembly system. Procedia Manufact. **11** (2017)
8. REFA: Methodenlehre der Planung und Steuerung Teil 2, 4th edn. Hanser, München (1985)
9. Bullinger, H.-J., et al.: Planung der Materialbereitstellung in der Montage. Teubner, Stuttgart (1994)
10. Golz, J.: Materialbereitstellung bei Variantenfließlinien in der Automobilendmontage. Zugl.: Berlin, Techn. Univ., Diss., 2013. Produktion und Logistik. Springer Gabler, Wiesbaden (2014)
11. Richter, M.: Adaptive Liefer- und Bereitstellungskonzepte für wandlungsfähige Montagesysteme zur Ausschöpfung der logistischen Leistungsfähigkeit. Logistikkonzepte für wandlungsfähige Montagesysteme. Schlussbericht, Hannover (2010)
12. Battini, D., et al.: "Supermarket warehouses": Stocking policies optimization in an assembly-to-order environment. Int. J. Adv. Manuf. Technol. **50**(5–8), 775–788 (2010)
13. Grünz, L.: Ein Modell zur Bewertung und Optimierung der Materialbereitstellung. Zugl.: Dortmund, Univ., Diss., 2004. Berichte aus der Logistik. Shaker, Aachen (2004)
14. Ranke, D., et al.: Materialbereitstellungsstrategien für die Montage. Auswahl nach logistikrelevanten Kriterien. wt Werkstatttechnik online **109**(6):447–451 (2019)
15. Heinz, K., et al.: Planung der Materialbereitstellung bei optimalen Kosten und Zeiten. wt Werkstattstechnik online **92**(10):531–535 (2002)

16. Augustin, H., et al.: Produktionslogistikplanung mit einer morphologiegestützten Planungssystematik. J. Eng. Manage. Oper. **1**, 111–120 (2019)
17. Kern, W., et al.: Planning of workstations in a modular automotive assembly system. Procedia CIRP **57**, 327–332 (2016)
18. Filz, M.-A., et al.: Analyzing different material supply strategies in matrix-structured manufacturing systems. Procedia CIRP **81**, 1004–1009 (2019)
19. Kern, W., Rusitschka, F., Kopytynski, W., Keckl, S., Bauernhansl, T. (eds): Alternatives to assembly line production in the automotive industry (2015)
20. Göppert, A., et al.: Frei verkettete Montagesysteme. ZWF **113**(3), 151–155 (2018)
21. Foith-Förster, P. et al.: Changeable and reconfigurable assembly systems – A structure planning approach in automotive manufacturing. In: Bargende, M., Reuss, H.-C., Wiedemann, J. (eds.) 15. Internationales Stuttgarter Symposium, pp. 1173–1192. Springer Fachmedien, Wiesbaden, (2015)
22. Adolph, S., et al.: Materialbereitstellung in der Montage. ZWF **111**(1–2), 15–18 (2016)
23. Hall, K.: Ganzheitliche Technologiebewertung. Deutscher Universitätsverlag, Wiesbaden (2002)

Smart Factory and the Unique Digital Order Twin

Wilmjakob Johannes Herlyn(✉) and Hartmut Zadek

Institute of Logistics and Material Handling Systems, Otto-von-Guericke Universität Magdeburg, Magdeburg, Germany
{wilm.herlyn,zadek}@ovgu.de

Abstract. Smart Factories are designed by Universities [e.g. 7, 1] and Research Institutes (e.g. IPA, IFF, IOSB, IFF, DFKI). Also, automotive companies are developing their 'own' Smart Factories [e.g. 3, 6]. The focus is on manufacturing processes, technical equipment, and control of technical processes but not on business and order processes [4, 5]. To run a Smart Factory also a concept for a "Digital Order Twin" (DOT) is needed to ensure that for each final product the right components are available in the right quantity at the right time and at the right place, especially if the product is configurated by a customer or dealer. But existing IT-Systems for Enterprise-Resource-Planning (ERPS), Material-Requirement-Planning (MRP) and Manufacturing Execution (MES) are not able to exploit the huge amount of acquired data 'in-time', because they are separated IT-Modules which are connected by batch-oriented interfaces only, the concept and algorithm base on the "Water-Fall-Model" without recursion to a preceding IT-System or IT-Module. In an autonomously controlled Smart Factory the right components must be referenced to each single final product instantly and in a more flexible way than the concept of pearl chain can do, instead of a unique DOT uses instantly data of RFID, QR-Code, Smart Devices and Cyber-Physical Objects and can interact in time with Systems and Modules.

1 The Idea of a Digital Order Twin

The idea of a 'Digital Twin Concept' (DTC) was introduced by John Vickers and Michael [13], it "contains three main parts: a) physical products in 'Real Space' b) virtual or digital products in 'Virtual Space' and c) the connections of data and information that ties the virtual and real products together" [12]. Based on their approach a logistic-oriented 'Digital Control Twin' (DCT) is designed [21]. This build the conceptional environment for the 'Digital Order Twin' that bases also on three pillars to regulate the order of final products and required material and is illustrated by a 'Big-Picture' (s. Fig. 1).

1. Reality: covers all production and material flow sections in a factory and all material items that flow thru. This includes production, transportation, storing, buffering, and sorting of final products and of all required components like modules, assembly groups, and single parts too. Reality includes also transporting means like trucks, loading means like bins or racks and all kinds of handling equipment for logistic purposes.

P. Weißgraeber et al. (Eds.): *Advances in Automotive Production Technology – Theory and Application,* ARENA2036, pp. 89–96, 2021. https://doi.org/10.1007/978-3-662-62962-8_11

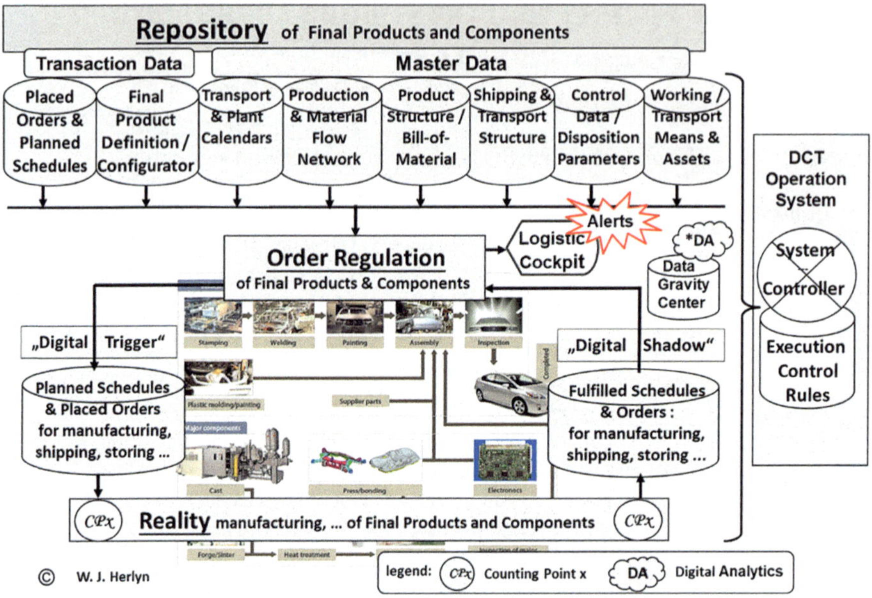

Fig. 1. Big picture of the 'Digital Control Twin' for a 'Digital Order Twin'

2. Repository: covers customer orders and all kinds of product schedules as well as order and call-offs of required components and other material flow items (**'Transaction-Data'**). It includes also the mapping of real production and material flow (PMF) and product-structure in specific databases as well as shipping and transportation structure and working and transportation means, transport and plant calendars and also control data for material dispatching (**'Master-Data'**).

3. Regulation: covers scheduling, ordering, and monitoring the flow of final products and all required materials. The task is to ensure that each single product is finished at the right time and for this the required material must be available at the right time, in the right quantity and at the right place. to achieve this, we use the method of 'closed-loop-control' for regulation whereby the target values are here the placed orders and planned schedules for the final products and required components named as **'Digital Trigger'**. And the actual values are here the fulfilled orders and schedules of real material flow objects named as **'Digital Shadow'**. The Digital Shadow should cover at a minimum the quantity and exact location of an exactly identified material flow item or a couple of identical flow items. Data can be acquired in different ways, supporting by different techniques and technical equipment and features, the chosen techniques and technical equipment depends on the kind of logistic process and handling environment and the required response time for regulation.

It is important to understand, that there is a general difference between an engineering and order oriented DCT. Target values in an engineering-oriented DCT like CAD/CAM-Data are fixed values that are replicated by instances [12, 1, 9]. In a DCT target values like order sequence and fulfillment time can be changed over time and must be balanced by actual values and recalculated. Therefor an order oriented DCT treats deviations of target-actual-values with other methods than an engineering oriented DCT. The order oriented DCT must ensure that the right components are available at the right place, right time and in the right quantity for a unique final product across the entire manufacturing and transportation network and requires a dynamic method of order regulation and material flow balancing [20, 21].

2 Repository – Main Databases for the Digital Order Twin

All required data of the DCT are stored in serval specific databases, which are called 'Repository', which map the virtual world of the real world. The core database of the Repository for order regulation are 'Production and Material Flow' (PMF) structure and the 'Product Structure' (PS) that are explained in the following.

2.1 Production and Material Flow Structure of a Smart Factory

The fundamental database are product and material flow (PMF) structure of the factory and the product structure stored as BoM. The material flow of discrete products is not arbitrary but linear oriented and follows the structure of the production process itself and the delivery network of components is depending on this. "Today most material flow systems are networks because the process is partly organized in series or parallel" [2]. We assume that the material flow of discrete products can be represented by a linear ordered chain of Boolean intervals which are closed open defined and meet the requirements of an ideal Boolean interval lattice. There are no jumps or overlaps between the next following intervals, no interval in the chain is missing or lies outside the entire interval whereby each interval is defined as closed-open. The beginning of an interval is defined by an 'Entry Point' (EP) which lays inside the interval and the end is defined by another EP which lays outside the interval and must be the beginning of the next-following interval. If the EP is referenced to a real existing 'Data Acquisition Point', we called this a 'Counting Point' (CP). The complete algebraic definitions of an ideal Boolean interval chain, subintervals and Boolean trees are omitted here, detailed information about can be found in [16, 15].

PMF-Structure of a Smart Factory without product data of a final products is an empty framework only and must be supplemented therefor by the product structure of required components that are normally stored in different kinds of 'Bill-of-Material'.

2.2 Product Structure and Bill-Of-Material in a Smart Factory

In a Smart Factory we need for each manufacturing interval a separate **'Bill-of-Material'** (BoM) which contains all components that are required in the concerned interval. Based on the BoM-data we can plan manufacturing, stocking, feeding, and delivering and we can calculate orders for all required components. Due to the complexity and configuration of a final product and production we use specific BoMs. For this we distinguish four different kinds of BoM that match with the PMF-structure and reflex typical techniques of manufacturing and fabrication. Each BoM relates to a corresponding manufacturing interval of the PMF-structure and as a result, we have got a PMF-structure where different kinds of BoMs which are linked to corresponding manufacturing sections so the complete product structure is transparent (s. Fig. 2).

1. **'Configurable BoM' for Final Product Assembling**: Configurable BOM for Final Products Assembling: In industries with extraordinarily complex products a product can be configured by the customer himself. Because of the huge amount of different product variants, it is not possible to define a separate BoM for each of billions of product variants. Instead of this a configurable BoM is used, also called Super-BoM or Complex-BoM. Final products are configurated by a combination of different options and the valid components can be chosen by the product options; for detail information see [19, p. 820 ff., 11].
2. **'Variant-BoM' for Module Assembling**: Modules are high-aggregated assembly groups, which are assembled according to a specified end-product. Because of their high variety and specific technical function for a product these modules are often produced in a specific pre-assembly line and delivered 'Just-in-Time' or 'Just-in-Sequence'. For modules with high variety a 'Variant-BoM'is used, which contains the variants of a specific type of module (e.g. 4-cyl. Engines, 6-cyl. Engines etc.).
3. **'Single-Level BoM' for Assembly Group Manufacturing**: Final products and modules incorporate assembly groups which can be used in different product and modules. Because of their frequently usage they can be produced in advance for better economic scale and quality. Normally these assembly groups can be mapped by a single-level-BoM which can be connected to another single-level BoM, so we can generate a multi-level BoM from the highest (top) assembly group to the lowest level of single parts or raw material.
4. **'List of Material' for Single Part Manufacturing**: Single parts which are described and documented in a CAD-system can often manufactured in different ways and steps. Because of the specific technologies and different kinds of the manufacturing process single parts can be produced in different ways with the consequence that these different stages of raw material or raw parts can be used; all alternatives must be described by alternative stages of raw material in the 'Material List' Depending on the actual situation in a factory the appropriate alterative must be chosen every day.

2.3 Unique BoM and Semantic Product Memory of Final Products

A final product must be exactly defined resp. completely configured by a customer. As soon as the customer order is stored in the Master Production Schedule (MPS) we can generate a unique BoM for each final product before manufacturing starts (s. Fig. 2). At a certain point in the early stage of production the final product passes the Order-Entry-Point (OEP) so we can transmit and store the unique BoM-Data in the Unit, that is attached at the final product. As soon as a certain manufacturing or handling operation is finished the actual data is registered/gathered and stored in the Chip e.g. the serial number of the mounted component, the ID of the used tool or results of a quality check. By this all mounted components are stored directly in the Chip and in the repository, which is especially crucial for items that must be documented because of laws. In addition, all acquired data are transferred to a 'Data-Gravity-Center' and can be analyzed by 'Data Analytic' tools (s. Fig. 1). The digital information, that are stored by the product is called the "semantic product memory" and contains relevant data that are gathered during the lifecycle [14]. The data can therefore be used not only during production but also for later product maintenance and repair processes in the after sales processes [10]. If an object is very small and frequently used like a nut or screw it make no sense or is not possible to fasten a RFID-Tag and if an object is very complex an RFID-Chip cannot store all data so a bigger Micro-Chip is needed; finally the physical item becomes a Cyber-Physical-Object (CPO) and can communicate via internet with other CPOs which provides high communication and interaction performance for the DOT and can be used for other purposes later on.

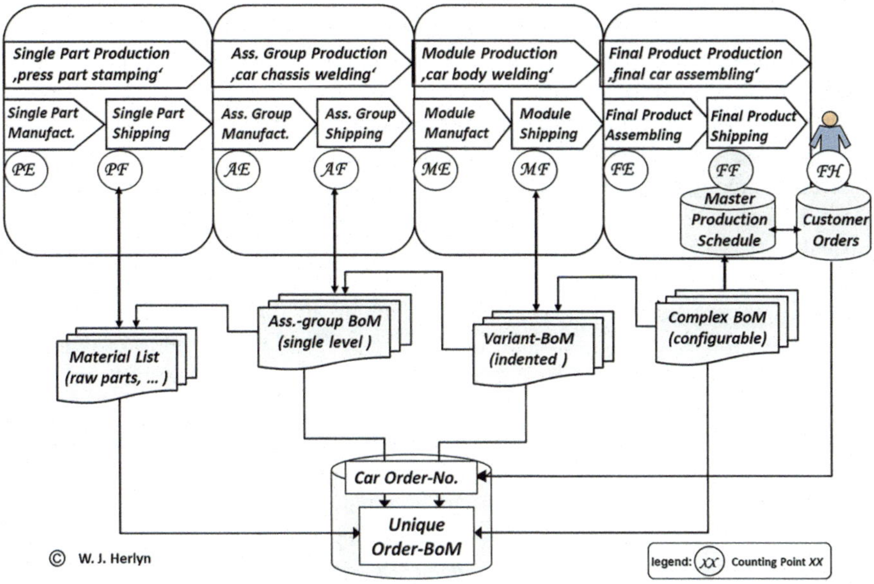

Fig. 2. Unique bill-of-material for an order of a unique final product

3 Regulation and Ordering of Final Product and Components

3.1 The Role of 'Order-Entry Point' in a Smart Factory

The CP where the costumer order is connected to the real physical product which is called the 'Order-Entry-Point' (OEP) also called as 'Order Decoupling Point' [20] is especially important for control and regulation of single customer order. And in addition, the OEP can be taken as interface of production DCT to the order DCT [9]. In car manufacturing normally the OEP lays in the area of car chassis welding where an ID-Plate or RFID-Chip is attached at the car chassis e.g. the side beam or cross beam of a vehicle chassis. In our example we take the CP 'AE' as the OEP and as soon as the car chassis passes this CP the Car-ID is related to the component, e.g. a gearbox and an engine. If the components are already manufactured, they are related temporary to this serial Car-ID and if a component will be produced later an order-trigger is generated. This procedure can be performed for each component, so we get a data grid for all required components at the CP's in the Smart Factory whereby we must know the lead-time (LT) which is the main logistic parameter to harmonized order and material flow of final products and components. LT is the real elapsed process time between two next-following CPs. Because of the ideality of PMF-Structure we can add up the LT of the next-following intervals, so the complete LT of an item is the sum of all passed CPs. For LT-calculation neither the kind of operation nor the physical length of an interval is relevant once and only the real elapsed time between two CP's is the decisive factor for all manufacturing areas with or without autonomous production control. Because of turbulences and change of sequences in material flow or autonomous control we distinguish different order relations between the final product and its components for flexible control of material items (s. Fig. 3. When a final product passes the OEP the order relation will be at first 'temporary', when the final product passes the CP 'BW' the order relation can be changed to 'committed' by exchanging the ID's of a dependent component. The relation is 'frozen' at a certain CP (e.g. "FE") and when the component is mounted into the final product the relation is 'fixed' and the serial ID's of items can be stored. In our example the CP 'AE' represents the Order-Entry-Point and at this CP the order relations to the engine and gearbox are temporary. When the final product passes the CP 'BW' the order is 'committed' while the engine and the gearbox enter a warehouse (see 'EWE' and 'GW'). When the final product passes the CP 'FE' the order relation for the gearbox and engine is frozen during the powertrain assembling. When the completed powertrain is mounted into the final product at CP 'mp' (called 'mariage') the order relation is fixed. Thereby the informational order relation of the final order number and the component number changed into physical item relation with the serial IDs for the final product ('Vehicle Identification Number') and the incorporated components. At the end of manufacturing serial numbers of all components are stored as unique DOT, the data can be used not only for official quality documentation but also for other purposes especially for after sales processes like maintenance and repairing.

Sometimes there are changes in customer orders or turbulences in manufacturing and delivering processes so the sequence of cars must be recalculated over time. Because of this it makes sense to change the references between Car-ID and Component-ID also. The best 'areas' to do this are warehouses, where the sequence of cars can be changed. When a car or a component leaves a warehouse, the references can be changed, and the status are changing from a temporary to a committed or frozen status. After a component is mounted into the final product the serial number of the component is fixed. At the end of the final assembly line all serial numbers of the components are stored for the required unique manufacturing and quality documentation of each single product and can be used for further Data Analytics.

4 Conclusion and Short Outlook

The presented concept of a DOT is designed to use the capabilities of Industry 4.0 specially to exploit the acquired data of material flow items that are available thru new

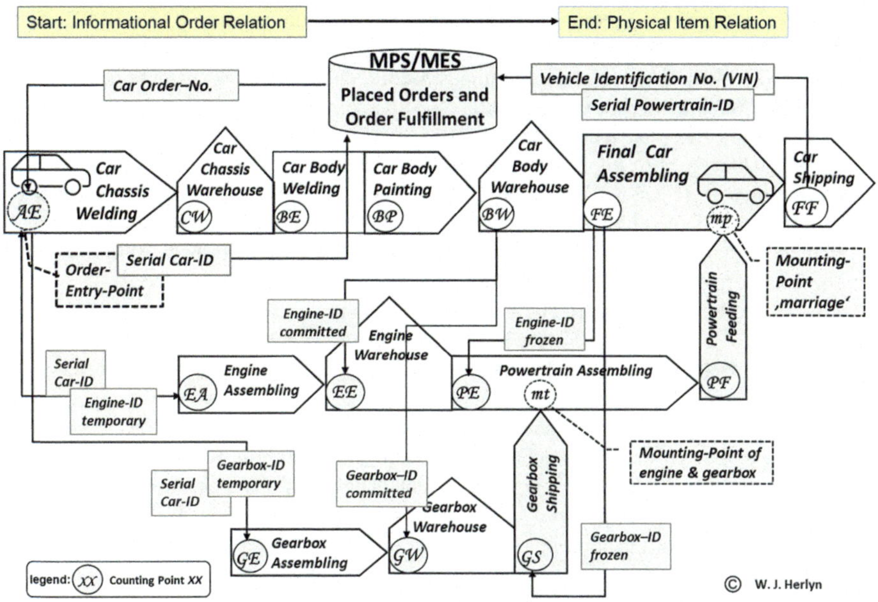

Fig. 3. Flexible order relation of final products and components of a unique DOT

technologies like RFID, QR-Code etc. The DOT is a logistic-oriented concept that is complementary to the engineering-oriented DT in a Smart Factory. The development of such a logistic-oriented DOT is still in its early stage and must be implemented in Smart Factories or Labs so the order process for products and required components can be virtually controlled, traced, and balanced. Existing engineering control applications should be connected to the DOT and the regulation tool must be integrated into the overall control software of the Smart Factory. Finally, each physical final product should have a unique DOT to harmonizes the order process of each single final product and its required components. The concept is not only useful for big international companies like enterprises in the automotive or machine tool industry it can also be applied to smaller companies which produce technical products.

The unique data of each single DOT cannot be used for manufacturing and quality documentation only, but also for repair, service, and maintenance in after sales processes, so the DOT exceeds the Smart Factory application and leads to a Smart Repair Shop and in addition all data is stored in a Data Gravity Center for Big Data Analysis.

References

1. Anderl, R.: A Digital twin perspective – A methodology to generate manufactured component instances from part types, German-Czech-Workshop. (2016) Accessed 8 June 2020
2. Arnold D., Furmans K.: Materialfluss in Logistiksystemen, 6th edn., p. 47 ff. Springer, Berlin. https://doi.org/10.1007/978-3-642-01405-5 (2009)
3. AUDI: "Smart Factory", Dialoge. AUDI AG, Kommunikation, Ingolstadt (2017)

4. BMBF: Industrie 4.0 Innovationen im Zeitalter der Digitalisierung, p. 16. BMBF, Berlin (2020)
5. BMBF: The ARENA2036 research campus, Berlin, published by www.bmbf.de (n. d.)
6. BMW: https://www.bmwgroup.com/en/innovation/company/industrie-4-0.html (2020)
7. Bauernhansl, T., et. al. (eds): Industrie 4.0 in Produktion, Automatisierung und Logistik. Springer Vieweg. https://doi.org/10.1007/978-3-658-04682-8 (2014)
8. Bauernhansl, T. (eds.): MANUFUTURE-DE, (Langfassung), Fraunhofer IPA. https://www.ipa.fraunhofer.de/de/Publikationen/studien/studie-manufuture-de.html (n. d.)
9. Eigner, M.: Digitaler Zwilling – Stand der Technik. ZWF Z. Wirtschaft. Fabrik. **115**(3–6), https://doi.org/10.3139/104.112300 (2020)
10. Erfle, R.: Alle Macht den META-Daten. TEKOM-Kommunikation **1**, 41–47 (2019)
11. Frischen, C., Marbach, A., Tichla, F., Mantwill, A.: Durchgängige Variantensteuerung mit Hilfe der regelbasierten Komplexstückliste. Tagungsband des 30. DfX-Symposium 2019. https://doi.org/10.35199/dfx2019.2 (2019)
12. Grieves M.: Digital twin – Manufacturing excellence through virtual factory replication, Whitepaper. https://www.apriso.com/library/white_papers.php June 2020 (2014)
13. Grieves M.: Virtually intelligent product systems: Digital and physical twins. In: Flumerfelt, S., et al. (ed.), Complex Systems Engineering: Theory and Practice, American Institute of Aeronautics and Astronautics, pp. 175–200. Reston (VA), US (2019)
14. Haupert J.: DOMeMan: Repräsentation, Verwaltung und Nutzung von digitalen Objektgedächtnissen, Dissertation Universität Saarbrücken (2013)
15. Herlyn, W.-J.: PPS im Automobilbau – Produktionsprogrammplanung und -steuerung von Fahrzeugen und Aggregaten. Hanser, München (2012)
16. Koppelberg, S.: Ch. 6 "Special Classes of Boolean Algebra". In: Monk, J.D., Bonnet, R. (ed.) Handbook of Boolean Algebra, vol. 1, North-Holland, Amsterdam (1989)
17. Sauer, O.: "The Digital Twin- a key technology for Industry 4.0". VisIT. https://www.iosb.fraunhofer.de/servlet/is/14330/visIT_1-26-03-2018_web.pdf June 2020 (2019) (Erstveröffentlichung 2018)
18. Schönsleben, P.: Integral logistics management: Operations and supply chain management within and across companies, 5th edn. CRC Press, Boca Raton, US (2016)
19. Wiendahl, H.-P.: Fertigungsregelung. Hanser, München (1997)
20. Zadek, H., Herlyn, W.: Der Digitale Steuerungs-Zwilling. In: ZWF, 115, special „Digitaler Zwilling", pp. 70–73, https://doi.org/10.3139/104.112338 (2020)
21. Zuehlke, D.: Smart factory – A vision becomes reality. DFKI, Kaiserslautern (n. d.)

Developing Technology Strategies for Flexible Automotive Products and Processes

Lukas Block[1(✉)], Maximilian Werner[1], Matthias Mikoschek[2], and Sebastian Stegmüller[2]

[1] University of Stuttgart, Institute of Human Factors and Technology Management (IAT), Allmandring 35, 70569 Stuttgart, Germany
lukas.block@iat.uni-stuttgart.de

[2] University of Stuttgart (Student), Allmandring 35, 70569 Stuttgart, Germany

Abstract. Flexibility provides a way to address technological progress and changing user needs in a vehicle's architecture. Yet, this has serious implications on the development and manufacturing process. It necessitates additional investments in long-lasting interfaces and changeable production design. Within this paper, we address the research question "which components of a vehicle should be designed flexible and which parts can remain static to streamline investments in flexibility?". We develop a methodology to support this decision by combining technology intelligence methods with comprehensive user acceptance research. Our approach builds upon a certain notion of function, which connects technological progress with customer utility. To this end, we analyze patents to evaluate technology dynamics and a user survey to detect functional needs. The research project *FlexCAR* designs a flexible vehicle architecture and was used for evaluation. Thereby, the methodology showed significant and sound results in its application and revealed flexible components with little effort. To the best of our knowledge, it is the first methodology of its kind.

1 Introduction

Flexibility in vehicle software, hardware and mechanical design provides a way to address constant technological progress and changing user needs. Connectivity, automation and electrification for example have increased the importance of adaption to latest technological progress. Studies suggest that for 36% to 80% of the users, connectivity positively stimulates the incentive to buy a vehicle [1]. Yet, innovation cycles in this area are significantly shorter than the total life cycle of a car. Thus, many automotive original equipment manufacturers have implemented over-the-air (OTA) software updates in their latest series [2]. It allows them to add new "versions" of software components or implement completely new functionality while the vehicle is on the road.

However, providing such flexibility in the long run has serious implications on the development and manufacturing process. On the one hand, it influences modularization and interface design because flexible modules and interfaces require a broader, future-oriented definition to last. On the other hand, flexible product design effects the

P. Weißgraeber et al. (Eds.): *Advances in Automotive Production Technology – Theory and Application,* ARENA2036, pp. 97–107, 2021. https://doi.org/10.1007/978-3-662-62962-8_12

assembly process and requires high changeability in manufacturing design [3]. Both challenges necessitate additional investments in flexibility. Costs occur. The research question arises: Which components of a vehicle should be designed flexible and which parts can remain static over the vehicle's lifetime to streamline investments in flexibility? Within this paper we address this research question by combining technology intelligence methods with comprehensive user acceptance research. The associated methodology was developed from scratch in the research project *FlexCAR* which develops a flexible architecture for future vehicles and its associated manufacturing processes. Thus, it faces a similar challenge in defining its flexible and static components.

2 Theoretical Background

In manufacturing and engineering design literature, flexibility is defined as a set of adaption options, which allow to react to changes without substantially changing the basic structure of the system and/or without causing major change effort [3–5]. Accordingly, a lot of literature addresses flexibility in product design: Suh et al. [4] as well as others (e.g. [6, 7]) utilize the design structure matrix or related systematic approaches (e.g. [8]) to guide architectural design decisions towards flexibility. A methodology called Uncertainty- or Change-Mode-and-Effect-Analysis systematically identifies possible changes and evaluates them in analogy to the Failure-Mode-and-Effects-Analysis (e.g. [9, 10]). However, the estimation of the change's dynamics (amplitude and frequency) is beyond the scope of these approaches.

Contrarily, technology intelligence addresses the early stage identification of new technologies or their potential impact. It thus serves the corporate ability to prepare for technological change (e.g. [11, 12]). To this end, basic activities in technology intelligence range from unfocused scanning for new information, over more targeted monitoring of technology areas, to detailed scouting of specific, technology-relevant information. Within these activities, the technology intelligence process comprises identifying, obtaining, evaluating and communicating information regarding technological innovations [12]. Thus, technology intelligence can serve the identification of crucial change spots, based on which flexibility requirements can be determined.

In technology intelligence, several tools and methods are well established to estimate technological change, such as patent or literature analyses, bibliometrics, scenario or Delphi method, and roadmaps which Reger [11] suggests to select based on the timescale to be considered and type of data available. Usually, the results are then transferred into a technology strategy to plan technology-specific activities and define corporate goals respectively [13]. Yet, Zhang et al. [14] show that differing type of results ask for a technology strategy development founded on different approaches. While technology roadmaps based on qualitative expert knowledge are widespread to schedule strategic R&D activities, Choi et al. [15] criticize more recent quantitative approaches being too limited to single keywords and propose to additionally derive functional correlations from patent information. Moreover, Eger and Schoder [16] argue that successful innovations rest on meeting consumers' needs and call for higher user-orientation in technology intelligence by applying text mining to user-generated content. In this context, novel text-mining approaches and algorithms for web-based data sources find increasing attention in technology foresight (e.g. [17, 18]).

However, to answer the research question present in this paper, it is not sufficient to identify possible changes. A rough estimation for the dynamics (frequency and amplitude) in a certain technological field is required. Technometrics for example is a method to measure technology dynamics based on historical data of certain key performance indicators [19, 20]. Moreover, text-mining methods are also suitable for estimating the frequency and amplitude of technological change. For instance, patent metrics such as citations and age (e.g. [21–23]) or similar evaluations on bibliometric indicators (e.g. [24, 25]) assess the development speed of technologies. Furthermore, Duwe et al. [26] suggest Google Trends to identify the diffusion of innovations. However, they state that it is only limited applicable to forecast new trends. Lenz and Winkler [27] calculate early indicators of technological change through automated information retrieval from technology-related news articles. Additionally, Ryu and Byeon [28] show a qualitative approach by proposing to evaluate technology dynamic levels using a Delphi survey.

To summarize, insights from technology dynamic evaluation can guide product architecture design, to incorporate the right degree of flexibility in the right place. Hartkopf's [3] technology roadmap approach already connects technology intelligence with flexible factory design. Yet, application to product design is limited as manufacturing design differs essentially. Thus, to the best of our knowledge, little research is known, which connects flexibility in product design to technology dynamic estimation. We close this gap and develop a new methodology to answer the research question.

3 Approach and Methodology

In the following, we will develop a new methodology to determine the right degree of flexibility for each component of a vehicle. It considers the results from the theoretical background (Sect. 2) as well as the project's requirements. It connects them to support the decision-making process concerning which vehicle components should be flexible. The new methodology is then applied to the *FlexCAR* project to define flexible and static components (Sect. 4).

The developed methodology consists of a procedure model (Fig. 1) and a decision canvas (Fig. 2). The procedure model of the methodology assesses different flexibility criterions for each component of the vehicle. The canvas then guides the decision for the degree of flexibility for each component based on these criterions. The methodology is derived by combining the output of technology dynamic research and the main statement of flexible product architecture design (see Sect. 2): A flexible architecture is only of value, if it enables effortless transitions towards new technologies, which reveal additional utilities for the user. Accordingly, flexible architectural design decisions must consider two aspects: Firstly, the degree and speed of the technological progress (technology dynamic), and secondly, the influence of the technological progress on user satisfaction and thus value of a change. Our notion of functions connects the utility-based perspective with technological progress in the decision canvas. We will elaborate this notion of function in the following.

In product development, a function is part of the requirements and describes a certain functionality the user expects or wishes the product to fulfil [29]. Utility and thus user satisfaction are directly dependent on the existence and the quality level of the

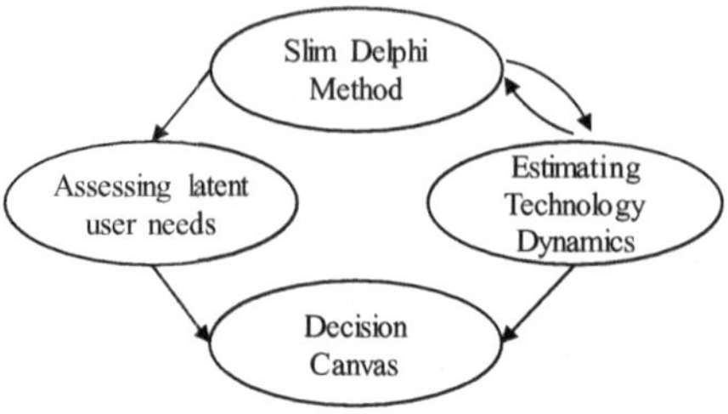

Fig. 1. Steps of the procedure model

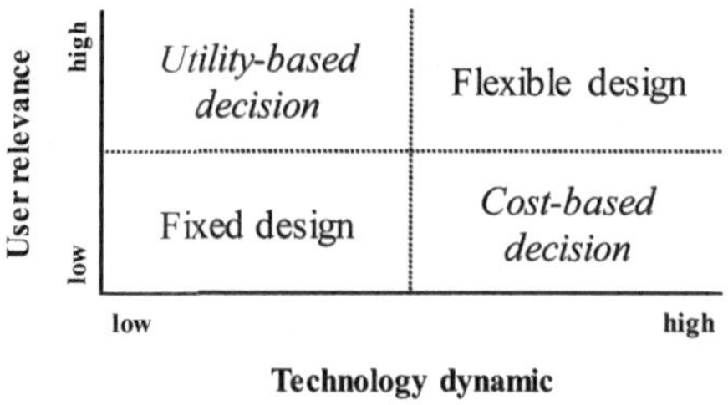

Fig. 2. Decision canvas for flexible product design

realized product function. As such, technology lays the foundation for the realization of a function. It influences its existence or its quality and thus the value for users [30].

This inter-dependency between function, utility and technology is depicted in the decision canvas (Fig. 2): Each point in the decision matrix represents one component and its function with its relevance for end users and its dynamic of the underlying technology. The horizontal axis describes the degree and speed of technological progress. The vertical axis denotes the relative sensitivity of utility in relation to the current state-of-the art technology, if technological progress happens. Thus, a component's degree of flexibility is derived from its function's position within the matrix.

A high sensitivity of user relevance and high technology dynamics indicate that a component should be designed as a replaceable and/or flexible part of the product architecture. User satisfaction and thus valuation of the product depends on the quality or existence of the function. Furthermore, technological progress happens foreseeably often. Thus, keeping modules up-to-date ensures customer loyalty. The inverse holds for the bottom left part of the matrix. Low user relevance and less technological progress imply a fixed or static module in the architecture. In this case, changes in technology happen seldom and the user is not interested in them, even if they happen.

For the remaining areas, mainly external factors determine the decision. A high user relevance and low technology dynamic can be seen as an opportunity to strengthen one's market position by increasing the products' utility. If technology progresses, users will require the new advantages. Technology leadership can be achieved if the module concerned is flexible enough to incorporate the change. Contrarily, this can also turn into a thread, if a technological leap is missed. Implementing flexibility for less relevant but highly dynamic functions is an economic concern (bottom right part). If the new technology provides additional value e.g. by enabling a cheaper manufacturing process, then incorporating flexibility for this module into the architecture makes sense.

Both latter choices focus on criterions beyond the user and technology dynamics. As such, we link to them in our approach, but see their evaluation beyond the scope of this paper. Consequently, the procedure model (Fig. 1) determines utility and dynamics of emerging functions and their respective technologies.

3.1 Assessing Latent User Needs of Future Product Functions

The decision canvas describes the function's user relevance as one criterion to assess. Therefore, a comprehensive understanding of user needs and requirements is necessary. Latest trends in product development apply open innovation approaches in order to involve users themselves in the assessment [31]. However, active forms of user integration are often limited to incremental innovations with functionalities already existing. The assessment of latent user needs during early stage innovation processes – of radical innovations in particular – is critical [32]. Consequently, we base our approach on a customized form of user surveys. The questionnaire is primed with expert knowledge and comprises a creativity enhancing question design. First, we use the professional expertise to develop hypotheses and to confine a solution space of future functionalities' ways of implementation (see Sect. 3.2) [33]. Hereupon, the questionnaire is formulated following a storytelling approach with narrative language. A user's perspective supports respondents in constructing their own vision of the future product [34, 35]. The quantitative survey results then represent a compilation of all respondents' product configurations based on their personal needs and requirements. Both, the general acceptance of a certain functionality as well as preferences regarding different ways of its technical implementations can be evaluated with this procedure.

3.2 Estimating Technology Dynamics

The decision canvas builds upon the critical notion of function-technology relationships. However, the selection of technologies to realize a user-visible function is at its core still left to development team's creativity [36]. Furthermore, visible or tangible functions are located at high levels of the function hierarchy [29]. Consequently, multiple technologies can contribute to the realisation of a user-defined requirement.

Thus, we apply a customized version of the Delphi method. It creates the function-technology mapping in the first step and generates the input to the methodology described in Sect. 3.1. In our Delphi method, experts give their input regarding possible mapping concepts and promising technologies to realize the functionality at hand. Afterwards, the dynamic of these technologies is investigated. Results of additionally interesting technology fields might occur during this investigation. These new technologies are fed back to the experts who evaluate them. They will then be added to the function-technology mapping or dismissed. This slim Delphi method is especially useful for applications with only few experts at hand (two to three). It delivers fast, approximate results, which are sufficient for the approach described in this paper.

Consequently, technology dynamic investigation starts with two or three expert interviews regarding promising current and future technologies to realize a function. Furthermore, they suggest whether the technology readiness is more in the direction of basic or applied research with respect to the given field. Accordingly, a bibliometric (basic research) or patent-based investigation (for applied research) is conducted. The patent-based approach in our case calculates the technology dynamic as follows:

$$Technology\ Dynamic = 100 \times \left(\frac{1}{3} \times \frac{MCC}{MCC'} + \frac{1}{3} \times \frac{TCT'}{TCT} + \frac{1}{3} \times \frac{MPA'}{MPA} \right) \quad (1)$$

The Mean Citation Count (MCC) is the average number of forward citations of all relevant patents within three years of publication [21]. It describes the growth rate of the technology. The Technology Cycle Time Indicator (TCT) is the average age of back-cited patents in a patent set [22]. The MPA indicator (Mean Patent Age) calculates the average age of the patent set under investigation [23]. It assumes that older technologies tend to develop more slowly and thus measures inventive activity. Finally, we normalize the values against a benchmark technology (MCC', TCT', MPA') because technological progress happens relative to the general progress in the applied field.

A combination of keyword and patent class search is applied to identify the set of patents, from which the technology dynamic is calculated. The experts initialize the patent set via a keyword description of the technology. The relevant patent class is then extracted and the results are narrowed down using another keyword set and Boolean operators (see e.g. [37]). We tested our described approach against two ex-post technology dynamic studies (both present in [23]). The high correlation (adjusted R^2 of 0.91) indicates a good approximation of future technology dynamics. The described process can be performed in a similar manner for bibliometric (e.g. research based) databases. For instance, Daim et al. [38] and Gupta et al. [39] describe such an approach.

4 Application and Results

The new methodology at hand (Sect. 3) is applied to the design of a flexible product line architecture for vehicle development. We show how it reveals the necessary flexible components with little effort. The evaluation took place in the research project *FlexCAR*, a joint initiative within the *ARENA2036* research campus. The *FlexCar* project consists of multiple work packages, each one addressing another product or manufacturing component. Furthermore, different research institutes and companies were involved in the component's development. As such, the challenge for the methodology was to be applicable with little effort and straight forward in usage. Evaluation criterions are whether the methodology's results are sound and significant, and whether the approach can be employed easily.

Soundness and significance are exemplary evaluated with the 'LiDAR' and 'seats' components of the *FlexCAR*. The components' relevance for users is based on an already existing, international user survey (see [40]). For instance, the importance of comfortable passenger seat design is used as a utility proxy for 'seats' and the willingness to use autonomous vehicles is a proxy for the 'LiDAR' function. The results show that user relevance of autonomous driving is fairly high across all respondents (Fig. 3). Comfortable seats however are rated less important by comparison (Fig. 4).

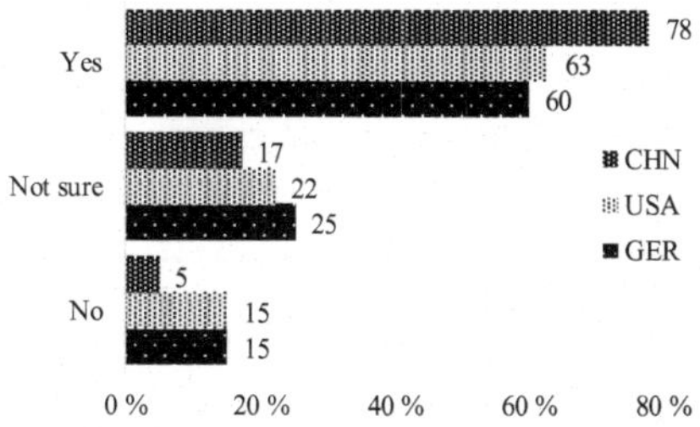

Fig. 3. Willingness to use an autonomous cab [40]

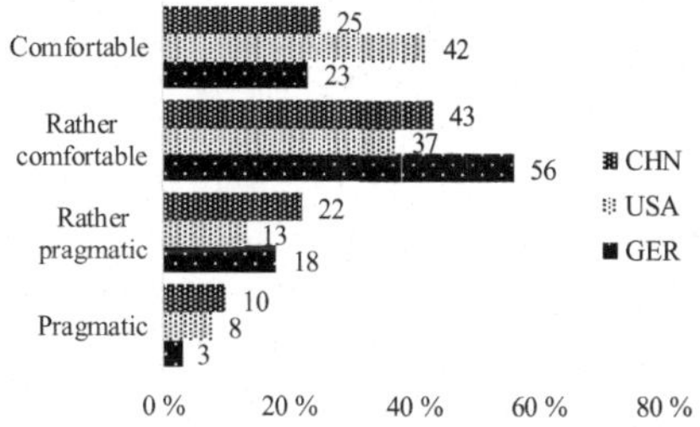

Fig. 4. Importance of seating comfort in autonomous cabs [40]

The evaluation of technology dynamics is realized through an automated search script that accesses the U.S. Patent and Trademark Office database. In accordance with the methodology in Sect. 3, expert knowledge from within the project was previously incorporated into the search strategy. We assessed the components under consideration in comparison to three other components – 'electrochemical energy storage', 'wheel assembly', and 'electric drive train' – with 'LiDAR' being the reference technology. Figure 5 exhibits the components' technology dynamic scores, determined with the methodology described in Sect. 3.2. Relatively, 'LiDAR' technology is developing fastest and 'wheel assembly' slowest. The components' technology dynamic and user relevance are now known. They can be located in the decision canvas (see Fig. 6).

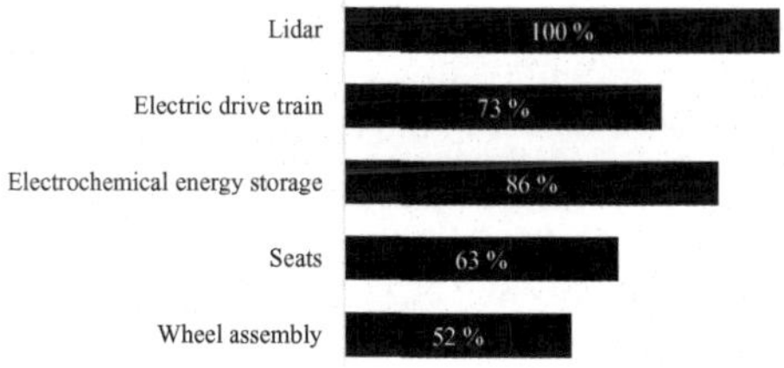

Fig. 5. Comparison of relative technology dynamic scores

Accordingly, it is recommended to pursue a relatively flexible module design for the implementation of 'LiDAR' technology due to high technology dynamic as well as user

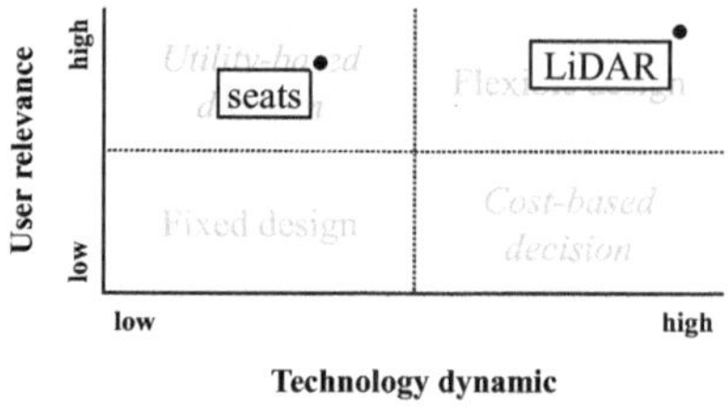

Fig. 6. The components 'LiDAR' and 'seats' located in the decision matrix

relevance. Seats still have relatively high user relevance but lower technology dynamics. Thus, their degree of flexibility is of utility-based, strategic importance. For example, a high-quality luxury OEM might design them with a certain amount of flexibility to maintain comfort and technology leadership over the vehicle's lifetime.

The overall results were consistent with expert opinions on the matter. As such, the methodology provides significant and sound results. Furthermore, application happened with little effort, because pre-existing information sources and automated information retrieval were used. The evaluation criterions are fulfilled with respect to the boundary conditions and requirements of the *FlexCAR* project.

5 Discussion and Conclusion

Within this paper, we developed a methodology to decide whether certain vehicle components should be designed flexible or fixed. The approach brings together methods from technology intelligence and comprehensive user acceptance research. To the best of our knowledge, it is the first approach for product design so far. Hartkopf's approach [3] for a flexible factory layout shows similarities in that it targets the same research question and also applies technology intelligence. However, the approach is much more fine-grained in that it takes operational, tactical and strategic uncertainties into account, when defining the flexible components of the manufacturing process. Yet, his methodology requires a higher effort and can only be carried out with a software tool. Our methodology requires less effort but assesses user valuation and technology dynamic only approximately. It is a lean approach that shall provide a basis for early stage decision making. Yet, the ease of application limits exactness in terms of the flexibility degree. The methodology cannot derive concrete flexibility requirements. Therefore, subsequent development phases must apply more advanced approaches in the field of technology and product strategies.

Most technology intelligence methods include a component to evaluate and derive the effect of a certain technology on a product (see e.g. [12]). However, this analysis happens in depth for each technology and again requires a higher investment. As such, we utilize more general technology dynamics measures. In the evaluation step for example, we applied patent analysis with sound results regarding effort and content.

Overall, the evaluation showed that the methodology is applicable within the *FlexCAR* research project. However, the methodology itself also offers a contribution to other vehicle development projects by answering the generic question, which components should be designed flexible. Therefore, future research should address broader field tests and application to validate the methodology's usefulness under different boundary conditions. Furthermore, already defined manufacturing strategies with their existing processes and facility designs must be kept in mind to define a flexible but economically viable product concept. As such, a firm connection of our methodology to flexibility approaches (e.g. [3]) in manufacturing design should be established.

To summarize, we suggest a methodological transfer, which transforms insights of technology intelligence to well-defined starting points for flexibility in product concepts. Subsequent approaches can then for example target the derivation of a flexible architecture from this information.

Acknowledgements. The research and development project *FlexCAR* is funded by the German Federal Ministry of Education and Research (BMBF) within ARENA2036 Research Campus (funding no. 02P18Q640 – 02P18Q649) and implemented by Project Management Agency Karlsruhe (PTKA). The authors are responsible for the content of this publication.

References

1. Vitale, J., Bowman, K., Singh, R., et al.: 2020 Global Automotive Consumer Study. Deloitte Development LLC (2020)
2. Truett, R.: Over-the-air vehicle updates can help dealership service departments. https://www.autonews.com/commentary/over-air-vehicle-updates-can-help-dealership-service-departments (2019). Accessed 15 Sept 2020
3. Hartkopf, M.: Systematik für eine kontinuierliche und langfristig ausgerichtete Planung technologischer und kapazitiver Werksentwicklungen. Dissertation, Univ. Stuttgart. Stuttgarter Beiträge zur Produktionsforschung, vol. 18. Fraunhofer, Stuttgart (2013)
4. Suh, E.S., de Weck, O.L., Chang, D.: Flexible product platforms: Framework and case study. Res. Eng. Design **18**, 67–89 (2007)
5. Westkämper, E., Zahn, E., Balve, P., et al.: Ansätze zur Wandlungsfähigkeit von Produktionsunternehmen. Ein Bezugsrahmen für die Unternehmensentwicklung im turbulenten Umfeld. Werkstattstechnik **90**, 22–26 (2000)
6. Giffin, M., de Weck, O., Bounova, G., et al.: Change propagation analysis in complex technical systems. J. Mech. Design**131**, 081001-1–081001-14 (2009)
7. Rebentisch, E., Schuh, G., Riesener, M., et al.: Assessment of changes in engineering design using change propagation cost analysis. In: Maier, A., Kim, H., Oehmen, J., Salustri, F., Škec, S., Kokkolaras, M. (eds.) Proceedings of the 21st International Conference on Engineering Design (ICED 17). Design Methods and Tools, pp. 69–78. Design Society, Red Hook (2018)
8. Jaring, M., Bosch, J.: Representing variability in software product lines: A case study. In: Chastek, G.J. (ed.) Software Product Lines. Second International Conference, proceedings, vol. 2379, pp. 15–36. Springer, Berlin (2002)
9. Rajan, P., van Wie, M., Campbell, M., et al.: Design for flexibility. Measures and guidelines. In: Folkeson, A., Gralen, K., Norell, M., Sellgren, U. (eds.) Proceedings of the 14th International Conference on Engineering Design (ICED 03). Design Society (2003)
10. Zhu, G.-N., Hu, J., Qi, J., et al.: Change mode and effects analysis by enhanced grey relational analysis under subjective environments. AIEDAM **31**, 207–221 (2017)
11. Reger, G.: Technologie-Früherkennung: Organisation und Prozess. In: Gassman, O., Kobe, C. (eds.) Management von Innovation und Risiko. Quantensprünge in der Entwicklung erfolgreich managen, pp. 303–329. Springer, Berlin (2006)

12. Wellensiek, M., Schuh, G., Hacker, P.A., et al.: Technologiefrüherkennung. In: Schuh, G., Klappert, S. (eds.) Technologiemanagement, vol. 10, pp. 89–169. Springer, Berlin (2011)
13. Schulte-Gehrmann, A.-L., Klappert, S., Schuh, G., et al.: Technologiestrategie. In: Schuh, G., Klappert, S. (eds.) Technologiemanagement, vol. 34, pp. 55–88. Springer, Berlin (2011)
14. Zhang, Y., Robinson, D., Porter, A.L., et al.: Technology roadmapping for competitive technical intelligence. Technol. Forecast. Soc. Chang. **110**, 175–186 (2016)
15. Choi, S., Kim, H., Yoon, J., et al.: An SAO-based text-mining approach for technology roadmapping using patent information. R&D Manage. **43**, 52–74 (2013)
16. Eger, M., Schoder, D.: Consumer-oriented tech mining: Integrating the consumer perspective into organizational technology intelligence. The case of autonomous driving. In: Bui, T. (ed.) Proceedings of the 50th Hawaii International Conference on System Sciences, pp. 1122–1131 (2017)
17. Kayser, V., Blind, K.: Extending the knowledge base of foresight: The contribution of text mining. Technol. Forecast. Soc. Chang. **116**, 208–215 (2017)
18. Silva, J., Vidal Pacheco, L.d.C., Negrete, K.P., et al.: Design and development of a custom system of technology surveillance and competitive intelligence in SMEs. Procedia Comput. Sci. **151**, 1231–1236 (2019)
19. Grupp, H.: Messung der technischen Eigenschaften von Innovationen (Technometrie). In: Grupp, H. (ed.) Messung und Erklärung des Technischen Wandels, pp. 101–141. Springer, Berlin (1997)
20. Coccia, M.: Technometrics: Origins, historical evolution and new directions. Technol. Forecast. Soc. Chang. **72**, 944–979 (2005)
21. Benson, C.L., Magee, C.L.: Quantitative determination of technological improvement from patent data. PLoS ONE **10**, e0121635 (2015)
22. Kayal, A.A., Waters, R.C.: An empirical evaluation of the technology cycle time indicator as a measure of the pace of technological progress in superconductor technology. IEEE Trans. Eng. Manage. **46**, 127–131 (1999)
23. Benson, C.L.: Cross-Domain comparison of quantitative technology improvement using. Dissertation, Massachusetts Institute of Technology (2014)
24. Boyack, K.W., Klavans, R.: Co-citation analysis, bibliographic coupling, and direct citation: Which citation approach represents the research front most accurately? J. Am. Soc. Inform. Sci. Technol. **61**, 2389–2404 (2010)
25. Small, H., Boyack, K.W., Klavans, R.: Identifying emerging topics in science and technology. Res. Policy **43**, 1450–1467 (2014)
26. Duwe, D., Herrmann, F., Spath, D.: Forecasting the diffusion of product and technology innovations: Using google trends as an example. In: 2018 Portland International Conference on Management of Engineering and Technology (PICMET), pp. 1–7, IEE, Honolulu (2018)
27. Lenz, D., Winker, P.: Measuring the diffusion of innovations with paragraph vector topic models. PLoS ONE **15**, e0226685 (2020)
28. Ryu, J., Byeon, S.C.: Technology level evaluation methodology based on the technology growth curve. Technol. Forecast. Soc. Chang. **78**, 1049–1059 (2011)
29. Pohl, K.: Requirements engineering, 2nd edn. dpunkt, Heidelberg (2008)
30. Bullinger, H.-J.: Einführung in das Technologiemanagement. Vieweg, Wiesbaden (1994)
31. Reichwald, R., Piller, F.T.: Interaktive Wertschöpfung, 2nd edn. Gabler, Wiesbaden (2009)
32. Carlgren, L.: Identifying latent needs: Towards a competence perspective on attractive quality creation. Total Quality Management & Business Excellence **24**, 1347–1363 (2013)
33. Lindemann, U.: Methodische Entwicklung technischer Produkte, 3rd edn. VDI-Buch, Berlin (2009)
34. Körber, M., Eichinger, A., Bengler, K., et al.: User experience evaluation in an automotive context. In: 2013 IEEE Intelligent Vehicles Symposium (IV), pp. 13–18. IEEE, Piscataway (2013)

35. Michailidou, I., Saucken, C. von, Lindemann, U.: How to create a user experience Story. In: Marcus, A. (ed.) Design, User Experience, and Usability. Design Philosophy, Methods, and Tools, pp. 554–563. Springer, Berlin (2013)
36. Menshenin, Y., Crawley, E.: A framework for concept and its testing on patents. INCOSE Int. Symp. **28**, 1564–1577 (2018)
37. Patent and Trademark Office, United States Department of Commerce: 7-Step U.S. Patent Search Strategy Guide. https://www.uspto.gov/sites/default/files/documents/7_Step_US_Patent_Search_Strategy_Guide_2015_rev.pdf (2015). Accessed 18 May 2020
38. Daim, T.U., Rueda, G., Martin, H., et al.: Forecasting emerging technologies: Use of bibliometrics and patent analysis. Technol. Forecast. Soc. Chang. **73**, 981–1012 (2006)
39. Gupta, B.M., Bhattacharya, S.: Bibliometric approach towards mapping the dynamics of science and technology. DBIT **24**, 3–8 (2004)
40. Stegmüller, S., Werner, M., Kern, M., et al.: Akzeptanzstudie Robocab. Autonome Mobilitätskonzepte aus Sicht der Nutzer, Stuttgart (2019)

Structured Information Processing as Enabler of Versatile, Flexible Manufacturing Concepts

Simon Komesker[1,2](✉), Wolfgang Kern[3,4], Achim Wagner[5], Thomas Bauernhansl[3,6], and Martin Ruskowski[1,5,7]

[1] TU Kaiserslautern, 67663 Kaiserslautern, Germany
simon.komesker@volkswagen.de
[2] Volkswagen AG, 30884 Wolfsburg, Germany
[3] GSaME, University of Stuttgart, 70569 Stuttgart, Germany
[4] AUDI AG,, 85045 Ingolstadt, Germany
[5] Deutsches Forschungszentrum für Künstliche Intelligenz GmbH, 67663 Kaiserslautern, Germany
[6] Fraunhofer Institute for Manufacturing Engineering and Automation IPA, 70569 Stuttgart, Germany
[7] Technologie-Initiative SmartFactory KL e. V., 67663 Kaiserslautern, Germany

Abstract. Automotive production systems face the challenge to produce models and brands with different drive concepts and individually configured equipment variants in a highly efficient way.

Studies on modular assembly systems in automotive industry have demonstrated potential for productivity gains through the implementation of an alternative, value-add-oriented process organization. The rigid concatenation of mechanical production processes is the limiting factor; firstly, for an efficient implementation of product individualization and secondly, for a highly available robust production which can optimize the overall factory production flow.

Rising degrees of freedom in material flow control associated with more flexible production flow increases the complexity of the overall production system. Decision support for humans by planning systems with integrated control logic is thus a decisive factor for mastering complexity. Currently, the overall performance of modular manufacturing processes is not sufficiently supported by the IT-architecture on factory level. The individually operating subsystems are not capable of supporting reactive manufacturing control.

As a basis for reactive manufacturing control, the information requirements towards modular manufacturing processes across different domains are defined in this paper. Furthermore, a cross-system information and communication matrix is proposed that structures information processing between individually operating subsystems. The application of broker-technology could subsequently enable holistic information-based control on factory level to support human experience-based decision making.

1 Introduction

The development of technology markets, especially in the field of digital data processing, is taking place faster than in almost all other areas and offers new possibilities for

P. Weißgraeber et al. (Eds.): *Advances in Automotive Production Technology – Theory and Application*, ARENA2036, pp. 108–116, 2021. https://doi.org/10.1007/978-3-662-62962-8_13

the demand-oriented control of production processes [1]. There is potential for optimization on the shop floor through the occurrence of unforeseen, short-term changes that require a high degree of coordination and control [2]. The technical possibilities of digital, flexible IT-networks offer great potential for increasing productivity for medium-sized businesses and industry. One potential of the increasing connectivity in production systems is the possibility of implementing alternative, modular manufacturing concepts. The challenge for automation technology is to support the process requirements for flexibility and autonomy in the production systems with comprehensive information and control technology [3].

In the following, a procedure will be presented which shows the corresponding information requirements and possibilities for interaction in modular production systems enabled by agent-based control systems.

2 State of the Art and Related Work

In the following section, the influence of alternative manufacturing concepts for the automotive industry is given as well as the development of Cyber Physical Production Systems (CPPS) and the associated challenge for the implementation of control concepts in production systems.

2.1 Efforts Towards Implementing Alternative Manufacturing Concepts for Modern Production Systems in the Automotive Industry

Current manufacturing systems in the automotive industry are based on the concept of continuous flow production for manufacturing tasks with a high proportion of identical parts and a uniform, high production volume [4]. With the implementation of product individualization in mixed model lines, the line concept is being questioned to be the most effective and efficient manufacturing concept. As a result of the wide range of variants, the production system has to support different process variants and thus a wide range of flexibility requirements are addressed with alternative methods [5, 6]. A freely interlinked material flow can meet the requirements for flexibility and dynamics in the production system and, among other things, increase the value-add time per resource as a whole [7].

In order to cope with the flexibility requirements in mass production, a change in the manufacturing concept with modular stations based on the model of decentralized fractals can be considered, which, by using intelligent technologies, can make the considerations on the organizational principles realizable [8]. For this purpose, design principles were developed which can support humans in the form of rule-based planning for a modular assembly concept in variant flow production [5, 9].

With the intended paradigm shift to an Industry 4.0 age, the complex cognitive work content for humans has increased [2]. A modular production system with an increased number of degrees of freedom requires intelligent software support to solve complex production problems.

2.2 Distributed Information Processing as Requirement for Implementing Modular Production Systems

Due to the leaps in development in the field of information and communication technology, electronic hardware components with intelligent software can be organized as Cyber-Physical Systems (CPS) or Cyber-Physical Production Modules (CPPM) in a Cyber-Physical Production System (CPPS). Industrie 4.0 components describe a physical object that provides a digital and actively communicating representation over its product life cycle – the asset administration shell (AAS) [10]. By means of an actively communicating representation of the actors in a CPPS, a self-organized, value creation-oriented coordination of production process steps and production resources is possible [11].

The state-of-the-art of control systems in automotive industry is a predominantly heterogeneous, historically grown system landscape according to ISA95. However, systems with several independent decisions that are distributed over time offer potential for agent-based control systems [12]. A dynamic reconfiguration with the consideration of information from several process participants requires, for example, parallel, decentralized and autonomous decision-making processes, which must be enabled by the control architecture. An agent-based control system can be used to integrate cyber physical systems into production systems [13]. Control architectures using middleware technology seem promising for providing cross-system communication and information integration for subsequent physical and logical reconfigurability in the system [14].

Approaches to the use of multi-agent systems (MAS) in production control rely on data from existing ERP-, MES- or SCADA-systems to integrate level-specific information [3, 13–15]. One method for developing control systems for MAS is the Designing Agent-based Control Systems (DACS) method, which is a suitable method for solving control problems in production based on shop floor decisions [13]. It designs agents based on decision clusters and interaction of the entities and their information flow. It is easy to apply in different domains and was extended by Vogel-Heuser et al. to include the possibility of integrating the field level [13].

Although there are several approaches for designing MAS in industrial use cases, implementing existing methods with the latest technology in both research and industry is missing. Existing approaches do not fit modern Industrie 4.0 standards such as RAMI 4.0, which leads to a limited deployment of such system, even if they have reached an industrial degree of maturity. There is a gap in the current industrial information structures for supporting reactive manufacturing control. Therefore the potential of modular processes in manfucaturing systems is still limited. The holistic consideration of distributed information from different systems needs to be supported by the information design of the manufacturing system.

3 Concept for Information Processing in Alternative Versatile Manufacturing Systems Implementation

In today's production systems in the automotive industry, humans are responsible for cross-divisional information processing. Information systems support them in the processing of planning and control tasks, but humans link the information to form an overall picture based on experience. A future system structure for modular production systems is required, which allows the user to combine the distributed experience-based process knowledge and make it available in the system in a sustainable way. This will allow decisions spanning different systems and areas to be made in an optimal way in the future. The following procedure is proposed for an application of a method for agent-based control of a modular manufacturing system:

1. analysis of the design principles [5, 9] in 3.1 regarding the information requirements for modular manufacturing systems and a resulting
2. definition of information requirements for modular production in the end of 3.1
3. informational relationships of level-specific interaction of process components in 3.2

3.1 Requirements for Information Processing in Modular Production Systems

The manufacturing system with a modular structure is intended to enable a flexible allocation of production resources and production process sequences. Depending on the order situation and the planned production program, this can lead to a need for negotiations between the actors in the system. Therefore, a planning methodology for demand-oriented design and a corresponding condition-based control is required. The previous principles according to Kern et. al [5, 9] are structured as follows: In the first step, the basic relationships between product and process are presented, in the second step, the requirements for a dynamic system network and in the third step, the extensions for the production system through the influences of the areas of quality, logistics and adaptations in the system are added.

A) Basic product and process information support: For a **variable station sequence**, the product-specific priority graph must be taken into account. This graph organizes the predecessor and successor relationships of the shoring process and determines the degree of freedom of the sequence. The workstation requires the ability to identify products and the ability to execute the required process steps. The station represents itself and its process capabilities determined by its resources and employees and is a production resource for the processing steps. On the other hand, the product provides the information about the **product-specific processing requirements** and process steps still to be completed. For the lead time of the product in the stations, no cycle-time-bound time specification applies, but rather the process time. The product and the station require a **self-description and identification capability**.

B) Dynamic system network: In a network of several workstations and products, status-based self-control and demand-dependent allocation of the stations is required as a logical component. The central assignment of products to process stations represents a recurring negotiation situation depending on the number of variants and production sequence. Optimum central process control can only be achieved by coordinating decentralized product needs and a central process control instance. The information about product sequence, product processing time and station availability in the status current priority graph must be transparent and available in "real-time". The reactions due to capacity and demand changes should be dynamic and flexible. One example is the higher-layer reaction to changes in customer demand and the related product mix in the production plan, as well as the lower-layer reaction to short-term disruptions in machine scheduling planning. This requirement for **logical reconfigurability** can be achieved by **real-time data exchange**.

C) Extension of the system behavior (integration of further control loops).
Integration of Logistics: The challenge of multi-model lines is to organize logistics expenditure in line with demand. A demand-oriented provision of components at the installation site requires real-time communication with the Automated Guided Vehicle's (AGV) central control system and the warehouse system. AGVs must be able to react locally to **safety-relevant external influences in real time**.
Integration of Quality: The modularization of the production process and the associated possibility of parallel processing of the same or similar materials at different locations requires the **flexible integration of quality control loops**. A comprehensive traceability of the individual and distributed shoring processes is necessary. For this purpose, the possibility of central storage of decentralized process results must be created. Evaluations in real time enable centralized knowledge for process optimization. Central coordination of decentralized quality information could even contribute to **predicted process and quality improvement** by appropriate use of technology.
Integration of Resource (CPPM/employee): The information about decentralized change or integration requirements of CPPM or employees is taken into account by the central process monitoring system and processed according to the production program. Components or entire stations can be temporarily blocked by setup processes or qualification measures and then be made available again in modified form. They require an adjustment of the capacity planning in the overall system for product, process, logistics and quality. In addition to the already mentioned logical reconfigurability, this also requires the ability of the system to consider **physical reconfigurability**.

Table 1 presents an overview of the requirements for decentralized (station-related) and centralized (network-related) control loops in the four information structure clusters. This serves, first of all, for rough planning and sets the premises to be observed for further detailed planning of the systems and information interactions.

3.2 Structuring of the Information Flow of Entities

Based on the material flow relationships, an overview of the interaction of all process participants in the production flow can be displayed. The aim is to derive the information relationships between the process participants from the process sequences of the freely linked material flow. In this way, the individual information relationships for the variant to be produced can be structured to show control decisions. This procedure makes it possible to integrate the requirements identified in step 3.1 into an interaction representation of the material and information flow. Figure 1 shows an example of centralized, decentralized and hybrid control structures in an interface matrix. By pointing out the process driven relationships for the levels of the production system, it can be shown which information-technical relationships exist between the actors of the respective level.

From Fig. 1, the level-specific control requirements can be derived and clustered with the interdependent information relationships. By applying the interface matrix,

Table 1. Information requirements for holistic production systems

Information Requirements of Modular Production Systems	Product		Process		Logistics		Quality	
	Central	Decentral	Central	Decentral	Central	Decentral	Central	Decentral
Self-descriptiveness	–	x	–	x	–	x	–	x
Identifiability	–	x	–	x	–	x	–	x
Real-Time Capability	x	x	x	x	x	x	x	x
Logical Reconfigurability	x	–	x	–	x	-	x	-
Physical Reconfigurability	x	x	x	x	x	x	x	x
Safety	–	x	-	x	–	x	x	x

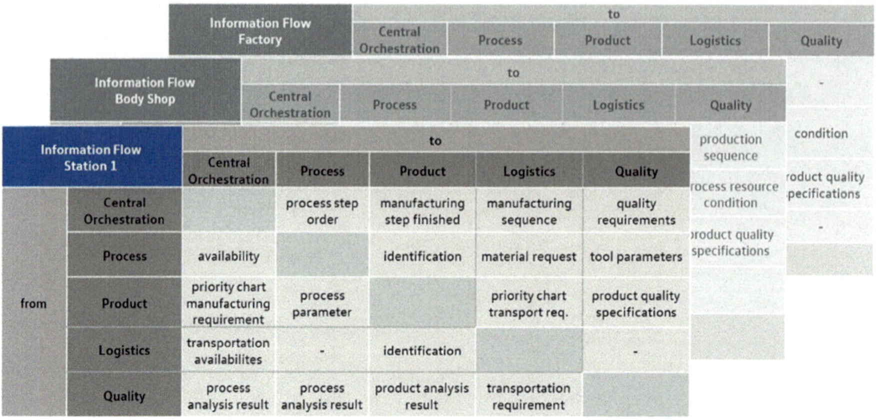

Fig. 1. Interface Matrix for the material flow driven interaction of entities and their information flow on system and station level

the information flow relationships of the actors and the central (horizontal and vertical connections) or decentral (diagonal connections) control needs become transparent. This can be used as a basis for designing control loops processing the relevant product, process, quality and logistics information for each level. After the basics for the product- and process-related design of modular assembly systems have been described, a holistic information support is required. The integration and consideration of the individual cross-process information of the corresponding process participants and other areas becomes challenging. Therefore, domain spanning MAS and intelligent middleware solutions can support handling this complexity. The interface matrix is also suitable for the analysis of existing processes which can be transferred into a modular structure.

4 Discussion and Conclusion

With the analysis of the design principles and the derivation of information requirements for modular assembly systems, a step towards a cross-system control for an application in modular production systems in the automotive industry has been taken. In interaction with the application of the interface matrix, the process participants can be identified and integrated according to their process requirements. The control requirements of modular assembly systems fulfill the factors for the agentification of control systems and the development of a distributed, agent-based control system therefore offers potential for further application.

With these decision clusters, the design for processing process, product, logistics and quality information can be pursued further, which enables an agent-based design for nested control loops in production systems. The computing effort in decentralized or centralized system structures for ensuring a robust and efficient operating system behavior must then also be taken into account. As no production systems with a modular structure have been operated in automotive industry so far, the described way of designing the information technology support with a cross-system approach will also be validated towards relevant KPIs.

In this article, the information flow requirements for modular assembly systems were defined and a novel procedure for structuring information relationships could be proposed. It is based on the product- and process-related manufacturing requirements and takes into account the information relationships with logistics and quality. By applying the procedure for structuring information relationships and considering the information requirements, flexible, dynamic production systems can be reconfigurable in terms of information technology. Further work will quantitatively validate the procedure and investigate the integration of further external system information using latest middleware architecture technology like OPC-UA, Apache Kafka, and BaSyx. The results can then be transferred into a sustainable system structure connecting different domains in order to deliver the informational basis to implement different manufacturing concepts in the automotive industry. Subsequently a strategy to design the information processing for central and decentral processing can follow in further research by using methods from the field of systems theory and engineering.

This seems to be a promising approach for the implementation of cross-system optima linking different abstraction layers by using latest technology. The aim is to deliver a broadly applicable system structure based on structured information processing of product, process, logistics and quality information for efficient and intelligent manufacturing systems.

References

1. Bauernhansl T., ten Hompel M., Vogel-Heuser, B.: Industrie 4.0 in Produktion, Automatisierung und Logistik, 1st edn. Springer Vieweg, Wiesbaden (2014)
2. Spath, D., et al.: Produktionsarbeit der Zukunft – Industrie 4.0. Fraunhofer-Verlag, Stuttgart (2013)
3. Block, C., Morlock, F., Kuhlenkötter, B.: Ganzheitliche flexible Vernetzung durch Erweiterung bestehender IT-Strukturen zu Serviceorientierten Architekturen mithilfe von Agentensystemen zur humanzentrierten Entscheidungsunterstützung. Ein Konzept zur RAMI Umsetzung. In: Schlick, C. (ed.) Megatrend Digitalisierung – Potenziale der Arbeits- und Betriebsorganisation. GITO, Berlin (2016)
4. Wannenwetsch, H.: Integrierte Materialwirtschaft und Logistik: Beschaffung, Logistik, Materialwirtschaft und Produktion, 4th edn. Springer-Lehrbuch. Springer, Berlin (2010)
5. Kern, W., Rusitschka, F., Kopytynski, W., Keckl, S., Bauernhansl T.: Alternatives to assembly line production in the automotive industry. In: Proceedings of the 23rd International Conference on Production Research (ICPR), Manila, Philippines (2015)
6. Bochmann, L., Gehrke, L., Böckenkamp, A., Weichert, F., Albersmann, R., Prasse, C., Mertens, C., Motta, M., Wegener, K.: Towards decentralized production: A novel method to identify flexibility potentials in production sequences based on flexibility graphs. Int. J. Autom. Technol. **9**(3), 270–282 (2015)
7. The Boston Consulting Group (BCG): https://image-src.bcg.com/Images/BCG-Will-Flexible-Cell-Manufacturing-Revolutionize-Carmaking-Oct-2018_tcm9-205177.pdf (2020). Accessed 20 May 2020
8. Warnecke, H.J., Hüser M.: Die fraktale Fabrik. Revolution der Unternehmenskultur [The fractal company. A revolution in corporate culture]. Rowohlt, Berlin (1996)
9. Kern, W., Rusitschka, F., Bauernhansl, T.: Planning of workstations in a modular assembly system. In: Proceedings of the 49th CIRP Conference on Manufacturing Systems (CIRP-CMS 2016), Stuttgart, Germany, 25.–27.05.2016, pp. 327–332 (2016)

10. Adolphs, P., Auer, S., Bedenbender, H., Billmann, M., SE, B., Coskun, G., et al.: Statusreport – Fortentwicklung des Referenzmodells für die Industrie 4.0 – Komponente. Struktur der Verwaltungsschale. VDI, ZVEI Düsseldorf, Frankfurt a. M. (2016)
11. Hermann, J., Rübel, P., Birtel, M., Mohr, F., Wagner, A., Ruskowski, M.: Self-description of cyber-physical production modules for a product-driven manufacturing system. 29th International Conference on Flexible Automation and Intelligent Manufacturing (FAIM2019), Limerick, Ireland. In: Procedia Manufacturing, Vol. 38, 2019, pp. 291–298 (2019).
12. Bussmann, S., Jennings, N.R., Wooldridge, M.: On the identification of agents in the design of production control systems. In: Goos, G., et al. (ed.) Agent-Oriented Software Engineering. pp. 141–162, Springer, Berlin (2001)
13. Vogel-Heuser, B., Göhner, P., Lüder, A.: Agent-based control of production systems – and its architectural challenges. In: Leitão, P., Karnouskos, S. (eds.) Industrial Agents: Emerging Applications of Software Agents in Industry. Elsevier, Amsterdam, pp. 153–170 (2015)
14. Trunzer, E., Calà, A., Leitão, P., et al.: System architectures for Industrie 4.0 applications. Prod. Eng. Res. Devel. **13**, 247–25 (2019)
15. Leusin, M., Kück, M., Frazzon, E., Maldonado, M., Freitag, M.: Potential of a multi-agent system approach for production control in smart factories. IFAC PapersOnLine **51–11**, 1459–1464 (2018)

A Novel Approach to Generate Assembly Instructions Automatically from CAD Models

Alexander Neb(✉) and Johannes Scholz

Fraunhofer Institute of Production Engineering and Automation IPA, Nobelstraße 12, 70569 Stuttgart Stuttgart, Germany
alexander.neb@ipa.fraunhofer.de

Abstract. For the distribution of consumer and industrial goods, every company is obliged to provide assembly instructions. For the consumer goods market, for example, the German Civil Code defines that defective assembly instructions must be declared as a material defect. However, the creation of comprehensible and defect-free assembly instructions is still a very time-consuming manual process, which must be determined in an extremely time-consuming procedure. Nonetheless, assembly instructions are more than just obligatory documents. They are also required in places where they are not prescribed. For example, assembly instructions are needed in production to pass on assembly knowledge to the assembly operators. Here, it often turns out that this knowledge is either not available or can only be used to a limited extent.

The two key elements of an assembly instruction are the assembly sequence and the visual illustrations. Currently, the assembly sequence is determined manually by the designers based on their personal experience, whereas illustrations are generated with costly software tools which are not even able to check the feasibility of the planned instruction. This work presents a novel approach to generate assembly instructions directly and automatically from CAD models of the designers. For this purpose, the commercial CAD software SolidWorks was extended by a macro Tool. All necessary data to generate an assembly instruction are extracted from the CAD model. The extracted data are assembly features, stability and geometric restrictions, subassemblies and assembly directions. Based on these data, the assembly operations are evaluated with a fitness function which includes the attributes like tool changing costs or distances of assembly paths. The whole assembly sequence optimization process was modeled as a Travelling Salesman Problem. After the ideal assembly sequence was found by the macro Tool, this tool also generated matching visualizations of the assembly operations based on the CAD model. The approach was validated by three different models, an assembly benchmark, a single-cylinder engine and a gear box.

1 Introduction

Planning of assembly sequences has always been difficult for both human and machine. For example, an assembly of 12 parts has 479.001.600 possible sequences to be assembled. Additionally, there are different possibilities to orient the assembly regarding the

P. Weißgraeber et al. (Eds.): *Advances in Automotive Production Technology – Theory and Application,* ARENA2036, pp. 117–125, 2021. https://doi.org/10.1007/978-3-662-62962-8_14

gravity direction, which increase the number of possible assembly sequences furthermore [1]. However, just a few of them are feasible. Therefore, an assembly instruction is needed which provides the information how a part has to be assembled. It also needs to provide the needed order and resources for their assembly.

The German civil code defines that a non-assembled product is only complete if it contains an accurate assembly instruction. If there is no assembly instruction or a faulty assembly instruction, it must be declared as a material defect. Therefore, the creation of an accurate assembly instruction is essential.

Although the product design is largely digitalized with CAD and PLM systems, the creation of the assembly instructions still requires a great amount of manual effort [2]. With high editorial effort and a huge amount of expert knowledge the assemblies have to be analyzed manually, because the CAD model itself do not provide information regarding the assembly sequence [3].

This leads to a great gap between designing and planning of assemblies, which becomes more and more difficult based on the shrinking product life cycles. However, an automated generation of assembly instructions from CAD models would support to bridge the gap between the designing and planning phase and reduce the required expert knowledge in the planning of assemblies.

The presented approach generates the required assembly information automatically out of the 3D CAD model and generates a feasible and optimized assembly sequence. Furthermore, the generated assembly sequence is visualized automatically by the help of a SolidWorks macro in Microsoft PowerPoint. The generated assembly instruction also includes the necessary resources that are assigned to the step based on the extracted data.

2 State of the Art in the Generation of Assembly Instructions

The aim of this work is to create an assembly instruction automatically from a CAD model. This requires both an automatic generation of the assembly sequence as well as an automatic generation of illustrations and description of the assembly steps.

To generate feasible and optimized assembly sequences three restrictions for an assembly sequence were defined. In an assembly sequence the next assembled part must always be in contact with one of the previous assembled parts. Furthermore, it must be possible to move the parts to their mounted position without any collisions and lastly each assembly state must be stable [1, 4].

Finding assembly sequences, which meet the described conditions from CAD models, was tried by using different data from the CAD model. In the past [5] and [6] just contact features were used. However, this approach showed that more data is required to generate assembly sequences automatically. In [7] a collision analysis, it is presented how to get the geometric restrictions and collision free assembly sequences. Additionally, this approach analyzes the stability by checking which parts support each other in a manually defined gravity direction. The determined restrictions are saved with a restrictions matrix like also presented in [8]. To determine the best possible assembly sequence within the restrictions, various optimization parameters are considered in literature. The most used ones are the number of reorientations and tool changes. Nonetheless, assembly operation in gravity direction and the stability of the assembly states should also be considered and not been forgotten [9].

An assembly instruction describes the required assembly operations best with a combination of illustrations and texts. For this purpose, a guideline exists which specifies several aspects of an assembly instruction. The DIN EN 82079-1 [10] specifies that an assembly step must be described clear and understandable. Therefore, it is best to use action diagrams to visualize assembly steps. The part which will be assembled next is moved against the assembly direction out of its assembled position for the illustration and the assembly direction can be visualized with an arrow [11]. Furthermore, the additional process information like required tools should be shown together with the illustration. The guideline recommends summarizing equal successive assembly steps, too. Therefore, it is necessary to detect all these steps.

Until today, research has not developed a program capable of automatically creating the illustrations and texts for assembly steps based on an automatically generated assembly sequence. Therefore, there exists a kind of documentation software, which requires a lot of manual work. All the parts must be moved manually in their position for the action diagram. Additionally, the required tools must be defined physically and inserted manually to the description.

This demonstrates the huge effort, which is needed to create assembly instructions. Furthermore, another challenge in this area is the trend of mass personalization, which results in assembly instructions with many variants. Based on the missing link between assembly instructions and CAD models, every assembly instruction of every variant has to be done manually in a repeating process.

3 Interface to CAD Model

As described in the previous chapter, for the generation of assembly sequences, various data must be extracted out of the CAD model. This approach uses an internal approach by accessing the CAD system SolidWorks though the API. First, the contact information is extracted based on the approach in [12]. This approach detects plane contacts, concentric restrictions, conical contacts and screwing connections. Additionally, for each contact feature, a vector direction will be derived.

For the description of the parts, the part properties like dimensions, weight and material information are also extracted from the CAD system through the API. Furthermore, norm parts, which are generated through the SolidWorks Toolbox, are detected by reading out the configuration.

The geometric restrictions are extracted with a collision analysis based on the approach in [7]. The parts are moved along their directions of the contacts and the assembly direction with the least number of collisions is chosen as assembly direction. This approach also defines subassembly groups based on the extracted contact information. Nonetheless, the approach is not able to guarantee that a subassembly is mountable. Therefore, a collision analysis for subassemblies was developed. The restrictions are saved in a restriction matrix as described in Chap. 2 [8]. The approach in [7] requires a manual gravity direction selection. To remove this manual step, the possible gravity direction is limited to the six possible directions parallel to the coordinate axis and the best ranked direction is selected automatically in the assembly sequence generation. For each of the six possible gravity directions, a support matrix is created through analyzing the contacts. Each of the six support matrices are matched with each other to the geometric restriction matrix. If the stability restriction is contradicting the geometric restrictions, the stability restrictions are deleted. All the data are generated by a macro accessing the SolidWorks API. All extracted data are stored as text files.

4 Assembly Sequence Generation

Creating the assembly instruction out of a CAD model requires the generation of an optimized assembly sequence. In Chap. 3 the data generation out of the CAD model is described. Based on this data, the optimal assembly sequence is generated using a developed Python 3.7 script, which is called by the SolidWorks macro. This also includes the best possible orientation of the assembly regarding the gravity direction.

The assembly group is modeled in a directed graph (see Fig. 1). The nodes are the parts. The edge from part 3 to part 4 represents the assembly operation if part 4 is the following part to part 3.

Each assembly operation is evaluated by three optimization parameters. For the reorientation, the assembly direction of the two parts of an edge is compared. For the tool changes each part is assigned to a tool based on its part properties. If the tool of the two nodes of an edge is different a tool change is required. The third parameter is the stability of the assembly group after an assembly operation. Assembling part 4 after part 3 in Fig. 1 is an instable step because part 4 has just a plane contact after the assembly. Assembling part 2 after part 1 the operation is evaluated as stable because one has a screwing connection after the assembly.

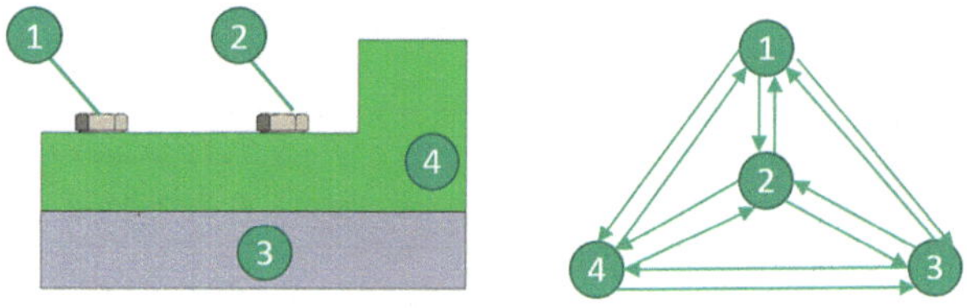

Fig. 1. Example assembly and the corresponding directed graph

Through the evaluation, each edge has its cost and so the best possible assembly sequence equals the cheapest path through the graph under consideration of the restrictions. Therefore, the optimization problem equals a Travelling Salesman Problem with Restrictions. The optimization problem is solved with a cheapest insertion algorithm. The restrictions are transformed in precedence tuples to make them readable for the solution procedure. For example if node 1 has to be assembled after node 4 the tuple is [1, 4] and the restriction is that the sum of all edges before node 4 has to be smaller or equal to the sum of all edges before node 1.

To guarantee the stability of the assembly sequence, the orientation regarding the gravity direction is determined automatically before the optimization. The approach in [7] defines a base part which has contact to the workspace during the hole assembly process. Therefore, this approach only considers gravity directions for which the base part is in contact with the work surface. This part will also be the first assembled part.

This approach allows to reorient the assembly group regarding the gravity direction once. A reorientation is required if the assembly of a part must be carried out through the work surface. For reorientation, all parts, which are assembled before, must have a concentric restriction. If there is a part which just has plane contacts before the reorientation, the reorientation must be carried out before this part is assembled. Because of the limitation to one reorientation, all parts must be mountable for the new gravity direction. To choose the best orientation for the assembly, the possible gravity directions is evaluated by the parameters number of reorientation, number of assembly steps not in gravity direction, gravity stability and size of the contact area of the base part with the work place. If no reorientation is needed, the corresponding restriction matrix is chosen. If there is a reorientation, all parts before the reorientation part are a precedence to the reorientation part.

5 Generation of Assembly Instructions

Before an assembly sequence for the main assembly is generated, the subassemblies will be assembled as "already assembled" parts. For this, a separate assembly sequence for each subassembly must be generated and included to the main assembly sequence. Until today, there are many approaches to create regular feasible assembly sequences. However, there is no approach that is able to visualize the necessary assembly steps, the subassembly steps and add the necessary process information to the assembly steps automatically. This step was also implemented in the presented approach by the help of the SolidWorks macro.

To create the assembly instructions, the mentioned main assembly sequence will be red by the SolidWorks macro. The automatically defined best possible gravity direction points from top to bottom. This means that the assembly must only be aligned exactly as shown in the visualization of the assembly step. For this purpose, the isometric view was chosen and the rotation axis and the angle are based on design rules. (see Fig. 2a).

To create the action diagram of an assembly step the next part, which must be assembled, is shown in SolidWorks and is moved against the assembly direction until it reached the bounding box of the assembly. This guarantees that the assembled part is not hidden by another part of the assembly. Furthermore, the minimum distance of 10 cm to the assembly position, as shown in Fig. 2b, ensures also that the assembled part does not cover its own assembly position in the visualization.

The black dotted line in Fig. 2b shows the assembly path for the part. If there is a concentric restriction, like in Fig. 2b, the assembly line is located on the axis of the restriction. For this purpose, a point on the axis of the cylindrical contact faces is read out. This point is moved by the distance between the part and the assembly against the assembly direction. If there is just a plane contact after the assembly of a part, the mass center of the part is used to locate the assembly line. This approach is also able to identify identical screws and visualize them in one assembly step, if they have a plane contact to the same part and they have the same assembly direction.

For each assembly step, also the visibility of the assembly location is checked. For this, the assembly direction is compared to the actual view. If the mounting direction points to the display plane, as in Fig. 2b, or along the direction of gravity, the mounting operation is visible. Otherwise, the assembly is rotated by 180° around the axis of the gravity direction.

The tools for the assembly steps are visualized with pictures from a database, which is shown in Fig. 2c. The considered tools are limited to hex wrench, wrench and circlip plier. With the described norm part detection, norm parts, like circlips or hexagon head screws, can be detected and matched to the required tools with the correct size automatically from a database. Therefore, if a circlip has to be assembled, also the required plier will be displayed in the assembly instruction next to the assembly step. In case of a nut the necessary wrench will be displayed with an additional information about its size in a text information. The same applies also to hexagon head screws and their wrenches.

Before assembling a subassembly, it is necessary to visualize all steps, which are needed to preassemble the subassemblies. Here it is required to assemble the parts in a different orientation than in the assembly. Therefore, the whole subassembly is transformed in the coordinate system of the base part. After all preassembly steps, the subassembly is inserted in the main assembly.

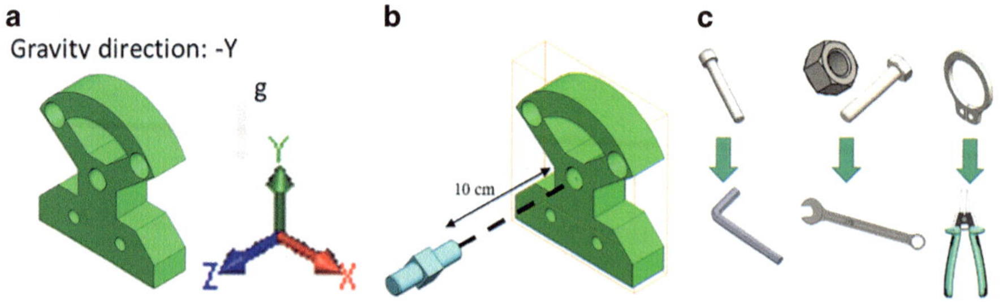

Fig. 2. a) Orientation of the view that shows the axis of gravity from top to bottom b) Generation of the action diagram for the assembly step c) Tool match based on the standard part type

6 Validation

The developed system was validated with two use cases. The first use case is a gearbox with two shafts; each of them was detected as base parts of a subassembly. The bearings as well as the gear wheels and circlips are parts of the subassemblies and have to be preassembled to the shafts. The whole gearbox consists of 36 parts.

The second use case is a single-cylinder engine with 59 parts and is much more complex. In Table 1 the computation times of the use cases are compared to each other. Noticeable is the big difference in computing times for collision detection. The engine takes 180 times longer than the gearbox and clearly shows the limitation of the approach regarding the computation time. Additionally also the calculation time for the assembly sequence generation requires much more time. This is caused by the exponential raise of calculation complexity through every single additional part. On the other hand, the generation of the visualizations for the gearbox requires more time than for the engine. This is based on the fact, that the generation of the visualizations is independent form the number of parts itself. The calcifications were performed using a workstation with an Intel Xeon W-2125 CPU, 32 GB RAM and Nvidia Quadro P200 graphics card.

Table 1. Computation times for the use cases separated by the major steps of the system

Use case	Contact analysis [s]	Collision detection [s]	Sequence generation [s]	Visualization [s]
Gearbox	284.7	528.2	1.14	6797
Engine	1597.1	95259	32.7	2684

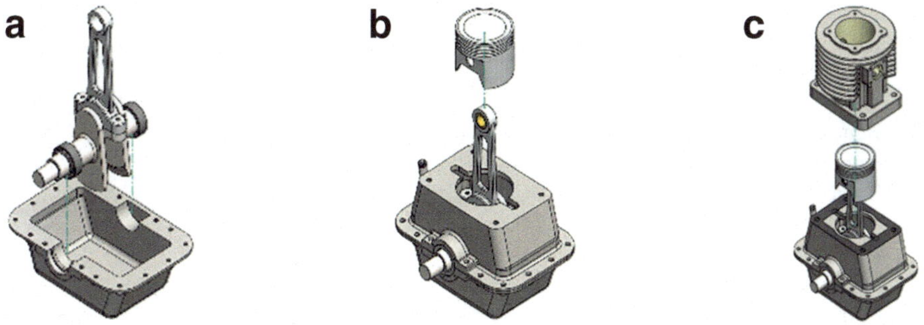

Fig. 3. Assembly sequence for the subassembly with the crankshaft as base part

In Fig. 3a the assembly of the subassembly with the crankshaft as base part is shown. Based on the subassembly detection the assembled parts are detected as subassembly and therefore will be documented in a separately. If the subassembly is not detected the parts are not mountable to the crankshaft. Afterwards the subassembly will be assembled into the main assembly. Each part of the subassembly, which is in contact with the already assembled parts gets a green line to visualize the assembly path as well as the contact points. In this example the two bearings have plane contacts to the oil pan.

Figure 3b shows the assembly step of the piston head. The connecting rod is already assembled, and the piston head is assembled on top of it. Afterwards the engine block has to be assembled (see Fig. 3c). Figure 3 displays a valid way to assemble an engine, however some experts would rather assemble the connecting rod first to the piston head and afterwards the piston head to the engine block from the top. The reason for this difference in our approach is the definition of the base part, which has to be in contact to the work surface. Therefore, just the oil pan in Fig. 3a can be the main base part. Some experts however, would start with different parts first. Nevertheless, the automatically generated assembly sequence is clearly a valid and possible solution.

7 Conclusion

In this work, a novel approach was presented to create assembly instructions out of a CAD model. The examined approach extracts necessary data out of the CAD model via the SolidWorks API with a developed macro. Based on these extracted data, the tool determines the best possible gravity direction for the assembly sequence. Here, the tool can also handle a reorientation of the assembly and is still generating a stable assembly sequence. Based on the generated assembly sequences, the tool generates an illustration of each assembly step via the SolidWorks API and adds the necessary tools to the instructions.

With this work, it is now possible to generate the assembly instructions just based on CAD models. The needed information for an assembly generation can be collected with less effort and there is no additional software for the generation of the instructions required.

However, a limitation of this approach is the collision analysis because it causes long computing times for big assemblies. The use of the internal approach limits the approach to the use of SolidWorks and the native data format of SolidWorks.

References

1. Bahubalendruni, M.R.A., Biswal, B.B.: A review on assembly sequence generation and its automation. Proceedings of the Institution of Mechanical Engineers, Part C: J. Mech. Eng. Sci. **230**, 824–838 (2015). https://doi.org/10.1177/0954406215584633
2. Westkämper, E., Decker, M.: Einführung in die Organisation der Produktion. Springer-Lehrbuch. Springer, Berlin (2006)
3. Ou, L.-M., Xu, X.: Relationship matrix based automatic assembly sequence generation from a CAD model. Comput. Aided Des. **45**, 1053–1067 (2013). https://doi.org/10.1016/j.cad.2013.04.002
4. Bourjault, A.: Contribution à une approche méthodologique de l'assemblage automatisé: Élaboration automatique des séquences opératoires. Université de Franche-Comté, Thése d'Etat (1984)
5. Viganò, R., Gómez, G.O.: Assembly planning with automated retrieval of assembly sequences from CAD modelinformation. Assem. Autom. **32**, 347–360 (2012)
6. Neb, A.: Review on approaches to generate assembly sequences by extraction of assembly features from 3D models. Procedia CIRP **81**, 856–861 (2019). https://doi.org/10.1016/j.procir.2019.03.213
7. Neb, A., Göke, J.: Generation of assembly restrictions and evaluation Generation of assembly criteria from 3D assembly models by collision analysis (in print). CIRP CATS (2020)
8. Choi, Y.-K., Lee, D.M., Cho, Y.B.: An approach to multi-criteria assembly sequence planning using genetic algorithms. J. Adv. Manuf. Technol. **42**, 180–188 (2009). https://doi.org/10.1007/s00170-008-1576-4
9. Abdullah, M.A., Ab Rashid, M.F.F., Ghazalli, Z.: Optimization of assembly sequence planning using soft computing approaches: A review. Arch. Computat. Methods. Eng. **26**, 461–474 (2019). https://doi.org/10.1007/s11831-018-9250-y
10. Deutsches Institut für Normung e. V. Erstellen von Gebrauchsanleitungen – Gliederung, Inhalt und Darstellung – Teil 1: Allgemeine Grundsätze und ausführliche Anforderungen 01.110(DIN EN 82079-1) (2018)

11. Agrawala, M., Phan, D., Heiser, J., et al.: Designing effective step-by-step assembly instructions. ACM Trans. Graph. (TOG) **22**, 828–837 (2003)
12. Neb, A., Schoenhof, R., Briki, I.: Automation potential analysis of assembly processes based on 3D product assembly models in CAD systems (in print). CIRP Design 20 (2020)

Selective Assembly Strategy for Quality Optimization in a Laser Welding Process

Manuel Kaufmann[1(✉)], Ira Effenberger[1], and Marco Huber[1,2]

[1] Fraunhofer-Institute for Manufacturing Engineering and Automation IPA, Nobelstr. 12, 70569 Stuttgart, Germany
manuel.kaufmann@ipa.fraunhofer.de

[2] Institute of Industrial Manufacturing and Management IFF, University of Stuttgart, Allmandring. 35, 70569 Stuttgart, Germany

Abstract. Selective Assembly is used to reduce the impact of geometrical variation of parts on the quality of their assembly. Therefore, matching rules for appropriate part combinations are determined in order to optimize the assembly quality. In this work, the geometrical features which are considered for selection are derived from the Virtual Assembly of the measured individual parts. Here the measurements are performed with a Computed Tomography (CT) system. In this new approach, the complete population of parts in a batch is optimized instead of a serial optimization procedure. This minimizes or even avoids the number of waste parts. In order to apply this new assembly strategy and to evaluate the achieved results, the assembly of the cover and the housing of a screen washing nozzle is studied, which is joined by laser welding. By means of global optimization using a genetic algorithm, the overlap in the welding seam, which determines the quality of the final product, is optimized. Deviations to the nominal welding seam are reduced up to 45% compared to a worst case assembly.

1 Introduction and Problem Setting

Major challenges for today's production systems are the high demands on product quality as well as the increasing complexity of processes and products. Often production processes are already near their manufacturing limits [1]. However, an increasing variety of product variants, fast innovation cycles and a high quality standard lead to a "First Time Right" maxim of avoiding loss. In contrast to that, companies underlie a challenging cost pressure due to the ongoing globalization. Thus, the management of geometrical deviations is an increasingly important activity in the product engineering process [2, 3]. Within the scope of this paper, we focus on the production optimization considering actual geometrical measurement data. Geometrical variations are a main cause for quality loss, as determined by [4] in context of the automotive white body production. The use of dimensional measurement systems is intensified, initiated by the *Industry 4.0* initiative. Especially in the field of optical metrology, new sensor concepts rapidly emerge that allow capturing areal scans with a high measurement point density. In industry, widely used optical sensors are triangulation sensors and also Computed

P. Weißgraeber et al. (Eds.): *Advances in Automotive Production Technology – Theory and Application,* ARENA2036, pp. 126–134, 2021. https://doi.org/10.1007/978-3-662-62962-8_15

Tomography (CT). In this paper, CT is considered. The term *Industry 4.0* comprises activities and recent progress in computer science, sensor systems as well as communication and information technology [5]. The concept constitutes new approaches for the *Geometrical Variations Management* (GVM), which is a generic term summarizing activities to minimize the effect of geometrical variations of parts on the product quality [6]. The GVM includes data-driven product and process optimization strategies and methodologies. These can be applied in the three domains of design, manufacturing, and assembly. In order to minimize the impact of geometrical deviations on the product quality, control loops establishing correlations between the final product and the specific domains are defined [3, 6, 7]. In the assembly domain, methods deployed before or during the physical assembly are differentiated [7]. In this paper, we focus on the physical compensation by Selective Assembly (SA).

1.1 State of the Art of SA and Deficits of Current Approaches

The SA approach aims at reducing geometrical variation by applying matching rules for the combination of parts. This approach is mainly used when the variation of the part does not ensure conformity with assembly tolerances. Variation and tolerances are linked by the process capability index C_{pK} according to

$$C_{pK} = \min(\mu - LSL;\; USL - \mu) \cdot (3\sigma)^{-1} \geq 1.33. \tag{1}$$

This index is a metric representing the exploitation of the tolerance interval $T = USL - LSL$ by a normal distribution with mean value μ and standard deviation σ, where USL and LSL are the upper and lower specification limits, respectively. It quantifies the ability of a serial production to manufacture parts while maintaining a maximum permissible scrap rate. SA determines individual, optimal combinations of parts that make less use of assembly tolerance specifications than random combinations based on Taylor's concept of part interchangeability. Hence, C_{pK} increases, because the first factor in Eq. (1) increases and σ decreases.

Relevant methods for SA can be classified into approaches considering two parts or more than two parts in the respective assembly. The former approach is mainly represented in the state of the art. Hereby, the conventional method to select combinations is the classification, which means that geometrical features of parts are attributed to certain classes. Combinations of parts are selected by matching parts from corresponding classes, see Fig. 1. The selection order within a class is arbitrary or depends on the chronology of selection.

For classification, mainly an equal class width is used. The smaller the class width, the larger the number of classes and the less information about the actual feature characteristic is neglected. However, a small class width requires a large number of parts, assuming a minimal required number of parts in each specific class [1, 3]. If the number of parts per class is unevenly distributed and the part variations are dissimilar, there will be a surplus of parts, because certain combinations occur less frequently. Therefore, advanced approaches for dissimilar variances were developed that aim for reducing the

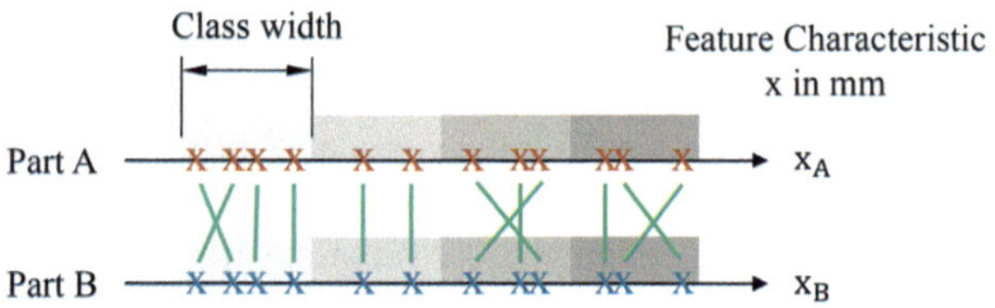

Fig. 1 Selective assembly by classification and arbitrary combination within classes. Feature characteristics x_A and x_B are quantitative manifestations of specific geometrical features.

surplus of parts. For similar variances, further approaches to either reduce scrap rates, to minimize tolerance variation, or to avoid mismatching are described in [3].

A classless matching approach is proposed in [8]. However, this concept demands individual part traceability, an efficient data processing and sufficient parts logistics. SA is a widely used measure to increase production efficiency by reducing geometrical variation, scrap rates, and production costs by maximizing quality yield. Since it is not possible to exchange any parts on site, SA is particularly useful when parts are bonded firmly or with form-fit. Its main application branch is the automotive industry in terms of large-scale productions. As described in [1], the manufacturing cost for an injector nozzle production could be reduced about 300,000 € per year by implementing SA. In [7], a reduction about 120,000 € is determined by simulation for a stator production of electric engines. In [9], the manufacturing costs of a nonlinear stack-up could be reduced by about 30% by introducing SA and additionally widening the tolerance specifications, enabled by an increased capability C_{pK}, as per Eq. (1). The authors assume a linear tolerance-cost function, which is common in most manufacturing processes [9]. In practice, most of manufacturing processes underlie a degressive tolerance-cost function [10], thus tolerance widening is more effective for small tolerances.

Deficits of Current Approaches. The information loss due to classification affects the quality compliance of a product. Thus, with an increasing class width, the loss increases for characteristics near the class limits, because the assembly tolerance is better fulfilled at the tolerance center. In order to reduce information loss, the class width should be minimized, which however demands a large number of individual parts considered. A large population number though is disadvantageous concerning storage and handling costs. As a conclusion, a classless selection strategy is required to overcome the mentioned shortcomings. Most SA strategies based on classification furthermore are sequential processes. These greedy selections are performed one after another chronologically. According to the selection criterion, first the best-quality combinations are determined, whereas at the end of the selection process, only combinations of barely compliant quality are produced. This first-in first-out effect causes a number of remainder parts that can be hardly or never combined. By a simultaneous selection process instead, where a complete population is selected at once, this deficit can be overcome.

Moreover, in this work the concept of Virtual Assembly (VA) is considered, as described in [11]. VA is based on the Skin Model Shapes (SMS) of the acquired measurement data. The concept of SMS is described in detail in ISO 17450-1. By VA, local form deviations of individual parts are considered in the assembly by recreating the physical workpiece contact virtually as shown in Fig. 2. On the one hand, the information content is increased, compared to the existing default datum definition as defined in ISO 5459:2011. On the other hand, advanced feature characteristics can be used for selection, which are directly derived from the virtually assembled SMS.

1.2 Objectives and Scope of this Paper

In this paper, SA is implemented as global optimization problem that simultaneously maximizes the quality of the entire batch. Minimal geometric deviations indicate quality. Here a batch is a set of parts of each assembly component that is considered in the SA at once. The global optimization achieves both zero scrap and a homogenous quality distribution across produced assemblies in case the number of parts is sufficiently large. The latter aspect prevents the production of only a few high-quality assemblies and many hardly acceptable assemblies. The concept of the VA for datum creation allows recording feature characteristics that correspond to physical assembly characteristics.

1.3 Description of the Use Case and Problem Setting

As Use Case, the assembly of two components, a housing and a cover is studied. The product is used as nozzle for screen washing in vehicles. Fig. 3 shows the housing, the cover as well as the VA from both components. Using CT, the holistic geometrical data of 24 housings and 24 covers was captured. The covers and housings are physically joined by a non-detachable connection through laser-welding. A possible functional restriction is caused by geometrical deviations of the welding seam. The virtual overlap in the welding seam is illustrated in Fig. 3, which shows the virtual intersection of both part volumes. In the physical assembly, this overlap is required as material reservoir for welding. If on the one hand the overlap is too low, too little material is available for the welding process, causing a mechanically unstable joining. If on the other hand too much material is available, a functional failure could possibly be caused.

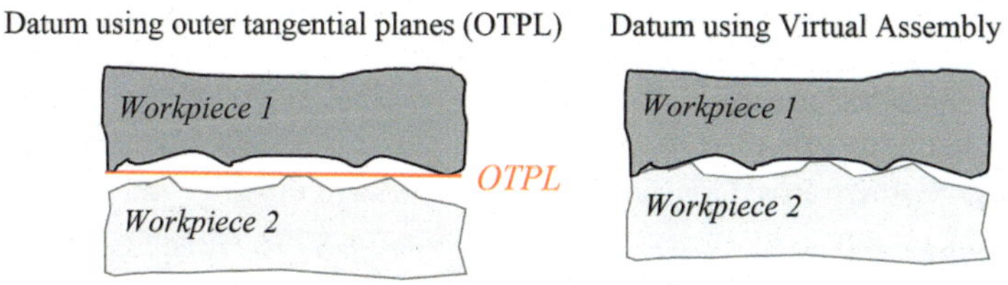

Fig. 2 Current datum definition using outer tangential planes (OTPL, left image). Function-oriented representation using Virtual Assembly (right image). The datum defines position and orientation of the SMS and thus the geometry propagation of the assembly state.

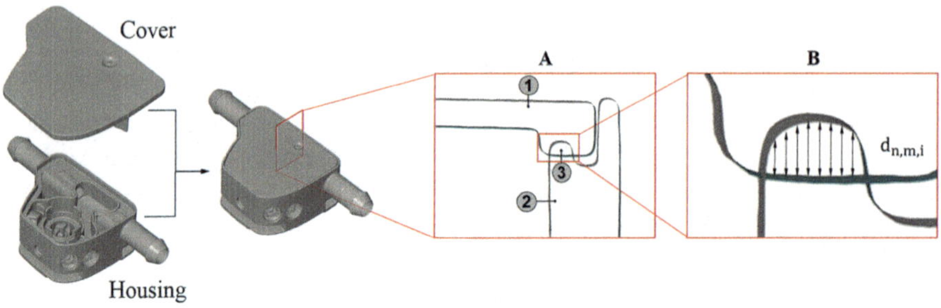

Fig. 3 Assembly of a screen-washing nozzle consisting of housing and cover. The cross-section A of the VA shows cover (1), housing (2) and welding seam (3). The cross-section B shows the welding seam detail and distances $d_{n,m,i}$ derived (parts n, m, point index i).

2 Experimental Realization of the Selective Assembly

2.1 Mathematical Statement of the Optimization Problem

For this Use Case the selection is formulated as global optimization problem.

$$\min_{j \in [1,2,\ldots,z!]} f_B(j), \tag{2}$$

with discrete optimization variable j and objective function f_B according to.

$$f_B(j) = \mathrm{med}\left(\underline{\mu}_{c_C, c_{H(j)}}\right) - t = \min!, \tag{3}$$

with med(.) being the median and $\underline{\mu}_{c_C, c_{H(j)}}$ being a tuple of mean values of distances $d_{n,m,i}$ towards the nominal distance t in the welding seam, whose k-th value is

$$\mu_{c_{C,k}, c_{H,k}(j)} = \frac{1}{L}\sum\nolimits_{i=1}^{L} d_{n,m,i} \,\forall \left\{d_{n,m,i} | \left(d_{n,m,p} < d_{n,m,i} < d_{n,m,1-p}\right)\right\}. \tag{4}$$

The variables n and m, with $n, m \in \mathbb{N}$ and $1 \le n \le N$, $1 \le m \le M$ are integer values representing a given cover and housing, respectively. The number of resulting assemblies is $z = \min(N, M)$. The tuples c_C and $c_H(j) = \underline{\pi}_j$ of cardinality $w = \max(N, M)$ describe the order of covers and the order of housings, respectively, where $\underline{\pi}_j$ is the j-th permutation of the order of housings. Each corresponding entry $c_{C,k}$ and $c_{H,k}$ describes the k-th individual combination, where $k \in [1, 2, \ldots, w]$. If $N > M$, i.e., the number of covers exceeds the number of housings, $\underline{\pi}_j$ are permutations of the set $\{1, 2, \ldots, M, 0, \ldots, 0\}$ comprising $N - M$ zeros. Conversely, if $M > N$, the entries $c_{C,N+1}$ to $c_{C,w}$ are filled with zeros. Combinations $(c_{C,k}, c_{H,k})$ for which one entry equals zero are ignored in the optimization. By doing so, the selection of z individuals from the larger set of components is determined. The number of possible permutations of $\underline{\pi}_j$ is $q = w!$ for this strategy of sampling without replacement, thus $j \in [1, 2, \ldots, q]$. The objective function f_B according to Eq. (3) includes the q-tuple $\underline{\mu}_{c_C, c_{H(j)}}$, which is composed from q mean values $\mu_{c_{C,k}, c_{H,k}(j)}$, occurring for a candidate combination $(c_{C,k}, c_{H,k})$. The nominal welding thickness t of 0.3 mm is subtracted from the median of $\underline{\mu}_{c_C, c_{H(j)}}$. By stating $n = c_{C,k}$ and $m = c_{H,k}$, the mean value $\mu_{c_{C,k},\, c_{H,k}(j)}$ of the distribution of i individual distances $d_{n,m,i}$ in the welding seam of cover n and housing m is calculated according to Eq. (4). Fig. 4 shows distances $d_{n,m,i}$ for combination (17, 10) as color-coded visualization after coarse-aligning to the CAD assembly. Here only the interval between the percentiles d_p and d_{1-p} with significance $p = 0.05$ is considered in order to avoid influence of outlier values. The objective function f_B as per Eq. (3) is nonlinear since part variation is discontinuous. The optimization problem is a bounded mixed integer nonlinear optimization problem. To assess the worst case combination, the corresponding objective function $f_W = -f_B$ with swapped sign is introduced. Generally in non-convex solution spaces, local optimization results depend on the initial search point and are only capable to find the nearest local optimum. The objective function described in this chapter is non-convex, thus requiring global optimization.

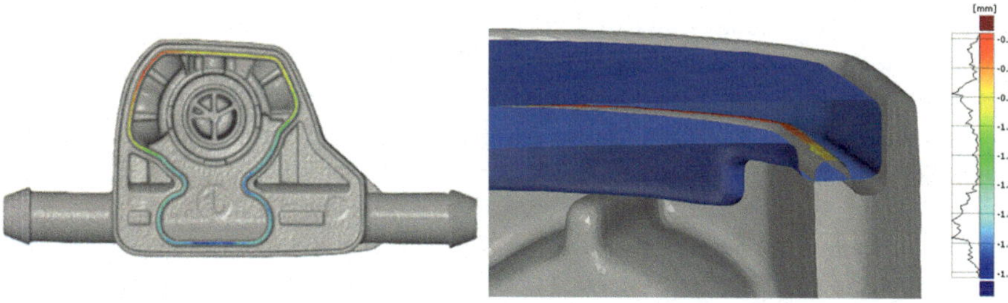

Fig. 4 Distances in the welding zone between cover n = 17 (blue) and housing m = 10 (grey).

Well-known solvers for global nonlinear optimization are Simulated Annealing, Artificial Bee Colony algorithm, Particle Swarm optimization, and *Genetic Algorithms* among others. Genetic algorithms (GA) are meta-heuristics in the field of evolutionary algorithms and thus in artificial intelligence, which are widely used in production optimization. GAs are sufficient to determine global optima of optimization problems with a short number of function evaluations and small computation times [1, 12]. Therefore, a GA implementation as described in [13] is utilized as promising method in this paper.

2.2 Optimization Settings and Parameter Study

The number of possible combinations is z!. Here, $6.2 \cdot 10^{23}$ combinations occur for 24 parts per component. Thus, by an exhaustive search, the computation time would take at least weeks with today's PC hardware. By using the GA, for the Use Case only approximately 300,000 function evaluations were performed, which endured about 30 s. (Intel i5-2430M, 2.4 GHz parallel processor) with the implementation provided by [13]. The optimization parameters used in this GA implementation are

- the population size P, defining the number of individuals (potential selection combinations) per generation (algorithm iteration), with P = {250; 500},
- the number of generations G = {150; 300; 450; 750},

- the crossover probability $p_c = 0.9$,
- the tournament size for tournament selection $s_T = 2$,
- the distribution index in the simulated binary crossover (SBX) operator $\eta_c = 20$ and
- the distribution index in the polynomial mutation $\eta_m = 20$ with $\eta_m \in [20, 100]$.

Please be referred to the article of Deb and Agrawal [14], where parameters and their quantities (except P, G) are defined. From a parameter study, best suitable optimization parameters were determined based on the metrics A_B, A_G, A_W, $R_{\bar{A}}$ and R_s, which are mentioned in the following Sect. 2.3. Thus, P = 500 and G = 300 was determined.

2.3 Results

As optimization result, both best and worst case-sequences were gathered using the objective functions f_B and f_W, respectively. Figure 5 shows the objective values for best and worst combinations as well as for a sequential (greedy) selection (see Sect. 1.1). In contrast to SA in classes, here also combinations occur, where the individual best combination is worse than the respective worst combination. However, referring to the complete population, the best combination is correlated to a smaller deviation towards the nominal welding thickness t, thus to a smaller geometrical variation in general. The area under the curves, denoted as A_B, A_G and A_W, is considered as metric to assess the exploitation of the objective functions. The resulting variation A_G for the greedy selection is about 10% smaller than the worst case result and 60% larger than the best case result. GAs use a random starting point, causing the optimization path to differ in each run. Therefore, for 25 runs of the optimization task, the average areas $\bar{A}_B$ and $\bar{A}_W$ and the standard deviations of their distributions, s_B and s_W, were determined. The ratios $R_{\bar{A}}$ and R_s are defined as.

$$R_{\bar{A}} = \left(1 - \frac{\bar{A}_B}{\bar{A}_W}\right) \cdot 100\,\% \rightarrow \max \text{ and} \tag{5}$$

$$R_s = \left(1 - \frac{s_B}{s_W}\right) \cdot 100\,\% \rightarrow \max. \tag{6}$$

The ratio $R_{\bar{A}}$ allows interpreting the SA result relatively to the worst case combination. Moreover, the ratio R_s is used to assess variance and robustness of the given results. Both ratios should be maximized in order to determine optimal combinations robustly. The ratio $R_{\bar{A}}$ equals 45% for P = 500 and G = 300, so that the deviation from t as defined in Eq. (3) is reduced nearly to the half of the worst case-deviation. The ratio R_s equals 11.1%, depicting that the variance of repeatedly estimated best combinations is reduced and that less extreme combinations occur.

3 Conclusion and Outlook

GA are highly feasible and computationally efficient to solve the underlying nonlinear mixed integer optimization problem. The formulation of the objective function is a highly problem-specific task for which no general approach can be provided. Here, it was derived from relevant weld seam geometries. By the shown reduction of deviations up to 45% and variation about 11.1%, either cost reduction or quality improvement can be obtained.

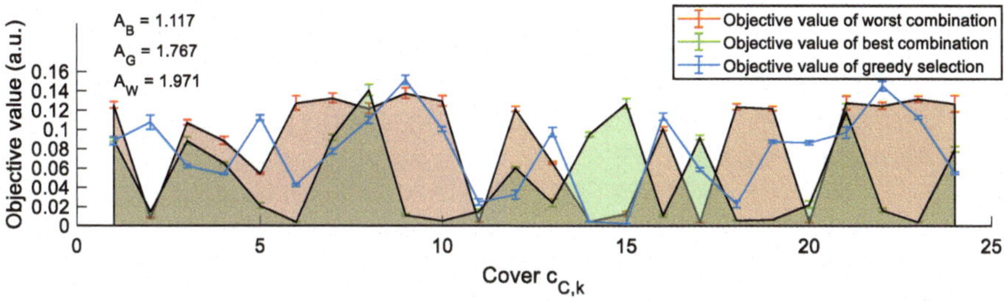

Fig. 5 Objective values for greedy selection (blue), best (green), and worst (red) combination, corresponding to objective functions f_B and f_W. A_W, A_G, and A_B denote the area under the worst, greedy, and best case-curves, respectively. The error bars indicate a $\pm 1 \cdot \sigma_{n,m}$ interval, where $\sigma_{n,m}$ is the standard deviation of deviations of combination (n, m), (N = 500, G = 300).

The assessment of economic potentials of SA is highly application-specific. A future challenge is to establish correlations between the ratios R, the initial tolerance value and variation reduction due to SA. Preliminary work in [9] regarding tolerance adaption and regarding cost models in [10] can be consulted to prospectively allows predicting cost reduction based on assessed geometrical variations reduction. The optimization of a batch of size z showed better results than a greedy selection. However, in order to find a sufficient tradeoff between the achieved quality and the number of parts to consider in a batch, as objective functions submodular cost functions can be studied.

References

1. Wagner, R., Haefner, B., Lanza, G.: Paarungsstrategien für hochpräzise Produkte. wt Werkstatttstechnik online **11**, 804–808 (2016)
2. Schleich, B., Anwer, N., Mathieu, L., Wartzack, S.: Skin model shapes: A new paradigm shift for geometric variations modelling in mechanical engineering. Comput. Aided Des. (2014). https://doi.org/10.1016/j.cad.2014.01.001
3. Kayasa, M.J., Herrmann, C.: A Simulation-based evaluation of Selective and Adaptive Production Systems (SAPS) Supported by quality strategy in production. Procedia CIRP (2012). https://doi.org/10.1016/j.procir.2012.07.004
4. Ceglarek, D., Jianjun, S.: Dimensional variation reduction for automotive body assembly. Manuf. Rev. **8**(2) (1995)
5. Monostori, L., Kádár, B., Bauernhansl, T., Kondoh, S., Kumara, S., Reinhart, G., Sauer, O., Schuh, G., Sihn, W., Ueda, K.: Cyber-physical systems in manufacturing. CIRP Ann. (2016). https://doi.org/10.1016/j.cirp.2016.06.005
6. Wärmefjord, K., Söderberg, R., Lindkvist, L., Lindau, B.: Inspection data to support a digital twin for geometry assurance. In: CIRP (ed.) Advanced Manufacturing. ASME 2017 Int. Mech. Eng. Congr. and Exp., Tampa, FL, USA, 03.11.2017. ASME (2017). https://doi.org/10.1115/IMECE2017-70398

7. Lanza, G., Haefner, B., Kraemer, A.: Optimization of selective assembly and adaptive manufacturing by means of cyber-physical system based matching. CIRP Ann. (2015). https://doi.org/10.1016/j.cirp.2015.04.123
8. Brecher, C. (ed.): Integrative Produktionstechnik für Hochlohnländer. Springer, Berlin (2011)
9. Chen, M.-S.: Optimising tolerance allocation for mechanical components correlated by selective assembly. Int. J. Adv. Manuf. Technol. (1996). https://doi.org/10.1007/BF01179810
10. Ehrlenspiel, K., Kiewert, A., Lindemann, U., Mörtl, M.: Kostengünstig Entwickeln und Konstruieren. Springer, Berlin (2014)
11. Schleich, B., Anwer, N., Mathieu, L., Wartzack, S.: Contact and mobility simulation for mechanical assemblies based on skin model shapes. J. Comput. Inf. Sci. Eng. (2015). https://doi.org/10.1115/1.4029051
12. SKS Labs: Single objective genetic algorithm. MATLAB central file exchange (2020)
13. Aderiani, A.R., Wärmefjord, K., Söderberg, R.: A Multistage approach to the selective assembly of components without dimensional distribution assumptions. J. Manuf. Sci. Eng. (2018). https://doi.org/10.1115/1.4039767
14. Deb, K., Agrawal, S.: A Niched-Penalty approach for constraint handling in genetic algorithms. In: Dobnikar, A., Steele, N.C., Pearson, D.W., Albrecht, R.F. (eds.) Artificial Neural Nets and Genetic Algorithms, vol. 29, pp. 235–243. Springer, Vienna (1999)

FlexPress – An Implementation of Energy Flexibility at Shop-Floor Level for Compressed Air Applications

Can Kaymakci[1,2](✉), Christian Schneider[1,2], and Alexander Sauer[1,2]

[1] Fraunhofer Institute for Manufacturing Engineering and Automation IPA, Nobelstr., 12, 70569 Stuttgart, Germany
can.kaymakci@ipa.fraunhofer.de

[2] Institute for Energy Efficiency in Production EEP, University of Stuttgart, Nobelstr., 12, 70569 Stuttgart, Germany

Abstract. One of the biggest challenges with the growing amount of renewable energy generation is the fluctuation in energy supply. In the case of Germany, renewable and volatile energy resources (wind and solar) are expanded. Industrial demand-side management, therefore, plays an important role for the automotive industry in Germany with its high energy demand. For a sustainable production manufacturing processes need to be more energy efficient and adaptable to volatile supply. The main goal is to synchronize manufacturing processes and their energy consumption with energy supply. While there are holistic concepts and ideas for a service-oriented platform for energy flexibility, a defined workflow for implementing energy flexibility signals at the shop-floor level is still missing. Our work proposes a method to adapt manufacturing processes with consideration to energy flexibility. The presented method aggregates data from a manufacturing process. Therefore, sensor data interoperates with flexibility signals on the shop-floor level where an intelligent controller can set process parameters by communicating through an OPC UA Server with the PLC. As an important cross-sectoral technology in the automotive industry, flexible compressed air is used for the validation of the presented method.

1 Introduction

The share of renewable energy sources is increasing and replacing energy generation from non-regenerative sources due to the energy transition in Germany [1]. However, the growing share of volatile energy sources is a challenge for grid stability due to the fluctuation in energy production of renewable sources [2]. Demand-side management (DSM) is seen as one possible solution to provide the power grid with necessary flexibility to help guarantee secure and resilient operation; thus offsetting the risks of a large share of volatile energy sources in the power grid [2]. Demand Response (DR) from industrial processes is not only relevant because of their high share in overall electrical consumption. Industrial sites, due to their complexity, can also cause a high level of stress and instability on the power grid, which could also be mitigated via flexibility [3].

P. Weißgraeber et al. (Eds.): *Advances in Automotive Production Technology – Theory and Application,* ARENA2036, pp. 135–141, 2021. https://doi.org/10.1007/978-3-662-62962-8_16

Demand-side Management is defined as the planning, implementation and monitoring of efficiency and flexibility measures, which enable industrial energy consumers to change their load profile depending on energy demand and supply [1]. Flexibility measures of demand-side management in the manufacturing industry can be grouped into four different categories proposed by Palensky et al. [2] – Energy Efficiency, Time of Use, Demand Response, Spinning Reserve. The focus of this paper is DR, which does not reduce energy consumption but changes the pattern of energy consumption. If the load change is based on signals from the energy system triggered by unplanned or irregular events, this is called DR.

Currently, one of the main challenges to accomplish meaningful industrial DR is the lack of a defined workflow or a method for implementing energy flexibility signals at the shop-floor level.

While there are holistic concepts and ideas for a service-oriented platform for energy flexibility [6] and a generic energy flexibility data model [4], a defined workflow for transferring information about processes and objects into the data model is still missing.

Therefore, we aim to answer the following research question: How can flexibility measures within compressed air systems be modelled using an existing energy flexibility data model? Our work proposes an easy-to-use method for implementing the workflow to adapt a manufacturing process to energy flexibility. Therefore, the presented method aims to facilitate the implementation of energy flexibility measures in various industries.

Our work is motivated by the fact that compressed air is described as one of the most expensive sources of energy required in a wide range of industries [4]. Therefore, a successful implementation of energy flexibility could be one measure to decrease energy costs for companies.

2 Related Work

One approach to synchronize energy demand from several industries and volatile energy markets is the usage of a holistic approach by integrating the entire process from the machine to the energy market. IT systems and digital platforms can support the process by implementing the data and information flow vertically and horizontally [5]. As presented by Schel et al. [6] the realization of a fully automated DR requires sound information technologies. The authors propose a concept of two logical platforms – a market-side platform and a company-side platform. For communicating energy flexibility between different layers and levels in manufacturing, the energy flexibility model was introduced by Schott et al. [7].

The energy flexibility data model (EFDM) defines a flexibility system as a collection of four different entity types or components – flexible load, flexible load measure, dependencies and storage. A **flexible load** contains the so-called key figures that describe the attributes of a physical entity (e.g. machine or actuator) with a flexible power consumption. This includes degrees of freedom and restrictions regarding the power curve and the associated costs. Specific attributes and the relations between the components are described in Schott et al. [7]. The **flexible load measure** is a time-dependent manifestation of the flexible load. **Storages** are objects that are capable of storing energy in industrial processes. This does not have to be electrical energy but can also be heat, cold and compressed air. A **dependency** describes the relations between different flexible loads. By modeling energy flexibility measures using the EFDM it is possible to communicate with different information systems and other participants of the process [6]. The usage of EFDM enables automatic trading of energy flexibility from machines to energy markets [7]. Nevertheless, the presented conceptual approaches of the EFDM are rarely used in practice. Only few simple examples are given by the authors of the data model.

A more general process for transferring data from machines to energy markets can be found in [8, 9]. Seitz et al. [9] define seven information flow steps based on the automation pyramid in manufacturing. The holistic approach starts with the long-term procurement of energy dependent on production planning. The next steps consider lower levels of the automation pyramid, where short-term adjustments of energy procurement but also ad-hoc flexibility can be identified and activated.

The presented platform [6], the data model [4] and the information flow [9] are necessary to synchronize energy flexibility between machine and energy market. Nevertheless, there is a gap between the concepts and actual usage. Especially the EFDM can be complex to implement. Therefore, a method for the combination of the EFDM, the company-side platform and the activation of energy flexibility for a pneumatic demonstrator is presented.

3 Case Study

3.1 Use Case

The use case presented in this paper considers an industry-related pneumatic demonstrator. Furthermore, the specific IT infrastructure of the aforementioned demonstrator is presented by explaining the different components, which guarantee an end-to-end communication. Within the demonstrator, different working scenarios for each actuator (with or without leakages to various degrees) can be simulated. Therefore, a high number of combinatorial scenarios can be simulated – ranging from a very inefficient to an efficient process. In this section, a brief introduction of the involved pneumatic demonstrator is presented.

A compressor is placed beside the demonstrator which delivers filtered, dried and non-oiled compressed air with a working pressure from 4 to 8 bar. The demonstrator simulates a compressed air intense production process represented by balls that are transported in a cycle. Within the demonstrator energy flexibility measures can be modelled and implemented with regard to the volume flow of the compressed air. Therefore, the proposed architecture of the energy synchronization platform [6] is used. The key components in the architecture are the company-side platform and the smart connector. The company-side platform has been implemented as a local instance on an Intel NUC and serves as an integration platform between different services and components like the smart connector. The company-side platform also communicates with the energy market and has access to the market platform [8]. The smart connector is an individually implemented software module for virtually representing a machine and the flexible loads related to the machine. The smart connector was used to build a direct connection from the OPC UA server of the PLC to the company-side platform. The consistent connection and synchronization from the market platform through the company-side platform to the smart connector is the enabler for commercializing industrial energy flexibility [8].

The flexibility-enabled compressed air demonstrator has the potential of executing two specific flexibility measures. The first flexibility measure is the short-term adjustment of different process parameters like the air flow rate. The technical and organizational restrictions are described by using different key figures given in the EFDM. Another flexibility measure could be the interruption or break of a manufacturing job after completing a cycle, which can also be modeled by using the EFDM approach. Technically it is also possible to interrupt the production immediately. However, it can be argued that such an immediate interruption can impair production quality.

3.2 Proposed Method for Flexibility Modeling

After the definition of the given architecture of the demonstrator including information systems and processes, it is necessary to derive the potential energy flexibility measures (see Sect. 3.1) into energy flexibility data models. For this reason an easy-to-use method has been implemented. The method identifies the relevant energy consumers or producers and transfers the given information into the EFDM and the specific key figures. The method has three subsequent steps.

(1) First, it is necessary to identify possible flexible measures [10]. In the beginning, the existing energy consumers or producers are identified. To minimize the complexity, the method focuses on energy consumers such as engines or compressors. Energy consumers are modeled as flexible loads. To determine the different key figures of the flexible load it is necessary to get specific information about the technical restrictions of the consumer. Therefore, the generic parameters of the consumer have to be determined.

(2) The second step combines the given information about flexible loads (e.g. energy consumers) like the nominal power or the modulation possibilities and connects it with other dimensions (e.g. production plan). Flexibility spaces are identified and transferred into an energy flexibility data model (EFDM). By specifying the validity (time dimension) and the possible power states (power dimension) derived from the production plan, multiple EFDMs are generated. A possible outcome of an optimized EFDM – according to the production plan, the flexible load is not in use from 12:00 am to 06:00 am (validity). Thus, the possible power state is defined from 0 kW to 4.2 kW. The EFDM can be extended by including more parameters – such as power gradients defined in Schott et al. [7] – when industrial processes are more interdependent and complex.

(3) The third step generates a specific flexibility measure related to the flexible load. The second step only defines the possible energy flexibility in different dimensions (e.g. different time and power). This is modeled as the flexible load. According to Schott et al. [7] there must be a specific flexibility measure (action) where the start time and the time-dependent power states of the measure have to be defined. This is called flexible load measure and is considered as the operation strategy or schedule of the flexible load.

In the next section, the proposed method is used to generate an EFDM for the compressed air system. Furthermore, it will be shown how to generate a specific flexibility measure (step 3).

3.3 Flexibility Model for Compressed Air

In this section, the generated energy flexibility data model (EFDM) is described by using the presented method for the use case of a flexible compressed air system. Flexibility or price signals from the energy market are simulated and handled as a black box to decrease the complexity of the implementation.

In Table 1 the flexible load "FlexPress1.Power" is described. Here, the power supply is modeled as a flexibility. All the necessary information for subsequent processes like production planning or job scheduling are included in the model. Two of the most important key figures in this example are *validity* and *power states*. The validity describes the valid time intervals where the flexibility could be active. Furthermore, the parameter attribute "total" sets the start and the end point of the flexibility load within the time interval (e.g. between 12:00 am and 06:00 am). One possible reason for modeling the flexibility load between 12 am and 6 am could be that the compressed air application is not used between these hours and thus cannot "consume" energy. Requirements from different planning scenarios of manufacturing execution or production planning systems can influence the validity of a flexibility load. The other important key figure *power state* describes the possible power at which the flexible load can run during each of the holding periods [7]. A positive algebraic sign means that the flexible load causes an increase of energy consumption whereas a negative sign represents a decrease in consumption. In the demonstrator the increase and decrease of energy consumption is possible from 0.2 to 4.8 kW.

Table 1. Energy flexibility data model for the flexible load

Key Figure	Type	Unit	Value
Flexible Load ID	String		FlexPress1.Power
Reaction Duration	$2^{\mathbb{R}_0^+}$	S	{0.1}
Validity	$2^{\mathcal{H}} \times$ {'start', 'total', 'end'}	S	([00:00, 06:00], "total"
Power States	$2^{\mathbb{R}}$	kW	{0,[0.2,4.8]}
Holding Durations	$2^{\mathbb{R}}$	s	∞
Usage Number	$2^{\mathbb{R}}$	–	∞
Modulation Number	$2^{\mathbb{R}}$	–	∞
Regeneration Duration	$2^{\mathbb{R}_0^+}$	s	0
Costs	$2^{\mathbb{R}}$	€	0

The flexibility measures "short-time adjustment of process parameters" and "pausing a manufacturing job" are on the manufacturing level [9]. Therefore, it is also possible to offer the specific flexibility on energy markets with smaller time and planning horizons like the day ahead market or spinning reserve.

Benefit of the proposed method for flexibility modeling is that it enables the comparison as well as the aggregation of different flexibility measures even between companies. Therefore, the defined workflow closes the gap between the shop-floor level and energy markets.

Considering a triggered flexibility signal for the short-time reduction of power from 1 am to 1:30 am by 3 kW, the flexibility signal arrives at 12:56 am. First, the implemented EFDM has to be evaluated whether validity and possible power states are aligned with the flexibility signal. The evaluation can be done manually from the production planner or automatically by using the capacities of the smart connector. In this case, the reduction of power (−3 kW) and the validity (1:00–1:30) are in the dimensions of the flexible load and the signal is transferred through the smart connector. In the energy flexibility data model, the "executing" is considered as a flexible load measure. The flexible load measures (dependent on usage number) for a given flexible load represent a load curve with time-dependent power states (holding durations) and a start time. The complexity level of planning and optimization increases when using more key figures that can be described by the EFDM (e.g. activation, deactivation or modulation gradient).

The compressed air use case underlines that energy flexibility can be modelled for this form of energy giving the potential to further decrease costs. However, the precise cost savings were not quantified in the context of this paper and are the objective of further research.

4 Results and Discussion

The use case of an industry-related pneumatic demonstrator validates the use of the EFDM and the proposed method of this paper. The energy carrier in the presented use case was compressed air, where the pressure flow is controllable. With the presented method it is also possible to take into account restrictions that further limit the space of available flexibility. Thus, the EFDM and its implementation is the basis for the marketing of DR with the main goal of synchronizing energy demand and supply. Another point is the integration of the potential costs and profits into the model to see the benefits of using DR in manufacturing companies.

Furthermore, the lessons learned from the pneumatic demonstrator can be used to derive a potential guideline for modeling and implementing demand response within manufacturing processes and industrial applications. These generalized guidelines can be used for effective EFDM modeling and extend the model by a method to implement it. The guidelines are part of further research and should be developed to support the manufacturing industry for implementing the EFDM to trade energy flexibility in a standardized way. It is necessary to analyze all systems related to the manufacturing process, such as the building automation, and not only the manufacturing system itself. In order to take other systems into account, the proposed method has to be extended. Additionally, the potential of the method is its capability to be applied to other energy sources such as electricity or gas.

5 Conclusion

In this paper a method for implementing flexibility measures in compressed air systems is introduced and implemented. The theoretical background of energy flexibility and the necessary information architecture is transferred into a more practicable solution. The key components for implementing energy flexibility into the demonstrator are the EFDM and the underlying energy synchronization platform for communicating flexible loads and flexibility signals from short-term energy markets. The work underlines that the existing framework for the generation of energy flexibility data models can be used for compressed air systems, demonstrating that the concept of energy flexibility can be applied to compressed air systems. With the presented link for implementing energy flexibility signals at shop-floor level, further research will focus on scheduling of several flexibility measures on the machine and plant level. Steps for extending the functionality of the demonstrator are the synchronization of the information flow with different energy markets (like day-ahead or intraday) by using the market platform. Additionally, other flexibilities like "change of production plan" or "change of energy supply" can be useful for this demonstrator.

References

1. Umweltbundesamt: https://www.umweltbundesamt.de/themen/klima-energie/erneuerbare-energien/erneuerbare-energien-in-zahlen#uberblick (2020). Accessed 7 July 2020
2. Alemany, J.M., Arendarski, B., Lombardi, P., et al.: Accentuating the renewable energy exploitation: Evaluation of flexibility options. Int. J. Electr. Power Energy Syst. **102**, 131–151 (2018)
3. Dulău, L.I., Abrudean, M., Bică, D.: Smart grid economic dispatch. Procedia Technol **22**, 740–774 (2016)
4. Mousavi, S., Kara, S., Kornfeld, B.: Energy efficiency of compressed air systems. Procedia CIRP **15**, 313–318 (2014)
5. Körner, M.F., Bauer, D., Keller, R., et al.: Extending the automation pyramid for industrial demand response. Procedia CIRP **81**, 998–1003 (2019)
6. Schel, D., Bauer, D., Haupt, L., et al.: IT platform for energy demand synchronization among manufacturing companies. Procedia CIRP **72**, 826–831 (2018)
7. Schott, P., Sedlmeir, J., Strobel, N., et al.: A generic data model for describing flexibility in power markets. Energies **12**, 1–29 (2019)
8. Roesch, M., Bauer, D., Haupt, L., et al.: Harnessing the full potential of industrial demand-side flexibility: An end-to-end approach connection machines with markets through service-oriented IT platforms. Appl. Sci. **9**, 1–26 (2019)
9. Seitz, P., Abele, El., Bank, L., et al.: IT-based architecture for power market oriented optimization at multiple levels in production processes. Procedia CIRP **81**, 618–623 (2019)
10. Zäh, M., Fischbach, C., Kunkel, F.: Energieflexibilität in der Produktion identifizieren. Z. Wirtschaft. Fabrik. ZWF **108**(9), 639–642 (2013)

Part B Smart Production Systems and Data Services

A Framework for Digital Twin Deployment in Production Systems

Ayman AboElHassan(✉), Ahmed Sakr, and Soumaya Yacout

Polytechnique Montréal, Montréal, QC H3T 1J4, Canada
ayman.aboelhassan@polymtl.ca

Abstract. Digital twins represent physical systems through dynamic adaptive digital replicas. These replicas are virtual images of the functionality and interactions of the physical system and its components. Digital twin provides real-time monitoring and decision-making support. These are essential pillars in the Industry 4.0 paradigm. On the system level, multiple architectures are established for digital twin concepts in the literature. However, the roadmap for deploying a functional digital twin has not been fully recognized thus far. In this paper, we propose a framework for the deployment of a digital twin in production systems. The framework covers both levels of virtualization; digital shadowing, and digital twining. It utilizes present technologies in network communication, data management, and knowledge extraction. We describe the main components of the digital twin, and their relation to the management schemes currently implemented in production systems. Additionally, we devise an approach for integration of available simulation and data analytic tools for dynamic modeling and system performance evaluation. Through our proposed framework, the current technologies and tools are capable of deploying a digital twin. Consequently, learning algorithms and monitoring procedures are exploited for dynamic performance improvement.

1 Introduction

Highly competitive markets and short product lifecycle push current production systems to adapt more proactive production plans. Computer-aided planning provides analysis capabilities for production systems in real-time and perform onsite optimization to meet production demands [11]. Moreover, customer-centric production lines need to schedule assembly-line operations for personalized products manufacturing. This need for real-time modification and resource allocation of production plans raises up the concept of cyber-physical production systems (CPPS) [3]. Digital Twin (DT) has been a highly active research topic lately because of its real-time analysis and managing capabilities. The emergence of information and communication technologies (ICT), and internet-of-things (IoT)

P. Weißgraeber et al. (Eds.): *Advances in Automotive Production Technology – Theory and Application,* ARENA2036, pp. 145–152, 2021. https://doi.org/10.1007/978-3-662-62962-8_17

in industrial manufacturing domain created potential economic value for the DT technology [11]. The role of DT in CPPS is to provide the real-time status of the physical system. This enables it to perform scheduling, simulation, and optimization to manage the production proactively.

DT term is usually used to define three different concepts [10]; product twin for product lifecycle management (PLM), production twin for customer-centric production systems, and system twin for CPPS.

1.1 Production Systems

Despite the massive focus on Industry4.0 topics in the academic domain, the industry is still studying the needed changes in their infrastructure to adopt the Industry4.0 technologies. In-service manufacturing systems still utilize the Purdue automation pyramid [1,8]. These businesses are lagging behind due to the high cost of upgrading the manufacturing lines. The upgrading costs include installation cost, production plan interruptions, and production line stoppages.

Large investments have been specified for DT technology. Newly constructed production facilities are best suited for Industry4.0 concepts as they utilize state-of-the-art equipment. However, small and medium-sized enterprises (SME) are more adaptive than standard large enterprises and complex businesses because of governmental initiatives [4].

Current production systems depend on central supervisory control and data acquisition (SCADA) server to collect information from shop-floor machines and send control commands to these machines. As production systems currently use industrial Ethernet, most equipment communicate with the SCADA server through OPC Unied Architecture (OPC-UA) protocol [6]. The central server stores the collected data into Historian database. Human operators interact with the machines via on-site terminals, or off-site connection to the SCADA server. System analysis is usually performed offline on the data collected and extracted from the historian database. Figure 1 shows the system architecture of data gathering, management, and planning in traditional production systems.

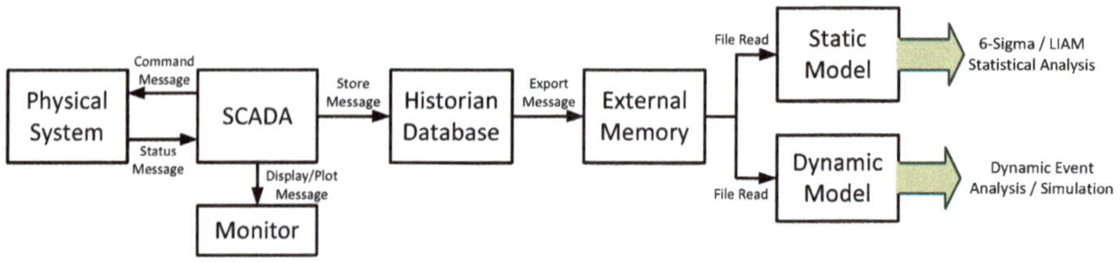

Fig. 1. Traditional approach for data gathering, management, and planning

1.2 Related Work

The number of publications are increasing in the field of Digital Twin. Most of the publications concentrated on DT applications [1]. While, some publications focused on the development of the DT framework [5,6,9,12].

The differentiation between DT and Digital Shadow (DS) terms is still an ongoing discussion. Two main groups of thoughts exist; The first opinion differentiate between the two terms according to the data-flow direction between the virtual and physical worlds [1,10]. In Digital Shadow, the physical world sends data to its corresponding virtual model. While in Digital Twin, the data-flow is bidirectional, from/to the physical world. This definition tends to describe DT as an upgraded version of DS. The second opinion uses Digital Twin term as a high-fidelity model, compared to the low-fidelity of Digital Shadow [1].

There is no clear separation between the roles of DS and DT in literature. Schuh et al. [7] defined DS a digitalization of the real manufacturing process, where a virtual copy of the factory is created using simulation models. However, they did not discuss the difference between DS and DT. Stark et al. [8] proposed an intersected functionality of DS and DT, for which DS term is used as the change in the digital model of the physical system over time. DT term refers to an instance of the digital model and its shadow, in which DT instance is used in system analysis and simulation.

1.3 Contribution

In this paper, we differentiate between DS and DT in terms of objectives and components. Digital shadow is defined as *a model-based virtual state of the physical world collected in real-time*. While digital twin is defined as *real-time decision-support sandbox based on digital shadow*. We use DS as a component of DT, which provides real-time state for DT functionalities. DS is connected to the physical world using current communication technologies. The main goal of using currently utilized communication technologies is to reduce the DT installation cost and encourage complex production systems to adapt Industry4.0 paradigm.

The rest of the paper is organized as follows; The proposed framework for DS and DT is presented in Sects. 2 and 3 respectively. The framework integration with the production system's components is discussed in Sect. 4. Finally, the conclusion is presented in Sect. 5.

2 Digital Shadow

Digital Shadow (DS) is defined as a digital replica of a physical component, system, or system of systems. This digital replica matches its physical counterpart's state in real-time. Digital Model (DM) describes the physical world's behavior in response to internal and external factors through time. Hence, we can define a DM as a mathematical model, behavioral model, data-driven model, or even a collection of data. The goal is to mirror the physical world in a digital form. We can summarize the objectives of a DS as follows;

1. Create an abstract model for the physical system.
2. Provide the state of the physical system in real-time.
3. Visualize the virtual state of the system.
4. Compare the virtual state to the physical state.
5. Update the physical system's configuration in the abstract model.
6. Update the abstract model to match the physical system's behavior.

DS is divided into components according to functionality. The DS architecture is developed as a service as shown in Fig. 2. Each component is a service that interacts with other components through a publish/subscribe messaging system.

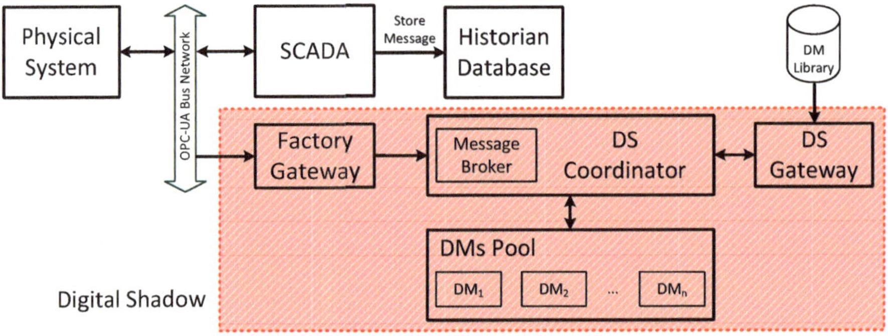

Fig. 2. A Digital Shadow architecture

Factory Gateway is an OPC-UA client application, which is installed on a small board. The gateway is connected to the SCADA server through a bus network topology. Upon receiving an OPC-UA message, the gateway converts the message into a set of <key, value> messages that is sent to the DS coordinator.

The use of passive data collecting technique allows the DS to read OPC-UA messages that the SCADA server receives. Accordingly, the SCADA server is not interrupted or requested to send data to the DS all the time. This approach reduces the overhead communication of passing all the manufacturing facility's information from the SCADA server to higher layers in real-time [2].

DS Coordinator manages the communication between the different DS components. The coordinator's message broker works as a middleware that relays messages from publishing components to subscribing DMs. The factory gateway publishes all received data onto the message broker, while DMs publish their virtual states. Each DM subscribes for the physical and virtual data that it needs. Moreover, the coordinator maintains a global state of the physical and virtual system from all the models in the DS pool. The global state is synchronized across all DMs using logical time synchronization. The DMs system time is synchronized by the coordinator using network time protocol.

DS Pool is a collection of all DM objects that are running in the DS framework. A DM object is initiated by the DM pool manager. On construction, the DM object subscribes to a list of required keys. Then, it sends a synchronization request to the coordinator. Upon synchronization, the DM runs the model in a real-time mode.

DM library is a database of all the DMs source models. A DM object is created from one of the models provided by the library. Moreover, the library tracks the version of each DM source model. This enables the modification of an initiated DM during runtime.

DS Gateway provides access to the global physical and virtual states. DT and user interface send data requests to the DS gateway using HTTP request protocols as a query for list of keys. The gateway then responds by the requested data from the global state provided by the coordinator. The requested data keys can refer to values published by the physical system or DMs.

3 Digital Twin

Digital Twin (DT) is defined as a real-time decision-support sandbox, which uses real-time digital representation of the physical component, system, or system of systems. By this definition, DS is used by DT to represent the physical world in real-time. While, DT provides a higher-level intelligence. We define intelligence in this context as any knowledge-based application that uses simulation, scheduling, searching, optimization, diagnosis, prognosis, and/or performance evaluation to help the decision maker in improving the efficiency of the production system. Intelligence applications can range from basic pre-programmed calculations, to artificial intelligence (AI) and machine learning (ML).

The separation between the virtual model in DS and intelligence in DT, eases the adaptability of the DT framework for new applications. DT applications read the state and model through a systematic approach. However, this generalization requires DMs to provide both real-time and simulation-time capabilities. This simulation-time speed differs from DS speed, which is required to match the physical world's pace. DT uses the DS real-time state to apply one or more of the following objectives:

1. Analyze the physical/virtual system's performance
2. Initialize simulation tools with the current virtual state.
3. Evaluate production policies.
4. Optimize shop-floor schedules.
5. Optimize maintenance plan.
6. Estimate an order price and delivery date.
7. virtual commissioning (VC) of equipment and/or product.

The DT framework is divided into components as shown in Fig. 3. Each component is an object that interacts with other components in a form of request/response messages.

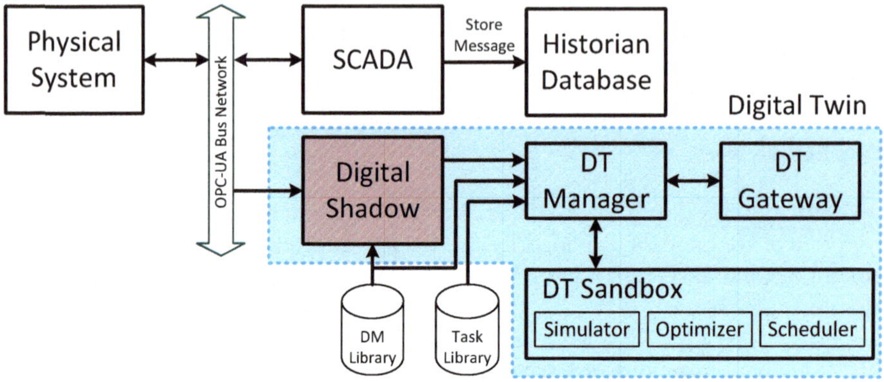

Fig. 3. A Digital Twin architecture

DT Manager coordinates between the different components of the DT's framework. The manager provides the needed data to the DT sandbox and passes its results to the user. The manager loads DMs from the DM library with their DS-configuration, in order to create a real representation instance for intelligence tasks.

DT Sandbox is a workspace for intelligence tasks. The sandbox provides necessary tools for these tasks as behavioral simulators, physics simulators, optimizers, and schedulers. Each tool is integrated into the sandbox using an API interface, which enables loading the tool with models and initial states. The sandbox initiates independent process for each task upon user request. Then, the tasks are initialized by real representation instances. Finally, task results are provided to the user through the DT manager.

Task Library is a set of intelligence scripts and procedures that use simulation, optimization, and machine learning to perform one of the intended objectives of the DT. A task script is executed by the DT sandbox. However, the task script is called from the DT manager. The task script specifies the tool to be used from the DT sandbox, the tool settings, and the DMs to be used in the run. Historical data can be used by the DT sandbox tools.

DT Gateway provides a user interface for the DT functionalities. Users access DT manager through a web interface using HTTP requests. This interface provides the user with the capability to view the real-time state, run intelligent tasks, review task results, and modify the DM library and task library.

4 Integration in Automation Process

In this section, we discuss two main aspects of the DT framework and its interaction with the other components of the automation pyramid; the feedback to the physical world, and the integration with MES/ERP.

4.1 Feedback to Physical World

The proposed DT framework does not explicitly define a way to control the physical system. However, the addition of a feedback connection to the SCADA server can be done through the DT gateway. In that sense, a DT system can be categorized as controlling or non-controlling. Non-controlling DT generates data insights and recommendations only. While, controlling DT uses a decision-maker module that controls the physical world through the SCADA server, as shown in Fig. 4.

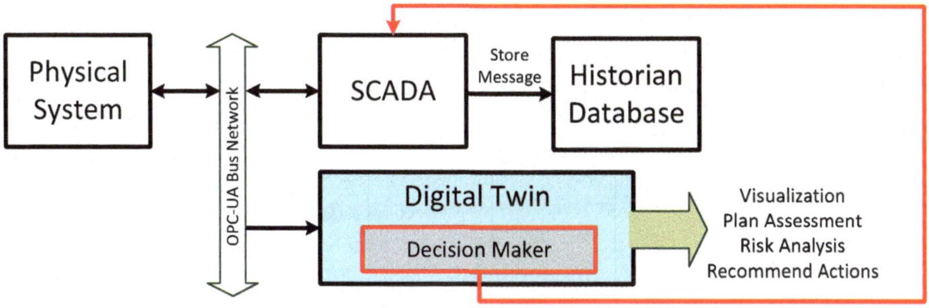

Fig. 4. A controlling Digital Twin

4.2 Integration with MES/ERP

DT differs from MES and ERP. It uses the real representation of the physical system with its current state. While MES/ERP uses nominal representation of the system, which induces higher abstraction error. The DT framework extends the MES/ERP by providing multiple operations as order dispatching, scheduling, virtual commissioning, maintenance planning, resource management, performance prediction, and estimation of price and delivery time. A MES/ERP module can directly command the DT framework through the DT gateway. The common scope between all DT applications is the use of the current state of the manufacturing facility as an initial state of the intelligence tool. Moreover, real representation of each component is used instead of nominal ones.

5 Conclusion

In this paper, we proposed the separation of functionalities between Digital Shadow and Digital Twin to describe a framework for DT deployment. Determination of roles provides a clearer view of the DT concept. Additionally, the separation of digital models from the higher intelligence allows wide range of DT applications to be applied on the same framework. We proposed a DT framework based on message-oriented communication between its components. The proposed framework provides real-time representation for users and high-level applications. Moreover, it provides a generic approach to extend the DT given new applications.

References

1. Cimino, C., Negri, E., Fumagalli, L.: Review of digital twin applications in manufacturing. Comput. Ind. **113**, 103130 (2019)
2. Eckhart, M., Ekelhart, A.: A specification-based state replication approach for digital twins. In: Proceedings of the 2018 Workshop on Cyber-Physical Systems Security and PrivaCy – CPS-SPC'18. (ACM), New York (2018)
3. Lu, Y., Xu, X.: Resource virtualization: a core technology for developing cyber-physical production systems. J. Manuf. Syst. **47**, 128–140 (2018)
4. Monsone, C., Mercier-Laurent, E., János, J.: The overview of digital twins in industry 4.0: managing the whole ecosystem. In: Proceedings of the 11th International Joint Conference on Knowledge Discovery, Knowledge Engineering and Knowledge Management. SCITEPRESS – Science and Technology, Setúbal (2019)
5. Park, H., Easwaran, A., Andalam, S.: TiLA: twin-in-the-loop architecture for cyber-physical production systems. In: 2019 IEEE 37th International Conference on Computer Design (ICCD). IEEE (2019)
6. Qamsane, Y., Chen, C.Y., Balta, E.C., Kao, B.C., Mohan, S., Moyne, J., Tilbury, D., Barton, K.: A unified digital twin framework for real-time monitoring and evaluation of smart manufacturing systems. In: 2019 IEEE 15th International Conference on Automation Science and Engineering (CASE). IEEE (2019)
7. Schuh, G., Kelzenberg, C., Wiese, J., Ochel, T.: Data structure of the digital shadow for systematic knowledge management systems in single and small batch production. Procedia CIRP **84**, 1094–1100 (2019)
8. Stark, R., Kind, S., Neumeyer, S.: Innovations in digital modelling for next generation manufacturing system design. CIRP Ann. **66**(1), 169–172 (2017)
9. Tao, F., Zhang, M.: Digital twin shop-floor: a new shop-floor paradigm towards smart manufacturing. IEEE Access **5**, 20418–20427 (2017)
10. Uhlenkamp, J.F., Hribernik, K., Wellsandt, S., Thoben, K.D.: Digital twin applications: a first systemization of their dimensions. In: 2019 IEEE International Conference on Engineering, Technology and Innovation (ICE/ITMC). IEEE (2019)
11. Wagner, R., Schleich, B., Haefner, B., Kuhnle, A., Wartzack, S., Lanza, G.: Challenges and potentials of digital twins and industry 4.0 in product design and production for high performance products. Procedia CIRP **84**, 88–93 (2019)
12. Yun, S., Park, J.H., Kim, W.T.: Data-centric middleware based digital twin platform for dependable cyber-physical systems. In: 2017 Ninth International Conference on Ubiquitous and Future Networks (ICUFN). IEEE (2017)

Assets2036 – Lightweight Implementation of the Asset Administration Shell Concept for Practical Use and Easy Adaptation

Daniel Ewert[1(✉)], Thomas Jung[1], Timur Tasci[2], and Thomas Stiedl[1]

[1] Robert Bosch GmbH, Renningen, Germany
Daniel.Ewert@de.bosch.com
https://www.bosch.com/research/

[2] Institute for Control Engineering of Machine Tools and Manufacturing Units, University of Stuttgart, Stuttgart, Germany

Abstract. ARENA2036 is a joint research campus incorporating production assets from different industrial and academic partners. To allow the implementation of cross-partner value streams and workflows, a common middleware for online date exchange and asset operation is mandatory. We implemented a generic and lightweight middleware which follows the concept of Asset Administration Shells as specified by the Platform I4.0. However, to allow for easy adaption and setup by a diverse range of partners we simplified modeling requirements and complexity of the actual data exchange. The result is a specification for the assets' self-descriptions in form of submodels (For many people, the term "submodel" implies the existence of a "model", which does not exist here. In order to have a consistent terminology with the standard of the Platform I4.0, we will use the term "submodel" here. For the future, we propose the term "aspect model" instead.) and a convention on how to map this onto MQTT. Additionally, we integrated means for online discovery and state monitoring of all connected assets.

1 Motivation

In the light of shortening product-life-cycles, production facilities must be able to change production faster while still being able to operate profitably also for smaller batch sizes. However, changeovers of a production site are still error-prone and troublesome. To improve this situation and to enable an efficient and changeable production technology are just two of the challenges the partners of the research initiative ARENA2036 currently address. ARENA2036 is a joint effort of established companies, startups, and academic institutes to develop common visions and solutions for the factory of the future.

P. Weißgraeber et al. (Eds.): *Advances in Automotive Production Technology – Theory and Application*, ARENA2036, pp. 153–161, 2021. https://doi.org/10.1007/978-3-662-62962-8_18

Easy and vendor-independent connectivity as well as establishing interactions between devices and services are key issues in this environment, while incompatible software interfaces are an industrial automation engineer's daily business. This domain particularly suffers from the need to connect the fieldbus layer with its focus on realtime and fast data exchange with the standard IT layer known for shorter innovation cycles and stronger innovative drive.

The German "Plattform I4.0" therefore works out the concept of the asset administration shell (AAS, [11]). The aim is to create a cross-vendor communication standard for I4.0-Components. It describes communication's syntax and semantics but omits to specify concrete technologies. Driven by practical needs and the requirements of different use cases on ARENA2036, we realized a lightweight implementation of the standard's main concepts that we present here.

2 Introduction and State of the Art

The Reference Architecture Model Industrie 4.0 (RAMI) describes a three-dimensional coordinate system or cube, which allows us to decompose complex interrelations into smaller, more manageable problems [6]. The dimensions are Life Cycle and Value Stream representing the product view, Hierarchy Levels representing the production view, and Layers representing the mapping of real resources into the digital world. To integrate a physical asset (e.g. machine or machine component) into this architecture, it has to be provided with an AAS. The tuple of asset and its AAS makes up the I4.0-Component. As a digital representation of the asset, its AAS offers the interfaces for the communication layer of RAMI.

An AAS contains a self-description of the asset consisting of its properties and functions allowing the virtualization of operation status and to communicate with the asset. Another characteristic of I4.0-Components and their corresponding AAS is the capability to order them hierarchically. The optimal level of granularity has to be identified w.r.t. the type of asset and its main use cases. This means, when we talk about I4.0-Components, we could mean a single drive, a machine or even a production line.

State of the Art In [9] formation and development of Cyber-Physical Systems (CPS) in the manufacturing environment is examined and it is determined, that CPSs are a driving force to reach Industry 4.0. In the ZVEI whitepaper [2] structural, functional, and information-related requirements to I4.0-Components are formulated. To simplify their usage in manufacturing, in [15] a hierarchical manufacturing system model is developed w.r.t. existing system structures. It also describes how to identify and localize I4.0-Components. [7] builds on that, identifies meaningful information about manufacturing, and assigns it to the hierarchical layers of RAMI. Both papers show the usage of I4.0-Components in real-world manufacturing environments, while not focussing on them.

In [4,5] the authors describe the digitized information of I4.0-Components using a semantic, knowledge-based approach. Tantik et al. [17] analyze the potential of AAS to support service-based economic solutions like remote access or data analysis. In [16] they describe an integrated data model offering flexible access to an I4.0-Component's information. However, they ignore the aspect of communication, which in our opinion is elementary.

For a definition of terms and better differentiation from related concepts like the digital twin, in [19] the role of the AAS during life cycle of an I4.0-Component is discussed and some bits of advice for implementation are given. Lueder et al. [8] present a prototypical implementation of an I4.0-Component based on AutomationML and OPC/UA.

In the recent past the term AAS is mentioned more and more often [3,13,14, 18,20]. However, those publications lack detailed descriptions or implementation details. In 2018 Plattform I4.0 published the an AAS specification paper [11]. The objective is to ensure the interoperability between partners along an I4.0 value chain. In contrast to the use cases we face on ARENA2036, the focus is put on the exchange of static information. Anyway, the report prepares the ground for further understanding and development of AAS.

3 Use Case and Derived Requirements

A central goal of Arena is to put innovative ideas into practice. Therefore typical use cases are:

Provisioning of runtime data for visualization and analytics The I4.0-Components of different partners should uniformly provide their runtime data so that tools for visualization and data analysis can easily access them.

Cross-vendor and cross-operator orchestration of assets and services In a dynamic, convertible manufacturing environment it should be possible to easily model processes making use of assets from different vendors and operators without the need for complex configuration.

From these typical use case patterns we derive the following requirements:

1. Uniform format of all data streams for distributed applications
2. Low barrier to entry for the communication infrastructure: Simple communication protocols, lean conventions, no dependencies to commercially not freely available software
3. Simple and self-explanatory syntax of communication
4. Capability for semantic descriptions: In addition to a self-explanatory syntax, also semantic description of interfaces should be supported

4 Concept and Implementation of Assets2036

4.1 Concept

To enable I4.0-Components to communicate, common language and common understanding of what to talk about is mandatory. In our concept, this is realized in the idea of submodels, which generically describe properties and capabilities. One submodel describes exactly one technical aspect. To make sure that this common understanding is as unambiguous as possible and is used by as many I4.0-Components as possible, submodels are published at a central location available to each partner.

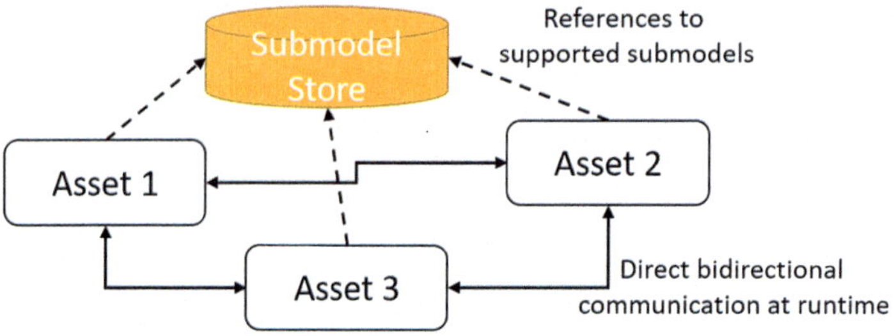

Fig. 1. Asset communication based on common submodels

If an I4.0-Component wants to offer its properties and capabilities to others, it introduces itself and references one or multiple submodels it supports (see Fig. 1). Herewith it describes its capabilities as the aggregation of all supported submodels, so others can use them.

Submodels One submodel specifies exactly one technical aspect of an I4.0-Component. To do so it describes the properties, events, and operations associated with this aspect:

Properties A submodel can describe an arbitrary set of properties. A property can be thought of as having a more static (e.g. engineering files) or a more variable nature (e.g. temperature). When a property is published, its value is to be assumed valid until a new value is published.

Each property has a submodel-wide unique name and a unique data type. The available data types are taken from JSON-Schema [12]. Non-primitive types can be further specified if needed using the JSON-Schema datatype "object".

Events Unlike properties, events have a strict temporal reference. Therefore they contain a time-stamp. Examples for events are the arrival of an AGV or the completion of an assembly process. Like properties events have a unique name. They are capable of carrying parameters. Each parameter is specified in analogy to a property.

Operations While properties and events describe outgoing communication, an operation is an offered capability which can be called from others. An example might be the triggering of an assembly process.

Operations are uniquely identified by the name in the submodel scope. Optionally call and return parameters can be specified. Parameters are specified in analogy to properties.

4.2 Submodel Description Format

Submodels are specified using JSON. Figure 2 shows a simple submodel of a signal light. It contains a property "light_on" describing the light's current state. The event "light_switched" will be emitted, if this state changes. The parameter "state" describes the light's new state. The operation "switch_light" offers the capability to change the light's state to the given target state in its only parameter "state".

```
{
    "name": "light",
    "revision": "0.0.1",
    "description": "describes a single switchable light source",
    "properties": {
        "light_on": {
            "type": "boolean",
            "description": "true if light is on"}},
    "events": {
        "light_switched": {
            "description": "triggered whenever the light is switched, parameter 'state' tells if on or off",
            "parameters": {
                "state": {
                    "type": "boolean"}}}},
    "operations": {
        "switch_light": {
            "description": "switches light into state state. Returns true if operation successful",
            "parameters": {
                "state": {
                    "type": "boolean"}},
            "response": {
                "type": "boolean"}}}}
```

Fig. 2. Submodel description for a signal light

4.3 Implementation of Communication Using MQTT and JSON

Until now we talked about communication contents, or: the "What". Now we will have a look onto the underlying communication technology, or: the "How".

For this concept we used MQTT [1] for data transport and JSON [10] for serialization. MQTT and JSON were chosen due to their wide distribution and simplicity. Anyway, in general, the usage of other technologies for this purpose is absolutely possible.

The described submodel concept is mapped to MQTT-topics as follows: <namespace>/<I4.0-Component>/<submodel>/<submodel element>.

- **namespace**: unique identifier to group together I4.0-Components e.g. by application, organizational affiliation etc.
- **I4.0-Component** namespace-wide unique name of I4.0-Component
- **submodel**: name of the submodel implemented by the I4.0-Component as defined in the submodel's description (see Sect. 4.2)
- **submodel element**: name of property, event or operation. For operations we additionally introduce the subtopics "REQ" and "RESP" to distinguish an operation's call and return message.

Via these topics messages in JSON-format are exchanged as follows:

- **Properties**: The message body is simply the JSON-serialized property value
- **Events** transmit a time-stamp describing when the event occured. The full message body as JSON-object is: {"timestamp": <time stamp according to ISO-8691>, "params": <JSON-object containing key-value-pairs of the event's parameters>}
- **Operations**: To separate call and response, sub-topics "REQ" and "RESP" are used. The caller sends a request message to the REQ-topic with {"req_id": <unique call-ID>, "params": <JSON-object containing key-value-pairs of the call's parameters>}
  The response sent to the RESP-subtopic contains the body: {"req_id":<req_id of the corresponding request>, "resp": <JSON serialized return value of the operartion call>}

4.4 Example

A signal light named "light_1" implementing the submodel as described in Sect. 4.2 would transmit the following messages:

Type	Topic	Message
New property value	arena2036/light_1/light/light_on	true
Emission of event	arena2036/light_1/light/light_switched	{"timestamp":"2020-03-12T07:20:34.837608", "params": {"state": true}}
Operation call	arena2036/light_1/light/switch_light/REQ	{"req_id": "266e8e766-4e6b-45a3-b1d8-3b574237e8b2", "params": {"state": false}}
Operation response	arena2036/light_1/light/switch_light/RESP	{"req_id": "266e8e766-4e6b-45a3-b1d8-3b574237e8b2", "resp": true}

As said before, each I4.0-Component has to publish the submodels it implements to enable other components to make use of it. This is done via a special "_meta"-property, which is implicitly part of each submodel. Each I4.0-Component has to publish the value of the _meta-property at least once for each submodel when it connects to the MQTT-broker. The value is in the form:

{

"submodel_url": <URL referencing the submodel description>,

"submodel_definition": <submodel description as JSON object>, "source": <name of the asset sending this information>

}

By "submodel_url", the referenced submodel specification is uniquely defined. By *optionally* sending the description directly in "submodel_definition", the interaction with so far unknown components is possible without the need to download external resources. The field "source" identifies the actual communication partner: If passive elements like primitive components or carriers are modeled, they do not communicate themselves. Instead, a service, we call it "endpoint", implements the communication of passive assets. This endpoint itself has its own AAS which is referenced in the "source"-field.

5 Conclusions

On ARENA2036 we witnessed some enthusiasm for our simple approach and experienced many different partners adapting it and participating very quickly. Here is what we think had a positive effect on this success:

Use lightweight, open and widespread standards like MQTT and JSON Both can be used without licensing, and in consequence, there are freely available software libraries in almost each programming language. Thereby the presented approach can be used on a wide range of I4.0-Components – from servers to embedded systems.

Low barrier to entry Even a partial implementation enables components to benefit from the whole. Using just an MQTT-client it is possible to observe topics and to focus on application, not on standards.

No need for proprietary software The presented approach only defines conventions, of how to use widely spread and open standards. Each partner can implement by himself with low effort. We provide some libraries for different languages that implement the convention. However, usage is optional.

Strict compliance with KISS-principle (keep it simple, stupid) The whole concept is based on properties, events, and operations encapsulated in submodels. If additional functionality is needed, it can be realized using these basic concepts. The concept itself remains small and simple.

Because of the strong focus on simplicity, some desirable aspects are ignored. For example, the concept lacks a security concept to protect the communication against unauthorized access and data manipulation. Those could be handled by securing the MQTT-traffic with TLS and a public-key-infrastructure, which would allow each I4.0-Component to identify itself reliably.

References

1. Banks, A., Gupta, R.: MQTT version 3.1. 1. OASIS Stand. **29**, 89 (2014)
2. Bedenbender, H., Billmann, M., Epple, U., Hadlich, T., Hankel, M., Heidel, R., Hillermeier, O., Hoffmeister, M., Huhle, H., Jochem, M., et al.: Examples of the asset administration shell for Industrie 4.0 components – basic part. ZVEI White Paper (2017)
3. Diedrich, C., Belyaev, A., Schröder, T., Vialkowitsch, J., Willmann, A., Usländer, T., Koziolek, H., Wende, J., Pethig, F, Niggemann, O.: Semantic interoperability for asset communication within smart factories. In: 2017 22nd IEEE International Conference on Emerging Technologies and Factory Automation (ETFA), pp. 1–8 (2017)
4. Grangel-González, I., Halilaj, L., Auer, S., Lohmann, S., Lange, C., Collarana, D.: An RDF-based approach for implementing Industry 4.0 components with administration shells. In: 2016 IEEE 21st International Conference on Emerging Technologies and Factory Automation (ETFA), pp. 1–8 (2016)
5. Grangel-González, I., Halilaj, L., Coskun, G., Auer, S., Collarana, D., Hoffmeister, M.: Towards a semantic administrative shell for Industry 4.0 components. In: 2016 IEEE Tenth International Conference on Semantic Computing (ICSC), pp. 230–237 (2016)
6. Heidel, R., Hoffmeister, M., Hankel, M., Döbrich, U.: Industrie4.0 Basiswissen RAMI4.0: Referenzarchitekturmodell mit Industrie4.0-Komponente, 1st edn. VDE Verlag GmbH and Beuth Verlag GmbH, Berlin (2017)
7. Hell, K., Hillmann, R., Lüder, A., Röpke, H., Zawisza, J., Schmidt, N., Calà, A.: Demands on virtual representation of physical Industrie 4.0 components. In: CIISE, pp. 65–71 (2016)
8. Lüder, A., Schleipen, M., Schmidt, N., Pfrommer, J., Henssen, R.: One step towards an Industry 4.0 component. In: 2017 13th IEEE Conference on Automation Science and Engineering (CASE), pp. 1268–1273 (2017)
9. Monostori, L., Kádár, B., Bauernhansl, T., Kondoh, S., Kumara, S., Reinhart, G., Sauer, O., Schuh, G., Sihn, W., Ueda, K.: Cyber-physical systems in manufacturing. CIRP Ann. **65**(2), 621–641 (2016)
10. n.a.: Standard ECMA-404: the JSON data interchange syntax. https://www.ecma-international.org/publications/files/ECMA-ST/ECMA-404.pdf (2017)
11. n.a.: Details of the asset administration shell: part 1 – the exchange of information between partners in the value chain of Industrie 4.0 (version 2.0) (2019)
12. n.a.: JSON schema. https://json-schema.org/, 9 (2019)
13. Prinz, F., Schoeffler, M., Lechler, A., Verl, A.: End-to-end redundancy between real-time I4. 0 components based on time-sensitive networking. In: 2018 IEEE 23rd International Conference on Emerging Technologies and Factory Automation (ETFA), vol. 1, pp. 1083–1086 (2018)
14. Profanter, S., Dorofeev, K., Zoitl, A., Knoll, A.: OPC UA for plug & produce: automatic device discovery using LDS-ME. In: 2017 22nd IEEE International Conference on Emerging Technologies and Factory Automation (ETFA), pp. 1–8 (2017)
15. Röpke, H., Hell, K., Zawisza, J., Lüder, A., Schmidt, N.: Identification of "industrie 4.0" component hierarchy layers. In: 2016 IEEE 21st International Conference on Emerging Technologies and Factory Automation (ETFA), pp. 1–8 (2016)
16. Tantik, E., Anderl, R.: Integrated data model and structure for the asset administration shell in Industrie 4.0. Procedia CIRP **60**, 86–91 (2017)

17. Tantik, E., Anderl, R.: Potentials of the asset administration shell of Industrie 4.0 for service-oriented business models. Procedia CIRP **64**, 363–368 (2017)
18. Tasci, T., Melcher, J., Verl. A.: A container-based architecture for real-time control applications. In: Conference Proceedings ICE/IEEE ITMC, pp. 1–9. IEEE, Piscataway (2018)
19. Wagner, C., Grothoff, J., Epple, U., Drath, R., Malakuti, S., Grüner, S., Hoffmeister, M., Zimermann, P.: The role of the Industry 4.0 asset administration shell and the digital twin during the life cycle of a plant. In: 2017 22nd IEEE International Conference on Emerging Technologies and Factory Automation (ETFA), pp. 1–8 (2017)
20. Wenger, M., Zoitl, A., Müller, T.: Connecting PLCs with their asset administration shell for automatic device configuration. In: 2018 IEEE 16th International Conference on Industrial Informatics (INDIN), pp. 74–79 (2018)

AutomationML in Industry 4.0 Environment: A Systematic Literature Review

Jiaqi Zhao[1,2(✉)], Matthias Schamp[1,2], Steven Hoedt[1,2], El-Houssaine Aghezzaf[1,2], and Johannes Cottyn[1,2]

[1] Department of Industrial Systems Engineering and Product Design, Ghent University, Technologiepark 46, 9052 Gent-Zwijnaarde, Belgium
Jiaqi.Zhao@UGent.be

[2] Industrial Systems Engineering (ISyE), Flanders Make, www.FlandersMake.be, Graaf Karel de Goedelaan 5, 8500 Kortrijk, Belgium

Abstract. AutomationML is an open neutral XML based data exchange format used in automation systems. It has come into the public for more than 10 years and is being used in many different areas in all kinds of manufacturing applications, such as digital twin, reconfigurable manufacturing systems, heterogeneous data exchange, etc. However, no comprehensive literature review on the research and application progress of AutomationML has been found since the initiation of AutomationML. Based on the study and analysis of AutomationML related publications, this paper gives a detailed illustration on the state-of-the-art of AutomationML. Firstly, the background and terminologies related to AutomationML are introduced. Secondly, the paper applies a methodology to collect AutomationML related publications, on which an analysis based on a multidimensional literature classification is conducted. Thirdly, according to the analysis results, current research status and whether AutomationML can meet the requirements for industry 4.0 environment are discussed. Finally, conclusion and outlook are illustrated in the end.

1 Introduction

In an industry 4.0 production environment, multiple different engineering disciplines are involved throughout the product life cycle [1]. The tools used by these disciplines are quite different, which leads to a broad "heterogeneous tool landscape" [2]. To efficiently integrate this landscape, data exchange between these tools is an obvious bottleneck waiting to be resolved [3].

To solve this problem, Daimler AG initiated the foundation of a consortium together with leading vendors and users of automation technology in 2006 [4]. The aim of the foundation is to develop a neutral data exchange format usable within the engineering process of manufacturing systems to exchange data among multidisciplinary tools. The consortium named this neutral data exchange format 'AutomationML', short for Automation Markup Language. The first version of AutomationML was brought to the public at the Hannover fair in 2008 [3]. Its original application scenario is the description of the static structure of a plant's shop floor [5].

P. Weißgraeber et al. (Eds.): *Advances in Automotive Production Technology – Theory and Application*, ARENA2036, pp. 162–169, 2021. https://doi.org/10.1007/978-3-662-62962-8_19

With the continuous development of industry 4.0 technology, manufacturing enterprises have to operate in a dynamic way to meet the diverse needs of customers [6]. This requires manufacturing systems to be flexible and reconfigurable enough to respond to orders from customers [7]. At present, the authors are trying to find a solution for a real-time reconfigurable digital twin system, in which the digital model can automatically change itself to reflect current situation of the production system according to the reconfiguration of the physical system. 'Digital twin' is an integrated multi-physics, multi-scale, probabilistic simulation of a complex product and uses the best available physical models, sensor updates, etc., to mirror the life of its corresponding twin [8]. This means that the digital model should always be consistent with the physical model. Due to the increasing flexibility of nowadays automation systems, having an up-to-date digital twin at all time is a great challenge. Therefore, an efficient way of bidirectional data exchange between the physical and its digital replica is indispensable. The authors believe that AutomationML could give an answer to this requirement.

This paper illustrates the state-of-the-art of AutomationML based on a systematic literature review. The question we are attempting to address is whether AutomationML can meet the requirements for industry 4.0 applications in general and for the generation of digital twins more specifically.

The remainder of the paper is structured as follows. In Sect. 2, a general overview of AutomationML is presented based on an example. Both the content and the architecture of AutomationML are discussed. In Sect. 3, a methodology which is utilized to collect AutomationML related publications is presented. An analysis method based on multidimensional literature classification is illustrated. Based on the analysis, the state-of-the-art of AutomationML is described and whether AutomationML can meet the requirements for making reconfigurable digital twin systems in industry 4.0 environment is discussed. In Sect. 4, conclusion and outlook are made according to the illustration and discussion above.

2 AutomationML Overview

AutomationML is an XML based data format, which is open, neutral and free [4]. The top level core of AutomationML is CAEX, which is utilized to interconnect all kinds of data formats. As CAEX is object oriented, all kinds of engineering information can be stored in AutomationML objects. These objects are called internal elements in AutomationML hierarchy. Typical categories of the information stored in internal elements are plant topology information, geometry and kinematics information, logic information, reference and relation information, and other data formats [9]. An example of AutomationML can be found on the right side of Fig. 1. The represented hierarchy is the topology information of a robot cell. From the hierarchy it can be directly seen that: (1) The robot cell is composed of a robot station and a safety fence, (2) The robot station has a robot and a conveyer, (3) The robot includes a robot machine and a gripper. Besides the topology information, the node "GeometryInterfaceCollada" contains the geometry information of a robot cell component, which is referenced to an external geometry file in the format of Collada, while the node "LogicInterfaceGantt" contains the logic information of the robot cell, which is referenced to an external logic file programmed in Gantt chart.

As AutomationML contains all kinds of multidisciplinary interrelated information, it is important to ensure the modeling efficiency. Therefore, three class libraries are introduced: the role class library (RCL), the interface class library (ICL) and the system unit class library (SUCL) [9]. An RCL is a container of role classes (RC), which define the semantics of internal elements (IE), and each IE has to refer to an RC. Similarly, an ICL is a container of interface classes (IC), and an IC can be used to interconnect between 2 IEs, or to refer to an external file. SUCL is a library of system unit classes (SUC), and it can be used by easily dragging and dropping during the modeling of the instance hierarchy. By using these libraries, the modeling efficiency of AutomationML is greatly improved.

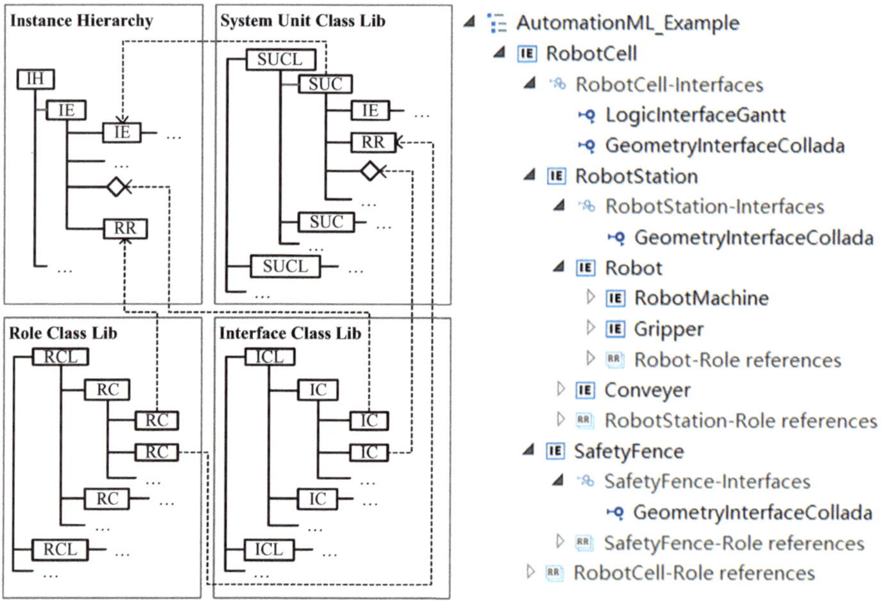

Fig. 1. AutomationML architecture and an AutomationML model example

3 Methodology and Discussion

This paper gives a systematic literature review on AutomationML. First, the authoritative website of AutomationML [4] is carefully checked. On this website, all AutomationML relevant research projects and technical documents are presented. Based on this, the authors have a clear understanding of AutomationML. Then, the authors use search engines Scopus, IEEE Xplore and Google Scholar to gather the publications containing the content of 'AutomationML'. In this way, 195 AutomationML related publications have been found. Among the papers, the amount of conference papers is 174, while the amount of journal papers is 18. The rest of the publications are 2 technical reports and 1 doctor thesis.

The amount of AutomationML related publications by year is shown in Fig. 2. It shows that the publication amount trend is mainly ascending from 2008 to 2018. It means that AutomationML gains more and more interest of researchers. The authors believe the publication amount in 2019 will be more than the amount in 2018, as some publications in 2019 are still waiting to become publicly available. The publication count for 2020 is similarly affected.

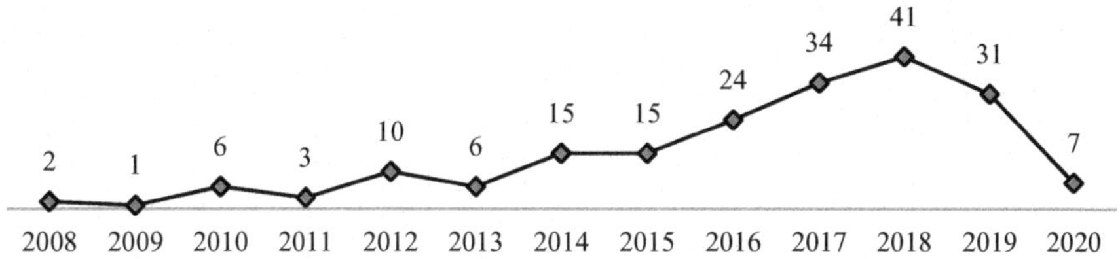

Fig. 2. The amount of AutomationML related publications by year until June 30, 2020

Figure 3 shows the percentage allocation of the research fields related to AutomationML. According to the role of AutomationML, the research fields are divided into 6 categories, which are: modeling, data exchange, concept, maintenance, integration, and evaluation. Modeling (29%) and data exchange (27%) cover more than half of the publications, which verifies that AutomationML based automation system modeling and AutomationML based heterogeneous data exchange are research hotspots. Furthermore, concept, maintenance and integration related publications have a considerable share of overall publications, respectively 16%, 14% and 11%. Evaluation related research publications come in the last, which cover only 3%.

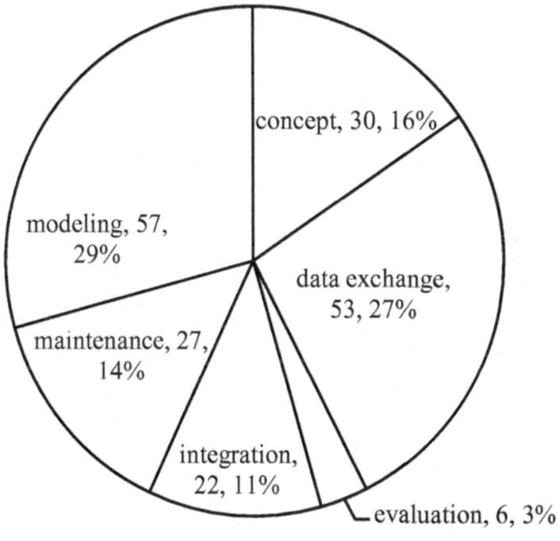

Fig. 3. AutomationML related research fields

- **Modeling:** The research field of modeling is using AutomationML based modeling method to express relevant information in AutomationML data format. The main question of this research field is what kind of modeling method should be used to describe the corresponding information. According to the publications, AutomationML based modeling methods are in many aspects, such as automation system modeling [10, 11], communication system modeling [12, 13], behavior modeling [14], process modeling [15], product-process-resource modeling [16], Asset Administration Shell (AAS) modeling [17], ISA-95 modeling [18], etc.
- **Data exchange:** This research field is focusing on doing data exchange based on AutomationML. This field is divided into two parts: (1) Data exchange between tools in the same field based on AutomationML, (2) Data exchange between AutomationML and other data formats. For (1), the data exchange paradigms include data exchange between 3D modeling software [19, 20], between logical programming software [21, 22], between simulation software [23, 24], etc. For (2), there are methods for AutomationML to do data exchange with OPC UA [25], SysML [26], RDF [27], OWL [28], PMIF [29], etc.
- **Concept:** This field is about new conceptual methods on AutomationML related engineering. Breckle et al. [30] introduce an approach of an evolving digital factory containing and visualizing all generated information based on an AutomationML metamodel. Kiesel et al. [31] present an approach of an AutomationML-based AAS, which is able to handle heterogeneous data. Wally et al. [32] use the information stored in a separate B2MML document to define an alignment of two industrial standards, ISA-95 and AutomationML.
- **Maintenance:** This research field focusses on the development of new approaches for working with AutomationML. Winkler et al. [33] present an AML-Review process approach towards reviewing AutomationML model elements with tool support. Wimmer et al. [34] propose a dedicated query language for AutomationML. Hua et al. [35] present a concept learning approach in AutomationML using the DL-Learner framework. Ananieva et al. [36] develop an approach to detect and repair inconsistencies in systems modeled with AutomationML.
- **Integration:** In this field, AutomationML is a component which is integrated in all kinds of digital automation systems. Panda et al. [37] develop a Plug & Play retrofitting platform based on AutomationML and OPC UA, where Industry 4.0 compliant sensor systems can be attached, detected, and configured automatically to the existing production environment.
- **Evaluation:** This field is about the evaluation of AutomationML related aspects. Meixner et al. [38] draft a novel, flexible evaluation framework in the context of AML model storage, modification, and retrieval and to evaluate two particular data storage paradigms.

From the classification of AutomationML related publications, we can conclude in several aspects. AutomationML, a data format which came to public just more than 10 years ago, is gaining more and more attentions from researchers. It is possible to model all kinds of manufacturing system information in AutomationML and to do data exchange based on the combinations of AutomationML, OPC UA, AAS, ISA-95, etc. We can say AutomationML is a promising data format which shows great potential for modeling rapidly changing automation systems in order to efficiently obtain and maintain an up-to-date digital twin.

However, the publications on AutomationML based reconfigurable digital twin systems are very limited. Only 6 papers describe the relevance of AutomationML in combination with digital twins. 3 of them are in the research field of modeling [39–41], 2 of them only describe a conceptual design [42, 43], and 1 paper illustrates a method of automatically generating the digital model based on AutomationML [44]. No reconfigurable digital twin system is indicated in literature.

The raising interest in digital twins and their applications in combination with the rise of mass customization, amplifies the need for a reconfigurable digital twin [45]. This may save a lot of effort for manufacturing enterprises, since the generation and maintenance of a digital twin is very time consuming. Therefore, the authors plan to make such a reconfigurable digital twin system. AutomationML will be used not only for digitally modeling the geometry and kinematic information, behavior and logical information, process information, etc. of the whole physical system, but also as a data exchange bridge between the digital model and the physical system. Plug-ins will be developed in both sides to ensure AutomationML based data transmission. AutomationML based real-time communication is also to be achieved.

4 Conclusion and Outlook

This paper gives a systematic literature review on the state-of-the-art of AutomationML. The authors classify AutomationML related publications into 6 categories according to the role of AutomationML. The majority of the publications is about what modeling method could be used for AutomationML (29%) and how to do data exchange based on AutomationML (27%). Furthermore, there are some publications that focus on conceptual descriptions (16%), maintenance of AutomationML files (14%) and the integration of AutomationML in real applications (11%). Only 3% of publications discuss the evaluation of AutomationML related aspects. According to a multidimensional literature classification, it can be stated that AutomationML is a promising data format to be used in an industry 4.0 environment. Continuously improvements due to current research will probably lift its potential even higher.

In the future, the authors are planning to develop an AutomationML based methodology to make a real-time reconfigurable digital twin system, in which the digital model can change synchronously along with reconfiguration of the physical model. Real time data exchange technology based on AutomationML will be used to ensure the equality between the physical system and its virtual replica.

Acknowledgements. This research is financially supported by the China Scholarship Council (CSC), which is a non-profit institution affiliated with the Ministry of Education of China.

References

1. Lüder, A., et al.: Validation of behavior specifications of production systems within different phases of the engineering process. In: ETFA, IEEE (2013)

2. Drath, R., et al.: Concept for interoperability between independent engineering tools of heterogeneous disciplines. In: ETFA, IEEE (2011)
3. Drath, R., et al.: AutomationML – the glue for seamless Automation Engineering. In: ETFA, pp. 616–623. IEEE (2008)
4. AutomationML Homepage. https://www.automationml.org. Accessed 30. June 2020
5. Wally, B.: Application Recommendation Provision for MES and ERP – Support for IEC 62264 and B2MML, 1st edn. AutomationML e. V. (2018)
6. Vogel-Heuser, B., et al.: Evolution of software in automated production systems: challenges and research directions. J. Syst. Softw. **110**(2015), 54–84 (2015)
7. Hoang, X., et al.: An Interface-Oriented Resource Capability Model to Support Reconfiguration of Manufacturing Systems. In: SysCon, IEEE (2019)
8. Kritzinger, W., et al.: Digital Twin in manufacturing: A categorical literature review and classification. In: IFAC World Congress, pp. 1016–1022. IFAC (2018)
9. AutomationML consortium: Whitepaper AutomationML Part 1 – Architecture and General Requirements. 2nd edn. AutomationML e. V. (2018)
10. Peres, R., et al.: GO0DMAN Data Model - Interoperability in Multistage Zero Defect Manufacturing. In: INDIN, pp. 815–821. IEEE (2018)
11. Najafi, E., et al.: Model-Based Design Approach for an Industry 4.0 Case Study: A Pick and Place Robot. In: ICMT, IEEE (2019)
12. Patzer, F., et al.: Towards the modeling of complex communication networks in AutomationML. In: ETFA, IEEE (2017)
13. Drath, R., et al.: Modeling and exchange of IO-Link configurations with AutomationML. In: CASE, pp. 1530–1535. IEEE (2018)
14. Brandenbourger, B., et al.: Behavior modeling of automation components using cross-domain interdependencies. In: ETFA, IEEE (2016)
15. Danny, P., et al.: An Event-Based AutomationML Model for the Process Execution of 'Plug-and-Produce' Assembly Systems. In: INDIN, pp. 49–54. IEEE, (2018).
16. Schleipen, M., et al.: AutomationML to describe skills of production plants based on the PPR concept. In: AutomationML User Conference (2014)
17. Drath, R., et al.: The AutomationML Component Description in the context of the Asset Administration Shell. In: ETFA, IEEE (2019)
18. Wally, B., et al.: IEC 62264–2 for AutomationML. In: AutomationML User Conference (2019)
19. Babcinschi, M., et al.: AutomationML for Data Exchange in the Robotic Process of Metal Additive Manufacturing. In: ETFA, IEEE (2019)
20. Fechter, M., et al.: From 3D product data to hybrid assembly workplace generation using the AutomationML exchange file format. In: CMS, CIRP (2019)
21. Hundt, L., et al.: Development of a method for the implementation of interoperable tool chains applying mechatronical thinking - Use case engineering of logic control. In: ETFA, IEEE (2012).
22. Estévez, E., et al.: A novel approach for flexible automation production systems. In: INDIN, pp. 695–699. IEEE (2017)
23. Bigvand, P.G., et al.: Concept and development of a semantic based data hub between process design and automation system engineering tools. In: ETFA, IEEE (2016)
24. Laemmle, A., et al.: Automatic layout generation of robotic production cells in a 3D manufacturing simulation environment. In: CIRP Design Conference, pp. 316–321. CIRP (2019)
25. Henßen, R., et al.: Interoperability between OPC UA and AutomationML. In: DET, pp. 297–304. CIRP (2014)
26. Berardinelli, L., et al.: Cross-disciplinary engineering with AutomationML and SysML. Automatisierungstechnik **64**(4), 253–269 (2016)

27. Hua, Y., et al.: From AutomationML to ROS: A Model-driven Approach for Software Engineering of Industrial Robotics using Ontological Reasoning. In: ETFA, IEEE (2016)
28. Hua, Y., et al.: Interpreting OWL Complex Classes in AutomationML based on Bidirectional Translation. In: ETFA, IEEE (2019)
29. Berardinelli, L., et al.: Integrating Performance Modeling in Industrial Automation through AutomationML and PMIF. In: INDIN, pp. 383–388. IEEE, (2016)
30. Breckle, T., et al.: The evolving digital factory – new chances for a consistent information flow. In: ICME, pp. 251–256. CIRP (2019)
31. Kiesel, M., et al.: AutomationML in a continuous products life cycle: a technical implementation of RAMI 4.0. In: AutomationML User Conference (2018)
32. Wally, B., et al.: Entwining Plant Engineering Data and ERP Information: Vertical Integration with AutomationML and ISA-95. In: ICCAR, pp. 356–364 (2017)
33. Winkler, D., et al.: AutomationML Review Support in Multi-Disciplinary Engineering Environments. In: ETFA, IEEE (2016)
34. Wimmer, M., et al.: From AutomationML to AutomationQL: A By-Example Query Language for CPPS Engineering Models. In: CASE, pp. 1394–1399. IEEE (2018)
35. Hua, Y., et al.: Concept Learning in Engineering based on Refinement Operator. In: ILP (2018)
36. Ananieva, S., et al.: Model-Driven Consistency Preservation in AutomationML. In: CASE, pp. 1536–1541. IEEE (2018)
37. Panda, S.K., et al.: Plug & Play Retrofitting Approach for Data Integration to the Cloud. In: WFCS, IEEE (2020)
38. Meixner, K., et al.: Investigating the Performance of selected Data Storage Concepts for AutomationML Models. In: IECON, pp. 2785–2791. IEEE (2019)
39. Schroeder, G.N., et al.: Digital Twin Data Modeling with AutomationML and a Communication Methodology for Data Exchange. In: IFAC World Congress, pp. 12–17. IFAC (2016)
40. Zhang, H., et al.: Information modeling for cyber-physical production system based on digital twin and AutomationML. The International Journal of Advanced Manufacturing Technology **107**(2020), 1927–1945 (2020)
41. Peng, G., et al.: Data Exchange of Digital Twins Based on AML in Space Science Experiment Equipment. In: IOP Conference (2020)
42. Um, J., et al.: Plug-and-Simulate within Modular Assembly Line enabled by Digital Twins and the use of AutomationML. In: IFAC World Congress, pp. 15904–15909. IFAC (2017)
43. Lou, X., et al.: An idea of using Digital Twin to perform the functional safety and cybersecurity analysis. In: INFORMATIK Workshops, pp. 283–294 (2019)
44. Spellini, F., et al.: Production Recipe Validation through Formalization and Digital Twin Generation. In: DATE, pp. 1698–1703. IEEE (2020)
45. Zhang, C., et al.: A Reconfigurable Modeling Approach for Digital Twin-based Manufacturing System. In: IPSS, pp. 118–125. CIRP (2019)

Generic and Scalable Modeling Technique for Automated Processes

Martin Karkowski(✉), Rainer Müller, and Matthias Scholer

Zentrum für Mechatronik und Automatisierungtechnik gemeinützige GmbH, Eschberger Weg 46, D-66121 Saarbrücken, Germany
m.karkowski@zema.de

Abstract. Modularity, adaptability and integration of new technologies like Human-Robot Cooperation (HRC) help in facing the major challenges posed by the increased product variants with shortened life cycles and fluctuating market conditions of the automotive industry. However, utilizing them requires strong software support and complicates the already demanding planning and implementation of an assembly system. The strong dependency on software creates a new void in the planning and implementation processes. Usually, the programmer, not the process owner, fills this void based on his knowledge. This results in frequent and resource-intensive adaptations during commissioning due to implicit knowledge and requirements during the development process. This paper presents a lean approach for implementing an adaptable assembly system. Our approach combines an abstract process description, a virtual model of assembly system and a standardized control system which enables the realization of an assembly system. Our modeling technique helps a process owner to develop robust assembly systems. Also, it enables the design of a process and supports in obtaining the corresponding boilerplate code needed to execute the process on standard hardware utilized by the industry. This is demonstrated and tested by means of a HRC underbody assembly process in vehicle assembly under realistic conditions in a demonstrator factory.

1 Introduction

The constant integration of new production technologies—such as HRC—into assembly systems as well as the increased requirements for mass customization require flexible and versatile assembly systems. This results in an increased implementation effort and complexity of the assembly system. This contradicts the demand for being capable of reacting quickly to fluctuating sales figures—for instance by exchanging production equipment. [1] Also, modification of the assembly system often requires fundamental adjustments to the software to ensure compatibility with the utilized equipment. To guarantee a higher level

P. Weißgraeber et al. (Eds.): *Advances in Automotive Production Technology – Theory and Application,* ARENA2036, pp. 170–177, 2021. https://doi.org/10.1007/978-3-662-62962-8_20

of system availability, malfunctions of the utilized equipment have to be analyzed to derive appropriate recovery strategies. The definition of those strategies necessitates the modeling and implementation of robust assembly systems suited for industrial applications even more.

In order to address these challenges, a comprehensive behavioral description of an assembly system based on Petri nets is presented. Section 2 presents the theoretical basis of the utilized modeling technique. Based on the currently published approaches to model an assembly system (Sect. 3), the research needs are identified (Sect. 4). After the presentation of the developed approach (Sect. 5), the application of the modeling technique follows, with an example from the automotive final assembly.

2 Theoretical Basis

2.1 Petri Nets

Petri nets are used to describe discrete event dynamic systems. To describe a system, places, transitions, and edges are utilized. Places define the states of a system or subsystem whereas state changes, e.g. by events or actions, are modeled by transitions. The explicit modeling of states simplifies the description and analysis of the system. Relations between states and transitions are modeled via edges. The structure of a Petri net—the number of places and transitions as well as their relations, modeled by edges—describes the system to be represented. A marking describes the distribution of tokens in the network, where tokens denote active states of a system. Tokens can represent elements of the system as well as abstract conditions. They are consumed, produced, changed or transported by firing transitions. Firing a transition changes the marking of the network. [2]

2.2 Refinement Action Engine

A Refinement Action Engine (RAE) uses facts, events, tasks and a set of interactions to control a system. Facts describe the state of the system and are utilized to select Refinement methods from the set of interactions to fullfill the desired tasks. Refinement methods specify the conditions to enable the interactions as well as their effects on the environment. The selected commands of the selected refinement methods are executed via an execution platform. This platform interacts with the environment and forwards events to the RAE, which may result in new tasks. In addition, tasks can be provided by other external components, such as users or planners. [3]

2.3 Utlized Control Hardware in Production Systems

In most production systems, process state changes are triggered by events. By now PLCs are utilized as economical solutions for controlling those production systems. However, machine manufactures are facing constraints due to the characteristics of PLCs. PLCs are characterized by their cyclic data processing and real-time execution, which results in a state based design for sequencing control. [4]

3 Approaches utilized by Research and Industry

3.1 Modelling Techniques in Industry

The Unified Modelling Language (UML) has established itself as the standard to model systems and combines different modeling techniques. For Example, activity diagrams and UML state machines can be used to describe system behavior. However, no execution model can be derived from this description automatically. Further UML follows a monolithic design approach. As a result, UML behavioral diagrams are often considered a less than optimal modeling solution for the composed systems. [5]

AutomationML (AML)—a neutral XML-based data exchange format—describes geometric, kinematic, logical and structural information of a production system. The behavior of the system is defined via the PLCopen-XML standard. [6] Hardware-oriented modeling and standardization approaches based on the IEC 61131-3 industry standard, such as PLCopen, only allow a low level of abstraction and are therefore only conditionally suitable for modeling composed system. [7]

3.2 Approaches in Research

"Plug and Produce" describes the vision of implementing structural changes to automated systems with or without minimal manual adaptation and reconfiguration of the software. [8] Different aspects such as "high-level" programming [8,9] or automatic programming [10] are considered in research projects. With the implementation of concepts such as skills and self-describing assets, an automatic configuration of the systems can be achieved (see e.g. [10]). These concepts use a stateless description of the assembly process. Relevant aspects for industrial applications—such as errors or disturbances in processes—are thus not explicitly modeled. Furthermore, these approaches are often not transferable to industrial hardware and software. Complex self-organized approaches based on several agents are not suitable for industrial use due to their hardly predictable behavior [11] and the high complexity of the software.

4 Research Needs

A major challenge is to increase the required adaptability and flexibility of today's assembly systems without increasing the complexity of the assembly system. Current approaches based on AML and UML are only partially suitable for modeling a modular assembly system running on industrial hardware. Relevant aspects of the logic cannot be represented in those approaches, or only in an implicit and simplified form.

The use of simple, predictable, flexible and maintainable task descriptions can reduce the implementation effort of the assembly process and system, and adaptations of the system during its life cycle can be simplified. The main research questions are:

- How can an assembly process and system be modeled to support a developer during the specification, configuration, (re-)implementation and adaptation of an assembly system?
- Is it possible to model industrial applications with such an approach?

5 Methodology

In previous work, an adaptable and modular control system with a Petri net-based modeling technique has been developed (see [12] and [13]). This approach is displayed in Fig. 1 and summarized in the following.

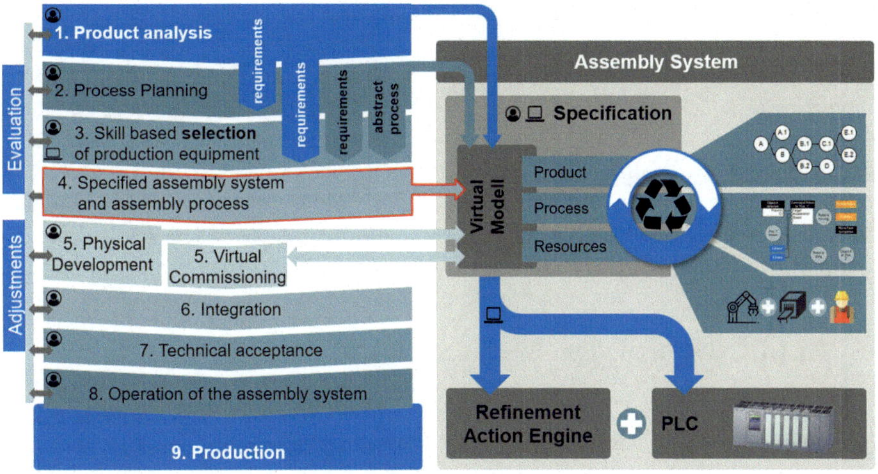

Fig. 1. Used methodology to implement industrial applications

In the beginning, a detailed product analysis is performed. This analysis results in requirements for the assembly process. Based on the requirements the assembly process is derived and its logical sequence is represented by Petri nets. A solution-neutral description of the assembly process is provided by matching the assembly operations of the process with skills. These skills are used to manually select appropriate resources from a resource construction kit. This derives the assembly system and results in detailing the assembly task by adding additional states and skills in an iterative designing process. To fully specify the process, the newly added states have to be analyzed. For this purpose, the resources of the resource construction kit provide behavioral models in the form of a Petri net, which has to be manually combined into an aggregated system behavior. The aggregated behavior contains various relations, restrictions, and requirements of the individual resources.

By detailing the assembly tasks, critical events of the task regarding the application and the current state of the system can be determined easily. This allows specifying actions to prevent those unwanted states of the system and helps the user to define safe and robust applications.

The resources implement the skills via services. The skills of the generic assembly task are automatically mapped with the corresponding services of the resources and an executable model of the system is derived. For this purpose, the flow of information between the services of resources and their mechanical, electrical as well as structural relations enhance the executable model. To validate and optimize the model of the production system, the specified model is transferred to the developed RAE (see [12]). Linking the RAE with a simulation environment (see [14]) allows adaptations of the system.

To carry out the defined assembly task, the RAE consists essentially of five elements. A dynamic logic interpreter implements the defined logic of the process, whereby the information flow is implemented via a volatile memory system. An extensible broker delegates the tasks determined by the logic to the corresponding operating resources via communication layers. These provide the functionalities as services. Standardized service interfaces ensure the easy exchange of resources.

After successful validation of the model with the non-real-time capable RAE, IEC-61131-3 compatible PLC-code can be generated and transfered to the hardware utilized in the assembly system. For simplification of the PLC code, the generated code utilizes the RAE for the execution of non-real-time critical tasks.

6 Application of the Methodology

The presentation of the application of the methodology in this paper is mainly limited to step four (specified assembly system and assembly process) and eight (operation of the assembly system) (see Fig. 1).

6.1 Description of the use Case

Mounting a cars underfloor paneling is a typical assembly task that has to be carried out on the vehicles underbody. Due to the poor ergonomics, the installation of the underbody paneling is stressful for the workers and is therefore considered problematic (see Fig. 2 left).

To improve ergonomic and economical aspects of the assembly process, a human-robot cooperation process has been developed in previous works (see [15]). A lightweight robot assists and relieves the worker from non-ergonomic tasks. The worker is in charge of feeding the underfloor paneling and align the component with the underbody of the car. Afterward the worker fits individual screws to ensure the proper arrangement of the paneling. The lightweight robot then performs the joining process automatically and tightens screws on the remaining joints. An automated alignment of the endeffector enables a robust process against disturbances like displacements or oscillations. After all screws have been tightened, the process starts over again.

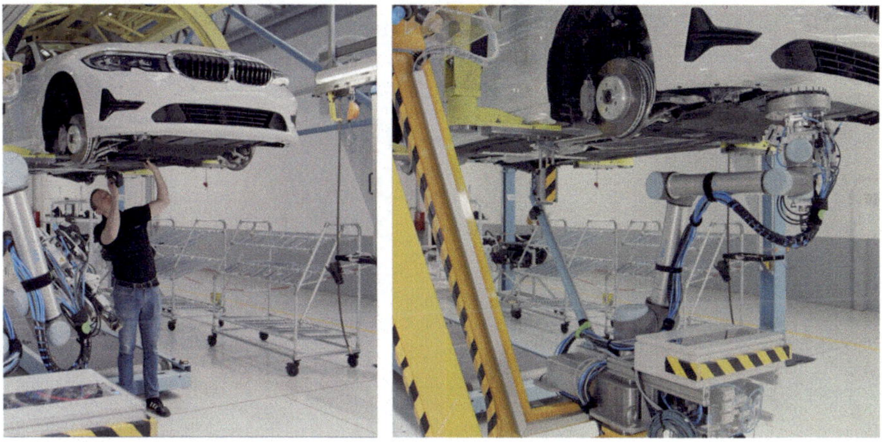

Fig. 2. Left: Worker tightening the underbody Panel. Right: the automatic system tightening the left joints

6.2 Modelling the System

To perform the automatic joining process the system consists of a lightweight robot, a specialized endeffector with an automatic screwdriver and a vision system, a feeder and a transport system, on which the robot is mounted (see Fig. 3). Modeling the presented system results in combing the behavior of the selected resources. Each resource provides at least one unique base behavior model.

The aggregated system is modeled by combining the base behavior models of the resources (see Fig. 3). To simplify the aggregation of the resources, the modeling technique adds additional arc types to classical Petri nets. These extensions enable relations such as:

- A resource is able to change its state if other resources are in a certain state or explicitly not in that state. However, the states of the other resources aren't modified.
- A state change of a resource by the occurrence of an event or the execution of action results in executing an action—thus fire a transition—in other resources.

Relations between the resources can be defined with these extensions without adapting the base behavior model of a resource. For a detailed description of the behavior, a distinction is made between event-based transitions, that cannot be influenced—such as the entry of a human into the workspace of a robot

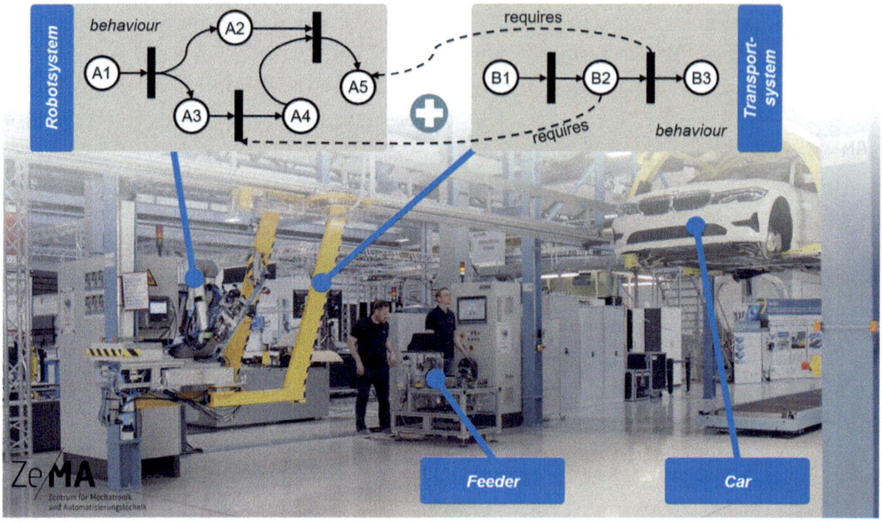

Fig. 3. Overview of the system, showing the aggregation of multiple resources

system—and planned transitions, called actions,—such as allowing the robot to move from position A to B. While event-based transitions fire directly once all conditions are met, actions require an additional release before firing.

6.3 Operation of the Assembly System

For the industrialization of the developed HRC process, the assembly system was evaluated in an endurance test with our partners. To optimize the system, individual modules and concepts of the system were tested at the beginning. Thereby different aspects like availability, process times, etc., were analyzed. The modeling technique supports this by enabling automatic analysis and quick adaptations of the behavior. On the basis of the optimized modules, the behavior of the assembly system was aggregated. The modeling technique supported the development of a robust assembly system.

7 Conclusion and Outlook

In order to reduce the implementation effort of flexible assembly systems, previous approaches presented a modeling methodology in combination with a generic modular control system. An enhanced Petri net-based approach is utilized to model the behavior of the assembly system. To optimize the behavior, the model can be used to control a digital twin in a simulation environment. Afterwards, the optimized model can be transferred to common automotive hardware via code generation. This significantly simplifies the commissioning process and reduces the time-to-process. The described approach was tested in a demonstrator factory with a previously developed HRC underbody process.

In future research, aspects of automatic planning will be considered to further simplify the implementation of an assembly system. Due to the great flexibility of the system, there are numerous possibilities to extend the functionality of the system.

Acknowledgments. This article is written within the project Mittelstand 4.0-Kompetenzzentrum Saarbrücken, as part of the Support Initiative "Mittelstand-Digital" of the BMWi. The nationally funded competence centers provide information to small and medium-sized companies about the opportunities and challenges of digitization. The authors are responsible for the content of the publication.

References

1. Michalos, G., Markis, S., Papakotas, N., Mourtzis, D., Chryssolouris, G.: Automotive assembly technologies review: Challenges and outlook for a flexibleand adaptive approach. **2010**(2), 81–91 (2010)
2. Reisig, W.: Understanding Petri Nets. Springer, Berlin (2013)
3. Ghallab, M., Nau, D., Traverso, P.: Automated Planning and Acting. Cambridge University Press (2016)
4. Olsson, G.: Programmable Logic Controllers. In Hristu-Varsakelis, D., Alur, R., Årzén, K.-E., Baillieul, J., Henzinger, T., Levine, W.S. editors, Handbook of Networked and Embedded Control Systems, Control Engineering, pp. 259–278. Birkhäuser, Boston
5. Jørgensen, J.B.: Coloured Petri Nets in UML-Based Software Development – Designing Middleware for Pervasive Healthcare (2002)
6. Hirzle, A.: AutomationML – Fachexperten erklären das Forma (2014)
7. Loskyll, M., Heck, I., Schlick, J., Schwarz, M.: Context-Based Orchestration for Control of Resource-Efficient Manufacturing Processes. **4**(3), 737–761 (2012)
8. Zoitl, A.: AutoPnP - Plug&Play für Automatisierungssysteme: Schlussbericht – Konsortialbericht
9. Antzoulatos, N., Castro, E., Scrimieri, D., Ratchev, S.: A multi-agent architecture for plug and produce on an industrial assembly platform. **8**(6), 773–781 (2014)
10. Danny, P., Ferreira, P., Lohse, N., Guedes, M.: An AutomationML model for plug-and-produce assembly systems. In 2017 IEEE 15th International Conference on Industrial Informatics (INDIN): University of Applied Science Emden/Leer, Emden, Germany, 24 26 July 2017 : Proceedings, pp. 849–854. IEEE
11. Wooldridge, M.J.: An Introduction to Multiagent Systems, 2. ed. Wiley (2009)
12. Müller, R., Scholer, M., Karkowski, M.: Increasing the Flexibility of Customized Assembly Systems with a Modular Control System. pp. 46–53 (2018)
13. Müller, R., Scholer, M., Karkowski, M.: Generic automation task description for flexible assembly systems. **81**, 730–735 (2019)
14. Illmer, B., Karkowski, M., Vielhaber, M.: Petri net controlled virtual commissioning – A virtual design-loop approach. p. 6. Elsevier B.V (2020)
15. Scholer, M.: Wandlungsfähige und angepasste Automation in der Automobilmontage mittels durchgängigem modularem Engineering -Am Beispiel der Mensch-Roboter-Kooperation in der Unterbodenmontage (2018)

On Automation Along the Automotive Wire Harness Value Chain

Marc Eheim(✉), Dennis Kaiser, and Roland Weil

IILS mbH, 72818 Trochtelfingen, Germany
eheim@iils.de, kaiser@iils.de, weil@iils.de
http://www.iils.de

Abstract. An ambitious reduction of wire harness design time and cost by at least 50% each can only be achieved by automation. In the ITEA2-project IDEaliSM, it was demonstrated that the automation of 3D wire harness generation can be achieved using graph-based design languages and the VEC (vehicle electrical container) as an open data standard for wire harnesses. All required data and process steps for generating the wire harness are described, as well as a wire harness assembly simulation.

1 Introduction

The wire harness represents to date the second most expensive part in a car and it is expected that its complexity grows even further. The associated value chain typically starts with the definition of the boundary conditions set by the original equipment manufacturer (OEM) with contractual specifications: A) the predefined electrical circuit diagram, B) the design space definition and C) the selection of dedicated components. It typically ends with the delivery of the designed and manufactured wire harness by the supplier to the OEM using just-in-time and just-in-sequence (JIT/JIS) logistics. Due to the high complexity and intensive coupling with other car parts only the design, manufacturing planning and organizational tasks have often remained in close geographic proximity to the OEM. Due to the price pressure in global competition, the manual wire harness assembly work has often been relocated to lower wage countries.

1.1 Current Wire Harness Design

The current status quo of wire harness design is a predominantly manual process chain and comprises the electrical and geometrical design process, which are performed concurrently. In the electrical design process an electric and electronic architecture solution is designed based on a choice of mechanical, electrical and electronic components, which need to be packed and routed inside a constrained installation space. In the geometrical design process, the OEMs usually provide

P. Weißgraeber et al. (Eds.): *Advances in Automotive Production Technology – Theory and Application,* ARENA2036, pp. 178–186, 2021. https://doi.org/10.1007/978-3-662-62962-8_21

the installation space as Digital Mock-Up (DMU) in a specific CAD data format. Positions of holes, tunnels, fixings, etc. can be derived from the DMU and serve as a part of the design requirements. The virtual 3D wire harness design is then typically created manually in a so-called *"cable workbench"* module in a CAD-system. Downstream, the wire harness manufacturing process steps lead to a full-scale 2D form-board plan for the manual wire harness assembly. Depending on the size and complexity of a wire harness, the entire development cycle may take up to 2 years and more (e.g. A380 wire harness).

2 Wire Harness Design Automation

During the manual process hundreds of design changes are likely to occur which need to be addressed in both process chains (geometrical and electrical) to avoid inconsistencies. In this section an automated approach to generate new wire harness designs is presented which is able to cope with these changes automatically.

2.1 Automation Approach

The automation approach based on graph-based design languages (GBDLs) is a novel technology to model and execute design processes of complex systems incrementally. GBDLs possess a vocabulary, rules and a production system [9] and allow to encode the design knowledge of a product in a (re-)executable way. This know-how can be re-executed with changing customer requirements, which enables the generation of harness design variants and enables in this way a systematic exploration of the design space. This greatly accelerates the design, reduces the amount of errors, reduces time-to-market as well as costs.

Similar to programming languages, GBDLs require a compiler to be compiled (i.e. translated) into an executable model. In this project the Eclipse-based software *DesignCockpit43* (DC43) of IILS [10] was used to compile the GBDLs, which are modeled in the Unified Modeling Language (UML) as the underlying meta-model. While the GBDLs describe the use case, the generic (use case independent) automated harness design process described in Sect. 2.4 is implemented in Java using Eclipse plug-ins for DC43. More information about the approach with GBDLs and DC43 can be found in the IILS white paper [10]. So far, GBDLs have been created for numerous engineering design and manufacturing applications, such as aircraft cabin design [8] and satellite design [2].

The wire harness design automation using GBDLs has been developed with the goal to be generically applicable to various cabling problems. This includes the ability to cope with highly detailed three-dimensional geometry. Various constraints have to be considered, like the physical properties of wires, restrictions for the placement and mounting, compliance to electro-magnetic interference, and clearances to hazardous areas like hot components. To overcome these challenges the approach makes use of advanced graph algorithms for pathfinding and collision detection. The automated wire harness has already been applied to several engineering domains ranging from the aerospace (launchers, satellites [11] and aircraft [8]) to the automotive domain (car cockpit wire harnesses [3]).

2.2 Use Case

The example of an car cockpit wire harness shown here in Fig. 1 is the public demonstrator of the European ITEA2-project IDEaliSM [3]. The key goal of IDEaliSM in the automation of electrical wire harness design was to reduce the development time and cost by at least 50%, while still improving product quality. The solution developed with GBDLs generates the wire harness automatically, allows easy reconfiguration in case of different requirements and offers the capability to be embedded in an optimization loop.

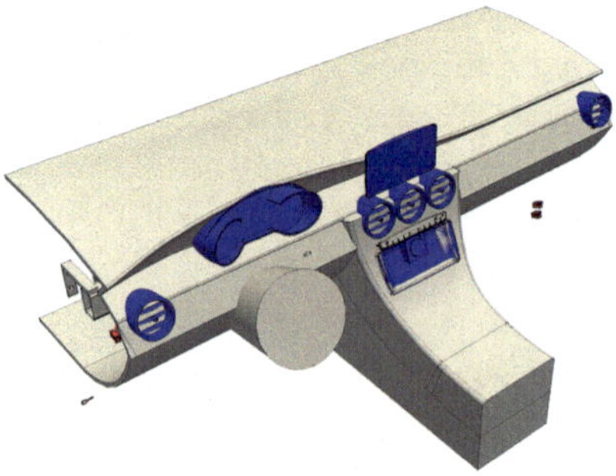

Fig. 1. Configurable automotive cockpit

In the IDEaliSM use case two versions of the cockpit are used, an industrial and an academic cockpit. The industrial cockpit with detailed geometry cannot be shown here for reasons of confidentiality. The academic cockpit variant [1] has a simplified geometry and was designed to show various cockpit configuration variants which lead to variants of the wire harness. The configurable components, highlighted in blue in Fig. 1, include different shapes (instrument cluster and ventilation outlets), different topology (number of ventilation outlet and integration of the display) and optional components (head-up display, analog clock, air conditioner outlets). Depending on the configuration, the wire harness consists of about 50 connectors of 8 types and up to 196 single wires.

2.3 Data

For the automation of wire harnesses, three different types of data are required, which have to be provided for the use case and need to be combined in a GBDL: 1) master data, 2) geometrical components and 3) electrical schematics.

The master data describe the selected electrical and non-electrical components which interact with or are part of the wire harness, e.g. connectors, fixings, ducts, grommets, terminals, splices, wires, etc. A surface mesh geometry of the components is required for collision checking as well as "qualified" or "electrified" information for these components. A so-called qualified connector additionally contains the exact point and the direction where the wire bundle leaves the connector. Figure 2 shows a such a connector (left) and a fixing (right). The geometry data can be imported via various formats like *stl*, *vtk*, *obj*, *stp* or *jt*. In the cockpit use case STEP files are used for the connector types. The additional electrical information for connectors and wire types (diameters, stiffnesses) was defined by a comma-separated values (CSV) file.

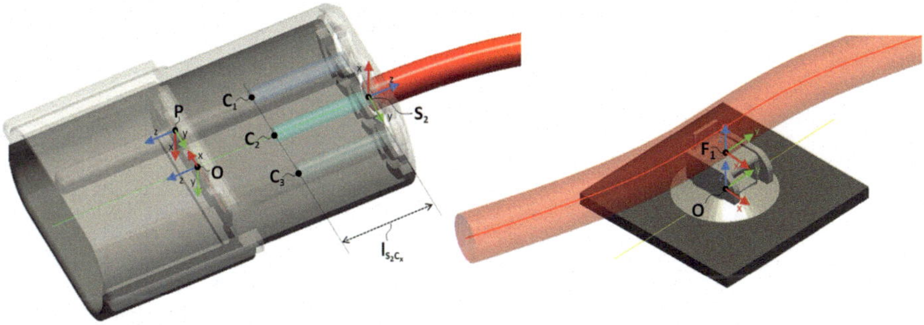

Fig. 2. Qualified connector (left) and fixing (right) geometry

Geometrical Components of A) the installation space which constrains the wire harness and B) the instances of connectors, fixings, etc. are required. Since the master data already defines the geometrical representation of these components, providing their positions and orientations as well as a unique reference ID will be sufficient. Usually, a "cloud" of connectors and fixings defines the positions of these components, e.g. in an CAD assembly as STEP file.

Electrical Schematics data defining the connection list must be provided. As in this project, the connection list can be imported e.g. via an Excel or CSV file. The unique reference IDs of the two wired connectors must have the same reference IDs in the geometrical data world.

The data of the separated design processes for geometry and electrical architecture must be imported and merged. The combination of data is easily possible with GBDLs expressed in UML, since the designer can define his own data format with the help of the GBDL vocabulary. More convenient is the integration of a widely used XML-based data standard. With the Vehicle Harness Container (VEC) [6] such a data standard is available for the holistic description of a wire harness along its life-cycle. Our automated approach offers full VEC integration and uses only in-built extensions of the VEC.

2.4 Process

Our approach is divided into several separated steps. Each step reads the available VEC data, applies a specific task on it and writes back the result by modifying the data in the VEC model. The VEC model is thus enriched with each process step and the degree of completeness of the data is increased. In the following, each of the process steps 1 to 6 in Fig. 3 is explained.

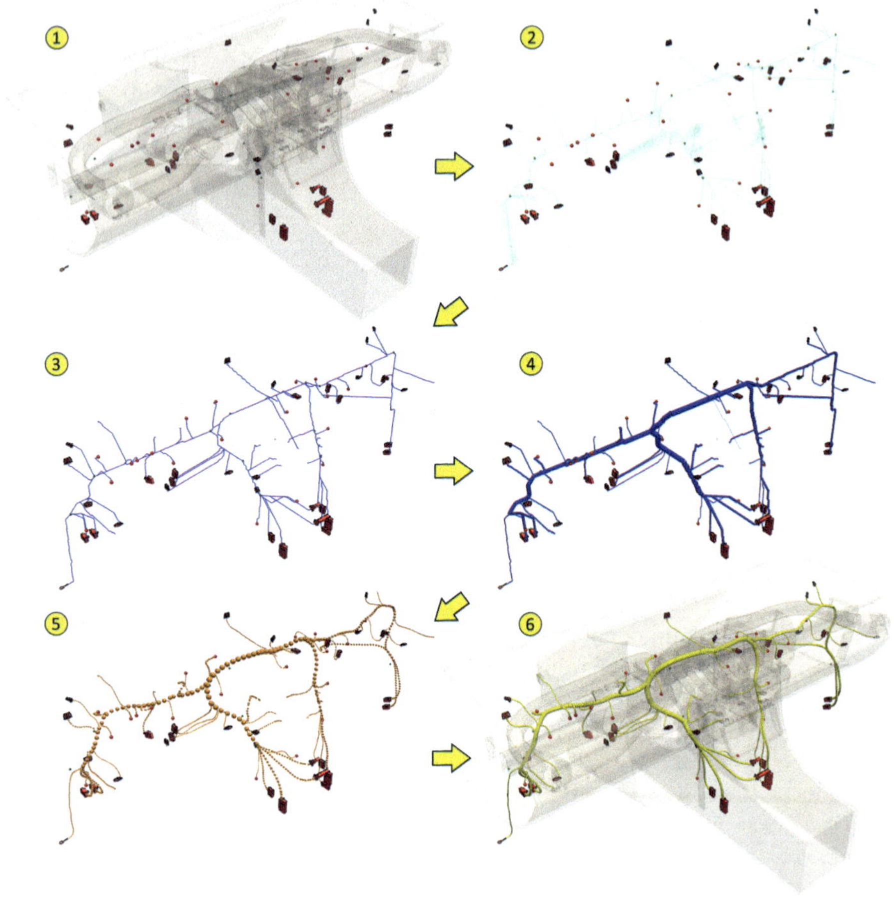

Fig. 3. Automated wire harness design process steps 1 to 6

1. Data Import The first process step builds up the holistic model incrementally by executing the design rules of the GBDL. The input data described in the last section is read and imported to the VEC model. Since this step depends on the available type of data (i.e. file formats) a data importer created for a specific OEM can read the data from a specific set of file formats and link the data together. Data merging works properly if the reference IDs of geometry and electrical schematics match. Inconsistent or missing data can be identified by a consistency check which is executed directly after the import. After this step the model holds the connectivity data, the harness components and their respective master data specifications. Figure 3 shows the connectors in the design space which will be connected to a wire harness in the next steps.

2. Pathfinding The pathfinding step can be characterized by finding the network topology of the wire harness by predefined sequential pathfinding tasks. This is done by using a logic to define the pathfinding sequence to generate the wire harness topology. A modified A* pathfinding algorithm finds the optimal paths on the discretized installation space between selected targets (e.g. path of connector A to a partial topology of the harness). In Fig. 3 the light blue areas are zones which are visited by the pathfinding algorithm. During this path search, a collision detection algorithm checks if the path is free from penetrations with the installation space geometry (i.e. the cockpit components) which serve as obstacles for the pathfinding algorithm. The topology and geometry of a wire harness is driven by the possibility where to attach the harness on the mountable structure geometry. In the cockpit use case, an intermediate topology of the wire harness is defined by waypoints along the cross beam, since it is possible to place fixings for mounting the wire harness there. To generate the final topology the connectors are routed to the intermediate topology via optimal paths.

3. Topology Generation This step collects the resulting paths from the previous task and assembles the network topology on which the single wires can be routed afterwards. The result is a topology which is mathematically a graph, represented by nodes (terminal points at connectors and break-out points) and edges (harness segments between the nodes). The harness topology (shown in Fig. 3) is written into the VEC model as well as placement information for the connectors and optionally occurring fixings on the harness topology.

4. Routing The routing step (see Fig. 3) routes the individual single wires on the harness topology created in the previous step. The routing is based on the shortest paths on the graph by minimizing cable lengths between the connectors which are connected by the wire. Inputs for this step are the topology, the harness components, contacting and placement information. When all wires are routed the diameters of the harness segments are calculated with a heuristic based on the information of included single wires and their types. The found routings for each wire of the connection list are added to the VEC model as well as the updated segment diameters.

5. Multi-Body Simulation So far, the topology of the harness is defined, but the geometry consists of mostly rectangular geometrical paths. A multi-body physics simulation is performed to smooth the harness segment paths considering the different diameters and stiffnesses to produce realistic segment curvature and break-out points. Required minimum bending radii can be taken into account. Figure 3 shows the discrete mass points of the simulation represented as spheres which are connected by constraints. Again, a collision detection algorithm ensures that there are no penetrations with the installation space geometry during the relaxation process. The result is a realistic geometrical representation of the wire harness which is written to the VEC model.

6. Data Export Since all harness data including the complete bill of material (BOM) is now available combined in a consistent model, it can be exported for downstream processes in form of an XML file conform to VEC or KBL (*Kabel-*

baumliste, the preceeding wire harness standard). For example, the generated geometry can be exported directly to various CAD systems where it can be used for installation space protection in the DMU (see Fig. 3).

2.5 Assembly Simulation

Before the wire harness is delivered to the OEM it must be ensured that the wire harness can be assembled into the constraining installation space. Some segments are long enough when the connectors are already plugged in, but must be lengthened to allow the connectors to be plugged in during integration. These segments can be automatically identified by performing the integration virtually in the simulation instead of a usual physical mock-up. A study of such an assembly simulation has recently been created [4] and applied to this use case for the assembly of a car radio into the foreseen slot in the cockpit. The multi-body physics simulation with collision detection mentioned before is used to ensure that the harness segments have sufficient lengths to plug the connectors.

2.6 Reduction of Design Time and Design Cost

The key goal of the ITEA2-project IDEaliSM [3] in the automation of electrical wire harness design was to reduce the development time and cost by at least 50%, while still improving product quality. Because of the fact that the exact savings identified in a dedicated campaign by industrial experts from the different project partners are strictly confidential, it can be clearly stated that the original claim of saving at least 50% was overfulfilled. As a further hint to a clear and full understanding of the true savings potential of the automation approach using GBDLs, a quote from the PhD abstract [2] is reproduced here: *"By the rigorous mapping of the design knowledge to a formal description, the consequent automated processing reduces the duration for a single … design to the execution times of the different algorithms. With regard to the chosen models and algorithms thus the theoretical lower limit for the duration of the design process is reached."* Such significant savings are frequently reported in the literature [7].

3 Summary and Outlook

The wire harness development process has been successfully covered from the point where the electrical architecture is defined until the final virtual wire harness is available. That wire harness is collision free in arbitrary constraining geometry. The automated process steps are repeatable with individual input data, can be executed automatically and used in optimization loops. For example, if a design change makes adjustments necessary, e.g. due to an offset or relocation of an equipment, the adjusted wire harness can easily be generated by clicking one button. The generation runtime of the described process in this use case was about 8 min on a regular desktop computer[1].

Along the wire harness life-cycle there are still some processes for automation conceivable. Upstream of the automated process, the electrical architecture must be designed. One task thereof is the decision about the grounding concept, i.e. splices have to be positioned. Downstream, before entering manufacturing, wire protections could be automatically selected and placed. For the manufacturing on a form-board, a harness flattening process is required for the form-board drawing. In the past, efforts were taken in digital factory simulations with GBDLs. This preliminary work could be applied to an envisioned future project, in which the manufacturing of wire harnesses with robots will be examined in a digital factory.

Acknowledgments. Parts of the research (i.e. wire harness automation) leading to these results were performed within the European ITEA2 research project IDEaliSM (13040) as part of the EUREKA cluster program. The authors would like to express their gratitude to the consortium members for their support and contributions (see https://itea3.org/project/idealism.html for details).

SPONSORED BY THE
Federal Ministry
of Education
and Research

Parts of this project (same as above) are sponsored by the Federal Ministry of Education and Research in Germany.

Clean Sky

Clean Sky 2 The project leading to this application has received funding from the Clean Sky 2 Joint Undertaking under the European Union's Horizon 2020 research and innovation programme under grant agreement No. 865044.

References

1. Bohn, J.G.: Eine Entwurfssprache zur Variantengenerierung von Automobilcockpits zur automatischen Verkabelung. Master Thesis, Institut für Statik und Dynamik der Luft- und Raumfahrtkonstruktionen, Universität Stuttgart (2016)
2. Groß, J.: Aufbau und Einsatz von Entwurfssprachen zur Auslegung von Satelliten. Dissertation, Institut für Statik und Dynamik der Luft- und Raumfahrtkonstruktionen, Universität Stuttgart (2014)

[1] Intel Core i5-2500K @ 3.3 GHz with 32 GB RAM.

3. IDEaliSM – Integrated & Distributed Engineering Services Framework for MDO. http://idealism.ifb.uni-stuttgart.de (2017)
4. Kaiser, D.: Entwicklung einer Einbausimulation für den Kabelbaum des Kleinsatelliten "Flying Laptop". Master Thesis, Universität Stuttgart (2019)
5. Motzer, M.: Integrierte Flugzeugrumpf- und Kabinenentwicklung mit graphenbasierten Entwurfssprachen. Dissertation, Institut für Statik und Dynamik der Luft- und Raumfahrtkonstruktionen, Universität Stuttgart (2016)
6. Project Group Car Electric of VDA Working Group PLM: Vehicle Electric Container (VEC), VDA Recommend. 4968. Verband der Automobilindustrie (2014)
7. Reddy, J., Sridhar, C., Rangadu, P.: Knowledge based engineering: notion, approaches and future trends. Am. J. Intell. Syst. **5**(1), 1–17 (2015)
8. Rudolph, S., Hess, S., Beichter, J., Motzer, M., Eheim, M.: Architectural analysis of complex systems with graph-based design languages. In: 4th International Workshop on Aircraft System Technologies (AST'13), Hamburg (2013)
9. Rudolph, S.: Übertragung von Ähnlichkeitsbegriffen. Habilitationsschrift , Fakultät der Luft- und Raumfahrttechnik, Universität Stuttgart (2002)
10. Schmidt, J.: Total Engineering Automation. Whitepaper, IILS Ing.-Ges. für intelligente Lösungen und Systeme. https://www.iils.de/#downloads (2017)
11. Weil, R.: Automatisierte Verkabelung des Kleinsatelliten Flying Laptop. Diplomarbeit, Institut für Raumfahrtsysteme, Universität Stuttgart (2012)

An ISA-95 based Middle Data Layer for Data Standardization—Enhancing Systems Interoperability for Factory Automation

Chen Li(✉), Soujanya Mantravadi, Casper Schou, Hjalte Nielsen, Ole Madsen, and Charles Møller

Department of Materials and Production, Aalborg University, Aalborg, 9220 Aalborg, Denmark
cl@mp.aau.dk, sm@mp.aau.edu, cs@mp.aau.edu, hn@mp.aau.edu, charles@mp.aau.edu
https://www.aau.dk/

Abstract. This paper presents a middle data layer that is designed and implemented based on the ANSI/ISA-95 industrial standard. The proposed middle data layer extracts key information on manufacturing operations from the control systems as well as the business systems. The proposed model enhances interoperability in Industry 4.0 by using a standardized and formalized data structure. We expand the above idea into three directions. Firstly, we analyze all the layers of the traditional automation hierarchy model of ISA-95 to get an overview of objects and activities that are needed for building a middle data layer and restructure the traditional hierarchy levels by introducing the ISA-95 based middle data layer. Secondly, we design the data architectures by categorizing the data source and explain the formalized and standardized data models. Finally, we use a Smart Manual Station of a production setup to show how to apply the proposed ISA-95 middle data layer to the real world case. The results indicate that the middle data layer enhances the interoperability of manufacturing systems and creates a universal standardized data structures for systems integration for factory automation.

1 Introduction

Industry 4.0 paradigm has been crucial in driving innovation in manufacturing, especially through its concept of "Smart Factory" [1,2]. A smart factory uses advanced digital technologies for production automation and is guided by the design principles of (a) Interconnection (b) Information transparency (c) Decentralized decision-making (d) Technical assistance [3]. Many companies are on their journey toward Industry 4.0 by improving their methods of data acquisition

P. Weißgraeber et al. (Eds.): *Advances in Automotive Production Technology – Theory and Application,* ARENA2036, pp. 187–194, 2021. https://doi.org/10.1007/978-3-662-62962-8_22

and processing. Over the years, manufacturing execution systems (MES) have been crucial in replacing paper-based production execution and management. It has been a core software to implement the functionalities of manufacturing operations management (MOM) set out by ISA-95 standard. However, the MES is challenged with low interoperability [4], causing problems due to its ambiguous data semantics and for difficulty of data exchange with other systems.

The manufacturing industry relies on automation of manufacturing operations in a factory to improve business and operational efficiency. For this, manufacturers need to integrate business and manufacturing processes quickly and efficiently [5]. One of the major challenges is how to smoothly reconstruct, aggregate, and standardize data from various systems, e.g, MES, Industrial Internet of Things (IIoT) platform, without interference with the existing business or manufacturing processes. Among other industrial standards, ISA-95 has been widely used to solve this problem [6–9]. ISA-95 [10] Level 3 layer acts as a bridge between enterprise resource planning (ERP) and the machines for automation of information exchange. Since most of the MES software that complies ISA-95 is supplied by a third party vendor, the data format (e.g, int, string or byte) and data structure (e.g, XML style or a relational database) vary from each other, causing low interoperability. Interoperability is imperative to achieve industry 4.0 shop floors as the implementation of reconfigurable manufacturing systems require a robust information management [11,12]. To address this problem and design information systems that are industry 4.0 ready, we designed a middle layer that holds all the production data and acts as a repository to retrieve the production data on demand.

Smart factory capabilities include manufacturing systems that are reconfigurable, autonomous, and modular. The digital transformation due to the IIoT technology promises extraction of production data from the machines and field devices. This data can be processed to run analytics as well as to build autonomous agent-based manufacturing entities based on learning algorithms. Recent studies have raised the concerns of enterprise integration and interoperability in manufacturing systems that hinder a cooperative information-driven environment required for Industry 4.0 [13]. On this premise, we propose a solution for enhancing interoperability of the systems through the ISA-95 middle data layer to support factory automation. We use Aalborg University (AAU) Smart Production Lab to build and test our design. For this, we used a manual work station for a case study and tested its application for providing the universal standardized data model for data exchange among different systems.

This paper is organized as follows: Sect. 2 presents the motivation for our problem and the experiment we conducted in 2019 to replicate an industrial problem of data exchange. Section 3 presents the research approach, which includes the data architecture and data models of the proposed middle layer. Section 4 describes the case study and analysis. Finally, we conclude the paper in Sect. 5.

2 Motivation and Problem Statement

In this section, we introduce the AAU Smart Manual Station (SMS), as an example, to indicate the problem.

AAU Smart Manual Station AAU SMS is part of AAU Smart Production Lab, which is an interdisciplinary platform for research and teaching within smart factories for Industry 4.0. It is located at Aalborg University in Denmark [14]. The AAU SMS is a modular, smart manual station, which can be quickly and easily configured for new production tasks. Figure 1 shows the setup of the AAU SMS.

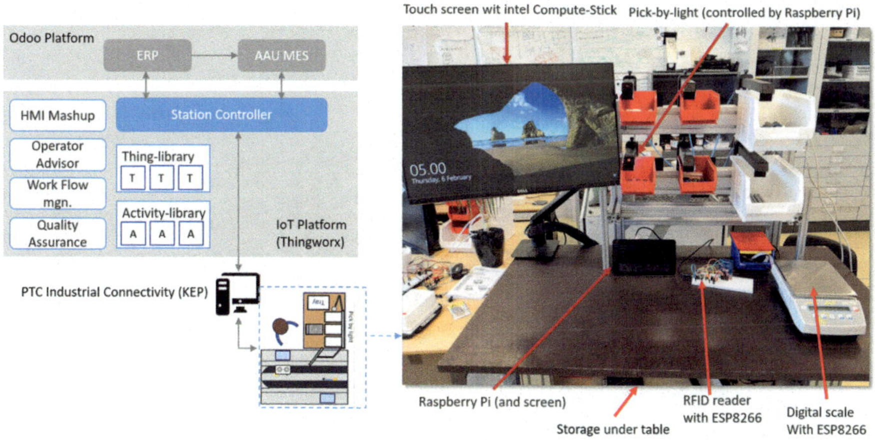

Fig. 1. AAU smart manual station

The right side of the Fig. 1 shows the physical setup of AAU SMS. It consists of worktable on wheels, numerous intelligent devices each with their own micro-controller and connectivity; including a pick-by-light system, an RFID scanner, a primary touch screen running ThingWorx (i.e., a IIoT platform) mashup through a webbrowser, a secondary 7-inch touch screen, a distance sensor and a digital scale. All the devices are independently connected to a PTC KEP server using OPC/UA or MQTT over a local Wifi. The central controller of the station is implemented in a locally hosted PTC ThingWorx cloud platform. By doing so, we facilitate easy data collection, data exchange and changes to the production control. Furthermore, we have in PTC ThingWorx implemented an equipment taxonomy (formal model) allowing the station control to operate on generic functionality rather than device specific syntax and implmentation. This is a crucial element in quick and easy configuration of the station. In AAU Smart Production Lab, Odoo [15] is chosen as the business platform providing various business modules, e.g, Sales and Purchase. The functionality of manufacturing operation is implemented by AAU MES [16] as an app in the Odoo framework (see the left side of the Fig. 1).

The setup of the AAU SMS described, requires integration between the ThingWorx platform running the station controller and the AAU MES running in the Odoo platform. In order to achieve a quick and efficient configuration of both existing and new shop-floor equipment, a formalized and standardized terminology and a consistent set of concepts and data models are required across the factory. As the common denominator on factory level typically is the MES or ERP system, one of these should provide such consistent modelling. However, the majority of these systems provide their own, vendor-specific modelling, further enhancing the "lock-in" to specific systems.

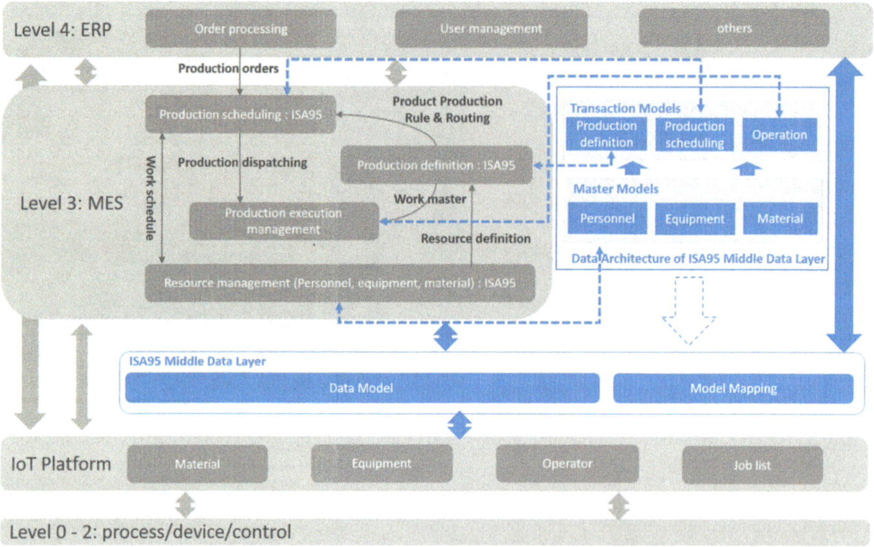

Fig. 2. Design logic of ISA-95 middle data layer

3 Approach

In response to the challenges highlighted in the problem statement, we propose an ISA-95 based middle layer which follows an industrially accepted and consistent modelling and terminology. It provides a guideline of what function and data flows should be considered within manufacturing organizations.

3.1 Main idea

The research work starts by analyzing the functional hierarchy model according to the ISA-95 standard. It provides a general picture of a role-based equipment hierarchy across different levels, i.e., business, manufacturing, control, and the actual process. The purpose is to have an overview of what kind of processes and functions should be considered for building the ISA-95 Middle data layer. The second step is to identify the core objects and activities by investigating the identified processes and functions. In order to build the data architecture of ISA-95 middle data layer, data source are extracted from the key objects and activities. By categorizing the data sources, two types of models are designed, master data model and transaction data model, as fundamental elements of the proposed data architecture. Figure 2 shows the general idea of our approach.

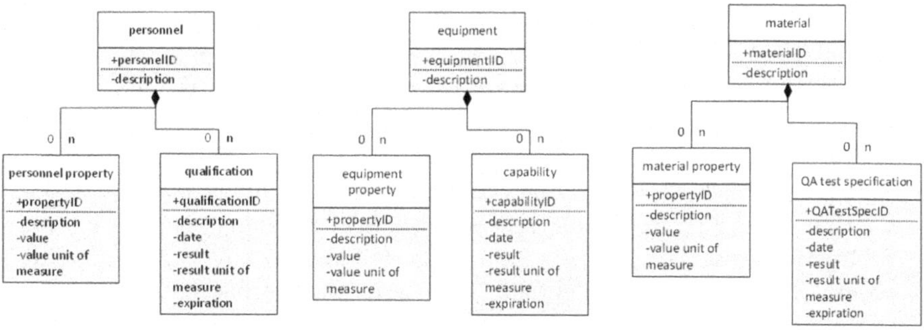

(a) Personnel Model (b) Equipment Model (c) Material Model

Fig. 3. Master data model

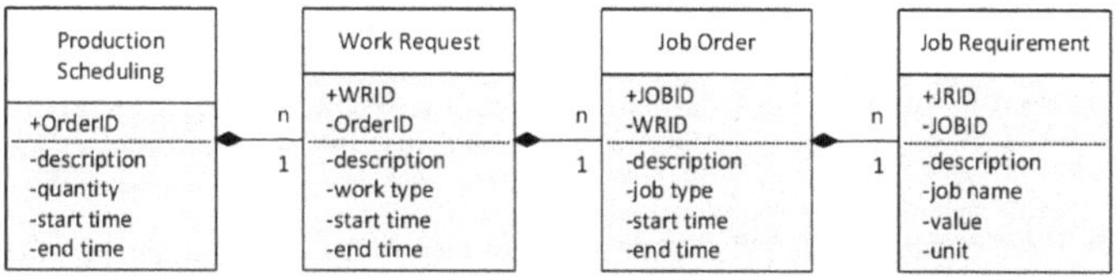

Fig. 4. Transaction data model

3.2 Data Architecture Design

Data architecture, as the core element of the proposed ISA-95 middle data layer, is composed of two types of data models, master data model, and transaction data model. It is classified based on the analysis of the data flow and information exchanging of the above activities (see Fig. 2).

- **Master data model**. The master data model, including personnel model, equipment model, and material model, represents the objects that contain the most valuable, agreed-upon information shared across a manufacturing company. Since the master model is directly related to resource management, it is also called the resource model. It provides the basic elements for fulfilling a production execution. 1) personnel model needs to answer the questions of what is an employee's identification and the special qualification he/she has (e.g., Level A certified). Such information are required in different levels, e.g., user management at level 4, detailed job order at level 3; 2) equipment model specifies the physical assets' capability and availability. It supports the detailed production scheduling; 3) the material model needs to provide the information of material class, quality test results, and inventory. Figure 3 shows the corresponding master data model.

- **Transaction data model**. The transaction data model is mainly composed of a product definition model, production scheduling model, and operation model that describes an event triggered during the manufacturing processing. It focuses on the events/actions happening during the manufacturing process, for example, production dispatching, product production rule execution, work scheduling. Those data information are specified in 1) product definition model (specifying the information of scheduling, material and production rules), 2) production scheduling model (defining the information of production resource management and production tracking) and 3) operation model (The operation model mainly concerns the production scheduling and product definition. The production scheduling can be mapped into a corresponding work scheduling model, e.g., work request, job list). See Fig. 4.

4 Case Study

In this section, we show how to use ISA-95 middle data layer to assist systems integration for factory automation. Due to the page limit, we provide two examples of master data and transaction data mapping between the middle data layer and the Thingworx platform. The production environment is described in Sect. 2.

4.1 Data Model Analysis

Data extracted from our case can be classified into two group master data and transaction data model. master data model defined in the scenario includes personnel (e.g., salesman, shop floor operator), equipment, and material (e.g., Lego brick). The production data mainly collected are the information of operators' qualification and availability, bill of materials (BOM), routing, quality test of material, etc. labor management should include providing a status of personnel, certification tracking, attendance reporting, etc, as well as order processing and product definition. The transaction data associated with production processes are collected, e.g., state of equipment, job order information.

4.2 Data Model Mapping

To integrate with the Odoo ERP, AAU-MES and Thingworx platforms, the data model mapping interface is provided by ISA-95 middle data layer. Figures 5 and 6

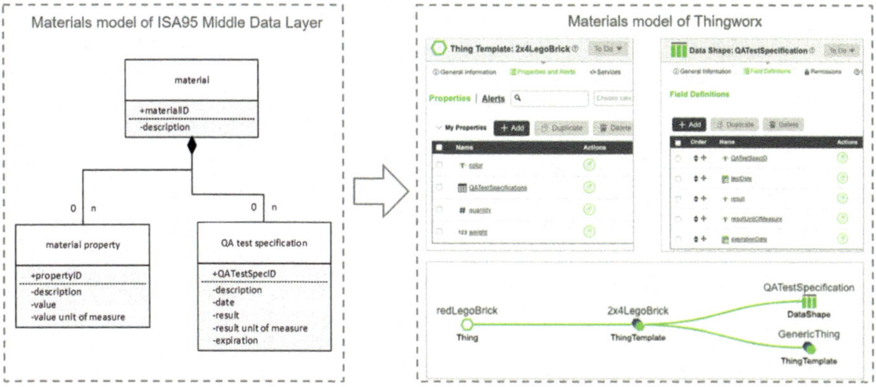

Fig. 5. Material model mapping

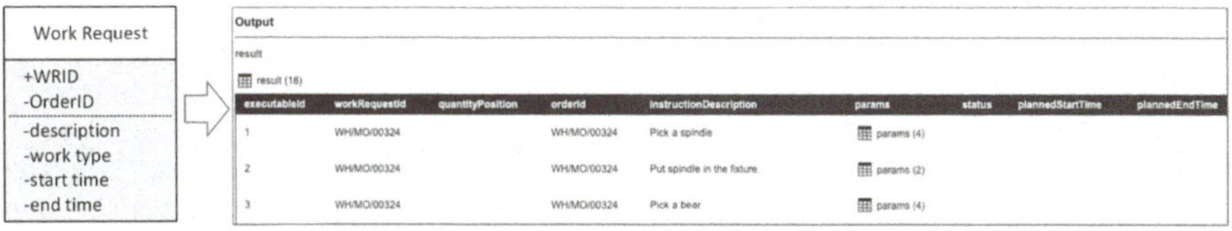

Fig. 6. Work request model mapping

are two examples of master data mapping and transaction data mapping respectively. In Fig. 5, the special property, QA test Specification of material model, is mapped to QATestSpecification as a data shape in Thingworx. The Thing in Thingworx is the actual instance of "Materials Class". Figure 6 describes how to map a work request into a work order in Thingworx.

In such way, the proposed ISA-95 middle data layer helps to smoothly transfer the data among systems without interfering the system execution. By using the proposed ISA-95 middle data layer, the data exchange among different systems can be easily achieved in a fast and smooth way. Therefore, it helps to enhance the system interoperability for automation of manufacturing operations.

5 Conclusions

To improve the degree of factory automation, we designed ISA-95 middle layer to reduce the cost and errors associated with implementing business and manufacturing operations systems. It helps to enhance the interoperability and systems integration. We provided a guideline to create a data architecture for building a flexible and scalable smart factory. By having an ISA-95 based middle data layer, the manufacturing companies do not need to struggle to reorganize the current data structure. It helps to standardize and formalize the data structure and smoothly integrates with the new standardized systems. Our approach suffers from some limitations, for example, extra efforts are needed for developing the data mapping interfaces. We aim to address this limitation in future work, as well as maintain high system performance after applying our middle layer.

Acknowledgments. This works is partially funded by the Manufacturing Academy of Denmark (MADE).

References

1. Zuehlke, D.: Smart Factory-Towards a factory-of-things. Annual Reviews in Control **34**(1), 129–138 (2010). https://doi.org/10.1016/j.arcontrol.2010.02.008
2. Kagermann, H., Wahlster, W., Helbig, J.: Recommendations for implementing the strategic initiative INDUSTRIE 4.0: Final report of the Industrie 4.0 Working Group. Acatech, München (2013)
3. Hermann, M., Pentek, T., Otto, B.: Design principles for industrie 4.0 scenarios. In: Proceedings of the Annual Hawaii International Conference on System Sciences, pp. 3928-3937. IEEE Press (2016). https://doi.org/10.1109/HICSS.2016.488
4. Wegner, P.: Interoperability. ACM Comput. Surv. (CSUR). **28**(1), 285–287, ACM, New York (1996)
5. Almada-Lobo, F.: The Industry 4.0 revolution and the future of Manufacturing Execution Systems (MES). J. Innov. Manag. **3**(4), 17 (2016)
6. Scholten, B.: The road to integration: A guide to applying the ISA-95 standard in manufacturing. ISA (2007)
7. He, D.Z., Lobov, A., Lastra, JL. M.: ISA-95 tool for enterprise modeling. In: Proceeding of the Seventh International Conference on Systems, pp. 83–87, IARIA (2012)
8. Emerson, D., Kawamura, H., Matthews, W.: Plant-to-Business (P2B) interoperability using the ISA-95 standard. Yokogawa Technical Report **43**, 17–20 (2007)
9. Govindaraju, R., Lukman, K., Chandra, D.R.: Manufacturing execution system design using ISA-95. Adv. Mater. Res. **980**, 248–252 (2014)
10. ISA95, Enterprise-Control System Integration. https://www.isa.org/isa95/ (2020)
11. Wang, S., Wan, J., Zhang, D., Li, D., Zhang, C.: Towards smart factory for industry 4.0: a self-organized multi-agent system with big data based feedback and coordination. Comput. Net. **101**, 158–168, Elsevier (2016)
12. Ray, S.R., Jones, AT.: Manufacturing interoperability. J. Intel. Manufact. **17**(6), 681–688, Springer (2006)
13. Panetto, H., Molina, A.: Enterprise integration and interoperability in manufacturing systems: Trends and issues. Comput. Indus. **59**(7), 641–646, Elsevier (2008)
14. Madsen, O., Møller, C.: The AAU Smart Production Laboratory for Teaching and Research in Emerging Digital Manufacturing Technologies. Procedia Manufact. **9**, 106–112, Elsevier (2017). https://doi.org/10.1016/j.promfg.2017.04.036
15. Ganesh, A., Shanil, K., Sunitha, C., Midhundas, A.: Openerp/odoo-an open source concept to erp solution, In: Proceedings of 6th International Conference on Advanced Computing (IACC), pp. 112–116, IEEE (2016). https://doi.org/10.1109/IACC.2016.30
16. Li, C., Mantravadi, S., Møller, C.: AAU Open Source MES Architecture for Smart Factories – Exploiting ISA 95, In: Proceedings of 18th International Conference on Industrial Informatics, accepted, IEEE (2020)

Deep Reinforcement Learning for IoT Interoperability

Sebastian Klöser[1,3(✉)], Sebastian Kotstein[2], Robin Reuben[1,3], Timo Zerrer[1], and Christian Decker[2]

[1] DXC Technology, 71034 Böblingen, Germany
robin.reuben@dxc.com, timo.zerrer@dxc.com
[2] Reutlingen University, 71034 Böblingen, Germany
sebastian.kotstein@reutlingen-university.de,
christian.decker@reutlingen-university.de
[3] haven association, Hamburg, Germany
sebastian.kloeser@dxc.com
https://www.dxc.technology

Abstract. The Internet of Things (IoT) is coined by many different standards, protocols, and data formats that are often not compatible to each other. Thus, the integration of different heterogeneous (IoT) components into a uniform IoT setup can be a time-consuming manual task. This lacking interoperability between IoT components has been addressed with different approaches in the past. However, only very few of these approaches rely on Machine Learning techniques. In this work, we present a new way towards IoT interoperability based on Deep Reinforcement Learning (DRL). In detail, we demonstrate that DRL algorithms, which use network architectures inspired by Natural Language Processing (NLP), can be applied to learn to control an environment by merely taking raw JSON or XML structures, which reflect the current state of the environment, as input. Applied to IoT setups, where the current state of a component is often reflected by features embedded into JSON or XML structures and exchanged via messages, our NLP DRL approach eliminates the need for feature engineering and manually written code for pre-processing of data, feature extraction, and decision making.

1 Introduction

Interoperability is the ability of two or more IT components to seamlessly communicate and interact, meaning they can *"'talk to and understand' each other"* [16]. The success of the Internet as an open foundation for applications and services is built upon a set of well-established open standards, protocols, and

P. Weißgraeber et al. (Eds.): *Advances in Automotive Production Technology – Theory and Application,* ARENA2036, pp. 195–204, 2021. https://doi.org/10.1007/978-3-662-62962-8_23

data formats granting interoperability for a large number of heterogeneous participants [5]. However, the recent developments around the Internet of Things (IoT) challenge this given interoperability: On the one hand, the IoT is expanding the Internet into new application fields and domains. On the other hand, both research and industry are adding new standards, protocols, and data formats to the Internet technology stack in order to meet the requirements of novel IoT applications. These new protocols, standards, and formats are often developed under use case specific criteria, resulting in a strong fragmentation of technologies into mutually incompatible groups [10]. For a developer, achieving interoperability and integrating a heterogeneous set of IoT components can be a time consuming-task marked by a high degree of manual effort.

Several efforts from research and industry have tried to address specific aspects of this interoperability challenge in IoT. Interestingly, only very few approaches adopt Machine Learning techniques (except of in the field of web service composition, e.g. in [18]). In particular, the use of Deep Reinforcement Learning (DRL) as a possible technique for autonomously learning and adapting complex communication interfaces as they are used in the field of IoT has not been examined so far.

In this work, we demonstrate that DRL algorithms, which leverage network architectures from Natural Language Processing (NLP DRL Systems from now on), can be used to autonomously learn complex behavior from data structures like JSON or XML, which are typically used for a variety of communication protocols in IoT as data representation formats on the application layer. Thus, our work addresses a specific aspect of interoperability, namely the automatic processing of input data to make autonomous decisions. In detail, with our approach we eliminate the need for feature engineering and manually written code for individual pre-processing of data, feature extraction, and decision making, and, therefore, we free a developer from these time-consuming tasks.

The full code used in this paper is publicly available on GitHub: https://github.com/SKloeser/DRL4IOT.

2 Background

DRL is an active field of research within the domain of Machine Learning that allows one to design systems that autonomously develop complex behavioral strategies without the need to supply explicit rules or expert data. Instead, DRL formalizes the idea of learning from experience gained in a stepwise interaction with the environment. The interaction is realized by an observation, action, reward sequence in which the agent receives and processes observations from the environment and decides on a concrete action. The action triggers a state transition in the environment that leads to a new observation together with an additional feedback signal (the reward) that represents the quality of the last action. With the help of DRL, some remarkable success has been achieved in a variety of domains, such as board games [14], computer games [11] or in industry robotics [7].

Most DRL environments are designed to require complex behavior, but to offer relatively simple control interfaces. In contrast, IoT components and systems often possess a complex business logic and additionally a complex control interface, which typically implements many nested communication protocols and data formats. For example, one such component could be a REST service that offers its functionality over a TCP/IP and HTTP-based communication interface with JSON or XML as a data exchange format. Integrating this service manually into an IoT setup requires developers to write a client adapter that is capable of creating and consuming HTTP requests and responses, respectively, as well as an integrated encoder and decoder for processing data in JSON or XML. Besides this, developers need to implement the client's logic for extracting the right features from a response payload, such that these features can be processed and interpreted to enable further decision making.

With our work, we investigate the automation of these time-consuming manual engineering tasks. Concretely, we present a framework that is capable of learning control strategies from observations that are provided as XML or JSON structures in an end-to-end approach. In this context, end-to-end refers to the property of the approach to learn two capabilities at once: The representation of the information in the data inputs (treated as strings) and the decision making from the input representation.

2.1 Related Work and own contribution

While end-to-end learning from visual observations is well established for DRL, textual inputs are much less investigated (see [9] for an overview). Most of the work in this domain focuses on text-based games (TBG) as benchmark environments [1,4] or textual dialog systems [8,13]. To the best of our knowledge, there is no research that applies techniques from NLP and DRL for adapting any kind of network-based communication interface.

During the last stages of our research, we came to know about the work of Woof and Chen who provide a framework for end-to-end learning on tree-structured data [19]. Their work possesses some parallels with ours, however differs on critical aspects. They mainly focus on supervised learning and use specifically designed network architectures tailored towards tree structured data formats. Our approach, however, relies on DRL, is more generic, and does not require any hierarchical order in the treated data formats.

3 Experimental Setup

Our main interest is to train NLP DRL Systems on observations provided as JSON or XML structures without applying any manually written code for individual pre-processing or adaptation of structured data, feature extraction, and decision making. In particular, we want to show that, by treating JSON or XML structures as strings, the policy architecture described in Sect. 4 is capable of learning suitable controls in an end-to-end fashion. To this end, we choose the standard Cartpole and Acrobot DRL environments provided by OpenAI Gym as a basis for our investigation [3] and equip them with wrappers that encode the original environments' outputs to JSON and XML.

In the original Cartpole environment, the agent needs to learn to balance a pole on a cart by moving the cart either to the left or the right. The Acrobot environment consists of a swinging rod system built from two links and one connecting joint. The goal is to swing it over a certain height threshold by applying a left, a right or no push to the joint. Both environments provide continuous numerical observations as shown in Table 1 and possess discrete action spaces. While Cartpole provides a dense reward signal with a maximum episode return of 200[1], Acrobot provides only a sparse reward once the final goal is reached. Since policy gradient algorithms struggle to learn from sparse reward signals regardless of the employed policy network, we add an additional term to the reward function that is associated with the current height of the system's lowest point. In this way, we obtain dense reward signals. To reflect the fact that this changes the structure of the environment, we name it *adjusted Acrobot* to avoid confusion.

The JSON and XML wrappers produce outputs of different complexity for the two environments, as shown in Table 1. It is especially noteworthy that with this approach, the numbers in the strings are treated as words and not as digits and we do not handle them any differently than other tokens in the string (see Sect. 4 for details). Since we use GloVe embeddings to map the string tokens to numerical vectors in the downstream (see below), our approach requires an additional adjustment: The original environments use float arrays as outputs. However, GloVe only contains integers and therefore we multiply the observations by 100 and cast them to integers before encoding them into JSON and XML.

The architecture of the wrappers is shown on the left of Fig. 1. In this work, we train both of the described extended environments with the policy network and the DRL algorithm detailed in Sect.4. As a benchmark, we additionally train a standard feed forward policy on the original environments with the same DRL algorithm and the same hyperparameters.

In addition to the sole learning capabilities we also investigate the transfer capabilities associated with the use of an NLP DRL System. To this end, we train a model on the JSON Cartpole environment, fix it and evaluate it without any further training on an observation structure that is incrementally changed towards an XML structure until we observe a drop in performance.

4 DRL Policy and Learning Algorithm

The NLP DRL System is inspired by the work around TBGs mentioned in Sect. 2.1. The right part of Fig. 1 shows the adopted architecture, capable of learning from textual observations in an end-to-end fashion.

[1] To make our results more reliable we increase the maximum number of steps within one episode such that, in our case, the maximum return is 500.

Table 1. Specifics of the wrapped environments and how they relate to the original Gym environments.

Cartpole

Original Output	Type	Example Output String	# Tokens
Cart Position Cart Velocity Pole Angle Pole Velocity	JSON	{"position": 2.0, "velocity": 1.0, "direction": "left"} {"angle": -3.0, "speed": -5.0}	34
	XML	<cart><position>-2.0</position><velocity>4.0</velocity> <direction>left</direction></cart><pole><angle>- 2.0</angle> <speed>0.0</speed></pole>	54

Adjusted Acrobot

Original Output	Type	Example Output String	# Tokens
Cos Angle Joint 1 Sin Angle Joint 1	JSON	{"joints":[{"id": 0, "torque": "idle", "cosine": 99.0, "sine": -1.0, "angular velocity": -3.0}, {"id": 1, "torque": "idle", "cosine": 99.0, "sine": -7.0, "angular velocity": -1.0}]}	77
Cos Angle Joint 2 Sin Angle Joint 2 Angular Velocity 1 Angular Velocity 2	XML	<joints><joint><id>0</id><torque>idle</torque> <cosine>99.0</cosine><sine>5.0</sine> <angular velocity>8.0</angular velocity></joint> <joint><id>1</id><torque>idle</torque> <cosine>99.0</cosine><sine>0.0</sine> <angular velocity>-2.0</angular velocity></joint></joints>	105

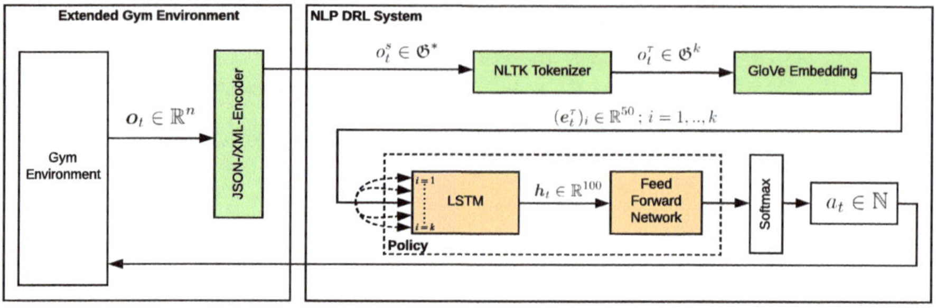

Fig. 1. Architecture of the used setup for the environment and the NLP DRL System. Here, $\mathfrak{G}$ refers to the set of all tokens available in the GloVe file that is used. The colored parts represent important constituents and the yellow parts are trainable

It consists of a static pre-processing unit built from a tokenizer, followed by a pre-trained embedding model. We use the tokenizer provided by NLTK [2] and a 50-dimensional GloVe representation[2] covering the tokens [12]. Our choice is based on the awareness that GloVe already provides good numeracy capabilities [17]. This is essential since we treat the numbers in the observation sequence as strings and not as digits.

The output of the embedding model is an array with a shape of $(seqlength, 50)$, which is then fed into the policy. It consists of an LSTM encoder [6] that processes the array sequentially to update its hidden state without producing any output. The hidden state is then used as an input to a standard feed forward network that maps to the discrete actions via a softmax function. With this approach, we only treat the observations as strings, but not the actions.

The policy is trained in an end-to-end fashion by applying a simple REINFORCE algorithm with a baseline [15], implemented in Tensorflow. As baseline we choose the median return of the current run. Given trajectories τ_i as observations and action sequences obtained from running the current policy $\pi_\theta(\mathbf{a}_t|\mathbf{o}_t)$, the policy gradient is estimated with the following loss function:

$$L \approx \frac{1}{N}\sum_{i}^{N}\left(\sum_{t=1}^{T}\log\pi_\theta\left(\mathbf{a}_t^i|\mathbf{o}_t^i\right)\left(\hat{Q}_{i,t}-b_i\right)\right) \quad \text{with} \quad \hat{Q}_{i,t}=\sum_{t'=t}^{T} r_{t'} \quad \text{and} \quad b_i = \text{median}_t\left(\hat{Q}_{i,t}\right)$$

as the reward to go and baseline.

All experiments are performed on a developer notebook (16 GB RAM, Intel Core i7-8850H processor, no GPU acceleration). For the hyperparameter configuration that is used in training, we refer to the code available on GitHub.

5 Results

Figure 2 shows the reward progress over policy updates for the three evaluated cases on both environments. In all cases, the NLP DRL System is able to learn

[2] 50-dimensional GloVe.6b

successful policies, as comparison with the benchmark shows. In the Cartpole scenario the systems learning progress is comparable for both JSON and XML with a slightly better performance for the XML structure. This is remarkable given the fact that the XML structure possesses many more tokens than the JSON structure. Equally notable is the fact, that both cases are learnt faster in terms of policy updates than the benchmark model. For the adjusted Acrobot environment the situation is opposite, here the benchmark model performs best while JSON is learnt faster than the XML structure. We interpret these results as follows: For the adjusted Acrobot environment the number of required policy updates correlates with the complexity of the observations. The benchmark model only possesses a four dimensional vector as observations while the JSON environment exports 77×50-shaped tensors, and the XML environment even has 105×50 big observations. Thus, the results are as expected.

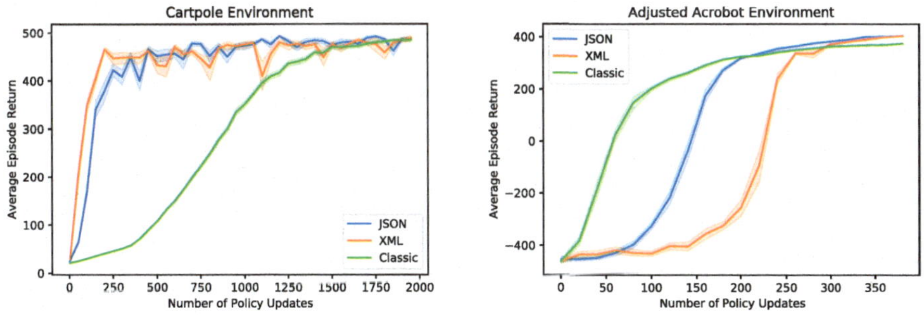

Fig. 2. Reward progress over policy updates for the Cartpole and the adjusted Acrobot environment. The orange line shows the progress for the XML structure, the blue line represents the JSON structure and the green line shows the results for the original environments used as a benchmark

For the Cartpole environment the situation is more interesting. First, it is important to note that the result for the benchmark model is obtained by choosing the same hyperparameters as for the NLP DRL System to ensure comparability. It is to assume that there are more suitable choices for the benchmark model in general, which will result in a better performance than for the JSON or XML environment. For the two NLP DRL System results the findings are opposed to expectation since also for the Cartpole environment XML is more verbose than the JSON structure. However, the difference in performance is not very dominant, and we rather interpret the results from the Cartpole environment as similar performance for both JSON and XML. Following this interpretation, we assume that there is a threshold below which the observation complexity only has subdominant effect on the performance. This assumption needs to be further investigated in future work.

Concerning the transfer capabilities, we find the most XML-like structure the JSON trained model can handle without a loss in obtained rewards to be:

$$< \text{cart:} >< \text{position:} > -2.0 < \text{velocity:} > 1.0 < \text{direction:} > \text{left} < /\text{cart} >$$
$$< \text{pole} >< \text{angle:} > -3.0 < \text{speed:} > -5.0 < /\text{pole} >$$

This structure is already far from an original JSON, and it is noteworthy that we did not apply any special techniques to obtain this robustness against changes in the observation structure. The only parts missing for the full XML structure are the deletion of the ':' that seems to be recognized as an important symbol for key-value paring during the JSON training and the closing </...>-brackets, which seems to confuse the model since we switch from a key-value structure to a key-value-key structure. However, the results look promising and motivate further investigation.

6 Conclusion and Outlook

In this work, we demonstrated that the presented NLP DRL System is capable of learning to control DRL environments from observations provided as JSON and XML structures by treating them as strings. This approach requires no individual code for preprocessing or feature extraction from the data formats as the system autonomously learns to interpret them. This holds especially for numerical information embedded into them. Since JSON and XML are common data structures in the field of IoT, our results indicate that NLP DRL Systems hold the potential to learn to control IoT components with significantly reduced manual effort that is otherwise required for feature engineering and the implementation of code for pre-processing of data, feature extraction, and decision making. The obtained models show robustness against changes in the data structure.

Currently, our results are limited insofar as we investigated cases of reduced complexity, both with respect to the environments and to the data formats. Moreover, our results are obtained by using the task dependent reward functions of the particular Gym environments. For broader applications of our approach, especially in the field of IoT, we need to enable task-independent learning of a communication interface. Therefore, in future work we plan to extend our approach by incorporating task-independent reward functions. Furthermore, we plan to increase the complexity and to further investigate the dependency between complexity and performance as well as techniques to increase the robustness of the learnt models even further.

References

1. Adolphs, L., Hofmann, T.: LeDeepChef: Deep Reinforcement Learning Agent for Families of Text-Based Games. In: 34th AAAI Conference on Artificial Intelligence, pp. 7342–7349. New York, USA (2020)

2. Bird, S., Klein, E., Loper, E.: Natural Language Processing with Python, 1st edn. O'Reilly Media, Inc. (2009)
3. Brockman, G., Cheung, V., Pettersson, L., Schneider, J., Schulman, J., Tang, J., Zaremba, W.: OpenAI Gym. arXiv preprint arXiv:1606.01540 (2016)
4. Côté, M.A., Kádár, Á., Yuan, X., Kybartas, B., Barnes, T., Fine, E., Moore, J., Hausknecht, M., El Asri, L., Adada, M., Tay, W., Trischler, A.: TextWorld: A Learning Environment for Text-Based Games. In: Computer Games, pp. 41–75. Cham, Switzerland (2019)
5. Fältström, P.: Market-driven Challenges to Open Internet Standards. Global Commission on Internet Governance Paper Series, Centre for International Governance Innovation (CIGI) (33) (2016)
6. Hochreiter, S., Schmidhuber, J.: Long Short-term Memory. Neural Computation **9**(8), 1735–80 (1997)
7. Kober, J., Bagnell, J.A., Peters, J.: Reinforcement learning in robotics: A survey. The International Journal of Robotics Research **32**(11), 1238–1274 (2013)
8. Lewis, M., Yarats, D., Dauphin, Y.N., Parikh, D., Batra, D.: Deal or No Deal? End-to-End Learning of Negotiation Dialogues. In: Proceedings of the 2017 Conference on Empirical Methods in Natural Language Processing, pp. 2443–2453. Copenhagen, Denmark (2017)
9. Luketina, J., Nardelli, N., Farquhar, G., Foerster, J., Andreas, J., Grefenstette, E., Whiteson, S., Rocktäschel, T.: A Survey of Reinforcement Learning Informed by Natural Language. In: Proceedings of the 28th International Joint Conference on Artifical Intelligence (IJCAI-19), pp. 6309–6317. Macao, China (2019)
10. Miorandi, D., Sicari, S., Pellegrini, F.D., Chlamtac, I.: Internet of things: Vision, applications and research challenges. Ad Hoc Networks **10**(7), 1497–1516 (2012)
11. Mnih, V., Kavukcuoglu, K., Silver, D., Rusu, A.A., Veness, J., Bellemare, M.G., Graves, A., Riedmiller, M., Fidjeland, A.K., Ostrovski, G., et al.: Human-level control through deep reinforcement learning. Nature **518**, 529–533 (2015)
12. Pennington, J., Socher, R., Manning, C.D.: GloVe: Global Vectors for Word Representation. In: Proceedings of the 2014 Conference on Empirical Methods in Natural Language Processing (EMNLP), pp. 1532–1543. Doha, Qatar (2014)
13. Sankar, C., Ravi, S.: Deep Reinforcement Learning For Modeling Chit-Chat Dialog With Discrete Attributes. In: Proceedings of the 20th Annual SIGdial Meeting on Discourse and Dialogue, pp. 1–10. Stockholm, Sweden (2019)
14. Silver, D., Huang, A., Maddison, C.J., Guez, A., Sifre, L., Van Den Driessche, G., Schrittwieser, J., Antonoglou, I., Panneershelvam, V., Lanctot, M., et al.: Mastering the game of Go with deep neural networks and tree search. Nature **529**(7587), 484 (2016)
15. Sutton, R.S., McAllester, D., Singh, S., Mansour, Y.: Policy Gradient Methods for Reinforcement Learning with Function Approximation. In: Proceedings of the 12th International Conference on Neural Information Processing Systems, p. 1057–1063. Denver, USA (1999)
16. Tsilas, N.L.: Open Innovation and Interoperability. In: DeNardis, L. (ed.) Opening standards: The global politics of interoperability, pp. 97–117. MIT Press, Cambridge, MA, USA (2011)
17. Wallace, E., Wang, Y., Li, S., Singh, S., Gardner, M.: Do NLP Models Know Numbers? Probing Numeracy in Embeddings. In: 2019 Conference on Empirical Methods in Natural Language Processing and 9th International Joint Conference on Natural Language Processing, pp. 5307–5315. Hong Kong, China (2019)

18. Wang, H., Zhou, X., Zhou, X., Liu, W., Li, W.: Adaptive and Dynamic Service Composition Using Q-Learning. In: Proceedings of the 22nd IEEE International Conference on Tools with Artificial Intelligence, IEEE, pp. 145–152. Arras, France (2010)
19. Woof, W., Chen, K.: A Framework for End-to-End Learning on Semantic Tree-Structured Data. arXiv e-prints arXiv:2002.05707 (2020)

Wireless Industrial Networks for Real-Time Applications

Jorge Luis Juárez Peña[1(✉)], Stefan Lipp[1], Andreas Frotzscher[2], and Frank Burkhardt[1]

[1] Fraunhofer IIS, Am Wolfsmantel, 33, 91058 Erlangen, Germany
{jorge.juarez,stefan.lipp,frank.burkhardt}@iis.fraunhofer.de
[2] Fraunhofer IIS, Zeunerstraße, 38, 01069 Dresden, Germany
andreas.frotzscher@eas.iis.fraunhofer.de

Abstract. One of the biggest challenges currently facing the industry is to make systems more flexible. Technological advances are enabling new application scenarios such as safe and efficient human-robot collaboration, which is based on the real-time availability of required data at the location where it is needed at that point in time. The crucial building block for this modernization of factories is the communication network between the automation components as it is the one enabler for constant and reliable exchange of information. In the future, an essential and ever increasing part of such a communication network will be wireless communication. It plays a major role in expanding the mobility and agility of sensors and actuators and in the retrofitting of rigid legacy structures. For smart production in particular, wireless technologies that are reliable and support deterministic cycle times in the range of less than 1 ms are required, which is barely addressed by current wireless developments based on IEEE 802.11 and 5G URLLC. This paper presents a novel real-time radio technology – Ultra Reliable Wireless Industrial Network (UWIN), developed by Fraunhofer IIS, which is currently in a preliminary pre-product development stage. In addition to explaining the UWIN concept, this paper provides a short overview on state-of-the-art real-time radio technologies.

1 Introduction and Motivation

In industrial automation systems, it is common for some of the sensors, actuators, and other system components to be installed on movable subsystems. Examples include tools mounted on robot arms, slide rails, or rotary and coordinate tables. These subsystems require a means of communicating with the central control system capable of operating at very short cycle times in the single-digit millisecond range. Nowadays, these connections are typically realized using cable drag chains, rotary feedthroughs or slide contacts. However, these inherently contain the disadvantages of heavy weight, susceptibility to wear and tear and the inflexibility with regard to changing the setup.

Table 1. Summarizes the requirements of typical applications in the context of motion- and closed loop control in factories.

P. Weißgraeber et al. (Eds.): *Advances in Automotive Production Technology – Theory and Application,* ARENA2036, pp. 205–212, 2021. https://doi.org/10.1007/978-3-662-62962-8_24

Table 1. Requirements of industrial applications.

	Motion control [1]	Machine tools [2]	ZDKI requirement profiles		IoT-COMMs requirement profile [4]
			HiFlecs Profile B [3]	HiFlecs Profile C [3]	
Isochronous cycle time	250 μs–1 ms	<500 μs	1 ms	0,5 ms	500 μs
Packet error rate	$<10^{-9}$	$<10^{-9}$	–	max. 1 Packet loss/day & link	<2 Packets lost in sequence
Telegram size	20–50 Byte	50 Byte	<50 Byte	17–37 Byte	<200 Byte
Motion speed	<10 m/s	–	<6 m/s	<10 m/s	<9000 rpm
Range	<50 m	–	<500 m	<250 m	<5 m

Curently, to connect these moving components, mostly wired real-time bus systems (e.g. Profinet RT/IRT, EtherCAT, Sercos III) are used in conjunction with cable drag chains, rotary joints and sliding contacts. However, the cabling limits the possible movements of the system's parts, the number of connectable automation components, and results in restricting later conversions or extensions.

2 Wireless Transmission Systems

Wireless transmission systems can achieve significant improvements for those mobile cases. In the following, some of these technologies are briefly described and a short summary of some current developments in real-time radio technologies is provided. This summary will take into account current 3GPP standards for 5G URLLC, ETSI (DECT Working Group URLLC) and IEEE (801.11be Extremly High Throughput). This three groups correspond ongoing activities in the field of mission-critical device control, the so-called ultra-reliable and low latency communication (URLLC), for applications in the context of motion- and closed loop control.

In the standardization organization 3GPP, the target latencies for 5G URLLC were described as 1 ms in the Work Item Description for URLLC on the physical layer [5]. However, this value is not normative. 3GPP specifies the standardized QoS classes of 5G networks in [6], which include the end-to-end one-way latency from the user equipment to the 5G core network. These standardized QoS classes in the specification [6] do not yet reflect the target value mentioned above. Available so far are– among others – the QoS classes with a latency of 5 ms and a packet error rate of $10 - 5$ (5QI value 85) and another QoS class with a latency of 10 ms and the corresponding packet error rate of $10 - 6$ (5QI value 80). Nevertheless, the actual achievable latency strongly depends on the network architecture (e.g. local campus network, nationwide network) and the selected 5G transmission parameters (FDD, TDD, frame configuration in TDD, subcccarrier spacing, HARQ etc.). The actually achievable latency on the radio interface (which is comparable with the UWIN technology as introduced in this paper) is significantly smaller than the end-to-end latency of the 5G system. The self-evaluation report [7] gives achievable latencies of down to 300 μs in the TDD configuration with frame structure DUDU and 60 kHz subcarrier spacing.

ETSI (DECT-2020 NR Working Group URLLC) also targets a transmission latency of less than 1 ms and packet error rates of less than $10 - 5$ [8]. This latency and the packet error rate are based on the URLLC requirements of ETSI and therefore their focus is not purely motivated by industrial applications but also based on requirements from within the media industry (e.g. professional audio streaming). The strengths of DECT-2020 NR, like 5G's, lie in the usage of reserved frequency bands as well as in the deterministic Time Division Multiple Access (TDMA) and Frequency Division Multiple Access (FDMA) in a Time Division Duplex (TDD) communication manner, seamless handover, and mobility support. The DECT-2020 NR Physical Layer uses Orthogonal Frequency Division Multiplexing (OFDM) with a cyclic prefix (CP) and modulation schemes of up to 1024-QAM. Both frame duration (10 ms) and slot duration (0.41667 ms) ensure coexistence with legacy DECT systems.

In IEEE, a task group for the upcoming standard 802.11be for "Extremly High Throughput" is currently defining key performance indicators and features. It targets at 30 Gbps peak transfer data rate and very low transmission latency. However, a fixed limit of the transmission latency has not been defined yet. 802.11be addresses various use cases, ranging from real-time gaming, cloud gaming, real-time video transmission and robotics, and industrial automation. Taking into account the addressed use cases, a target value for the transmission latency of <5 ms as well as a delivery ratio of <10–4 can be expected [9]. In order to achieve these target values, various technical features have been proposed, e.g. priority tagging, time-aware shaping, parallel transmission on multiple links, admission control between neighboring APs and channel reservations.

However, none of the radio technologies described above meets the requirements listed in Table 1.

3 Ultra-reliable Wireless Industrial Network

In various projects research is being conducted on radio technologies to fulfill the requirement profiles shown in Table 1. Ultra-Reliable Wireless Industrial Network (UWIN) as result of some of these projects (cf. Sect. 6) is a new radio technology that is designed to meet the closed-loop requirements profile of Table 1. The technology can be used as a wireless extension or even as a replacement for wired fieldbuses or real-time ethernet.

The system architecture of UWIN consists of a base station (base) and so-called field devices (devices), which form a star topology with the base at the center. The communication of the devices with the base is scheduled by the controller within the base. Additionally, the time synchronization of the devices is also controlled by the base in accordance with the requirements of the Precision Time Protocol [10]. Figure 1 shows the network topology together with the internal structure of the field devices. The communication stack is shown schematically, referencing the layers of the OSI (Open Systems Interconnection) model. In the figure, the lowest layers (PHY, MAC and network layer) are shown. In addition, the base station contains a field bus gateway between its controller and the infrastructure, while the devices have interfaces to connect sensors and actuators. Coexistence management is distributed over the base station and the devices as they all contain modules for coexistence analysis. Additionally, the base contains functions for cooperative self-organizing network management.

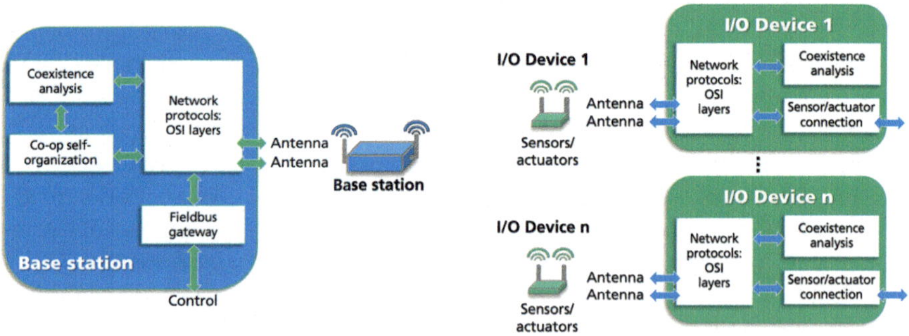

Fig. 1. Network topology and architecture of UWIN base station and I/O devices.

The coexistence analysis is used as a novel solution for some of the numerous challenges presented by industrial automation. For example, there are severe radio-unfriendly environmental conditions. These are characterized by strong and unpredictable multipath reception, interference with other radio systems, and interference from other electrical systems. This basically requires a robust waveform on the physical layer, with which the extreme and conflicting requirements of reliability and low latency can be reconciled within the specifications of sensor-actuator networks. The use of multiple degrees of freedom in diversity (space, time and frequency) enhances this robustness and the reliability of the system. On top of this the analysis of the actual interference/coexistence situation offers a new degree of freedom. The system becomes able to decide further actions based on the type and behavior of a coexisting device. This behavior can be a simple power adaption but an active, synchronized switching of carrier frequencies within the system becomes possible as well, before a communication link would result in a transmission error.

With such robust communication links, it becomes possible to uphold and guarantee the core requirement of control applications in the factory: the cycle time. This time is inherent to the currently used field buses and represents the time frame in which an update of a dataset is expected to be received, either by a controller (e.g. PLC) or an actuator. If that timing is not fulfilled, the application will stop after a certain number of cycles if no data has been received. This is accompanied by further strong requirements regarding the temporal synchronicity of the whole system.

In essence, the radio transmission represents an additional latency for data transmission. If the latency of the radio system is much smaller than the cycle time of the control system, the system is able to operate correctly. If either the transmissions and/or the internal calculation takes too long the cycle time will not be met and the system's performance degrades which may lead to a complete halt of the system.

3.1 Physical Layer

MIMO-OFDM modulation is used as the basis for the UWIN radio signal waveform. The use of MIMO (Multiple Input Multiple Output) serves to increase transmission reliability by exploiting spatial diversity. The OFDM (Orthogonal Frequency Division Multiple Access) signal is based on 802.11a [9], which was specially adapted for the 5 GHz U-NII (Unlicensed national information infrastructure) band. The bandwidth of 20 MHz per channel, together with a short symbol duration, are the basic building blocks to achieve very short packet lengths of 32 bytes, which can be decoded individually. OFDM modulation is especially suited to cope with the massive multipath reception occurring in industrial environments by utilizing suitable equalization algorithms and fitting guard times. Table 2. Summarizes the parameters of the OFDM waveform and the channel coding of UWIN.

Table 2. OFDM waveform parameters of UWIN.

Parameter	Variable	Value	Parameter	Variable	Value
Carrier frequency	f_{RF}	5.150–5.725 GHz	Occupied bandwidth	BW	16.5625 MHz
Sample period	T_S	50 ns	Subcarrier distance	Δf	312.5 kHz
FFT length	N	64	# of data subcarrier	K	38
OFDM symbol duration	T_{OFDM}	3.2 μs	# of pilot subcarrier	K_P	14
CP length	N_{CP}	0.8 μs	Modulation		Up to 64-QAM
# OFDM symbols per user	N_{OFDM}	2	Channel code		Conv. Code (base R = 1/2)
Burst duration	T_{Burst}	8 μs			

3.2 Media Access and Coexistence Management

The UWIN radio system implements a logical star topology. To achieve the simultaneously required reliability, the radio system uses two radio channels simultaneously. To support the required isochronous real-time capability, the master organizes the media access of the radio nodes according to a time-slot procedure. It is based on a TDMA frame structure (Superframe), which also realizes a temporal separation of uplink and downlink (Time Division Duplex). The various transmissions between the base station and the field devices are assigned dedicated time slots on the two radio channels, which are dynamically adapted at runtime according to the current channel characteristics and environmental conditions.

The dynamic scheduling of media access requires information about the status of the radio connections in the network and the radio environment. The coexistence management has the task of providing this information. To monitor the status of the various radio connections, the master node collects statistical parameters to assess the transmission quality of its received signal (including bit error and packet error rates, reception power). In addition, the use of the radio frequency bands, and adjacent frequency bands is monitored. For this purpose, the received signal is spectrally evaluated, radio packets are detected in the frequency spectrum and the respective radio technology is identified. Statistical parameters are derived from this information, including packet frequencies, reception power distribution and patterns in the packet sequences. The network management uses this information to assess the current network quality and, if necessary, triggers the change of a channel in the radio system. Since UWIN uses two channels simultaneously, the channel switching can be handled completely seamless.

4 UWIN Implementation and Prototypes

To demonstrate the performance, the radio system is implemented in hardware. Therefor was build a demonstrator based on a commercial PLC which is integrated with UWIN. Besides, UWIN was also implemented in an industrial environment within the scope of the research project IoT-COMMs [4]. Here, UWIN was integrated on a circular PCB (see Fig. 2) to establish a real-time connectivity between a drill head and its control unit. Currently, work is being done on the implementation of further features to achieve an even more reliable performance, objective is Packet error rate $<10^{-8}$.

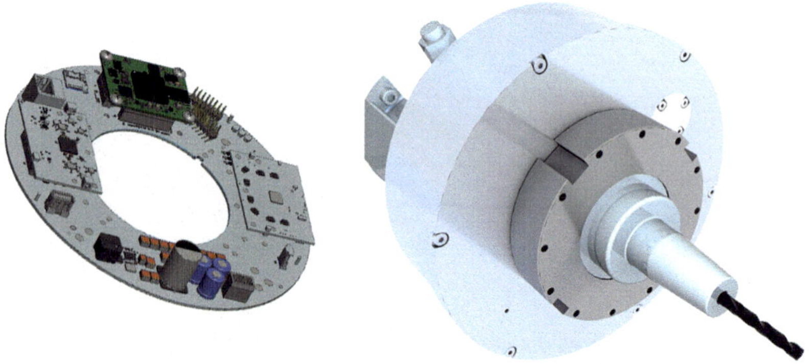

Fig. 2. Integration UWIN industrial prototype for IoT-COMMs Project in circular PCB.

The core parameters of UWIN are summarized in Table 3.

Table 3. Core Parameters of UWIN.

Parameter	Value	Comment
Latency	<100 μs	
Cycle time	125 μs–10 ms	Adaptable to requirements of the application
Number of devices per system	1–100	Adjustable according to latency and data rate requirements;
Range	Up to 20 m	Orientation values: 6 radio nodes (@ 125 μs), 100 radio nodes (@ 2ms)
Reliability	Packet error rate $<10^{-6}$	Depending on transmit power and frequency band
Data rate per device	≤2 Mbps	Depending on the number of deployed nodes and latency requirements
Payload per device	≤32 bytes/cycle	Maximum data rate per slot and frame
Bandwidth per cell	1 x 20 MHz or 2 x 20 MHz	Higher payload through slot aggregation
Frequency band	2.4 or 5 GHz + others	Depending on latency and reliability requirements and on the number of deployed nodes
Deployability	EU, USA, Asia	Use of license-free resp. Available frequency bands
Plug and manufacture	Self-organization according to application or user requirements, RF environment (coexistence management)	

5 Summary and Outlook

Industrial applications such as motion control and closed loop control have stringent requirements which cannot be met with wireless solutions such as the upcoming standards IEEE 802.11be and 5G URLLC. In this paper, a novel radio technology – UWIN – was presented, which is particularly suitable as a wireless extension of real-time field bus systems. Compared to the scope of other current developments, the key performance indicators of UWIN are significantly closer to the requirements of the production industry. Currently, UWIN radio technology is being verified in different application scenarios of industrial automation within the projects IoT COMMs [4] and IC4F [11]. In parallel, an evaluation-kit for UWIN is being implemented as well and will be offered to interested partners by the end of 2020. The scope of this evaluation kit is to provide the means to easily verify the KPIs of UWIN on-premise. Furthermore, it is indented to build a consortium with interested partners to standardize UWIN.

6 Acknowledgements

UWIN technology has evolved from various research projects. The projects fast real-time [12] and fast automation [13] have been supported by the German Federal Ministry of Education and Research in the context of the program "Region Zwanzig20" under reference numbers 03ZZ051 OA und 03ZZ0504D. Project IC4F [11] is supported by the German Federal Ministry for Economic Affairs and Energy in the technology program PAiCE (Platforms/Additive Manufacturing/Imaging/Communication/Engineering) under reference number 01MA17007A.

References

1. Aktas, I., Bentkus, A., Bonanati, F., Dekorsy, A., Dombrowski, C., Doubrava, M., Golestani, A., Hofmann, F., Heidrich, M., Hiensch, S., Kays, R., Meyer, M., Müller, A., ten Brink, S., Petreska, N., Popovic, M., Rauchhaupt, L., Saad, A., Schotten, H., Wöste, C., Wolff, I.: Funktechnologien für Industrie 4.0. VDE ITG, Frankfurt a. M. (2017)
2. Frotzscher, A., Wetzker, U., Bauer, M., Rentschler, M., Beyer, M., Elspass, S., Klessig, H.: Requirements and current solutions of wireless communication in industrial automation, IEEE International Conference on Communications Workshops (ICC), Sydney (2014)
3. Bittner, F.: Funksystem für industrielle Anwendungen – HiFlecs Homepage, https://www.vde-itg-kommunikationsnetze.de/images/stories/Fachtagungen/zdn2017/vortraege/ZDN2017_HiFlecs_Bittner.pdf. Accessed 30 Sept 2020
4. IoT-COMMs Homepage: https://www.cit.fraunhofer.de/de/zentren/fiot.html. Accessed 30 Sept 2020
5. Huawei, H.: Contribution title. In: RP-191584, 3GPP Work Item Description, Physical layer enhancements for NR ultra-reliable and low latency communication (URLLC) (2019)
6. 3GPP: TS 23.501 v16.5.0 System architecture for the 5G System (5GS) (2020)
7. 3GPP: TR 37.910 V16.1.0 Study on self evaluation towards IMT-2020 submission (2019)
8. ETSI: Digital Enhanced Cordless Telecommunications (DECT); Study on URLLC use cases of vertical industries for DECT evolution and DECT-2020, ETSI TR 103 515 V1.1.1 (2018)
9. The Institute of Electrical and Electronics Engineers (eds.): IEEE standard for information technology – Telecommunications and information exchange between systems Local and metropolitan area networks – Specific requirements – Part 11: Wireless LAN Medium Access Control (MAC) and Physical Layer (PHY) Specifications, IEEE Std 802.11 (2016)
10. The Institute of Electrical and Electronics Engineers (eds.): IEEE standard for a precision clock synchronization protocol for networked measurement and control systems, IEEE Std. 1588–2002 (2002)
11. G. F. M. o. E. A. a. Energy, PAiCE – Brief summary of the funding projects Homepage https://www.bmwi.de/Redaktion/EN/Publikationen/Digitale-Welt/paice-brief-summary-of-the-funding-projects.html. Accessed 30 Sept 2020
12. Fast realtime Homepage: https://de.fast-zwanzig20.de/basisvorhaben/fast-realtime/. Accessed 30 Sept 2020
13. Fast automation Homepage: https://de.fast-zwanzig20.de/industrie/fast-automation/. Accessed 30 Sept 2020

A Novel 'Automated Hardware Upgrade Service' for Manufacturing Systems

Christian Schneider[1,2(✉)], Martin Reisinger[1,2], Thomas Adolf[2], Nicolas Heßberger[1,2], and Alexander Sauer[1,2]

[1] Institute for Energy Efficiency in Production EEP, Nobelstr., 12, 70569 Stuttgart, Germany
christian.schneider@ipa.fraunhofer.de

[2] Fraunhofer Institute for Manufacturing Engineering and Automation IPA, Nobelstr., 12, 70569 Stuttgart, Germany

Abstract. Industrial energy efficiency is widely acknowledged as an important and effective way to mitigate industrial greenhouse gas emissions. However, so far, implementation rates of energy efficiency measures are disappointing. Industry 4.0 has been described as a way to increase energy efficiency and, therefore, barriers to energy efficiency. Here, we share the concept of a digital service providing assistance for investments in the replacement of energy consuming physical assets (Hardware) by continuously calculation the life cycle costs of the underlying assets and more energy efficient replacement versions (Upgrade). We present the specification for this digital after-sales service and a basic implementation in the context of a use case applied in the context of a compressed air demonstrator.

1 Introduction

Industry 4.0 is seen as a step forward towards a more sustainable industrial value creation including an efficient allocation of resources such as energy, products and materials [1]. Within the context of Industry 4.0 manufacturing systems are upgraded to an intelligent level taking advantage of advanced information and manufacturing technologies [2]. The term Cyber-Physical Systems "reflects a vision of a technology that deeply connects our physical world with our information world" [3]. The objective of this article is to utilize main features of Industry 4.0 to develop a digital service dedicated to increase awareness and transparency of possible energy efficient upgrades of components. The research question can be summarized as follows: What are the properties of a digital service that can support the user to identify energy efficiency measures within industrial machines utilizing a life cycle costing approach?

In order to increase awareness and transparency of available energy-efficient upgrades we present the concept of a so called "Hardware Upgrade Service". The service enables the automated and continuous benchmarking of available "upgrades" to energy consuming machine components or machine modules based on the continuous monitoring of a machine. The concept derives from commonly known software updates and hardware upgrades for personal computers. Transferred to manufacturing, the replacement of a worn-out component with the same or a comparable new component would

P. Weißgraeber et al. (Eds.): *Advances in Automotive Production Technology – Theory and Application*, ARENA2036, pp. 213–220, 2021. https://doi.org/10.1007/978-3-662-62962-8_25

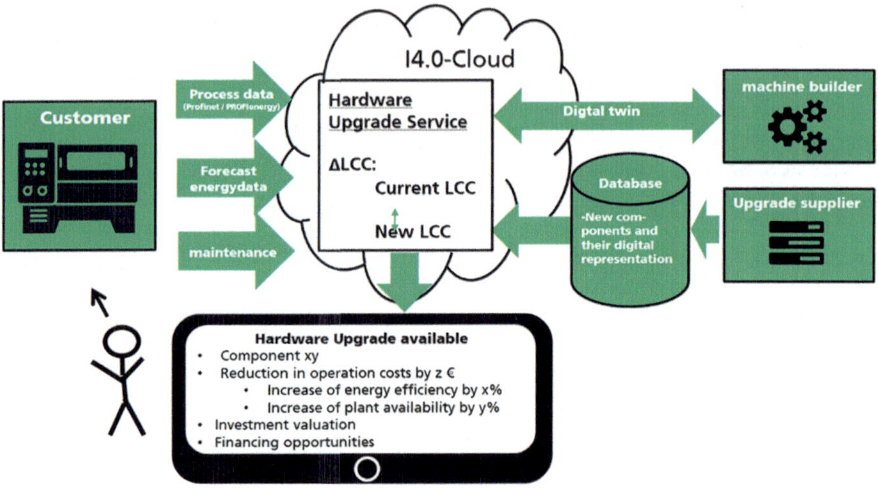

Fig. 1. Concept of the hardware upgrade service

be a "Hardware Update". The "Hardware Upgrade" aims to improve energy efficiency through replacing the worn-out component by a more energy-efficient variant.

This paper is structured as follows: In the 'State of Research' section the research questions and the scientific foundation is presented. We further present our approach depicted in Fig. 1 in the 'Automated Hardware Upgrade Service' section. Additionally, a proof of concept implementation is presented. Further research potential and conclusions on standardizations are drawn in the 'Conclusion and Outlook' section.

2 Current State of Research

We propose that the Hardware Upgrade Service can be assigned to the research field of upgradability in the context of product service systems [4]. The core method is an automated Life Cycle Costing (LCC) analysis on component level. Therefore, in the first subchapter we give an introduction to the term LCC in manufacturing. In the second subchapter we describe the state of research relevant for the service, focussing on LCC in manufacturing and sum up the challenges. In the third subchapter we focus on the state of research for automating LCC.

Introduction to LCC

LCC is an instrument for strategic cost management with the aim to enable the quantification and control of costs during the life cycle phases of a product. Application areas in a manufacturing company are in the procurement, controlling and logistics as well as the production and maintenance and the development and sales departments [5]. In practice, there is often no clear differentiating definition of the terms LCC and Total Cost of Ownership (TCO) [6]. One definition is, that LCC is used in industry for the procurement of investment goods because the investment and operational costs are dominating the business case. However, TCO also integrates more transaction cost, for example the contracting of a supplier, and therefore is often used for the procurement of consumables or services [6]. From this perspective, LCC is often described as a subset of TCO [5]. Therefore, the following literature research included LCC and TCO in its scope.

LCC in Manufacturing and Challenges

A decade ago, experts predicted an increased relevance of LCC for the products of mechanical and systems engineering industry and their end-user [7]. Survey results of the Energy Efficiency Index of German Industry, surveyed by the EEP of the University of Stuttgart, show, that for 80% of the 950 participating companies in Germany LCC is a relevant decision criteria for investments in energy efficiency [8]. However, the survey results just indicate a broad awareness about LCC in the context of tools for investment decision, not the application rate of LCC in reality and also not the challenges when doing so.

For understanding the challenges and how the concept of this paper tackles them we present the current state of art of LCC in context of manufacturing in Germany. Technical Guidelines like VDMA 34160 [9] or VDI 2884 [10] outline the application of LCC for the procurement of production equipment. [11, 12] presents comprehensive listing and comparison of these guidelines. These guidelines are mainly hosted by industry organizations and are quite generic. Specifications for the procedure of determining information for all individual LCC cost components are missing [13]. Reconfigurations on component level, for example the adaption of a gripper to a larger product, are not considered in these LCC methods [14]. Furthermore, there are technology-specific LCC models in form of calculation tools [15]. Most relevant in the context of the case study of this paper is the LCC modell for assembly and handling technology [16]. Furthermore, during the engineering process of pneumatic actor systems, LCC is used in branch-specific planning tools for the assessment of components [17]. However, currently these calculations are a forecast for an estimation [15] in the engineering phase. Calculations are only verified with real data after the ramp up of a production system in the field of maintenance cost. In case this is part of the purchasing contracts [18]. The central challenge for all tools is the high effort for the data acquisition for the cost inventory of the models [13, 19]. All described tools focus on the engineering phase and the procurement perspective, and a service-oriented perspective from the after sales unit of a supplier is not described. An approach with a focus on replacement investments on component basis could be identified in literature research.

Automation of LCC

[20] describes an approach to optimize the planning process of a reconfiguration of manufacturing processes. The scope of the method is to enable the reconfiguration of manufacturing systems for adapting to new production requirements and extending their use phase instead of investing in new systems. The method is tested on laboratory scale [14] and TRL 4–5 can be assumed. However, the execution of the method is time-consuming and in the latest publication, the service perspective is added to the method [21], however, only in form of a generic framework for a service-oriented reconfiguration. This highlights the relevant of "Hardware Upgrade", which can be seen as a first concrete application in the context of this framework.

Another central requirement for the automation of LCC is data exchange and energy transparency on component level. In the following, we describe the current state of research. Manual data acquisition is common, since only the biggest companies have implemented data interfaces for automating the LCC modell application [18]. The need for a structure to map energy consuming components of machines [12] and a data exchange for field data between machine operators and supplies [22] has already been described in research. Also in up-to-date research on TCO with a focus on automating the calculation of the maintenance cost component, this data exchange is still not based on automated data interfaces [23]. However, the current development in standardization in the field of Industry 4.0 could now enable the large-scale rollout of this data exchange [24]. Also, the quantification of the current energy consumption on component level is one of the central requirements for the Hardware Upgrade. The use of mobile measurement equipment is time-consuming and a fixed measuring concept on component level needs high investments. However, the actual state of the art of nonintrusive load monitoring enables the disaggregation of the electric energy demand of manufacturing machines on component level based on only one measurement point per machine [25]. For the ex-ante calculation of the energy consumption in upgrade scenarios there are research prototypes available, based on the assessment of measured comparable upgrades or based on simulation [26].

3 Automated Hardware Upgrade Service

In order to present our idea we utilize the "5C" architecture as a guideline for the CPS development for industrial applications [27]. The 5C level architecture presented by Lee [28] is described as a step-by-step guide for the development of cyber-physical systems (CPS) and services. The framework aims to define how to construct a CPS or service from the initial data acquisition, to analytics, to the final value creation differentiated by the so called 5Cs. We depict our classification of the proposed hardware upgrade service in Fig. 2 and further outline it below.

Within the *connection level,* necessary sensors and types of data have to be considered. In order create the digital twin representation necessary data for the LCC calculation within the Hardware Upgrade Service from the PLC as well as MES and energy management data are to be included.

The *conversion level* is described as the step that brings self-awareness to machines. With regard to the presented service this represents a possible integration of predictive maintenance in order to refine the hardware upgrade model as well as LCC calculation. However, this is not incluced at the current level of the implementation.

[28] describe the *cyber level* as central information hub in this architecture, stating that similarities between machine performance and previous assets (historical information) can be measured to predict the future behavior of the machine. In our presented service,this step contains the digital twin necessary for the benchmarking of possible hardware upgrades provided for the considered machine as well as possible hardware upgrades provided by the producer of the machinery and external providers.

The *cognition level* is also described as a decision support system (DSS). Similarly, in this step hardware upgrades are benchmarked in the digital twin and those meeting the requirements individually with regard to financial as well as technical dimensions will be submitted to the user.

Finally, the *configuration level* provides assets with self-configuration capabilities. In the described use case, the digital representation of the machine is self-updated in order to represent the implemented hardware upgrades.

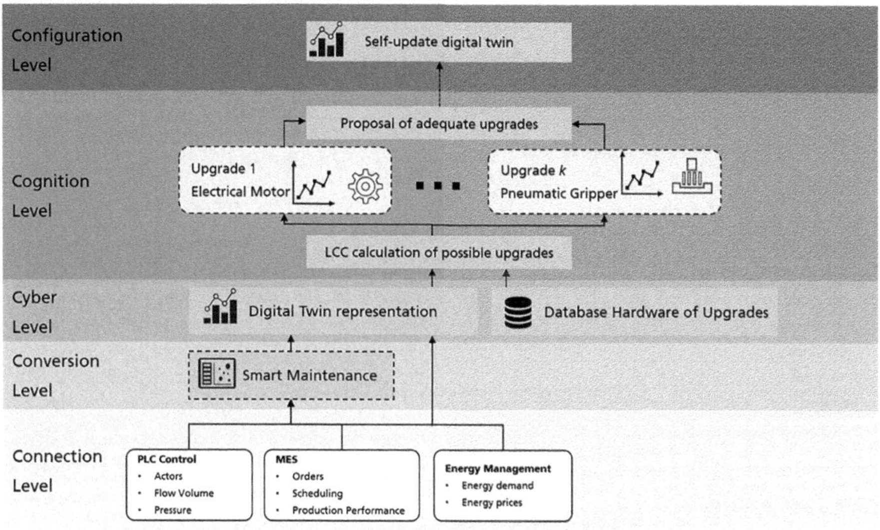

Fig. 2. Structure of the hardware upgrade service

4 Case Study

In order to validate our presented approach an industry-related pneumatic demon-strator was used in this paper. Within the demonstrator different working scenarios for each actuator scenarios (with or without leakages to various degrees) can be simulated. Therefore, a high number of combinatorial scenarios is possible ranging from a very inefficient to an efficient pneumatic system. The Proof-of-Concept (PoC) service was developed in a MATLAB environment. A parallel gripper was considered as the object to discuss the methodology. The goal was to investigate and compare the economic influence of similar pneumatic grippers with regard to their life cycle costs.

Accounting to the general LCC model, a mathematical expression has been formulated with modification for the parallel gripper. The derived model was based on the assumptions below:

- The calculation focuses on the economic performance from the perspective of the customer/equipment owner.
- The end-of-life phase is estimated by a fix residual value.
- Labor costs to conduct the upgrade relevant activities are estimated.

- Electricity and compressed air price is considered as being constant.

With these assumptions and restraints, the generalized LCC model can be modified to, $LCC = Ca + Co + Cd$ in which, $Ca = Acquisition\ cost$, $Co = Operation\ cost$, $Ce = end\text{-}of\text{-}life\ cost/disposal\ cost$. All costs are calculated in Euros.The energy price was collected from local energy supplier.

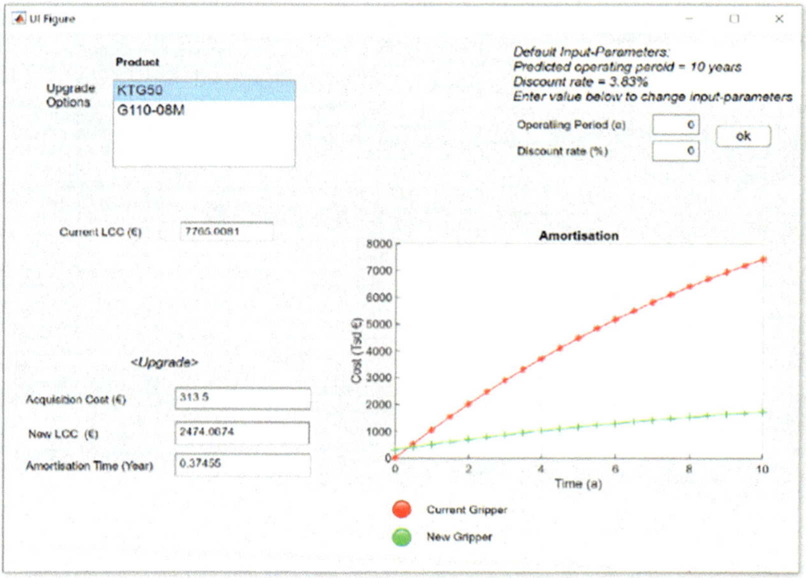

Fig. 3. Upgrade options

Hardware upgrade result suggested two options which could decrease the LCC. Figure 3 shows the LCC analysis result with proper visualization methods. The payback time was calculated with an assumed energy consumption and retailed price from the manufacturers. Other considered grippers were manually preselected by their similiarity to the currently deployed gripper. For instance, they all have the similar stroke and they can all hold and release the target items designed for the demonstrator.

According to the LCC analysis there were two hardware options for the upgrading, namely "KTG50" and "G110-08M". The displayed results changed automatically corresponding to user's selection in the "Upgrade Option" list.

5 Conclusion and Outlook

This paper proposes a smart service for proposing possible hardware upgrades for a continuous life cycle costing, the "Automated Hardware Upgrade Service". The major contribution of this study is to curve out the vision of the proposed service as well as a thorough representation of the state of research. The predictability of the cyber-physical system comes from the predictive analytics for the main components in critical assets, so that unexpected downtime can be reduced and users can act in time to increase productivity. In order the to scale this methodology within the automotive production a research project involving several partners is aspired ranging from equipment manufacturerand users as well as a possible third party upgrade provider.

Due to the importance and experience with compressed air system a first real-life use case can be applied in this field. With regard to the inclusion of predictive maintenance into the approach one solution is an encapsulated approach with an interface as a mean for a standardized data exchange to various predictive maintenance solutions. However, the detais of this interface are yet to be determined and scope of further research.

Currently, the hardware upgrade concept is developed with the aim to enhance energy efficiency in production. Further work should consider the integration of the hardware upgrade concept into reuse or remanufacturing concepts. For example, through delivering information and end-of-live hardware for circular economy value recovery concepts [28]. Furthermore, an integration of predictive maintenance into the Hardware Upgrade Service will be developed.

References

1. Stock, T., Seliger, G.: Opportunities of sustainable manufacturing in industry 4.0. Procedia CIRP**40**, 536–541 (2016)
2. Zhong, R.Y., Xu, X., Klotz, E., Newman S.T.: Intelligent manufacturing in the context of industry 4.0: A review. Engineering **3**(5), 616–630 (2017)
3. Lee, E., Ashford, S., Arunkumar S.: Introduction to embedded systems. A cyber physical systems approach, 2nd edn. MIT Press, Cambridge
4. Khan, M.A., Mittal, S., West, S., Wuest, T.: Review on upgradability – A product lifetime extension strategy in the context of product service systems. J. Clean. Prod. **204**, 1154–1168 (2018)
5. Geißdörfer, K.: Total Cost of Ownership (TCO) und Life Cycle Costing (LCC). Einsatz und Modelle: Ein Vergleich zwischen Deutschland und USA. Dissertation, European Business Schoool. LIT, Münster (2009)
6. Weissenberger-Eibl, M., Koch, J.: Fraunhofer ISI. Gesamtkostenabrechnung TCO Abschlussdokumentation. Fraunhofer ISI, Karlsruhe. https://vda-qmc.de/fileadmin/redakteur/Fraunhofer_ISI_-_Gesamtkostenrechnung_TCO_Abschlussdokumentation.pdf (2011). Accessed 13 July 2020
7. Schweiger, S.: Lebenszykluskosten optimieren. Paradigmenwechsel für Anbieter und Nutzer von Investitionsgütern, 1st edn. Gabler, Wiesbaden (2009)
8. EEP: Der Energieeffizienz-Index der deutschen Industrie. Ausgewählte Ergebnisse EII, 2. Halbjahr. https://www.eep.uni-stuttgart.de/dokumente/EEI-Winter-2018_19/2018_II_Ausgewaehlte-Ergebnisse.pdf (2018). Accessed 13 July 2020
9. VDMA: 34160: Forecasting Model for Lifecycle Costs of Machines and Plants. Beuth, Berlin (2007)
10. VDI-Fachbereich Fabrikplanung und -betrieb: VDI 2884: 2005–12. Purchase, operation and maintenance of production equipment using Life Cycle Costing (LCC). VDI-Gesellschaft Produktion und Logistik, Düsseldorf, Germany (2005)
11. Bünting, F.: Lebenszyklusbetrachtungen bei Investitionsgütern. In: Schweiger, S. (ed.) Lebenszykluskosten optimieren. Paradigmenwechsel für Anbieter und Nutzer von Investitionsgütern, 1st edn., pp. 34–50. Gabler, Wiesbaden (2009)

12. Abele, E., Dervisopoulos, M., Kuhrke, B.: Lebenszyklusanalysen bei Werkzeugmaschinen. In: Schweiger, S. (ed.) Lebenszykluskosten optimieren Paradigmenwechsel für Anbieter und Nutzer von Investitionsgütern, 1st edn., pp. 51–80. Gabler, Wiesbaden (2009)
13. Müller, A., Choudry, S., Bornschlegl, M., Mantwill, F.: Der modulare Lebenszyklus-Baukasten. ZWF **112**(7–8), 501–505 (2017)
14. Karl, F., Reinhart, G.: Reconfigurations on manufacturing resources: Identification of needs and planning. Prod. Eng. Res. Devel. **9**(3), 393–404 (2015)
15. VDMA: Calculating lifecycle costs. Manual fort he excel calculation tool. Business Management. VDMA Business Management, Frankfurt am Main, Germany (2012)
16. VDMA: LCC-Excelblatt Montage- und Handhabungstechnik. https://www.vdma.org/v2viewer/-/v2article/render/15237352. Accessed 13 July 2020
17. Pastewski, N., Krichel, S., Posselt, G., Linzbach, J., Plank, M.: Approach for achieving transparency in the use of compressed air in manufacturing as a basis for systematic energy saving. In: Thiede, S., Hermann, C. (eds.) Eco-factories of the Future, pp. 173–198. Springer International Publishing, Cham (2019)
18. Albrecht, V., Wetzel, P.: M-TCO – Daimler AG. In. Schweiger, S. (ed.) Lebenszykluskosten optimieren. Paradigmenwechsel für Anbieter und Nutzer von Investitionsgütern, 1st edn., pp. 81–96. Gabler, Wiesbaden (2009)
19. Ellram, L.: A framework for total cost of ownership. Int. J. Logist. Manag. **4**(2), 49–60 (1993)
20. Karl, F.L.: Bedarfsermittlung und Planung von Rekonfigurationen an Betriebsmitteln. Reconfigurations of Manufacturing Resources: Identification of Needs and Planning. Dissertation, Technische Universität München (2014)
21. Boucher X., Cerqueus, A., Delorme, X., Gonnermann, C., Paul, M., Reinhert, G. et al.: Towards reconfigurable digitalized and servitized manufacturing systems. Conceptual framework. In: Farhad, A., Stecke, E., von Cieminski, G., Krtisis, D. (eds.) Advances in Production Management Systems. Production Mangement for the Factory of the Future, pp. 214–222. Springer Nature, Cham (2019)
22. Boge, C.: LCC/TCO aus Sicht eines Werkzeugmaschinenherstellers. In: Schweiger, S. (ed.) Lebenszykluskosten optimieren. Paradigmenwechsel für Anbieter und Nutzer von Investitionsgütern, 1st edn., pp. 117–134 Gabler, Wiesbaden (2009)
23. Stähr, T., Lanza G.: Ausfallanalyse von Werkzeugmaschinen. Total Cost of Ownership 2.0 in Bezug auf den Anwendungsfall. wt Werkstattstechnik online **107**, 507–510 (2017)
24. ZVEI: Details of the Asset Administration Shell. Art. 1 – The exchange of information between partners in the value chain of Industrie 4.0. https://www.zvei.org/fileadmin/user_upload/Presse_und_Medien/Publikationen/2020/Januar/Details_of_the_Asset_Administration_Shell_Version_2-0/Details_of-the_Asset_Administration_Shell_Version_2.PDF. Accessed 13 July 2020
25. Gebbe, C.: Disaggregation of the electric energy demand of manufacturing machines. Dissertation, Technische Universität München, Munich, Germany. https://d-nb.info/120587982X/34 (2020). Accessed 13 July 2020
26. Matthes, K., Schröter, M.: Wirtschaftlichkeitsbewertung: Bewertung der wirtschafltichen Potenziale von energieeffizienten Anlagen und Maschinen. Fraunhofer ISI, Karlsruhe, Germany (2012)
27. Lee, J., Bagheri, B., Kao, H.-A.: A cyber-physical-systems architecture for industry 4.0-based manufacturing systems. Manufact Letters **3**, 18–23 (2015)
28. Cong, L., Zhao, F., Sutherland, J.W.: Integration of dismantling operations into al value recovery plan for circular economy. J. Clean. Prod. **149**, 378–386 (2017)

Deep Learning-Enabled Real Time In-Site Quality Inspection Based On Gesture Classification

Ioan-Matei Sarivan[1,2](✉), Stefan Andreas Baumann[1], Daniel Díez Álvarez[1], Felix Euteneuer[1], Matthias Reichenbach[1], Ulrich Berger[2], Ole Madsen[2], and Simon Bøgh[2]

[1] Mercedes-Benz AG, HPC F150, 71059 Sindelfingen, Germany
[2] Robotics & Automation Group, Aalborg University, Fibigerstræde 16, 9220 Aalborg Øst, Denmark
ioanms@mp.aau.dk
https://vbn.aau.dk/en/persons/144483

Abstract. In this paper we present a novel method for performing in site real time quality inspection (QI) and consequently, digitalization of manual processes performed by human workers. It complements and improves our previous work in this area, which makes use of telemetry gathered from a smartwatch to classify manual actions as successful or unsuccessful. This new methodology provides the worker with a real time capable, robust and more accurate quality inspector. This work enhances the existing system through the elimination of input from the user by making use of a BIOX bracelet that detects gestures. The signal processing and classification methods are simplified and optimised by using assembled neural networks thus merging together the data gathered from multiple signal sources. Consequently, the overall QI system is improved with around 70%, thus furthering the necessary development needed to have a system ready to be used on a production environment.

1 Introduction

The development presented in the hereby paper drastically improves an existing on spot quality inspection system named QCApp, thus making it faster and more reliable. The system uses acoustic and accelerometer signals gathered at the moment of electronic plug connections. The improvement is made on three different parts of the QCApp system. First, the process of starting and stopping the telemetry gathering process, which otherwise needs manual input, is fully automated by using an external triggering device. Then, the preprocessing of the gathered data is optimised by reducing the complexity of the used methods.

P. Weißgraeber et al. (Eds.): *Advances in Automotive Production Technology – Theory and Application,* ARENA2036, pp. 221–229, 2021. https://doi.org/10.1007/978-3-662-62962-8_26

Finally, merged neural networks are used to obtain decision fusion regarding the classification of the plug connection action.

The paper is structured as follows: Sect. 2 contains information about related work in the quality inspection field and about the previously developed system (QCApp) on top of which the development presented hereby was made. Section 3 presents the overall framework of the QCApp system with details about the triggering device that has been incorporated in it for starting and stoping the telemetry gathering. Sections 4 and 5 present details about the improvements made on the data prepossessing and classification components of the QCApp system. Section 6 contains details about the experiments made and the results obtained following the implementation of the methods presented hereby. The paper is concluded with suggestions for future work in the final section.

2 Related Work

The use of wearable devices for human machine interaction has met growth especially in healthcare applications. As P. Visconti et al. present in their paper about the technical features and functionalities of the Myo armband, the Myo armband is an optimum candidate for controlling various types of prostheses. [10]. The hand gesture recognition of the Myo armband is investigated by Tanasanee Phienthrakul [7] by applying various machine learning algorithms on the raw EMG signals extracted from the armband. Their conclusion on the ability of the Myo armband of detecting hand gestures is positive as long as it is not used for time critical applications and the complexity of these applications is relatively low.

The BIOX armband is a device designed and developed at Aalborg University, Denmark and it was initially meant for active exoskeletons control. In the paper published by M.R.U Islam and S. Bai [5], the armband is described as a device making use of force sensing resistors (FSR) that can detect the muscle deformation in the user's forearm. The FSR based telemetry is highly relevant for the QI application as it has comparable performance with similar technologies (e.g. electromyography EMG) and as opposed to those, it is not invasive and no prior skin preparation is required, while low requirement for signal processing ensures real-time control of involved systems [4]. The bracelet comes with a gesture classification software that is able to detect how the user grabs different objects. Moreover, the BIOX armband can be used to estimate the payload which the worker is trying to pick up with their hands [3].

The use of acoustic signals for quality assurance related applications is relatively common in the manufacturing industry. The sound emitted by machines or various components of the item while being assembled can be easily correlated with the proper execution of the manufacturing process itself. Bo-Si Kuo and Ming-Chyuan Lu present in their publication about sound analysis for quality monitoring in laser welding applications, the results obtained when employing an AI based classification system for detecting proper and faulty weldments which has an accuracy of up to 100% [6]. The acoustic signals are gathered using a microphone placed in the proximity of the welding device.

Following the investigation of the state of the art technologies at the moment of writing this paper, it was concluded that the development of wearable devices in the context of performing quality inspection is limited. Performing quality inspection at a rapid pace however, it is indeed a pressing matter which is addressed through various approaches in literature. Two technologies were investigated to make possible the automated control of the quality inspection system, the Myo and the BIOX armbands. The BIOX armband was chosen to proceed with being implemented in the already existing QCApp system due to its ability of reliably detect hand gestures and much faster signal processing, as concluded from the studied literature.

3 Overall Framework

The previously developed system for performing QI on-site makes use of a smartwatch that runs a specially developed application named QCApp. The previous work conducted by the authors is documented in the "Enabling Real-Time QI in Smart Manufacturing Through Wearable Smart Devices and Deep Learning" paper [8]. The app handles the necessary telemetry by sending it further to a computer that has binary classification algorithms based on convolutional neural networks for classifying the action of connecting the plug as successful or unsuccessful. The telemetry gathered by the smartwatch contains an audio signal sampled from the incorporated microphone, and XYZ acceleration signals sampled from the inertial measurement unit available on the device for tracking motions. A major advantage of the system is that it is highly portable, easy to use and the workers have no problem getting acquainted with it. Moreover, the equipped output devices (display, speaker and vibrating alert motor) allow for feedback to be provided to the user on multiple channels: visual, auditory and haptic, depending on the classification of the action as successful or not.

There are however a few drawbacks which are addressed in this paper. One of them is that the user needs to actively start and stop the telemetry gathering by touching a button on the screen of the smartwatch. This perturbs the flow of the process and it is highly desired for these steps to be removed through automation by integrating the BIOX armband into the system. Another one is that the selected signal processing method, Mel Frequency Cepstral Coefficients (MFCC) is too complex for the scope of the system. It is therefore desired to heavily simplify and therefore optimise the pre-processing method and improve the reliability of the classification system. The hereby presented project is based on a patent published in January 2021 [9] and owned by Daimler AG.

3.1 BIOX Bracelet Implementation

As previously stated, the user input necessary to start and stop the telemetry adds extra time to the plug connection process. Therefore, a BIOX bracelet is included in the overall framework in order to make the recording start/stop automatic. This is possible by having the BIOX system detect when the worker is grabbing a plug with the intention of connecting it into a socket. In Fig. 1, the software available with the BIOX bracelet can be observed. In subfigure a of Fig. 1, the BIOX bracelet detects the relaxed posture of the worker's hand. As soon as the worker grabs a plug (as observed in subfigure b), the BIOX system detects their intention to make a connection.

Having the BIOX software for detecting hand gestures developed using the Python programming language, it is accessible to integrate it in the rest of the QCApp system running on the computer that is making the classification using the telemetry gathered from the smartwatch. The telemetry gathering process is started when a "grabbed" gesture is detected by sending a start signal to the smartwatch through Wi-Fi. When the worker's hand is relaxed again, the recording is stopped. To make sure that a recording is not accidentally triggered by random gestures made by the user, any recordings shorter than one second are ignored. In average, a connection of a plug takes 1.5–3 s to perform.

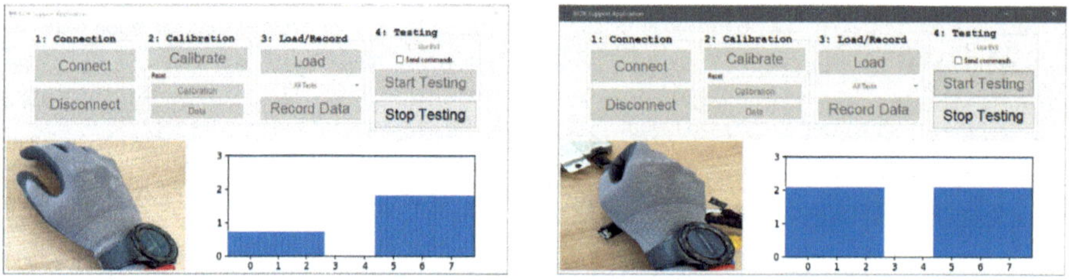

(a) BIOX GUI "Relaxed Mode" (b) BIOX GUI "Recording Mode"

Fig. 1. Screenshots of the BIOX GUI while the worker is relaxed (subfigure a) and while the worker intends to plug in the connector (subfigure b).

It is important to state that the system is meant to assist the worker and not monitor them, therefore recordings are to be made only when the worker is performing the relevant task for classification. Any other time, no telemetry should be gathered, as the sound recording is considered to be sensitive data in regards to the worker's privacy. By implementing the BIOX bracelet in the QCApp system, the start and stop of the telemetry gathering is fully automated without requiring the worker's direct input for starting and stopping it.

4 Data Pre-processing

Previous methods like [8] for audio quality inspection use the MFCC approach, but computing the Short-Time Fourier Transform(STFT) simplifies the process. The STFT is applied to obtain a frequency-domain representation of the recorded audio data by splitting the audio signal into multiple overlapping segments for which the result of the Fast Fourier Transform is computed. The resulting output represents the intensity of the multiple frequency bands for each segment and improves the results for transient classification problems [1, S. 26].

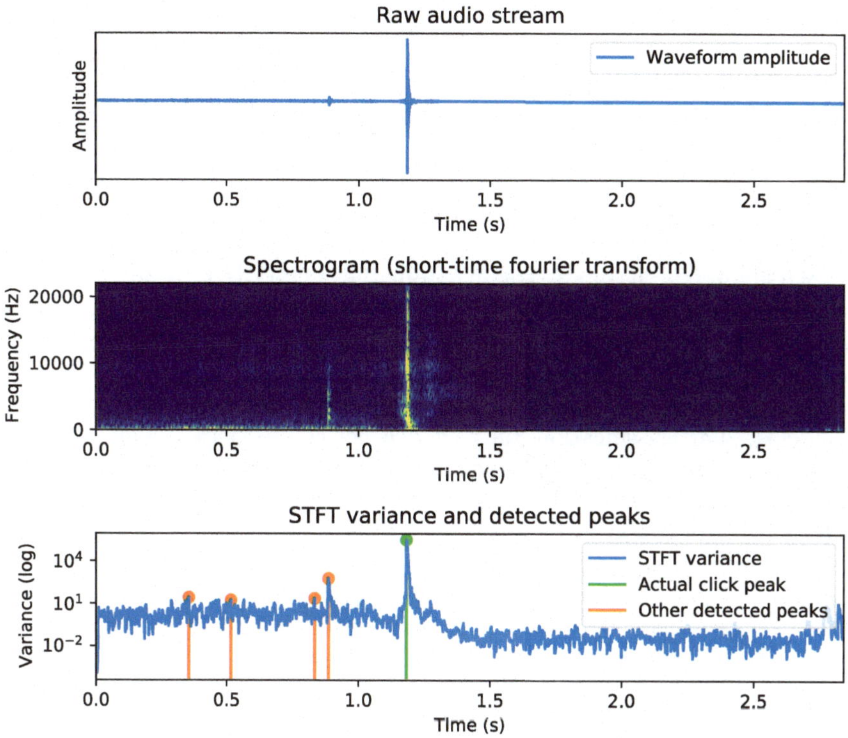

Fig. 2. Side-by-side representations of the audio input signal waveform, the generated spectrogram and the calculated variance with the detected peaks.

High energy components of both audio and acceleration signals are correlated and only these specific parts of the recording are processed through the AI algorithms. As shown in Figure 2, the events are extracted by detecting local maximum in variance across the time-dimension in the data generated by the STFT using a moving variance calculation which computes the local variance in a 3-sample range from each band. To reduce the computational complexity, the 1024 frequency bands from the STFT are consolidated into 8 bands from which the variance along the time-axis is computed. Compared to processing the input signal only, the frequency band based approach has the advantage of also being able to detect sudden changes in the frequency spectrum composition as opposed to just detecting sudden changes in overall signal energy.

Using the detected events as reference central points, signal windows are created around them. For the audio windows, a section of 200 ms in length is extracted from the STFT data. The event itself is in average 40ms long so the longer window allows the neural network to consider the background sounds that were present before the representative sound was produced. The total energy of these STFT windows is then normalized to prevent the network classifying sounds by their loudness which makes the model hardware-agnostic and robust to changes in the hardware or microphone. For the acceleration data, a similar approach is used, where a window of 750ms in length is extracted from the original signal with the representative event located in the middle of the window.

5 Deep Learning Model for Classification

For classifying the events extracted from the STFT data, a network utilising convolutions across the frequency and time axes is used. For training and validation of the AI model, a database of approximately 1200 recordings was used. The recordings are split into three categories: 15% correct connections sounds, 15% sounds similar with correct connections (e.g. mouse clicks), 70% environment noise. For classification, only the correct connections sounds are labelled as "correct" and the rest of the data as "incorrect". A relatively larger quota of "incorrect" data is used for ensuring the reduction of false positive classifications.

The network architecture consists of three sections—at the start there are two blocks which each consist of two bi-dimensional convolution layers of diminishing kernel size, followed by a max-pooling layer. For the third section, an alternating pattern between fully connected and dropout [2] layers is used with the network ending in a fully connected output layer with softmax activation to get probabilities for each class. For this structure, different hyperparameters were evaluated: different kernel sizes, filter counts, different neuron counts for the fully connected layers and different dropout rates.

For classifying the events using acceleration data, an artificial neural network was created using the same concept as the audio model. As the input data only has a single dimension, only a 1-dimensional convolution was used. It is to be mentioned that the accelerometer data is much simpler than the STFT audio data. Therefore, good results were obtained using a simpler network composed of one convolution-pooling block and only one fully connected layer. Multiple values were tested for the hyperparameters of the network (same as for the audio processing network), which resulted in the final used architecture.

The correlation between the audio and acceleration events is made during the classification performed by AI and it is possible by assembling together the two neural networks described above. This is made possible by cutting off the final layers of the original separate networks. A section that processes the output from both audio and acceleration networks together is added composed of a couple of fully connected layers on which dropout is applied with the RELU activation function and one final Softmax layer.

6 Results

The system was tested during the training of the AI model. Three different approaches of how to train which part of the network have been evaluated:

- Train the whole network at once without any separate pretraining of the audio and acceleration sections
- Pretrain the audio and acceleration networks independently, then construct the combined model using the pretrained weights and training the new layers only using the same data as in the pretraining
- Pretrain the audio and acceleration networks independently, then construct the combined model using the pretrained weights and allowing the training process to change all weights using the same data as in the pretraining

In our testing, the first and second approaches did not lead to better results compared to the audio-only model. The third approach significantly outperformed the audio-only approach, achieving perfect classification accuracy in some training rounds. The average model was able to classify noise and fail samples correctly and only miss-classified 2% of the success samples as unsuccessful.

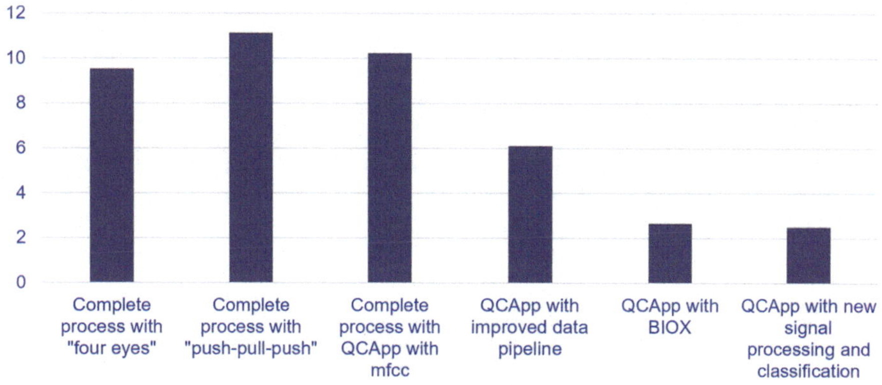

Fig. 3. Measured time needed to complete the plug connection with on spot quality inspection following the implementation of the above presented methods.

The new approach has reduced the required time for plug connection and on-spot quality inspection with more than 70% when compared with our previous version (QCApp). The testing was made using work benches similar with those used in production. Figure 3 showcases the time reduction from classical two-worker methods (first two measurements) to QCApp (next two measurements) and to QCApp+BIOX (last two measurements). The necessary time is drastically shortened by improving the data transfer process between the smartwatch and the computer, and by integrating the BIOX system for automatic recording triggering. Moreover, not only time is improved but also the accuracy of the new system QCApp+BIOX improves when compared with the initial version.

7 Conclusion

With satisfying results regarding the time needed to perform the manual task using the improved system, it is desired to turn the system into an actual product that can be implemented and used on the production lines at Mercedes-Benz AG. Future work will however be performed by analysing the possibility of adding an extra telemetry source, namely the raw signals from BIOX bracelet. The correlation between the gathered signals and the force applied by the worker at the level of their finger tips is planned to be investigated. Attempts on creating a system able to perform regression using these signals were made, it was however concluded that the FSR sensors used on the available version of the BIOX bracelet are not sensitive enough. This development will continue with the emergence of a new version of the BIOX bracelet.

Acknowledgments. This work was supported by Mercedes-Benz AG, ARENA2036 e. V. (Stuttgart, Germany) and Department of Materials and Production, Aalborg University (Aalborg, Denmark).

References

1. Brandt, Anders: Noise and vibration analysis: signal analysis and experimental procedures. John Wiley & Sons (2011)
2. Ilya Sutskever Geoffrey E. Hinton Alexander Krizhevsky and Nitish Srivastva. *System and Method for Addressing Overfitting in a Neural Network*. patentUS. [Patent: US9406017B2]. Aug. 2016
3. Muhammad R. U. Islam and Shaoping Bai. "Payload estimation using forcemyography sensors for control of upper-body exoskeleton in load carrying assistance". In: *Modeling, Identification and Control* 40.4 (2019), pp. 189–198. doi: https://doi.org/10.4173/mic.2019.4.1.
4. Muhammad R. U. Islam, Kun Xu, and Shaoping Bai. "Position Sensing and Control with FMG Sensors for Exoskeleton Physical Assistance". In: *Wearable Robotics: Challenges and Trends*. Ed. by Maria Chiara Carrozza, Silvestro Micera, and José L. Pons. Cham: Springer International Publishing, 2019, pp. 3–7. isbn: 978-3-030-01887-0
5. Muhammad Raza Ul Islam and Shaoping Bai. "Intention Detection for Dexterous Human Arm Motion with FSR Sensor Bands". In: *Proceedings of the Companion of the 2017 ACM/IEEE International Conference on Human-Robot Interaction*. HRI '17. Vienna, Austria: Association for Computing Machinery, 2017, pp. 139–140. isbn: 9781450348850. https://doi.org/10.1145/3029798.3038377
6. Bo Si Kuo and Ming Chyuan Lu: Analysis of a Sound Signal for Quality Monitoring in Laser Microlap Welding. English. In: Applied Sciences **10**(6), 1934 (2020)
7. T. Phienthrakul. "Armband Gesture Recognition on Electromyography Signal for Virtual Control". In: *2018 10th International Conference on Knowledge and Smart Technology (KST)*. 2018, pp. 149–153
8. I.-M. Sarivan et al. "Enabling Real-Time Quality Inspection in Smart Manufacturing Through Wearable Smart Devices and Deep Learning". In: *Procedia Manufacturing* 51 (2020). 30th International Conference on Flexible Automation and Intelligent Manufacturing (FAIM2021), pp. 373–380

9. I.-M Sarivan et al. (2021) Verfahren zur Überprüfung mindestens eines durch eine Person durchgeführten manuellen Arbeitsschritts (Daimler AG, Stuttgart, DE. Patent No. DE 10 2020 006 056 A1), German Patent and Trade Mark Office
10. Visconti, P., et al.: Technical features and functionalities of Myo armband: an overview on related literature and advanced applications of myoelectric armbands mainly focused on arm prostheses. International Journal on Smart Sensing and Intelligent Systems **11**(1), 1–25 (2018)

Detection and Monitoring for Anomalies and Degradation of a Robotic Arm Using Machine Learning

Hussein A. Taha(✉), Soumaya Yacout, and Lionel Birglen

Polytechnique Montréal, Montréal University QC, H3T 1J4 Montreal, Canada
hussein.taha@polymtl.ca

Abstract. Robotic arm performance varies due to normal and abnormal events. Normal events may include degradation of equipment, motors, mechanical system joints, and gears, while abnormal events may occur such as faulty episodes. In this paper, we address positional performance degradation that can be stopped and redressed if suitable required action is achieved. The Tool Center Point (TCP) position measurement devices are expensive, hence unavailable to every robot. Some industrial processes are critically sensitive to target tool position such as assembly, pin and past, and material handling. We propose a data driven artificial intelligence tool to detect anomalies and degradation of the robotic arm for a positional health assessment without the need for special advanced sensors. TCP deviation is predicted using deep machine learning models that train on time series of historical data of the robot's performance. Statistical thresholds are calculated to detect the robotic arm's degradation and anomalies by performing residual analysis. An alarm system is built by applying the proposed monitoring tool online.

1 Introduction

Industrial robots represent an important part of industrial manufacturing and machining process. They are critical to industrial processes, such as assembling, and are sensitive to robotic arm tool displacement. A small displacement or deviation leads to line stop, product reworking, and process down time. Robotic arms are challenged with significant variability and positional uncertainties that frequently result in robot position failure. Malfunctioning robots cost industrial plants more than 20 K$/min[1]. Robotic arm Position Health Assessments (PHA) have been addressed in the literature to solve displacement problems and enhance overall accuracy [1–4]. However, this requires highly accurate advanced measurements that are difficult to be found in each manufacturing plant. Meanwhile, it is not easy to send these robots to be calibrated every working day.

In recent years, researchers have worked to address fault detection of industrial robot arms. Their work is categorized into two groups, a model-based [1, 3, 5, 6] and historically based model [7, 8]. The model-based methods use mathematical equations to model the physics of the robotic arm and errors in the inverse kinematic matrices. Manish Goel, et. al. [3] used a model-based method to study unified locked-joint failure. These

P. Weißgraeber et al. (Eds.): *Advances in Automotive Production Technology – Theory and Application*, ARENA2036, pp. 230–237, 2021. https://doi.org/10.1007/978-3-662-62962-8_27

authors considered one type of manipulator fault of a locked joint. The robot arm was analyzed under a faulty state overall feasible space the robot could reach. This method was used to model 2 Degrees of Freedom (Dof) and 3 DoF simple robotic arms. However, this method needs significant computational power, thus it is difficult to be applied to the online operation of a 6 DoF robot arm. Guixiu Qiao, et. al. 2017 [5] developed a mathematical model to detect the positional error using an inverse kinematics model for a 6 DoF robot. This method has an industrial limitation, since it requires the measurement and the recording of the robot feasible working space using advanced 7-D measurements before operation. Yizheng Zhang, et. al. 2019 [1] used QR codes and an advanced camera to detect robotic arm joints that have had a failure for pick-and-place experiments. The proposed method needs the presence of cameras, hence a clean environment to obtain precise measurements.

The other detection methods are based on analyzing the robotic arm's stored data. Costa, Marcelo, et. al. 2019 [7] developed a hybridization of a boosting classifier to detect joint failure for a 6 Dof industrial robot. They assumed that the data fit a logistic distribution. This proposed method was a failure and it did not address the robot degradation anomalies. Qibo. Y, et. al. 2020 [8], studied the fault diagnosis for ball screw industrial robots. The current fault's signature was characterized to diagnose two types of faults. The developed diagnostic model used short-time Fourier transform (STFT) signal analysis and logistic regression classifiers to detect a faulty state. STFT signal analysis outperforms the wavelet decomposition. Current ball screw data consists of 20 samples for each fault and 20 samples for motor health status.

The main contribution points of this paper are: (1) the development of an online data-driven PHA tool for robotic arms. This tool monitors and detects displacement anomalies regardless of measurements' correlations. (2) The development of a time series Artificial Intelligence (AI) model to predict robotic arm displacements before incident or failures. (3) Cost savings, since the developed tool does not need special advanced displacement measurements in a robot's operational environment.

Table 1. Descriptions of data

Variable Index	Variable name	Variable type
1–6	Target joint [1–6] positions (angle)	Setting
7–12	Actual joint [1–6] positions (angle)	Measurement
13–18	Target joint [1–6] velocities (m/s)	Setting
19–24	Actual joint [1–6] velocities (m/s)	Measurement
25–30	Target joint [1–6] currents (m/s)	Setting
31–36	Actual joint [1–6] currents (A)	Measurement
37–42	Target joint [1–6] accelerations (m/s^2)	Setting
43–48	Target joint [1–6] torques	Setting
49–54	Joint control [1–6] currents (A)	Measurement
55–57	Actual Cartesian coordinates (x, y, z) of the TCP (m)	Response

2 Data Description

The proposed data based method ingests a time-series data for a six degree of freedom Universal Robot UR5 with six joints. The robot has a payload of 5 kg and joint ranges ±360°. The speed of all joints is 180°/S, the tool speed is 1 m/s, and the repeatability is ±0.1 mm. Table 1 shows the description of input and response variables that were used for the proposed supervised learning model. This robotic arm data is from laboratory data and was acquired by the National Institute of Standards and Technology's (NIST) with a frequency of 125 Hz for six position episodes [5]. The data response variables are the Tool Center Point (TCP) Cartesian coordinates (x, y, and z) and it was measured with laser sensor that has an index of [55–57] in Table 1.

3 Methodology

To have a robotic arm PHA, we developed a tool that consists of a TCP displacement prediction model and multivariate process monitoring. The combination of Deep Long Short-Term Memory (DLSTM) and Regression Adjustment Multivariate (RAM) shows up the strength points of each technique to have a value-added tool.

TCP Displacement Prediction

TCP displacement is considered the main position assessment factor for robotic arms.

The DLSTM model predicts robotic arm tool displacement, since the robot data is a time series. Our proposed DLSTM architecture consists of two main Long Short-Term Memory (LSTM1, LSTM2) layers, two dropout layers, and a Full Connected dense Neural network (FCN) layer. Figure 2 demonstrates the schematic of the Proposed DLSTM-RAM model. Its input is the robotic arm's joint data variables that have an index of [1:54] in Table 1 with time window 400 m sec (50 samples). Meanwhile, the output is the predicted TCP displacement for the 51st sample in the next time window. It is the Euclidean distance given in Eq. (1) and it depends on the response variables that were indicated in Table 1.

$$TCP\_d = \sqrt{(TCP\_x)^2 + (TCP\_y)^2 + (TCP\_z)^2} \quad (1)$$

Where, TCP_d is the Euclidean distance from the robotic arm base joint, TCP_x, TCP_y, *and* TCP_z are the measured tool center point displacements for x, y, and z Cartesian coordinates, respectively. This DLSTM prediction model eliminates the special need of TCPs for expensive position sensors. It achieved high accuracy, as will be shown in the results section.

Figure 1 plots the robotic arm Tool Center Point displacement (TCP_d) Auto-Correlation Function (ACF). ACF is the correlation between TCP_d observations as a function of time lag between each observation and a prior one [9]. Response variables are correlated variables, as given in Fig. 1. ACF for TCP_d was calculated with the Pearson's correlation coefficient, which is a dimensionless value between [−1, 1] to describe the positive and negative correlation, and zero ACF means zero correlation [9].

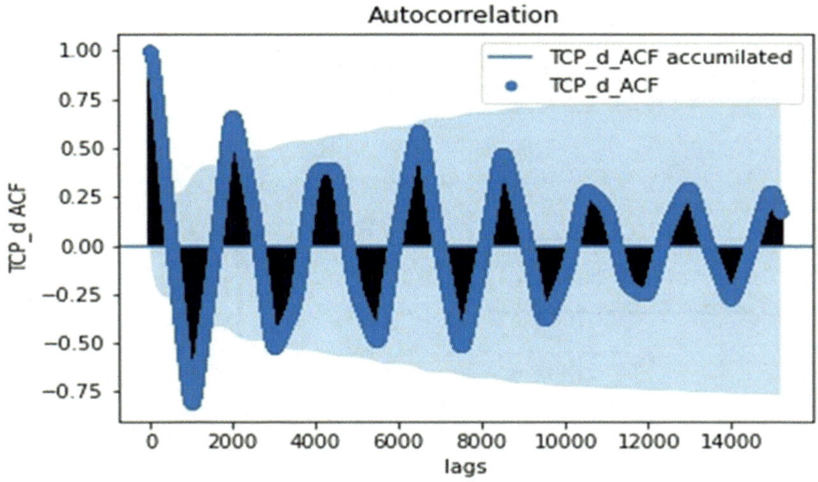

Fig. 1. Tool center point displacement autocorrelation

Multivariate Process Monitoring

Most of the industrial process has several output variables that affect process performance. Accordingly, multivariate process monitoring was developed. This was an extension of univariate quality control to avoid false alarms or failures of anomaly detection [10]. In multivariate, the monitoring data variables x is a matrix [p x n] given by $[x_1, x_2, \ldots, x_p]$, where n is number of data points and p is number of data variables. In our case, p = 54 variables while x is the data variables that have a variables index of [1:54] given in Table 1. The monitoring variables means is $\mu = [\mu_1, \mu_2, \ldots, \mu_p]$ and the covariance matrix (Σ) is [p × p] diagonal matrix of variables' standard deviation. Therefore, the multivariate normal probability density function is given by Eq. (2).

$$f(x) = \frac{1}{(2\pi)^{p/2}|\Sigma|^{1/2}} e^{\frac{1}{2}(x-\mu)'\Sigma^{-1}(x-\mu)} \quad (2)$$

The robotic arm TCP displacement response variable has a correlation, as shown by Fig. 1. Regression Adjustment for Multivariate (RAM) data is effective with correlated variables [11]. The RAM control chart uses linear regression to predict response variables. RAM models show the residual analysis of the predicted TCP position in comparison with the desired one. The residual error is the difference between the predicted TCP displacement and the target value. The residuals are uncorrelated, even though the original TCP data was correlated [10, 11]. In this paper, we develop a RAM and DLSTM prediction model (DLSTM-RAM) for robotic arm monitoring. The proposed DLSTM-RAM model architecture is given by Fig. 2. RAM model has inputs of TCP target settings and DLSTM predicted values to detect the positional error anomaly with the internal parameters of residual mean and standard deviation. The DLSTM-RAM internal parameters tuning and model capabilities are discussed in results section.

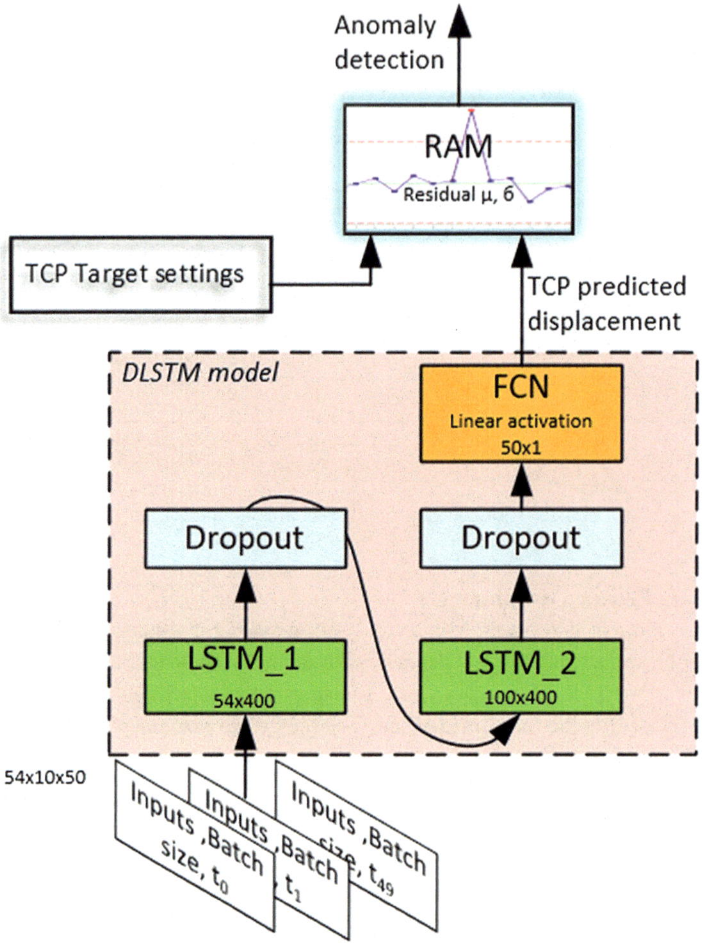

Fig. 2. Schematic of the proposed DLSTM-RAM robotic arm anomalies detection.

4 Results

Our DLSTM prediction model is trained with 200 training epochs, 10 input batches, 15206 training samples, and 481-validation samples. The training Mean Absolute Error (MAE) loss was 52 x 10^{-4}. Training ended at the 30th epoch when the model was trained with early stopping. Early stopping is one of the methods that we used to prevent overfitting [12]. The MAE was 27 x 10^{-4} with 30 epochs of training. Figure 3 depicts training and validation MAE losses during training with early stopping, and it shows the leaning stability of the proposed model. The lowest loss values were achieved at the 30th epoch.

The proposed model achieved testing MAE at 0.57% when it is tested with 8989 data samples. Figure 4 shows the predicted TCP displacement and the actual values with testing data.

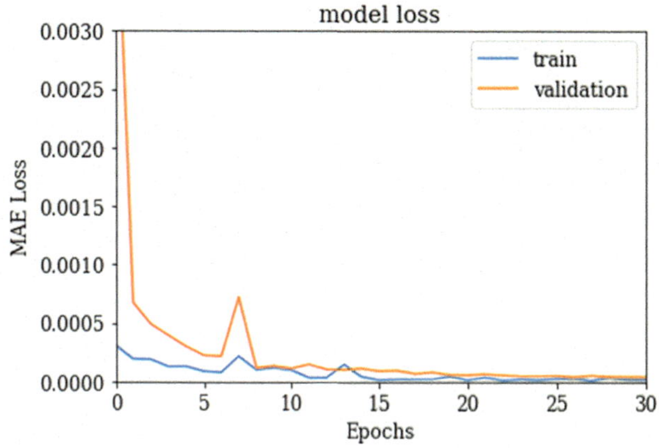

Fig. 3. DLSTM prediction model training and validation losses with early stopping

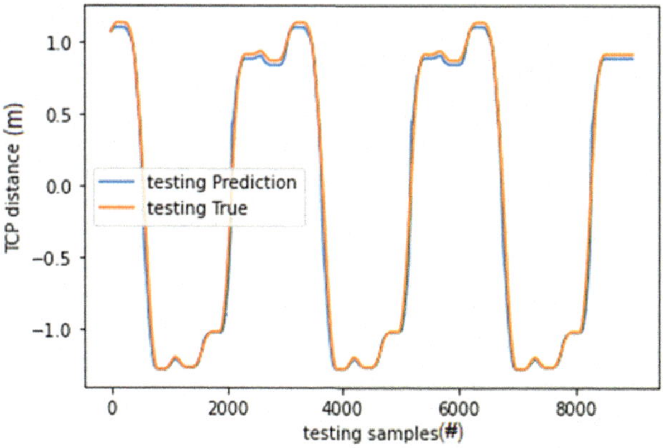

Fig. 4. DLSTM model testing to predict the TCP distance.

For comparison, a linear regression model was designed with the same training and testing data that we used for the DLSTM model. Table 2 represents the MAE of our proposed model and the linear regression model for training and testing data.

Table 2. DLSTM model vs. Linear regression prediction performance

	DLSTM	Linear Regression
MAE training (15241 samples)	27×10^{-4}	9×10^{-3}
MAE testing (8989 samples)	3.95×10^{-2}	13.98×10^{-2}

RAM detects the observation as an anomaly if the corresponding residual difference is more than the upper control limit or less than lower control limit. The control limits are defined as a residual (μ) ± 3* the standard deviation of б, which were calculated for the first 2000 samples data samples. Figure 5 shows the RAM residual control chart with testing observation data at rated speed and full payload. The number of anomaly points was 359 observations out of 6519 testing observations. The detected anomaly points are related to two positions out of the six that formulate a complete robotic arm operational cycle.

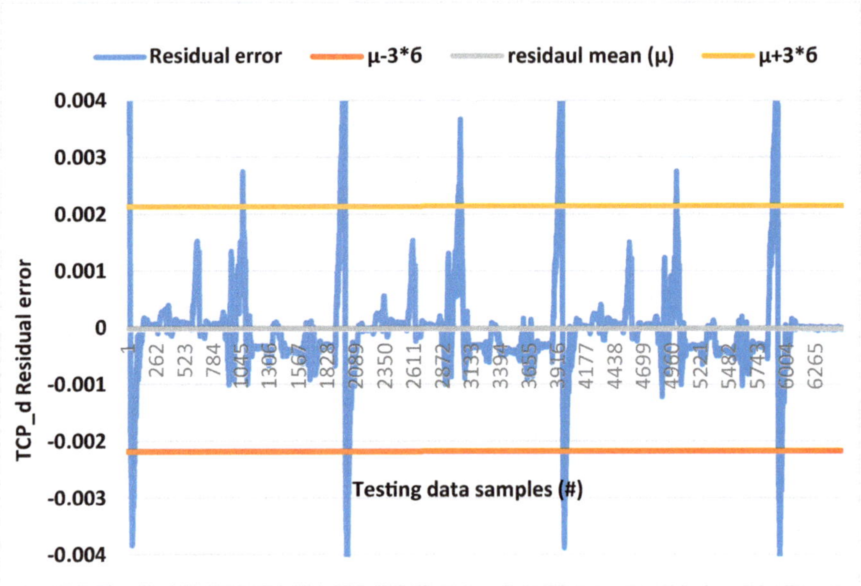

Fig. 5. RAM Control chart for robotic arm testing data with rated speed and full payload.

The chart in Fig. 5 shows a residual difference that is repeated periodically three times. Each one represents a repeated robotic arm complete operation. Therefore, the anomalies are periodically repeated. Hence, the robotic arm anomaly samples are affected by the same cause. For further research, analysis of this cause will be addressed. The time lag of detection and the self-healing actions will also be addressed. Once anomalies are detected, self-adjustment actions can be implemented during robot operations before the actual failure or incident.

5 Conclusion

This article presents a novel approach for robotic arm positional health assessments. Industrial robots are confronted with several environmental and field variability challenges that lead to robot failures. A combination of DLSTM-RAM models detects robotic arm TCP displacement anomalies. Our DLSTM model outperforms the traditional quality chart prediction model that uses a linear regression. It has lower MAE when testing approximately 10%. While the TCP sensor's measurement is very expensive and is not available to all manufacturing plants, the proposed DLSTM-RAM detects position anomalies based on measurements taken of indicators and flags an alarm signal without the need for additional special sensors. DLSTM-RAM was tested and verified with data obtained from a robotic arm that is working at rated speed and full payload. For further research, we will improve the DLSTM-RAM model to include the position anomaly's causes and use this information to perform self-healing actions.

References

1. Zhang, Y., Zhu, W., Rosendo, A.: QR code-based self-calibration for a fault-tolerant industrial robot arm. IEEE Access **7**, 73349–73356 (2019)
2. Qiao, G., Weiss, B.A.: Accuracy degradation analysis for industrial robot systems. In: ASME 2017 12th International Manufacturing Science and Engineering Conference collocated with the JSME/ASME 2017 6th International Conference on Materials and Processing (2017)
3. Goel, M., Maciejewski, A.A., Balakrishnan, V.: Analyzing unidentified locked-joint failures in kinematically redundant manipulators. J. Rob. Syst. **22**(1), 15–29 (2005)
4. Qin, J., Léonard, F., Abba, G.: Real-time trajectory compensation in robotic friction stir welding using state estimators. IEEE Trans. Control Syst. Technol. **24**(6), 2207–2214 (2016)
5. Qiao, G., Schlenoff, C., Weiss, B.A.: Quick positional health assessment for industrial robot prognostics and health management (PHM). In: 2017 IEEE International Conference on Robotics and Automation (ICRA) (2017)
6. Piltan, F., Kim, C.-H., Kim, J.-M.: Advanced adaptive fault diagnosis and tolerant control for robot manipulators. Energies **12**(7), 1281 (2019)
7. Costa, M.A., et al.: Failure detection in robotic arms using statistical modeling, machine learning and hybrid gradient boosting. Measurement **146**, 425–436 (2019)
8. Yang, Q., et al.: Fault diagnosis of ball screw in industrial robots using non-stationary motor current signals. Procedia Manuf. **48**, 1102–1108 (2020)
9. Berthold, M.R., Höppner, F.: On clustering time series using euclidean distance and pearson correlation. arXiv preprint arXiv:1601.02213 (2016)
10. Montgomery, D.C.: Introduction to statistical quality control. Wiley (2007)
11. Lowry, C.A., Montgomery, D.C.: A review of multivariate control charts. IIE Trans. **27**(6), 800–810 (1995)
12. Prechelt, L.: Automatic early stopping using cross validation: Quantifying the criteria. Neural Netw. **11**(4), 761–767 (1998)

Using Deep Neural Networks to Separate Entangled Workpieces in Random Bin Picking

Marius Moosmann(✉), Felix Spenrath, Manuel Mönnig, Muhammad Usman Khalid, Marvin Jaumann, Johannes Rosport, and Richard Bormann

Fraunhofer Institute for Manufacturing Engineering and Automation IPA, Stuttgart, Germany
marius.moosmann@ipa.fraunhofer.de
https://www.ipa.fraunhofer.de

Abstract. Entanglements can cause robots to pick multiple parts within random bin picking applications. Previous approaches cope with this problem by shaking the gripped workpiece above the bin. However, these methods increase the cycle time and may decrease the robustness of the application. Therefore we propose a new method to separate entangled workpiece situations by using deep supervised learning. To generate annotated training data for a convolutional neural network we set up a simulation scene. In this scene, bins are filled with different amounts of sorted workpieces in several entangled situations. Each workpiece is then moved into different directions to path poses which are evenly distributed along the surface of a hemisphere. The emerging dataset consists of cropped depth images of entangled workpiece situations and several path poses. A serial connection of convolutional neural networks is trained on this dataset and proposes a sequence of poses yielding the general departure path. Finally, the performance of this method is validated on simulated data. To the best of our knowledge, our proposed method is the first systematic approach to find the best extraction strategy to separate entangled workpieces in a pile while decreasing the effective cycle time for gripping entangled workpieces and increasing the robustness significantly.

1 Introduction

In production environments workpieces are mostly stored chaotically in bins. In many cases, the workpieces have complex shapes which potentially can entangle. Entangled workpiece situations within bin picking can cause increasing cycle times and a loss of robustness in production. To cope with these situations, we previously developed a method to detect and avoid entangled workpiece situations in a pile [1]. If however, there are no unentangled workpieces left, another

P. Weißgraeber et al. (Eds.): *Advances in Automotive Production Technology – Theory and Application*, ARENA2036, pp. 238–246, 2021. https://doi.org/10.1007/978-3-662-62962-8_28

method is needed to separate those entanglements to be able to empty the bin completely. In this paper we propose a learning-based approach for finding possible path points to separate entangled workpieces in a pile.

In summary, the main contributions of this paper are:

- a method to validate the effort of separating entangled workpieces,
- three different separation strategies and a comparison thereof,
- a serial connection of convolutional neural networks to predict path points for separating entangled workpieces, and
- an evaluation of the method on simulated test data samples.

2 State of the Art

Some of the most challenging tasks in random bin picking are motion planning of a robot and failure detection due to entanglements. Motion planning aims to find a collision-free robot path and to determine a trajectory for the gripper after grasping a workpiece. Ellekilde et al. [2] and Iversen et al. [3] focus on finding a trajectory by avoiding collisions with the environment. Spenrath et al. [4] proposed a method with fixed approach and depart poses that are checked for collisions with the point cloud. Considering coping with entangled workpiece situations in a pile, Moosmann et al. [1] focused on a model-based approach to detect those situations and to avoid them while bin picking. Matsumura et al. [5] proposed detecting gripping points in a pile that can lead to a failed grasp. The first approach not only detecting entangled workpiece situations but also separating them is presented by Leao et al. [6]. The robot trajectory to separate the entanglement is calculated by guessing the size of the workpiece. The robot movement itself takes place only in the x-y-plane. Other approaches focus on detecting entanglements with force-torque sensors [7,8]. To the best of the authors' knowledge, the approach presented in this paper is the first supervised learning approach to systematically separate entangled workpiece situations in a pile in random bin picking with six degrees of freedom motion planning.

3 Proposed Separation Methods

This section presents a method to validate the effort of separating entangled workpiece situations in a simulation scene, and three different separation strategies to generate trajectories for a robot developed with the simulation software CoppeliaSim. Through to the approach presented in this section, we are able to generate a dataset for training a serial connection of convolutional neural networks.

3.1 Method to Validate the Effort of Separation in Simulation

To validate the effort of separation in simulation, we propose a method considering the displacement of the entangled workpieces. First of all, we use the cropped

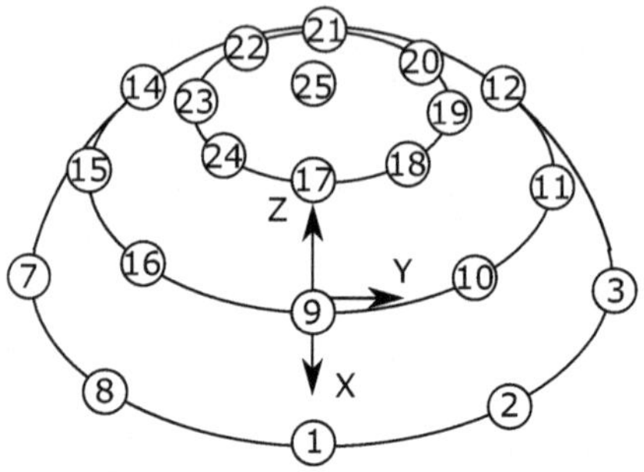

Fig. 1. Hemipshere that is spanned over a depth map crop with 25 possible path points

depth image from [1] which is oriented along a certain workpiece coordinate axis and span a hemisphere with the base in the workpiece plane over this depth map crop. To limit the number of iterations in the simulation scene, we designed a hemisphere consisting of 25 possible path points, which can be seen in Fig. 1. These path points are distributed over the surface of the hemisphere according to the following mathematical conditions. The origin of the gripped workpiece and therefore the origin of the depth map crop is defined as $M_{workpiece}$. With respect to the number E of point levels in the hemisphere, the radius r of the hemisphere, and the calculated distance between those levels d_E in Eq. (1), the Euclidean height of each level h_E can be calculated as shown in Eq. (2) with $n \in \{0, 1, ..., E\}$.

$$d_E = \frac{r}{E} \tag{1}$$

$$h_E = n \cdot d_E \tag{2}$$

When the workpiece is moved to one of those possible path points distributed over the surface of the hemisphere, it is also oriented with the angle α calculated in Eq. (3) around the z-axis of the workpiece coordinate system with $P \in \{1, 2, ..., 25\}$. There is also a rotation of γ around the x-axis. The angle γ is equal to the angle between the x-y-plane and the vector from the origin of the hemisphere to the possible path point P.

$$\alpha = (P - 1) \cdot \frac{1}{4}\pi \tag{3}$$

Under consideration of (1), (2), (3), the radius r of the hemisphere and the origin of the workpiece $M_{workpiece}$, the position (P_X, P_Y, P_Z) of each path point along the surface of the hemisphere can be calculated as shown in Eqs. (4), (5) and (6).

$$P_X = \sqrt{r^2 - h_E^2} \cdot \cos\alpha + M_X \tag{4}$$

$$P_Y = \sqrt{r^2 - h_E^2} \cdot \sin\alpha + M_Y \tag{5}$$

$$P_Z = h_E + M_Z \tag{6}$$

In the simulation scene, all 25 path points on the hemisphere are tested by moving the workpiece to each point. After the movement to one of those proposed path points, the Euclidean distance $|\overrightarrow{\mathrm{PQ}_k}|$ of all workpiece origins k in the scene except the gripped object are calculated before (q_X, q_Y, q_Z) and after the movement (p_X, p_Y, p_Z) as shown in Eq. (7) and the situation of the workpieces is reset to default.

$$|\overrightarrow{\mathrm{PQ}_k}| = \sqrt{(q_X - p_X)^2 + (q_Y - p_Y)^2 + (q_Z - p_Z)^2} \tag{7}$$

After reaching every possible path point we normalize the Euclidean distances with respect to the current situation and thereby annotate 25 labels for each depth map crop to generate a training data set for a convolutional neural network. Therefore, the effort of separation is described by the normalized total of all workpiece movements in the scene and the most suitable path point is the one with the smallest normalized total of Euclidean distances.

3.2 Three Proposed Separation Strategies

To predict a full separation trajectory instead of only one single path point, we propose three different separation strategies which take the method from Sect. 3.1 as a basis and can be seen in Fig. 2. We define three hemispheres for each separation strategy. The strategies differ by the origins of the hemispheres and the variation of their radii. In strategy A the radius of the hemispheres does not change but the origin changes. The most suitable path point of the previous hemisphere calculated by the validation method explained in Sect. 3.1 forms the origin of the following hemisphere. Strategy C resembles strategy A in changing

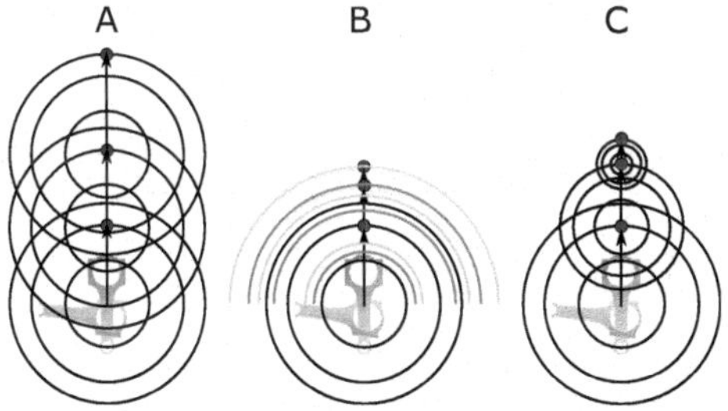

Fig. 2. Separation strategies; A: same radius of hemispheres with changing origins; B: same origin with increasing radius of hemispheres; C: decreasing radius of hemispheres with changing origins

the origin, however the radius of the hemisphere is decreasing. In strategy *B* the origin of the hemisphere remains in the origin of the entangled workpiece and the radius increases. An evaluation of the different strategies is presented in Sect. 5.1.

4 Training of a Serial Connection of CNNs

For the proposed method a series of three neural networks is used to predict scores of all point triplet combinations in the hemispheres to result in a separation. As seen in Fig. 3 every network gets the depth image as input. Network *II* and *III* additionally receive the outputs of the previous networks as indices of the predicted path points. So, for network *II* and *III* the output of the image classification is extended with the prediction of the previous networks. The final prediction for the hemisphere is then done by three final fully connected layers. The selected architecture for image classification resembles a DenseNet architecture in a TensorFlow framework with a depth of 22 layers over six dense blocks. For hyperparameter optimization we used a keras tuner with a novel bandit-based algorithm [9]. DenseNet is a neural network architecture first proposed in [10]. The approach to connect every layer to all previous layers in a dense block accelerates training and allows for less connections in total without the need to learn redundant feature maps. For training the serial connection of CNNs, we use a dataset generated with the method of Sect. 3.1 and the separation strategy A of Sect. 3.2 with 13,000 training data samples. Each CNN is trained for itself with the descriped input data and a binary crossentropy loss function. The networks are trained on an Nvidia Tesla V100 PCIe GPU over 50 epochs with around 13,000 samples, a batchsize of 128 and a dropout rate of 35% over the whole network to be able to not only predict one possible path point and to avoid overfitting.

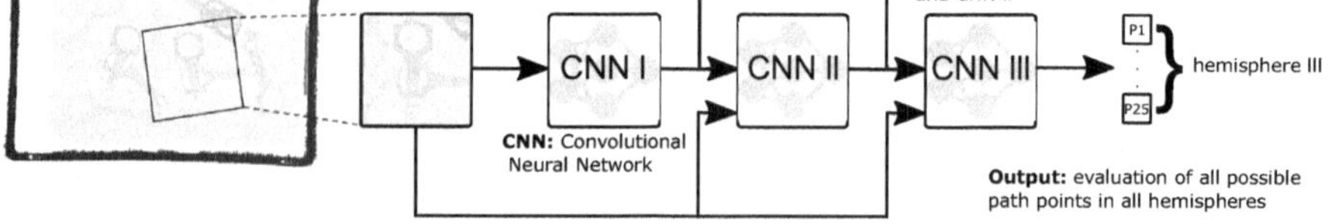

Fig. 3. Scheme of the serial connection of CNNs for trajectory generation

5 Results

5.1 Selection of Proposed Separation Strategy

To compare the three separation strategies presented, we divide our validation dataset into three different entangled workpiece situations, two with connecting rods and one with hooks as shown in Table 1. 100 entanglements were considered in each situation and strategy. Concerning the connecting rods, all three separation strategies *A*, *B*, and *C* are able to separate the entanglement. In comparison, strategy *A* performs best in the average number of hemispheres needed to separate the entanglement and the covered distance for separation. Regarding the hook, strategy *A* performes best in the average count of hemispheres to reach the separation and only strategy *A* is able to separate all 100 data sample entanglements. In the further course, we therefore focus on strategy *A*.

Table 1. Comparison of the proposed separation strategies A, B, and C in two entangled workpiece situations of connecting rods and in one with hooks

Strategy	Connecting rod						Hook		
	A	B	C	A	B	C	A	B	C
Average of hemipsheres	2.04	2.27	2.07	2.00	2.10	2.36	2.97	5.44	5.58
Distance of path [mm]	107.8	133.6	113.1	100.0	112.5	145.2	254.6	471.7	232.0
No path found [%]	0.0	0.0	0.0	0.0	0.0	0.0	0.0	80.0	83.33

5.2 Validation of Predicted Path Points with Strategy *A*

To validate the predicted path points and trajectories through the serial connection of CNNs, we set up a simulation scene. The entangled workpiece situation is modeled and the workpiece is moved along the predicted path points. In Table 2 we present the results of 3984 entangled workpiece situations. With the help of our proposed method, we separated 3917 entangled workpiece situations.

Figure 4 shows the variability of predicted trajectories to separate the entangled workpiece situation with strategy *A*. A perspective illustration of the predicted robot path is shown in Fig. 5.

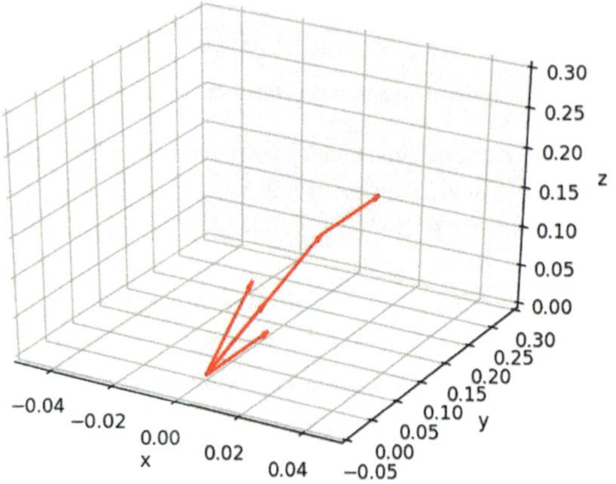

Fig. 4. Variability of possible predicted trajectories

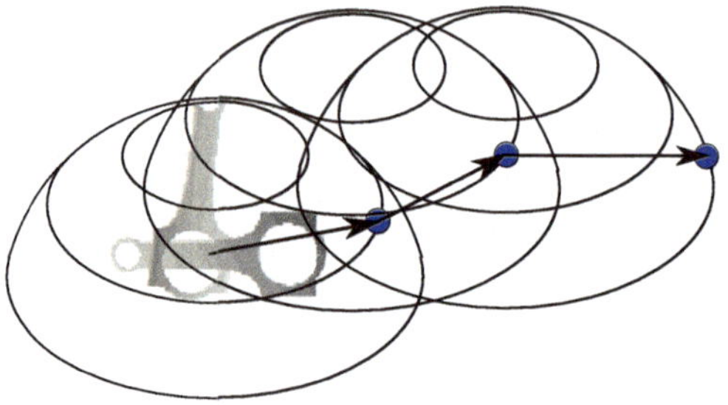

Fig. 5. Visualization of the predicted path

Table 2. Successful separations in different numbers of workpieces in the simulation scene and success rates

Number of workpieces in scene	Number of entangled situations	Number of successful separations	Success rate in [%]
2	1249	1212	97.04
4	2735	2705	98.90

6 Integration in Bin Picking Application

To cope with entanglements in bin picking, the proposed method needs to be integrated in an overall bin picking system. Figure 6 visualizes such an integration and especially the interaction with the contained grasp planning presented in [11]. The grasp planning approach represents all elements, which have to be considered for bin picking, as nodes in a search tree starting from the localized workpieces. Each node has a heuristic function, which estimates the likelihood to reach a solution from this node. A heuristic search algorithm is used to find a solution which represents a suitable grasp, including the motion path for the robot. Using the entanglement score from [2] in the heuristic function of the workpiece nodes, the search already prefers workpieces which are not entangled. If however an entangled workpiece has to be picked, the method presented in this paper can help to separate it from other workpieces. Since the grasp planning algorithm supports multiple paths by adding alternative path points into the search tree, the method can easily be integrated. The output of the proposed neural networks, i.e. the path point scores, can simply be used in the heuristic function of each path point to chose the ones that are most likely to solve the entanglement. During the evaluation of each path point it is checked for potential problems like collisions or singularities to make sure that the alternative paths are still safe to execute. This shows that the method can be seamlessly integrated in a full bin picking system to increase the robustness and lower the effective cycle times.

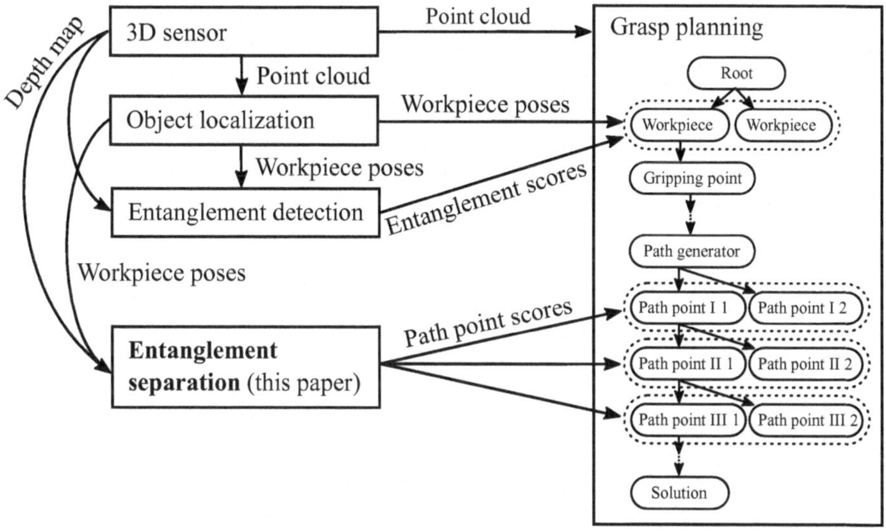

Fig. 6. Integration of the proposed method into a bin picking application

7 Conclusion and Outlook

In this paper, we proposed a method for separating entangled workpiece situations in random bin picking by using deep neural networks. First, we set up a method to validate the effort of separation of entangled workpiece situations in simulation and presented three separation strategies to generate a dataset. A serial connection of convolutional neural networks is trained and the success of separating entangled workpieces is presented. In the end, we put our paper in context of a bin picking application. In summary, this method helps to increase the robustness of random bin picking and decrease the effective cycle time for gripping. In future work, we will implement the proposed method in a real world bin picking application and try to train the serial connection of the convolutional neural networks concerning novel objects.

References

1. Moosmann, M., Spenrath, F., Kleeberger, K., Khalid, M.U., Mönnig, M., Rosport, J., Bormann, R.: Increasing the robustness of random bin picking by avoiding grasps of entangled workpieces. In: CIRP CMS, Chicago (2020)
2. Ellekilde, L.P., Petersen, H. G.: Motion planning efficient trajectories for industrial bin-picking. Int. J. Robot. Res. 991–1004 (2013)
3. Iversen, T.F., Ellekilde, L.P.: Benchmarking motion planning algorithms for bin-picking applications. Ind. Robot 189–197 (2017)
4. Spenrath, F., Spiller, A., Verl, A.: Gripping point determination and collision prevention in a bin-picking application. In: 7th Germ. Conf. on R. (ROBOTIK) (2012)
5. Matsumura, R., Harada, K., Domae, Y., Wan, W.: Learning based robotic bin-picking for potentially tangled objects. In: IEEE/RSJ IROS (2019)
6. Leao, G., Costa, C.M., Sousa, A., Veiga, G.: Detecting and solving tube entanglement in bin picking operations. Appl. Sci. (2020) (10.3390)
7. Kanoulas, D., Lee, J., Caldwell, D.G., Tsagarakis, N.G: Center-of-mass-based grasp pose adaptation using 3D range and force/torque sensing. J. Humanoid Robot. (2018)
8. Moreira, E., Rocha, L.F., Pinto, A.M., Moreira, A.P., Veiga, G.: Assessment of robotic picking operations using a 6 axis force/torque sensor. In: IEEE Robotic and Automation Letters (RAL), pp. 768–775 (2016)
9. Lisha, L., Jamieson, K., DeSalvo, G., Rostamizadeh, A., Talwalkar, A.: Hyperband: a novel bandit-based approach to hyperparameter optimization. J. Mach. Learn. Res. **18** (2018)
10. Huang, G., Liu, Z., van der Maaten, L., Weinberger, K.Q.: Densely connected convolutional networks. In: IEEE CVPR (2017)
11. Spenrath, F., Pott, A.: Gripping point determination for bin picking using heuristic search. In: 10th CIRP ICME, vol. 62, pp. 606–611 (2017)

Automatic Grasp Generation for Vacuum Grippers for Random Bin Picking

Muhammad Usman Khalid[1(✉)], Felix Spenrath[1], Manuel Mönnig[1], Marius Moosmann[1], Richard Bormann[1], Holger Kunz[2], and Marco F. Huber[1,3]

[1] Fraunhofer Institute for Manufacturing Engineering and Automation IPA, Stuttgart, Germany
muk@ipa.fhg.de, muhammad.usman.khalid@ipa.fraunhofer.de, fps@ipa.fhg.de
[2] FORMHAND Automation GmbH, Braunschweig, Germany
[3] Institute of Industrial Manufacturing and Management IFF, University of Stuttgart, Stuttgart, Germany

Abstract. In random bin picking, grasps on a workpiece are often defined manually, which requires extensive time and expert knowledge. In this paper, we propose a method that generates and prioritizes grasps for vacuum and magnetic grippers by analyzing the CAD model of a workpiece and gripper geometry. Using projections of these models, heatmaps such as the overlap of gripper and workpiece, the center of gravity, and the surface smoothness are generated. To get a combined heatmap, which estimates the probability for a successful grip, all individual heatmaps are fused by means of a weighted sum. Grid-based sampling generates prioritized grasps and suggests the most suitable gripper automatically. This approach increases the autonomy of bin picking significantly.

1 Introduction

In manufacturing industry, workpieces are often stored chaotically in bins. However, for many processing steps, the workpieces have to be separated, e.g., to be fed to a production machine. This is still mostly done manually, especially for large workpieces, but robots are nowadays used extensively for this task, which is called *bin picking* [1]. Bin picking is a highly researched topic and there are two different approaches: model-free and model-based. Model-free approaches search for possible grasps in the sensor data of the filled bin without using a model of the workpiece [2]. Since the pose of the workpiece is not known, these approaches struggle with a precise placement of the workpieces. In contrast, model-based approaches use a model of the workpiece to determine the pose of each workpiece inside the bin [3]. After the pose has been determined, a collision free robot path has to be calculated to extract each workpiece, e.g., by using a heuristic search algorithm [4,5]. To calculate the appropriate robot movement, the system has to know how the workpiece can be grasped. Since the orientation of the part is

P. Weißgraeber et al. (Eds.): *Advances in Automotive Production Technology – Theory and Application,* ARENA2036, pp. 247–255, 2021. https://doi.org/10.1007/978-3-662-62962-8_29

arbitrary in bin picking scenarios, many different grasp candidates are required. For working with new variants of workpieces, the definition of such grasps (target point on workpiece and approaching vector) can still be defined manually. However, it requires expert knowledge and manual labor for each workpiece, which is unpractical for a large variety of different workpieces. This paper proposes an approach to automatically generate these grasps for suction grippers, focusing on form-flexible suction grippers like FormHand [6]. While the goal of this paper is to generate grasps for bin picking, the technology can also be used for any other handling tasks, e.g., conveyor picking.

After a review of the state of the art in Sect. 2, the calculation of a heatmap that represents the probability of grasping success is developed in Sect. 3. In Sect. 4, we describe a method to generate discrete grasps based on this heatmap. Finally, the results are compared with physical experiments in Sect. 5. The main contributions of this paper are as follows:

- Generation of a heatmap that represents the probability of grasping success based on projections of the workpiece. This heatmap considers several factors like overlapping between the workpiece and the gripper, the surface smoothness, and the center of gravity.
- Generation of discrete grasps on the workpiece and the selection of the most suitable gripper for a workpiece.

2 Related Work

While there is a considerable amount of work on generating suitable grasps for clamping grippers [7–9], suction cup grippers are not studied that much. Kraft et al. [10] generate grasps for clamping grippers and suction cup grippers by randomly distributing the grasps on the workpiece and calculating a value for the grasp quality. The generated grasps are validated and prioritized using simulation to learn which grasps are more successful than others. Mahler et al. [2] generate grasps for suction cup grippers by using a quasi-static spring model. However, most of these approaches make the assumption that the grippers can not grip if there is even a small hole in the gripping area, which is quite common, e.g., with sheet metal parts. However, there are more flexible suction grippers, like FormHand, which can grip, even in the presence of holes [6]. This paper therefore focuses on generating grasps for such a robust suction gripper on generally flat parts like sheet metal parts.

3 Probability Matrix of a Successful Grasp

3.1 Projection of Workpiece and Gripper

Instead of calculating all the property matrices of gripper and workpiece directly from a CAD model, we use 2D projection matrices for both gripper and workpiece. For a gripper, we project it only from its top view so that the contact

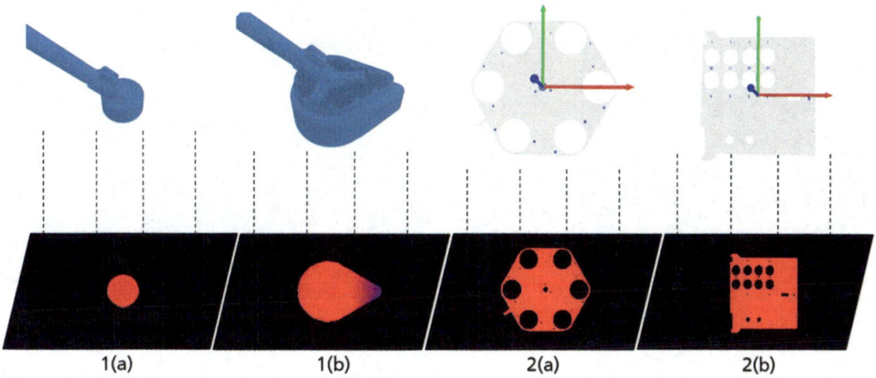

Fig. 1. Gripper and workpiece projections on 2D images. 1(a) Projection of a small gripper. 1(b) Projection of a large gripper. 2(a) Projection of a hexagonal type workpiece. 2(b) Projection of a rectangular workpiece.

surface of gripper can be projected to an image plane as shown in Fig. 1(a,b). This projection P_g works for all types of vacuum grippers consisting of any number of suction cups and also for magnetic grippers. In the similar manner, the CAD model of the workpiece is projected to a 2D image as shown in Fig. 1(c,d). This projection values represent the depth value of workpiece in each pixel. As our scope of work consists of flat workpieces, we project the workpiece from all its 6 sides. However, to speed up the process of grasps generation, some projections can be discarded at the start of the process by using the valid pixels V_p in the top projection. By comparing the valid pixels in each projection with V_p and using 60% as threshold, some projections can be discarded at the start of the process. In our test set, threshold was found by physically examining the geometry of workpieces.

3.2 Overlap Between Workpiece and Gripper

To find the contact surface area between gripper and workpiece, an overlap heatmap is calculated using projections of gripper and workpiece. The center pixel of gripper projection P_g is taken as an anchor point p_a. With respect to the anchor point, the gripper projection P_g is traversed pixel by pixel over the workpiece projection P_w. For each traversal, the number of pixels of workpiece for which gripper pixels have non-zero value are calculated. This defines the heatmap H_o of overlap between gripper and workpiece, as shown in Fig. 2(b). To calculate the overlapping, the gripper and workpiece projections are binarized:

$$b_w(x,y) = \begin{cases} 1 & \text{if } P_w(x,y) \neq 0 \\ 0 & \text{otherwise} \end{cases}, \quad b_g(x,y) = \begin{cases} 1 & \text{if } P_g(x,y) \neq 0 \\ 0 & \text{otherwise} \end{cases} \tag{1}$$

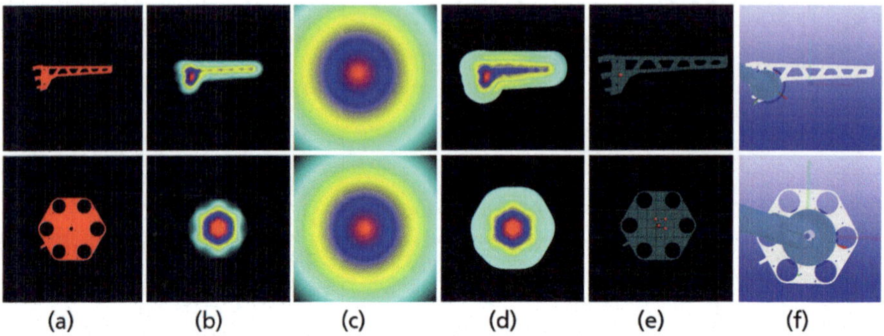

Fig. 2. Calculation of grasp reliability matrix and generation of gripping points for W5 and W6. (a) Projection of workpiece. (b) Overlapping heatmap of workpiece and gripper pixels. (c) heatmap of centre of gravity. (d) Weighted combination of heatmaps. (e) Grid based discrete grasp pixels. (f) Generated grasps.

For a specific point (x, y) in the workpiece projection P_w, the overlap $H_o(x, y)$ at this point is then calculated as the convolution of b_w with b_g:

$$H_o(x,y) = \sum_{i=1}^{w_g}\sum_{j=1}^{h_g} b_w(x - c_x + i, y - c_y + j) \cdot b_g(i,j) \\ \text{for} \quad c_x < x < w_w - c_x, \quad c_y < y < h_w - c_y \tag{2}$$

with the width w_g, the height h_g, and the center pixel (c_x, c_y) of the gripper projection and the width w_w, the height h_w of workpiece projection. To calculate the effectiveness of overlapping with respect to workpiece and gripper sizes, overlap heatmaps are normalized with respect to gripper H_{og} and workpiece H_{ow} by dividing overlap heatmaps with number of non-zero pixels in the gripper and workpiece projection, respectively.

3.3 Calculation of Center of Gravity Heatmap

It is desirable to pick the workpiece from its centre of gravity because it ensures that the weight of part is evenly divided and with firm gripping, the workpiece is less likely to drop. Therefore, we also introduced a centre of gravity heatmap into the calculation of grasp reliability matrix. For calculating the centre of gravity, we assumed that the part is uniformly thick and made of the same material. The centre of gravity pixel p_{cog} is calculated by adding all the x and y valid values in workpiece projection P_w and dividing it by total number of non-zero pixels. For a pixel at (i, j), heatmap for centre of gravity $H_{cog}(i, j)$ is calculated as

$$H_{cog}(i,j) = \max(H_{cog}) - \sqrt{(x_i - x_{cog})^2 + (y_j - y_{cog})^2}\,, \tag{3}$$

where $\max(H_{cog})$ is the maximum value in centre of gravity heatmap. Figure 2(c) shows the heatmap for centre of gravity.

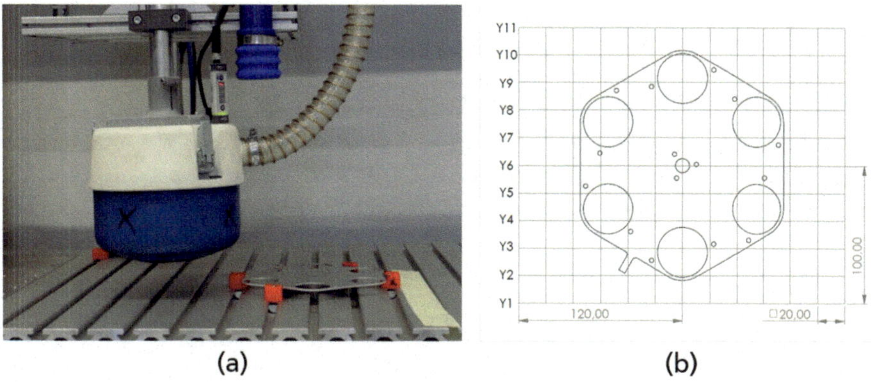

Fig. 3. Measurement of vacuum pressure with gripper. (a) Setup for measuring vacuum pressure. (b) Grid for measuring at different points on workpiece.

3.4 Calculation of Surface Smoothness

Some workpieces have grooves on their surface and when a vacuum gripper grasp the workpiece at that point, leakage of vacuum occurs and the workpiece is dropped during grasping. Therefore, it is important to find a matrix that provides the information of grooves or humps. We calculate a surface smoothness heatmap to consider these properties in our calculation of grasping points. Similar to the calculation of overlap heatmap, the heatmap for surface smoothness H_s is calculated by traversing pixel by pixel over workpiece projection P_w and calculating the sum of differences between z-surface value and z-value in workpiece projection.

3.5 Weighted Combination and Normalization of Matrices

To incorporate the influence of all property matrices of gripping, a weighted combination of heatmaps is calculated. To get the uniform influence of heatmaps in the combined heatmap, each heatmap is normalized with respect to its maximum value. The combined heatmap H_c is calculated as

$$H_c = w_{og} \cdot \tilde{H}_{og} + w_{ow} \cdot \tilde{H}_{ow} + w_{cog} \cdot \tilde{H}_{cog} + w_s \cdot \tilde{H}_s \tag{4}$$

Here, w_{og} tends to change the sensitivity of heatmap w.r.t the gripper. Heatmap H_{og} is defined in a way that a smaller workpiece with smaller gripper has higher heatmap H_{og} values as compared to a smaller workpiece with a larger gripper. Moreover, w_{ow} tends to change the sensitivity of heatmap w.r.t the workpiece. Heatmap H_{ow} is defined in a way that a smaller workpiece with larger gripper have higher heatmap H_{ow} values as compared to a smaller workpiece with smaller gripper. By adjusting the weights w_{og} and w_{ow}, a suitable gripper for a workpiece can be selected.

4 Grasps Generation and Gripper Selection

The heatmap is a continuous representation of grasp regions. However, for the generation of grasps, the heatmap has to be discretized. The heatmaps are discretized in a manner that the grasps are evenly distributed over the image with a certain distance in between. Moreover, we have extended our approach to suggest the best gripper from selection.

4.1 Gripper Selection

For some workpieces, particularly having holes, it is not easy to decide which gripper would be the most suitable one for that workpiece. We extend our approach of grasp probability matrices to suggest the suitable gripper from the selection. The heatmap H_{og} defines how well the gripper surface is overlapped with workpiece one. The greater the value of H_{og}, the greater is the contact surface of gripper on workpiece. On the other hand, heatmap H_{ow} defines how much surface area of workpiece is overlapped by the gripper. The gripper that receives the maximum score for the combination of these heatmaps is selected for a workpiece.

4.2 Generation and Prioritization of Grasps

For the generation of grasps from heatmap H_c, we use grid based discretization. Based on the bounding box of workpiece, a grid is generated with fixed width g_w and height g_h. Within each grid cell, the grasp pixel with maximum score is calculated. However, these grasp pixels can be very close to each other and not uniformly distributed over workpiece surface. Therefore, we applied a two step filtration of grasp pixels. In the first step, the grid pixels are sorted based on their heatmap scores. In the second step, a grid pixel is checked sequentially with its neighboring grid grasp pixels. If the distance between neighbouring grasp pixels is less than half of the grid width, the grid pixel with lower score is discarded. After the filtration of closely neighbored grasp pixels, grasps $p_g(x, y, z, \alpha, \beta, \gamma)$ are generated in which (x, y) are taken from pixel position of the heatmap, z is taken from depth value of grasp pixel in workpiece projection P_w and approach vector (α, β, γ) is taken from normal vector to the grasp pixel plane. Grasps are then converted to a transformation matrix T_p^g of grasp in pixel coordinate system, which is then transformed to its object coordinate system as $T_o^g = T_o^p \cdot T_p^g$. Here T_o^p is the transformation of object in pixel coordinate system as defined during workpiece projection W_p.

5 Experiments and Results

5.1 Hardware Testing Setup

We have arranged an experimental setup in our lab, as shown in Fig. 3. In our hardware setup, we fix the workpiece slightly elevated above from the base table

Table 1. Evaluation of grasp reliability heatmaps and gripper selection on test set

Workpiece	Gripper	Correlation	Maximum pressure	Maximum overlapping
W1	Small	0.53	244	312.99
W2	Small	0.76	214	452.83
W3	Small	0.71	211	384.59
W4	Small	0.73	210	202.81
W5	Small	0.73	291	459.24
W6	Small	0.81	**360**	**459.24**
W6	Large	0.75	270	436.56
W7	Small	0.60	421	459.03
W7	Large	0.74	**458**	**459.13**
W8	Small	0.76	379	459.26
W9	Small	0.95	242	440.22
W10	Small	0.71	242	454.83
Average		**0.73**		

to allow for free airflow through cut-outs in the workpiece simulating the situation after picking. The form-flexible suction gripper is positioned on the workpiece in a grid by a linear axis system and the resulting vacuum pressure [mbar] is measured. We chose a grid size of 10–20 mm resulting in up to 70 measurement points depending on the workpiece. The measurement was repeated five times at every position. The points where we get higher vacuum pressure are likely to be better points for the gripper to grasp that object.

5.2 Evaluation of Grasp Reliability and Gripper Selection

To validate our generated grasp reliability heatmaps, we compare the overlap heatmap H_o with the vacuum pressure recorded in physical experiments, described in Sect. 5.1. We calculated overlap score at each point defined in the grid for each workpiece, shown in Fig. 3(b). Pearson correlation coefficient is calculated for all the points of overlap with vacuum pressure and recorded in Table 1. A higher correlation coefficient closer to 1 shows that our algorithm is well correlated with grasp reliability in physical experiments. As the workpiece was fixed in our experiments, centre of gravity did not play any role in the measurements. Therefore, the correlation coefficient was calculated only with overlap heatmap.

Moreover, for the generation of combined heatmap of grasp reliability, we use $w_{og} = 1.8$, $w_{ow} = 0.01$, $w_{cog} = 1.5$, and $w_s = 0.5$. These weights were found empirically by comparing the heatmaps with the measured heatmaps of Sect. 5.1 of a subset of the test set. The weights w_{og}, w_{ow} are more sensitive for the selection of a suitable gripper and the weights w_{cog}, w_s are more sensitive with respect to the geometry of the workpiece. Also, as most of the workpieces have smooth surfaces, therefore, w_s has very little effect on combined heatmap. By fixing these weights, combined heatmaps for the complete test set are recorded and listed in Table 1. A total of 10 workpieces are tested with a small gripper and $W6$ and $W7$ also with a larger gripper. In physical experiments, it was observed that the small gripper is suited for $W6$ and a larger one for $W7$. We observe the similar trend in our combined reliability matrix. This result shows that our algorithm can choose the best suitable gripper from the available options. Also, we have applied our generated grasps successfully in the experimental setup of bin picking, details are provided in the supplementary material.

By using the speedup option of valid pixels V_p defined in Sect. 3.1, the average time for grasp generation is around 5.5 s on CPU Core i5-5200 @ 2.20 GHz, 8 GB. As our algorithm is deterministic and does not depend upon randomness, the same grasps can be generated repeatedly in the setup when weights and grid size are fixed in settings.

6 Conclusion

We presented a novel approach to generate grasps for a form-flexible suction gripper on generally flat workpieces like sheet metal parts. These grasps can be used for bin picking applications, where a lot of different grasps are needed due to the arbitrary placement of each workpiece. The results show that this approach can generate suitable grasps and determine an overlapping score which correlates well with physical experiments. Future work may include extending the approach to 3d objects and applying it to model-free grasping system.

References

1. Kleeberger, K., Bormann, R., Kraus, W., Huber, M.F.: A survey on learning-based robotic grasping. J. Curr. Robot. Rep. (2020)
2. Mahler, J., Matl, M., Liu, X., Li, A., Gealy, D., Goldberg, K.: Dex-Net 3.0: computing robust vacuum suction grasp targets in point clouds using a new analytic model and deep learning. In: IEEE ICRA (2018)
3. Khalid, M.U., Hager, J.M., Kraus, W., Huber, M.F., Toussaint, M.: Deep workpiece region segmentation for bin picking. In: IEEE (CASE) (2019)
4. Spenrath, F., Pott, A.: Gripping point determination for bin picking using heuristic search. Procedia CIRP **62**, 606–611 (2017)
5. Spenrath, F., Pott, A.: Using neural networks for heuristic grasp planning in random bin picking. In: 2018 IEEE (CASE) (2018)
6. Löchte, C., Kunz, H., Schnurr, R., Langhorst, S., Dietrich, F., Raatz, A., Dröder, K., et al.: Form-flexible handling and joining technology (formhand) for the forming and assembly of limp materials. Procedia CIRP **23** (2014)
7. Miller, A.T., Allen, P.K.: Graspit! A versatile simulator for robotic grasping. IEEE Robot. Autom. Mag. **11**(4), 110–122 (2004)
8. Jørgensen, J. A., Petersen, H.G.: Usage of simulations to plan stable grasping of unknown objects with a 3-fingered Schunk hand. In IEEE/RSJ (2008)

9. Mahler, J., Pokorny, F.T., Hou, B., Roderick, M., Laskey, M., Goldberg, K., et al.: Dex-net 1.0: A cloud-based network of 3d objects for robust grasp planning using a multi-armed bandit model with correlated rewards. In: IEEE ICRA (2016)
10. Kraft, D., Ellekilde, L.P., Jørgensen, J.A.: Automatic grasp generation and improvement for industrial bin-picking. In: Accelerating Cross-Fertilization Between Academic and Industrial Robotics Research in Europe. Springer, Cham (2014)

Flat Knitted Sensory Work Glove for Process Monitoring and Quality Assurance

Sarah Kim[1(✉)], Paul Hofmann[1], Hermann Finckh[1], Röder Uwe[1], Albrecht Dinkelmann[1], Michael Haupt[1], and Götz T. Gresser[1,2]

[1] German Institutes of Textile and Fiber Research (DITF), 73770 Denkendorf, Germany
sarah.kim@ditf.de

[2] Institute for Textile and Fiber Technologies (ITFT), University of Stuttgart, 70569 Stuttgart, Germany

Abstract. Flat knitted data gloves can be used as an assistance system for the detection of touches and gripping forces in logistics and production processes. By being able to determine, whether an action was carried out in accordance with the specifications, errors can be reduced, efficiency increased, and process documentation facilitated. This work focuses on the development and characterization of knitted sensor structures, which can be integrated during the production process of a completely knitted work glove. Compared to the integration of conventional rigid sensors into textiles, a fully textile sensor has advantages in terms of elasticity, breathability, draping properties, and flexible design. As it is a single step production process there is a high potential for very low production costs. Gripping forces, biomechanical requirements as well as knitting requirements were defined. Yarn material was fabricated for both the electrodes and the sensory functional layer. Different sensors were then manufactured and subsequently evaluated in regards to electromechanical properties such as sensor sensitivity and reproducibility of the sensor signal. Since the glove is stretched differently when worn, the behavior towards stretching was investigated. When subjected to dynamic forces while pre-stressed by different tensile forces, a clear distinction between loaded and unloaded states is possible.

1 Introduction

As the trend progresses towards small batch sizes and customized products, work is becoming more demanding, more diverse and more complex. These changes lead to the fact that a worker has to carry out different activities instead of consistently repeating the same task. This development causes a greater risk of mistakes accompanied by the increasing need of process monitoring for quality assurance. Errors in production require time-consuming reworking and can lead to costly recalls.

Camera-based systems are often used for process monitoring in production [1, 2] but come with limitations: If the fingers are concealed by an item, the system is unable to detect the accurate position of the fingers. In addition, the recorded image quality is strongly dependent on lighting conditions and the background. Besides that, tactile

P. Weißgraeber et al. (Eds.): *Advances in Automotive Production Technology – Theory and Application,* ARENA2036, pp. 256–263, 2021. https://doi.org/10.1007/978-3-662-62962-8_30

information such as the time and type of contact with an object is missing but might be of great importance for process monitoring. A camera system by itself is therefore not sufficient for a constant and reliable detection of hand movements.

Humans carry out most tasks with their hands. Since workers in production sites usually wear work gloves, it is evident to use these gloves as carriers of sensory and electronic assistance systems to gather information for interaction and control. The use of assistance systems such as a tactile data glove facilitates repetitive or complex tasks. By being able to determine whether an action has been carried out in accordance with the specifications, errors can be reduced and efficiency in logistics and production processes increased. In addition to improved process safety, assistance systems allow automated process documentation and consequently enable an increased transparency in production [3].

Even though there is a variety of tactile data gloves described in literature, most of them do not go beyond their use in research. They merely serve to evaluate their feasibility or as prototypes for studies [4, 5]. Poor stability of the data gloves, as well as technical challenges, high costs, and the lack of reliable production techniques that can be implemented in the existing textile production chain, represent crucial barriers for the transition from research to production. These challenges make these data gloves unsuitable for everyday use, especially for the use in an industrial environment.

Often, commercially available flexible foil-based sensors are integrated into the gloves via gluing or sewing, which have been extensively studied and show a very good reproducibility of the measurement signal [6, 7]. However, they are still very stiff, do not adapt to the finger shape, reduce the comfort, and limit the natural movement of the hand. In addition, flexible foil sensors are too expensive to be integrated into a (low price) work glove, which is usually replaced after an 8 h shift.

The development of purely textile sensors is therefore the better alternative. Completely textile pressure sensors have already been produced in the work of e.g. [5, 8, 9]. Disadvantages of previous solutions are complex and expensive manufacturing processes [5], expensive material such as CNT [8], and a structure of different textile layers that require fabrication or gluing together [5, 9]. Furthermore, integration of the sensors requires additional process steps. There are also textile sensors that are already integrated during the manufacturing process. However, these can only be found in other fields of application such as Structural Health Monitoring in lightweight construction [10, 11].

The aim of this work was to transfer the sensor principle of piezo-resistive foil-based pressure sensors to a knitted sensor structure (Fig. 1). An advantage of piezo-resistive sensors over piezo-electric or capacitive sensors is their insusceptibility to interferences. The sensory textile structure can be integrated automatically into a knitted work glove during its manufacturing process. Thus, the textile does no longer solely serve as substrate for the sensor, but rather the textile itself acts as the sensor. Due to the complete textile structure, textile aspects such as draping properties, elasticity, and breathability are retained. Moreover, the knitted sensor adapts to the shape of the wearer's hand and noise signal due to slippage with the skin is minimized.

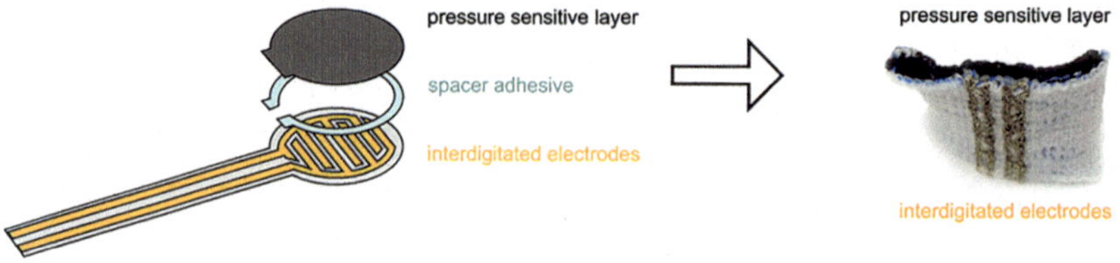

Fig. 1. Schematic view of a foil-based force sensor (left) and knitted force sensor (right)

2 Materials and Methods

In general, piezo-resistive foil-based force sensors, as depicted in the schematic view on the left in Fig. 1, consist of two electrodes arranged side by side in one layer (e.g. arranged as an interdigitated comb) and a functional (piezo-resistive) material layer located on top. The functional material is bridging the two electrodes. The total resistance of the sensor R_{Total} results from the contact resistance R_C between the electrodes and the functional layer and the volume resistance R_V of the functional material. The higher the force, the better is the contact between the functional layer and the electrodes and thus, the lower R_C. A compression of the functional material leads to a reduction of R_V. Consequently, R_{Total} decreases through a load application. This correlation is illustrated in Fig. 2.

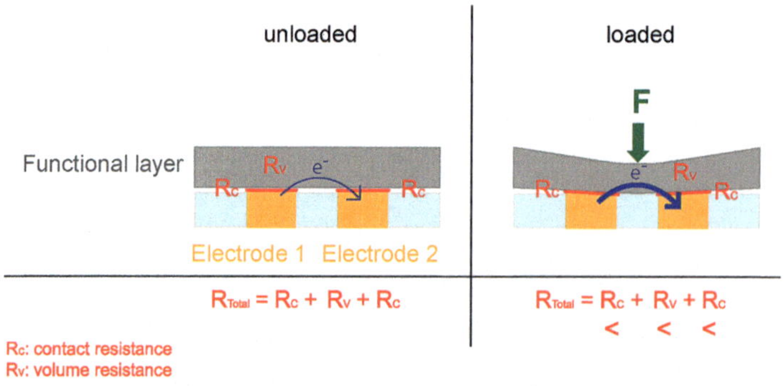

Fig. 2. Schematic cross section view of the functional principle of piezo-resistive sensors

2.1 Sensor Construction

Knitting is an established manufacturing process for gloves, as the entire glove can be produced in a single production process. Therefore, production costs and manufacturing time can be reduced, in contrast to the cut and sew production method. Moreover, flat knitting machines fulfil the requirements of a production process suitable for mass production while being highly customizable at the same time [12].

To create a knitted force sensor, yarn material for the electrodes and the functional layer is needed. We used the silver coated polyamide (PA) yarn Statex Shieldex® 235/36 dtex HC+B for the electrodes and developed polymer yarns filled with carbon black for the functional layer with piezo-resistive properties. Furthermore, we used polyester as base material for the glove. Instead of polyester, different materials can be used, depending on the requirements of the glove and its field of application, e.g. aramid for additional cut protection.

To produce seamless knitted gloves, each finger is knitted as an individual tube. For the sensor configuration described in Fig. 2, the functional layer is integrated as a third intermediate layer. The selection of alternate needles is required to produce this third (intermediate) knitted layer on the empty needles of the front and back needle bed (cf. the red yarn in Fig. 3) [13].

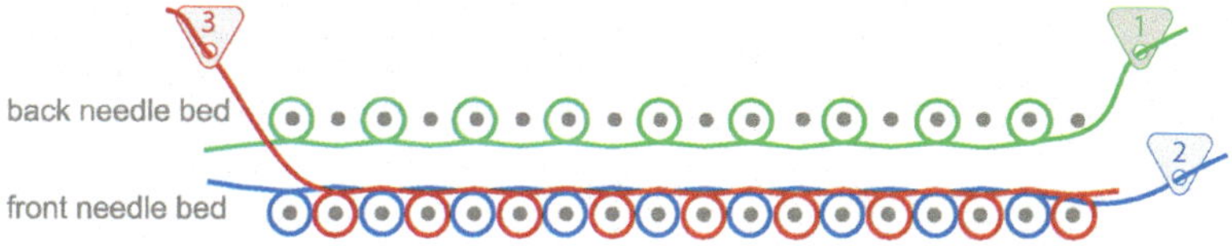

Fig. 3. Schematic view of the fabrication of three-layer knitted structures (green: back layer, blue: front layer, red: functional layer)

However, alternate needle selection results in a reduction in stitch density and elasticity, which needs to be compensated by using elastic yarn material [12]. This can be achieved by introducing the non-elastic conductive yarn together with an additional elastic yarn or by using an elastic conductive yarn. Fig. 4 depicts the difference between two knitted electrode layers: On the left, a fabric knitted with silver coated PA-yarn plated with Lycra® (blue); on the right, a fabric knitted with the conductive elastic core-spun yarn shown in Fig. 5. The conductive elastic core-spun yarn was developed together with Fein-Elast Umspinnwerke GmbH, Zeulenroda-Triebes, Germany. Two of the silver coated PA-yarns are wound around a Lycra core in Z and S direction, in order to reduce the shifting of the turns on the core and the tendency to curl.

As shown in Fig. 4 the developed core spun yarn improved the conductivity of the knitted electrodes by increasing and equalizing the proportion of conductive material on the surface. In addition the use of a core-spun yarn has technical advantages: There is no need for additional elastic yarn, so that the number of yarns in total as well as the corresponding feeding elements can be reduced. Therefore, more sensors could be integrated into one glove.

A Stoll flat knitting machine (CMS ADF 32 BW) with Autarkic Direct Feed (ADF) technology was used to produce the sensors. Instead of being pulled along by the carriage, the ADF technology enables the yarn feeders to move independently. Due to the individual control of the yarn feeders both horizontally and vertically, it is possible to switch between intarsia technique and plating without changing the feeders.

The two electrodes are integrated into the finger tube via intarsia technique (Fig. 6). The functional layer is knitted alternately on the empty needles of the front and rear needle beds, as described above (cf. Fig 3). A binding yarn binds the stitches of the

Fig. 4. Conductive yarn with additional elastic yarn (left) and elastic conductive yarn (right)

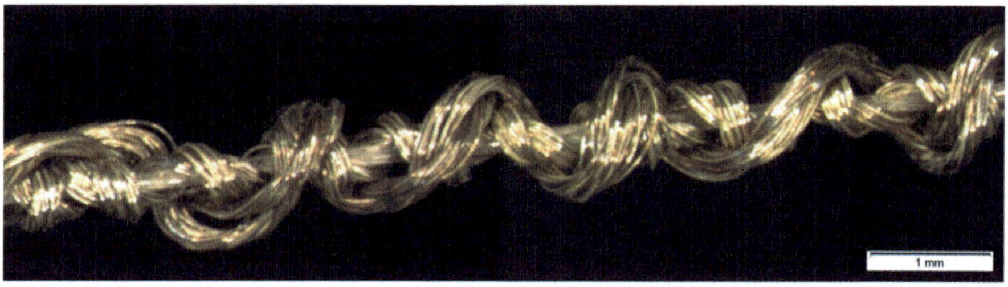

Fig. 5. Conductive core-spun yarn

functional layer to the front of the tube but not to the electrodes. There, the binding yarn floats on the reverse side, creating a small air cushion as seen in Fig. 6. Previous tests revealed that the less strongly the electrodes and the functional layer are attached to each other, the higher is the change in contact resistance R_c when a force is applied and thus, a higher relative change in resistance is measured.

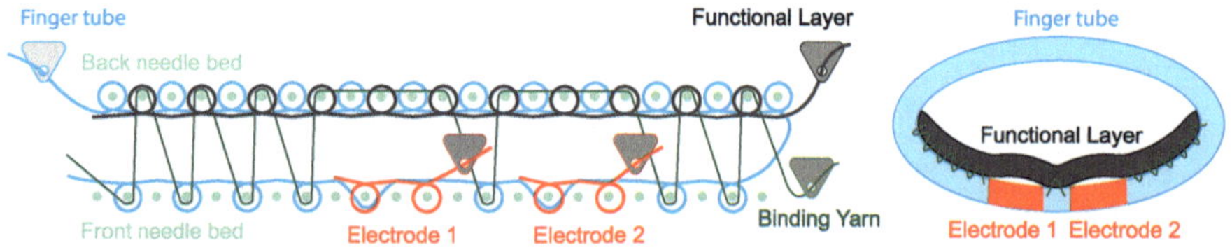

Fig. 6. Stitch notation (left) and schematic view of the cross section (right) of a flat knitted finger tube with integrated sensor

2.2 Sensor Characterization

We characterized the knitted sensors in regards to their electromechanical properties such as sensor sensitivity and reproducibility of the sensor signal.

According to literature [14–16] grip forces depend on the object to be gripped (shape, weight, surface material), the way something is gripped, the size of the hand as well as gender and age of the person. Therefore, it is difficult to identify specific forces that occur during gripping in industrial environments in general. In [15] forces from 0,2 N to 3 N are measured when gripping different lightweight objects (6 g – 200 g). As the gloves are intended to be used under normal physical activities, we subjected the sensors in this work to forces from 0 to 10 N with reference to [15, 16].

Experimental Setup for Sensor Characterization. In order to evaluate the sensor characteristics under dynamic load, several load cycles were continuously recorded using a test device developed at the DITF (Fig. 7). The device allows compression of the sensors at constant speed with an accuracy of 1 μm. Thus repeated loading and unloading of the sensors can be measured in order to analyze the reproducibility. Besides compression force and displacement, resistance of the knitted sensor is measured by a 0–10 V signal using a signal converter adapted to the resistance range of the knitted sensor. The samples were subjected to defined forces from 0 to 10 N. The circular test stamp has an area of 100 mm^2. Since the glove is stretched differently when worn, the behavior of stretching was also investigated. To examine the influence of tensile strain, the knitted sensor was automatically stressed by different tensile forces (0 cN/150 cN/300 cN) during the compression test. The samples were tested 10 times at each strain level to examine the reproducibility of the sensor signal.

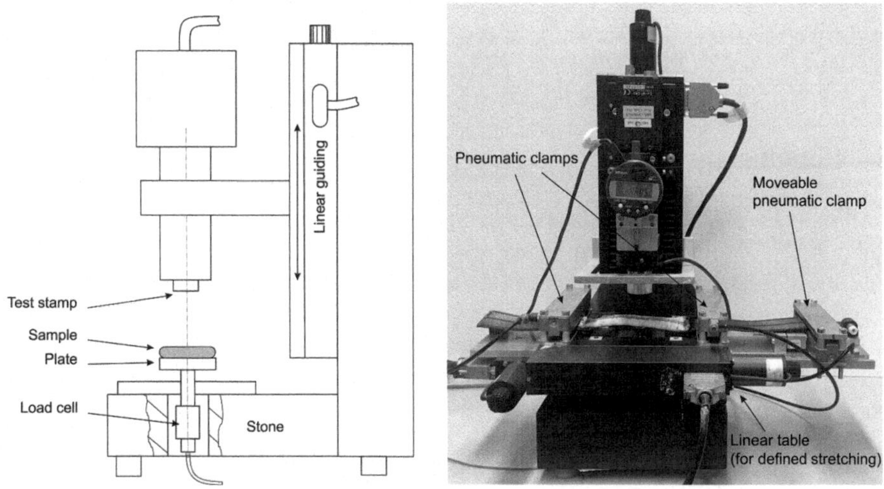

Fig. 7. Schematic view (left) and picture (right) of the test device

Results and Discussion. The graphs in Fig. 8 reveal a clear compression force dependent sensor signal. The diagrams show the dependence of the resistance R_{Total} to the applied force. Overall, when subjected to a dynamic force from 0 to 10 N the saturation range of the sensors is not reached. As a result, the knitted sensor provides distinguishable signals for compressive forces up to 10 N. The sharp drop of the curve in the beginning suggests excellent sensitivity of the sensors, especially when detecting low forces up to 3 N. Furthermore, Fig. 8 shows that, despite being pre-stressed by different tensile forces, a clear differentiation between loaded and unloaded states is possible. The sensor signal is unfiltered.

The graphs in Fig. 8 show that in particular only the curves of the first test cycles slightly deviate from the rest. This can be explained by the fact that the samples, due to the textile haptics, are not yet completely in contact with the test stamp prior to the first loading procedure. Therefore, a higher resistance value is measured. In general, the force-resistance curve of each subsequent cycle is usually slightly lower than the previous cycle. The reason for this is that the knitted fabric does not fully recover before the next load.

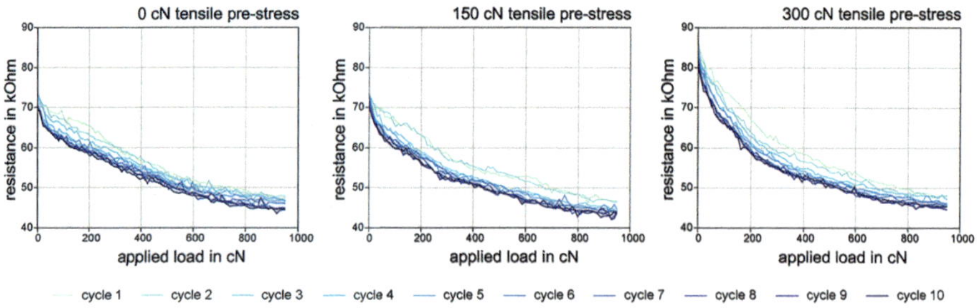

Fig. 8. Force-resistance curves of a knitted sensor sample during ten subsequent compression cycles pre-stressed by different tensile forces

3 Conclusion

In this work knitted sensors were developed which can be integrated during the production of completely knitted gloves without requiring additional process steps. The developed fully knitted pressure sensors meet the requirements of a dynamic usage due to good sensitivity (especially for the detection of forces up to 3 N) and high reproducibility of the sensor signal. A reliable detection of pressure is possible even when pre-stressed by different tensile forces. For these reasons, the fully knitted pressure sensor has been proven to be highly suitable for the integration into a fully knitted data glove.

Such glove can be e.g. used to detect how long an object is held or pressed. Applications such as press-in movements can be recognized as peaks on the sensor curve. In combination with acceleration sensors or camera systems, positioning or pick and place movements can be reliable recorded. Depending on the required application, a suitable number and the position of the sensors on the fingertips and palm must be identified.

The knitted force sensors provide great potential on expanding its use to a variety of other sectors such as logistics, non-productive industries, and public health. Even the possibility of applying such data gloves as a household device or for leisure, as an intelligent and multifunctional interaction device between people and their environment, appears very promising.

In order to protect the sensors from external influences such as moisture and dust, the knitted glove will be subsequently dip-coated, e.g. with rubber.

The sensory knitted structure is submitted by the DITF as the patent "Knit product and process for its preparation", DE102017127319A1.

4 Acknowledgement

We thank Fein-Elast Umspinnwerke GmbH, Zeulenroda-Triebes, Germany for producing the elastic electrode yarns. We thank the German Federal Ministry for Economic Affairs and Energy (BMWi) for funding our research within the framework of ZIM (Project-No.: ZF4060019DB6).

References

1. Lenz C., et al.: Human workflow analysis using 3D occupancy grid hand tracking in a human-robot collaboration scenario. IEEE/RSJ International Conference on Intelligent Robots and Systems, pp. 3375–3380. San Francisco (2011)
2. Mueller, B.C., et al.: Motion tracking applied in assembly for worker training in different locations. Procedia CIRP **48**, 460–465 (2016)
3. Huber, W.: Industrie 4.0 kompakt – Wie Technologien unsere Wirtschaft und unsere Unternehmen verändern. Springer Fachmedien, Wiesbaden (2018)
4. Ye, Q., et al.: Force-sensing glove system for measurement of hand forces during motorbike riding. Int. J. Distrib. Sens. Netw. **11**(11), 545643 (2015)
5. Büscher, G.H., et al.: Flexible and stretchable fabric-based tactile sensor. Rob. Auton. Syst. **63**, 244–252 (2015)
6. Sadun, A.S., et al.: Force Sensing Resistor (FSR): A brief overview and the low-cost sensor for active compliance control. First International Workshop on Pattern Recognition, p. 1001112. Tokyo (2016)
7. Vecchi, F. et al.: Experimental evaluation of two commercial force sensors for applications in biomechanics and motor control. 5th Ann. Conf. of Int. FES (2000)
8. Kim, G., et al.: Single-layer pressure textile sensors with woven conductive yarn circuit. Appl. Sci. **10**(8), 2877 (2020)
9. Kirthika, S.K., et al.: Fabrication and comparative study on sensing characteristics of soft textile-layered tactile sensors. IEEE Sens. Lett. **1**(3), 1–4 (2017)
10. Hofmann, P., et al.: Utilization of the textile reinforcements of fiber reinforced plastics as sensor for condition monitoring. Compos. A Appl. Sci. Manuf. **126**, 105603 (2019)
11. Hofmann, P., et al.: Woven piezoelectric sensors as part of the textile reinforcement of fiber reinforced plastics. Compos. A Appl. Sci. Manuf. **116**, 79–86 (2019)
12. Choi, W., Powell, N.B.: Three dimensional seamless garment knitting on V-bed flat knitting machines. J. Text. Apparel Technol. Manage. **4**(3), 1–33 (2005)
13. Weber M.O., Weber, K.-P.: Wirkerei und Strickerei: Technologien Bindungen Produktionsbeispiele, 6th ed. Dfv Mediengruppe Fachbuch, Frankfurt, a. M. (2014)
14. Hall, C.: External pressure at the hand during object handling and work with tools. Int. J. Ind. Ergon. **20**(3), 191–206 (1997)
15. Hiramatsu, Y. et al.: Control of precision grip force in lifting and holding of low-mass objects. PloS one **10**(9) (2015)
16. Westling, G., Johansson, R.S.: Factors influencing the force control during precision grip. Exp. Brain Res. **53**(2), 277–284 (1984)

Predictable and Real-Time Message-Based Communication in the Context of Control Technology

Timur Tasci(✉), Marc Fischer, Armin Lechler, and Alexander Verl

University of Stuttgart, Institute for Control Engineering of Machine Tools and Manufacturing Units, Seidenstr. 36, D-70174 Stuttgart, Germany
timur.tasci@isw.uni-stuttgart.de

Abstract. The current market trends demand a flexible production technology, but today's control systems are monolithic and therefore not flexible. The monolithism can be broken by using container-based virtualization and a microservice architecture. In a microservice architecture, messaging is used to achieve independence of services. Furthermore, deterministic real-time communication is necessary due to the control environment. Since messaging emerged from the IT world, real-time capability is not given. Therefore, in this work, a concept for real-time messaging is developed, implemented and validated on a test system.

1 Introduction

Due to the changing requirements of the markets and the constantly growing performance of electronics and computers, the rigid architecture of embedded real-time systems must be transformed into a more flexible system to fulfil those needs [9]. Today's predictable real-time systems, such as the Programmable Logic Controller (PLC) in the field of control technology, are designed in a monolithic manner. The Monolithic application describes a single-tiered software application in which different components combined into a single program from a single platform. Due to the tight coupling of functionalities, reconfiguration and making changes are complicated [9]. The use of a microservice architecture combined with a method of virtualization can achieve a loosely coupled, scaleable, exchangeable and hardware-independent system. This architecture requires a reliable communication method and often realized by messaging. However, it is difficult to use this type of architecture (microservices & messaging) due to the requirements of reliability and real-time in the field of production engineering, which are not supported by these technologies. In previous works, a container-based microservice architecture was designed for the PLC, taking reliability and real-time requirements into account [11,12]. The approaches for communication between container services in compliance with real-time requirements and the principle of loose coupling are still lacking. Within the scope of this paper, it is therefore examined whether messaging is a feasible approach.

P. Weißgraeber et al. (Eds.): *Advances in Automotive Production Technology – Theory and Application*, ARENA2036, pp. 264–271, 2021. https://doi.org/10.1007/978-3-662-62962-8_31

2 Related Works

Since the microservice architecture and messaging evolved from the office IT, there is no work that satisfactorily supports the requirements of real-time for control technology. In control technology a real-time system must respond to events within precise time constraints. The time between the event and the result is called the latency τ. The deviation of the latency $\tau$is referred to as jitter J [1,3]. Only the three following publications deal with the implementation of real-time messaging.

Bauer et al. [2] present a high performance and scalable pub/sub communication system called Tempo. The system is tested with TCP and UDP and achieves the timings $\tau = 380$ ms and $J = 10$ ms. Temporal requirements can be assigned to the messages which are check by a cyclic round trip time measurement. Tempo supports a broker-based and a broker-less implementation of the pub/sub pattern and has significant latency advantages with the latter. The scheduling of sending and receiving is cyclic and is implemented with the real-time scheduling RMS (rate monotonic scheduling). The message failure is not taken into account.

Liang et al. [8] implement a UDP-based protocol for a pub/sub communication system in RocketMQ to achieve better latency values in the range of $\tau = 60$ ms and $J = 180$ ms. They integrate a protocol with periodic acknowledgement of receipt. With the acknowledgement they can recognize a message failure. However, they make no statement about the scheduling.

Wang et al. [13] present an adaptive pub/sub communication system that is real-time capable and stores data from real-time critical systems in an adaptive real-time database with timings of $\tau = 21$ ms and $J = 3$ ms. The focus of the work is the integration of a communication system into the adaptive real-time database. The required communication is realized with an real-time pub/sub communication. However, the communication system is rather an asynchronous client-server architecture. Scheduling is solved by interrupts on the subscriber side and by cyclic scheduling on the publisher side.

The three publications only consider partial aspects for real-time messaging. Some works only consider the scheduling of sending and receiving, others only a failure detection. However, no publication shows how a predictable communication can be implemented holistically and how latency and jitter consist in the communication system. Furthermore, none of the publications shows a realization that meets the control engineering requirements of a latency below 250 μs and a jitter less than 10 μs.

3 Concept to fulfill real-time requirements

For the conception of a real-time messaging system with adherence to time limits τ and J, the process flow of a message transmission is considered first. For transferring a message from one application to another, the message is *created* and filled with the desired data, whereupon the sender *sends* the message by adding it to a channel. At *delivery* the message is delivered to the recipient and the recipient *receives* the message by reading the channel. Afterwards, the receiver *processes* the message and extracts the data. All steps have an impact to the real-time capabilities of the entire system. Every design decision has its advantages and disadvantages and have different consequences on the real-time capabilities. Therefore, the following sections provides an overview of the design decisions which we made.

Execution model: Two types of execution can be distinguish: *cyclic* and *event-based* execution. With *cyclic execution*, the data can be sent *during* processing as soon as the data is available, or the data can be sent *collectively after processing*. The distinction is necessary as it has to be defined who is responsible for keeping the data. Real-time is achieved when the control program runs completely within the cycle and the message arrives at the receiver during a given time-frame. When using *event-based execution*, an event is sent at any time. It can be differentiated whether data is included in the event or sent separately on a data bus. Real-time means that the response to the sent event takes place at the right time.

Scheduling: We compare two common scheduling algorithms which are supported in our system, which is based on Linux. The EDF (earliest deadline first) is based on the scheduling parameters period T_i, run time r_i, deadline $\tilde{d}_i$ and executes the tasks cyclically. The FPS (fixed priority scheduling) uses a fixed priority P_i and has the two variants FIFO (First-in first-out) and Round-Robin. If several tasks with the same priority are willing to run, FIFO executes the tasks in the order of the request time without interruption. With Round-Robin the tasks are executed in time slots one by one.

Both scheduling methods can be used to implement cyclic tasks, but EDF is recommend as the jitter J_{sched} is limited, overlapping task must not be considered and EDF is by design cyclically. For FPS, cyclic tasks can be implemented either with a timer interrupt request or with blocking for the remaining cycel time. The use of EDF is recommended.

For non-periodic tasks, the jitter is determined by J_{sched}. For the FIFO variant of the FPS it is $J_{\text{sched}} = \max(\sum r_{i,a})$ where $r_{i,a}$ is the runtime of the tasks that is ready to run. With Round-Robin the jitter is calculated according to $J_{\text{sched}} = \frac{r_i}{t} n_{\mathfrak{T},a}$ with the duration of the time slot t and the number of tasks that are currently ready to run $n_{\mathfrak{T},a}$.

It emerges that the jitter depends on the number of currently running tasks. The choice of scheduling influences the execution behavior and thus the jitter. To run the application independently of the system load, the maximum jitter must be taken into account for the latency.

Create and process: The time for creating τ_{create} and processing τ_{proc} the messages should be constant. Therefore, serialization and deserialization must be executed in constant time which can be satisfied by a constant length of the messages. Due to the constant length of the messages, it is useful to aggregate data if the length of the data is less than the message length. We distinguish *static* and *dynamic* aggregation. *Static* aggregation is defined and integrated into each service at development time and is therefore statically integrated into the compiled program. *Dynamic* is the aggregation at runtime.

Send and receive: The scheduling of sending and receiving can be *cyclic* or *event-based*, independently of the execution model. The scheduling of sender and receiver sides can also be selected independently.

If both sending and receiving as well as the execution model are cyclic or event-based, *synchronous* sending or receiving can be used. Synchronous means that sending or receiving is triggered from the user application. Consequently, the communication layer does not need its own execution context. In this paper, the execution context is considered as a task with all the necessary information that is required to execute a task. With *asynchronous* sending, the user application forwards the message to the execution context of the communication layer. With *asynchronous* receiving, the user application retrieves the message from the execution context of the communication layer. This means that the communication layer is responsible for the time of sending and receiving, which takes place independently of the user application flow.

In all cases, the latency for the transmission is at least the time for putting or taking a message from the transport medium. For cyclic and asynchronous processing the additional latency between sending request and the actual sending is at worst T_i and in the best case 0. With cyclic processing, the additional time for receiving a message is between T_i and 0, too. With asynchronous processing, a jitter occurs due to scheduling J_{sched}, whereas in synchronous processing, there is no jitter between communication and application. However, the scheduling of the user application also has a jitter, which is not taken into account, since it is in the scope of the responsibility of the programmer.

Transport: The realisation of the channel should be flexible so different transport mechanisms can be used but must fulfil all real-time requirements. Different approaches exist for communicating between internal systems and external systems. For the communication between external systems, time-sensitive-networking (TSN) can be used. For communication on a single system-environment, there are three ways to send messages between processes in real-time: shared memory, IPC sockes and the loopback adapter.

Pub/Sub realization Since no broker should be used due to higher latency [11], there are two ways in which pub/sub communication can be technically implemented. For the first approach, publishers and subscribers *know each other*. The subscriber registers to the publisher and thus subscribes to the publisher's provided data. Sending of messages is technically realised by a 1:1 connection per subscriber by sending a copy of the message to each subscriber. Within the second possibility the publisher and subscriber do *not know each other*. Sending and receiving refers to a channel in a data-oriented way. This can technically be done by a multicast communication, in which the publisher sends the message via multicast and the subscriber filters out the messages intended for him.

The advantage of a 1:1 connection is that messages are only sent if at least one subscriber has subscribed to a publisher. However, a copy of the message must be sent to each subscriber, resulting in a proportional increase of the time for putting the messages on the transport channel. If the number of subscribers varies, a constant latency can be achieved by a maximum number of subscribers and a compensating timeout.

The advantage of multicast communication is the independence of sender and receiver. The latency for a message sent is lower, since no additional copies must be created. The disadvantage of multicast communication is that a publisher must always send his message, even if no recipient is active.

Timing violations and incorrect messages: The time violation check is done by a *time stamp*, which is generated within the request of creating a message. The receiver checks the timing defined by the latency limit. Incorrect messages can be detected by methods like cyclic redundancy check or error sum. These methods are well established and thus not discussed here [10].

Message loss: Detection of message loss depends on the processing within the communication layer. With cyclic processing on the receiver side, a message loss can be detected using *heartbeats*. In each cycle a heartbeat is exchanged with the number of messages sent so far. The recipient can compare the number of actual received messaged. Since the transmitter must send a heartbeat cyclically with the total number of sent messages, there must be synchronisation between sending the heartbeat and sending the messages.

A combination of *sequence number and heartbeat* can be used to bypass the necessary synchronisation. An ascending sequence number is assigned to each message, which is checked on the recipient side. If a message is lost, this is recognised on the receiver side by a larger sequence number than expected. In the time span of no message sending, a heartbeat is used to inform the receiver. The maximum duration until a message loss is detected corresponds to the cycle time of the heartbeat.

In a purely event-based processing, in which no cyclic execution contexts are used, a message failure can only be detected by an **acknowledgement**. If the recipient receives a message, it sends an acknowledgement back to the sender. If the acknowledge is missing, the sender can assume a message failure. The maximum duration until a message loss is detected is the timeout for the acknowledge, which depends on the latency of the transmission and on the processing on the sender and receiver side. In addition, the receiver has to know the transmitter in order to send an acknowledge back to it.

4 Validation

The concept is evaluated on an Intel Core i5 3350P 3,1 GHz system with a Linux Preempt-RT. A scenario with running ten cyclic control services is constructed which represents a modularized control system. Each service represents a cyclic control service with a cycle time of 10 ms. These services are connected in a chain and communicate likewise with each sevices sending seven messages to the next service in line. The messages are grouped in one channel.

First, for the messaging, the concept of asynchronous processing with cyclic execution is used. Therefore, sending and receiving task are added to the services with a cycle time of 300 μs. For the scheduling of the services and messaging taks, EDF is used. The response time is measured in a two-minute time frame of τ_G in Fig. 1a, wherein τ_G of the message transmission is displayed as a cumulative distribution between each individual service. τ_G is the time between initiating the sending of a message in a service and receiving the message in the next service. The global minimum is 77 μs and the global maximum is 618 μs, which will achieve the expected latency values. The stepwise behavior is due to the asynchronous call and the time interval between the request.

For long-term validation, only the individual time violations of messages, control tasks and communication tasks are recorded. Due to the non-periodic system jitter of the hardware, time violations are expected for the test system, which are presented in Fig. 1b. Time violations of the tasks occur permanently, with the majority of the values below 50 μs. The time violations of the communication tasks result in values of up to 10 ms and longer. The time violations of the messages occur both frequently and infrequently. The different number of time violations of the messages in the measurement can be explained on the one hand by a changed execution sequence of the EDF, which is caused by the delays of the jitter. On the other hand, the different number of time violations can be attributed to the influence of the non-periodic system jitter. Depending on whether the non-periodic jitter falls into the execution of the real-time tasks, additional delays occur.

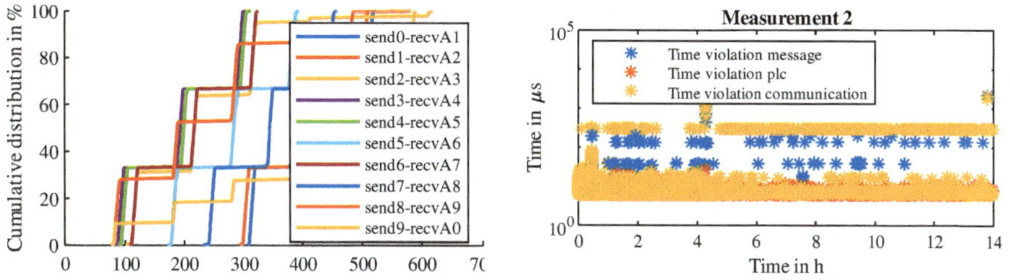

(a) Measurement of τ_G

(b) Long-term timing behavior

Fig. 1. Asynchronous execution

The jitter of the test system is so large that the parametrization used for this system cannot be used in critical areas due to too many time violations. On the other hand, it was shown that the system runs stable during a long-term test and that time violations are detected as expected.

To be able to use this hardware despite its poor real-time characteristics, it is examined whether synchronous processing improves the time behavior. Synchronous processing changes the latency τ_G of the messages to the time between the request of sending and receiving of the next service. The duration is between 0 and $T_{i,\text{st}}$ since the messages are sent from one service to the next in line. The maximum occurring latency considering the jitter is $\hat{\tau} = T_{i,\text{st}} + J_G = 10015.75\,\mu\text{s}$.

Here, the jitter is equal to the jitter of the asynchronous processing if the jitter of the control program is taken into account. Furthermore, the execution duration of the service tasks is set to $r_{i,\text{st}} = d_{i,\text{st}} = 200\ \mu\text{s}$ because sending and receiving requires additional time.

In Fig. 2a the time behavior of τ_G is analyzed by a two-minute measurement. In the cumulative distribution, constant latencies occur, which is noticeable a slight jitter.

Furthermore, the behavior of the time violations is examined in a 6 h test, which is shown in Fig. 2b. The measurement shows permanent time violations of the service tasks, whereby permanent time violations of the messages are also registered. This shows that adapting the design decisions can reduce the number of time violations, but the influence of the non-periodic jitter of the test system can not be prevented. For a communication with permanently fewer time violations in all versions, the use of other hardware with less jitter is necessary.

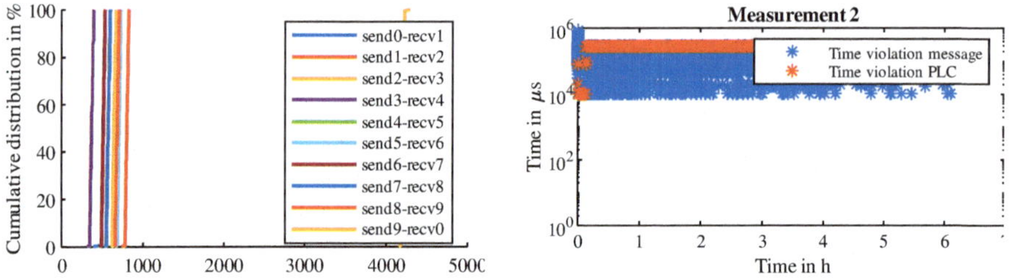

(a) Measurement of τ_G (b) Long-term timing behavior

Fig. 2. Sync execution

5 Conclusion

The related works only consider partial aspects in the implementation of real-time in messaging with a lack of a comprehensive approaches. For this reason, a holistic concept for a real-time capable and message-based communication was presented by combining the antagonistic requirements in the best possible way. Cyclic scheduling is often used in today's control technology. In the event-based execution model, the scheduling methods must still be thoroughly investigated in order to determine under which conditions a sufficient statement about the time behavior can be made and whether this time behavior satisfies the control applicaton. The event-based execution model depends strongly on the use case as well as on the number and frequency of events that occur. Therefore, an event-based execution model, in which communication is taken into account, must first be implemented.

Synchronous or asynchronous processing has a significant influence on the latency, which was shown in the implementation example. Even with an event-based execution model and event-based communication at the same time, the additional tasks must be taken into account in scheduling. Consequently, it can be said: The more global and predictive the scheduling, the better the real-time behavior.

Furthermore, the influence of the hardware on the real-time behavior must be considered. The system parameters must be determined to make a statement about the real-time behavior of the given hardware. In addition to a permanent jitter in some computer systems, non-periodic jitter also occurs, which cannot fulfill the temporal requirements for real-time. However, synchronous processing can improve the time behavior, as the validation shows. The validation showed that the concept works and behaves as expected. A validation with suitable hardware is still outstanding, in order to be able to use the system in a more time-critical manner.

References

1. Altenberg, J.: "Using the realtime preemption patch on ARM CPUs," *11th Real-Time Linux Workshop*, pp. 229–236 (2009)
2. Bauer, D., Garcés-Erice, L., Rooney, S.: "A Simple Time-Sensitive Messaging System," *1st IEEE Workshop on Autonomic Communications and Network Management* (2007)
3. Buttazzo, G.C.: Hard real-time computing systems. Springer, New York (2005)
4. Dragoni, N., et al.: Microservices: Yesterday, Today, and Tomorrow. Present and Ulterior Software Eng. **27**, 195–216 (2017)
5. Fowler, M., Lewis, J.: "Microservices: a definition of this new architectural term,". https://martinfowler.com/articles/microservices.html. Accessed 7 Nov 2018
6. Heinrich, B., Linke, P., Glöckler, M.: Grundlagen Automatisierung: Sensorik. Regelung, Steuerung, Wiesbaden, Springer Vieweg (2015)
7. Hohpe, G., Woolf, B., Brown, K.: Enterprise Integration Patterns: Designing, building, and deploying messaging solutions. Addison-Wesley, Boston (2004)
8. Liang, Y., Tang, X., Bing, L., Chen, D.:"Study on Service Oriented Real-Time Message Middleware," *11th International Conference on Semantics, Knowledge and Grids*, pp. 207–211, Beijing (2015)
9. Roth, A. (Ed.): *Einführung und Umsetzung von Industrie 4.0.*, Springer, Berlin (2016)
10. Schnell, G., Wiedemann, B.: Bussysteme in der Automatisierungs- und Prozesstechnik. Vieweg+Teubner, Wiesbaden (2006)
11. Tasci, T., Melcher, J., Verl, A.: "A Container-based Architecture for Real-Time Control Applications,". IEEE International Conference on Engineering, Technology and Innovation, pp. 1–9 (2018)
12. Tasci, T., Walker, M., Oliver, R.: "From monolith to microservices: A container-based en 61131-3 architecture,". 48'th International conference on computers & industrial engineering, pp. 1–9 (2018)
13. Wang, Q., Xu, J.-g., Wang, H.-a., Dai, G.-z.: "Adaptive real-time publish-subscribe messaging for distributed monitoring systems,". 2nd IEEE International Workshop on Intelligent Data Acquisition and Advanced Computing Systems–Technology and Applications, pp. 412–417 (2007)

Part C Advances in Manufacturing Processes and Materials

A New Adjustable Hemming Die for Automotive Body Construction: Simulation, Design and Experiments

Moritz Nowack[1] and Arndt Birkert[2(✉)]

[1] Hochschule Heilbronn, Max-Planck-Str., 39, 74081 Heilbronn, Germany
moritz.nowack@hs-heilbronn.de
[2] inigence gmbh, Bernbachstr, 36, 74626 Bretzfeld, Germany
arndt.birkert@inigence-gmbh.de

Abstract. In the present investigation a conception of an adjustable hemming die for the assembly of a roller hemmed engine bonnet in automotive industry is presented potentially permitting time and cost savings in serial production. Numerical simulations are performed revealing that the fitting accuracy of the roller hemmed part can, in principle, be drastically improved. Based on the results, a mechanical construction of the hemming die is developed, a realized prototype of which is experimentally explored. Tests demonstrate that the fitting accuracy of the assembled part can be manipulated significantly.

1 Introduction

The construction of a dimensionally accurate car body is a huge challenge in automotive industry, involving diverse manufacturing processes such as sheet metal drawing, cutting, hemming, mechanical joining and gluing [1]. In practice, the complexity of the individual steps, as well as their interaction, hardly admit a precise physical modelling or a valid prognosis of dimensional errors and, frequently, the production routines require to deal with errors of other production stages in an ad-hoc manner.

As an example, the doors and closures of a car body mostly consist of sheet metal stampings, typically an inner structural and an outer skin part, being joined by roller hemming. Already in the forming process of these stampings spring-back effects occur causing both dimensional and form deviations compared to the nominal components [2]. Such undesired deformation of the single components in combination with further shortcomings in the subsequent hemming process can affect the dimensional accuracy of the final assembly significantly [3], resulting in varying gap widths and transition offsets between the respective parts and their neighbors (Fig. 1) [4]. Reworking of the stamping tools to eliminate these defects (Fig. 2) is an extremely time and cost-consuming procedure, rather being considered a last resort in practice [2]. However, such a procedure is often inevitable owed to the fact that the classical hemming beds, being used in industry, lack sufficient tool adjustment options and therefore are inappropriate to cope with errors of previous production stages.

P. Weißgraeber et al. (Eds.): *Advances in Automotive Production Technology – Theory and Application,* ARENA2036, pp. 275–285, 2021. https://doi.org/10.1007/978-3-662-62962-8_32

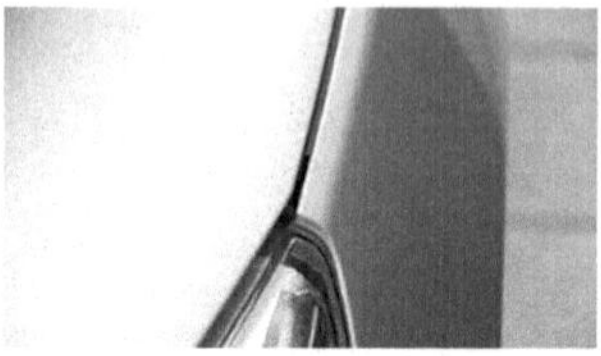

Fig. 1. Transition Offset

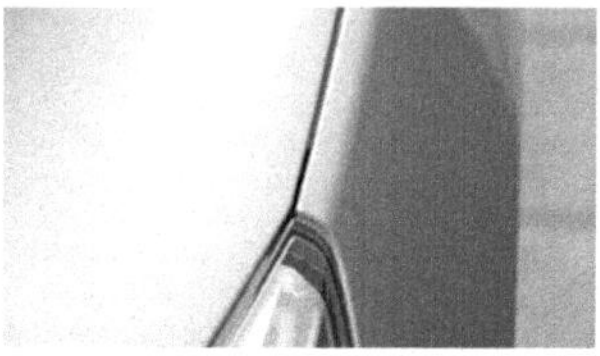

Fig. 2. No Offset

The geometry of assemblies which only consist of structural parts is normally adjusted in the body shop by varying the position of the parts prior to the joining process [5], a procedure known as *matching* in industry. The fitting accuracy of nearly every assembly can be influenced by matching [3, 6] and, therefore, it is only appropriate trying to extend the concept of matching to the hemming process as well. In this spirit the present investigation explores the possible design and simulated behavior of an *adjustable* hemming die which is, via adequate tuning of the active surface, potentially capable to improve the fitting accuracy of roll-hemmed doors and closures directly in the body shop. To the best knowledge of the authors this is the first attempt to explore such an adjustable hemming die.

2 FE-Based Optimization of Hang-On Parts by Modification of the Active Surface of a Virtual Hemming Die

In the following investigation the concept of an adjustable hemming die is considered for the manufacturing process of a specific body part, namely an engine hood/bonnet, being assembled from an inner structural and an outer skin component (Fig. 3, 4); both consist of 6016 aluminum alloy with a sheet thickness of less than 1.0 mm. For this purpose the complete manufacturing process, involving both the stamping of the single components and the assembly by roller hemming, is simulated via the commercial FE software AutoForm^{TM} R8[1]; for similar investigations based on other software, see also, [7] and references therein.

In practice the dimensional accuracy of the stamping parts is compromised by springback effects which can, due to different requirements of the surface quality, more easily been compensated for structural inner parts than for outer skin parts. Thus, in the present context, the structural stamping part is assumed to correspond to the target geometry

[1] For drawing and hemming simulations the standard parameters of AutoForm^{TM} R8 are used.

(Fig. 4) while the outer skin part exhibits spring-back deviations of more than 5 mm being determined by forming simulation (Fig. 3); these deviations between the spring-back and the target geometry have been measured in normal direction to the latter while both geometries are positioned according to a best-fit approach.

However, despite these deviations, according to industry standards, the conception of the hemming die would rely on the target geometries of the single parts only, which explicitly means that the active surfaces of the hemming die correspond to the target geometries of the single parts. Based on these assumptions a first hemming simulation has been conducted with the deformed single parts, leading to the result in (Fig. 5). The deviation, measured in the same way as above, is less than 2 mm in magnitude which is significantly less compared to the deviation of the single outer part. Obviously, a positive compensation effect occurs as a result of the hemming process which is, however, insufficient to achieve an aspired fitting tolerance of ±0.5 mm. In summary, the simulation result of the hood based on a classical hemming die shows that 72% of the transmission areas of the roller hemmed hood, namely those areas that will be adjacent to other hang-on parts in the final assembly, are outside the tolerance.

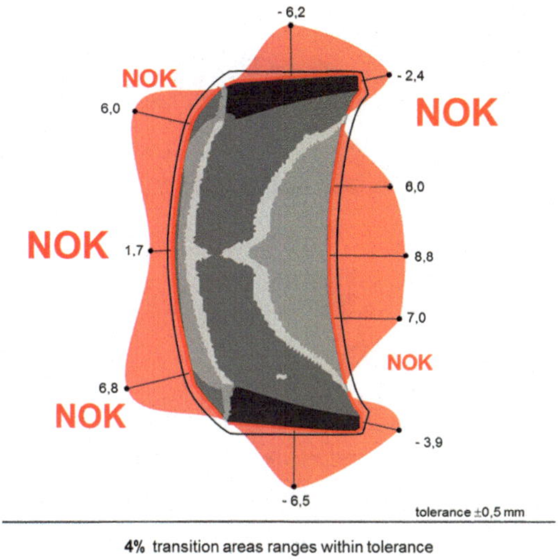

Fig. 3. Fitting accuracy of the outer part

To improve the fitting accuracy of the hood, the active surface of the hemming die has been iteratively updated based on geometrically non-linear elastic calculations via the simulation software ANSYS^{TM}. For this purpose, the deviation between the previous geometry, obtained by the conventional hemming process, and the ideal target geometry is measured and used to adapt the virtual active surface of the hemming die. For the present investigation, the virtual adjustment of the active surface is done in a continuous way and otherwise not subjected to any limitations, except that area and arc length measures always correspond to those of the target parts. After repeated simulation of the hemming process, using the adapted active surface, the deviation is measured again

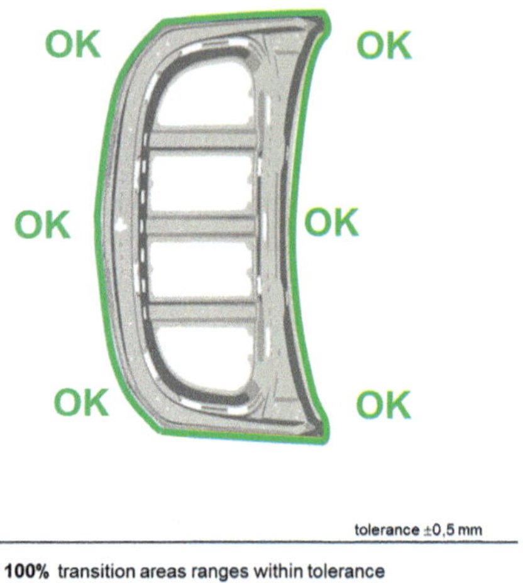

Fig. 4. Fitting accuracy of the inner part

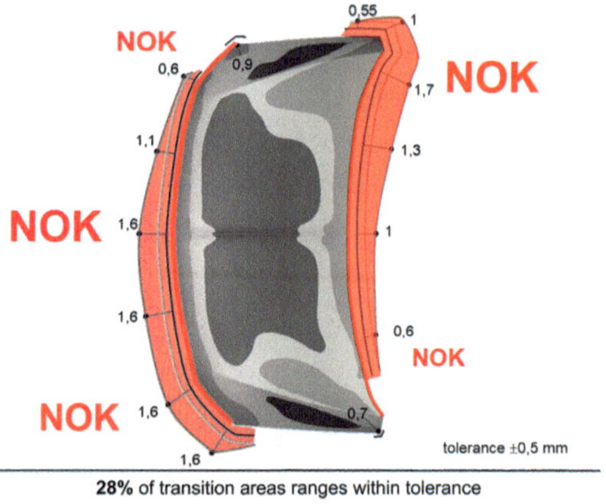

Fig. 5. Fitting accuracy using a classical hemming die

and the result employed for another optimization iteration and so on. This procedure is similar to the spring-back compensation of stamping parts which is performed in an analogous iterative fashion [8].

After four virtual iterations with adjustments of the active surface the simulation result satisfies the fitting tolerance of ± 0.5 mm apart from the – less relevant – area near the wind screen (Fig. 6). In the course of the iteration, the active surface has been adjusted in a range between ± 2.20 mm. For selected positions, corresponding to those of possible setting units of a realistic hemming die, the displacement of the modified active surface in normal direction to the initial configuration is depicted in (Fig. 7). A local deformation measure (▲) for the modified active surface, which has been numerically obtained taking into account a point ($\times$) and its respective neighbor points (Fig. 9, dashed

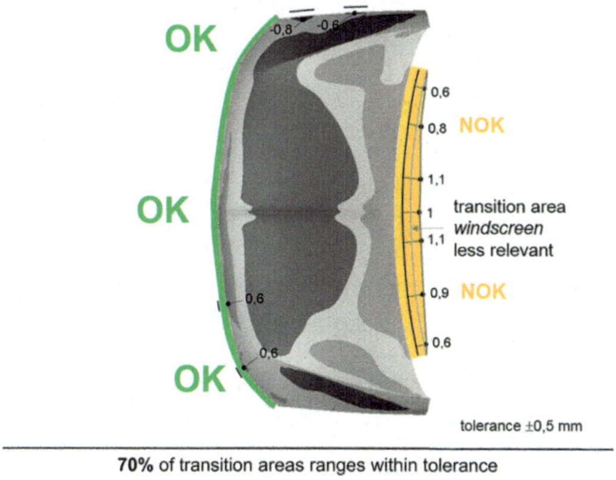

Fig. 6. Fitting accuracy using an adjustable hemming die

line), leads to values of 0.4 mm on the average (Fig. 8, 9 and Fig. 10). The critical area with respect to this local deformation measure is located on the right side between the A-pillar and the indicator, showing values up to 0.7 mm.

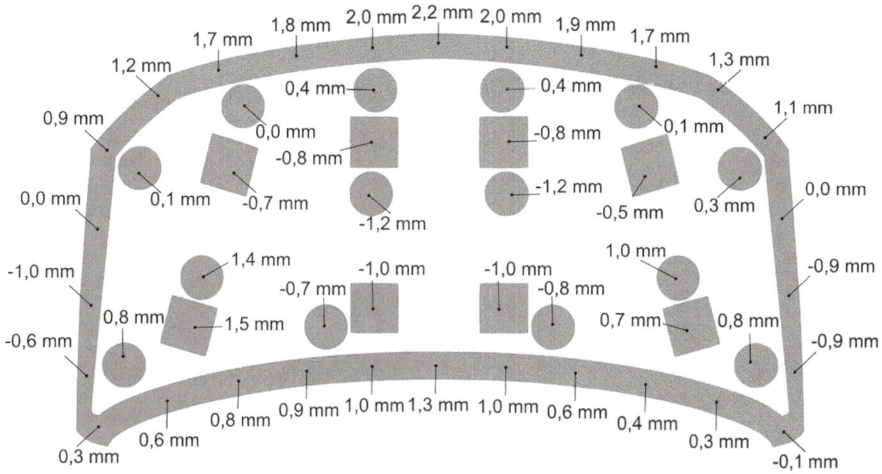

Fig. 7. Required hemming bed adjustment to improve the fitting accuracy

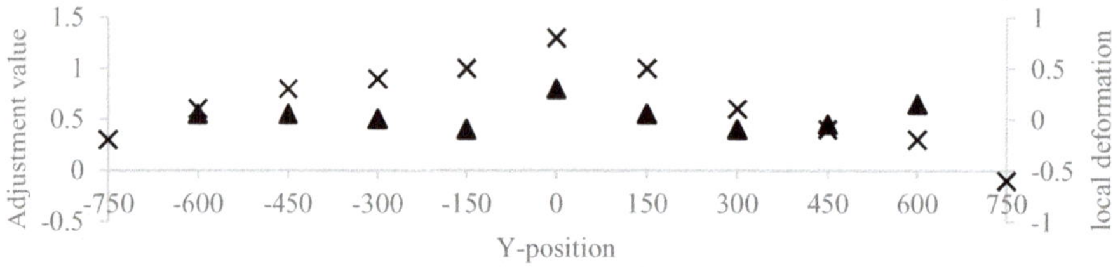

Fig. 8. Local deformation of the hemming crown along the windscreen

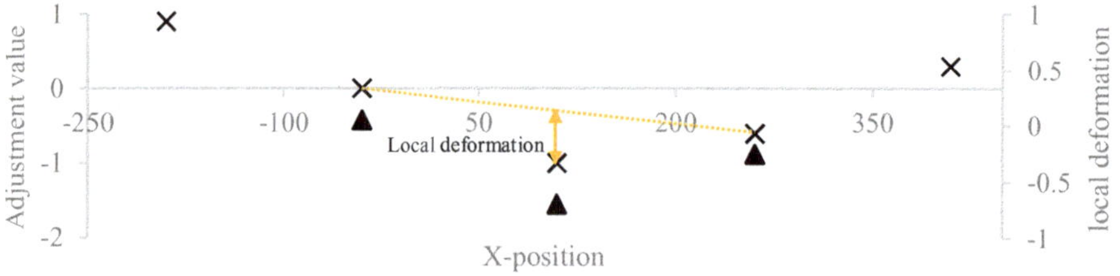

Fig. 9. Local deformation of the hemming crown along the fender (right side)

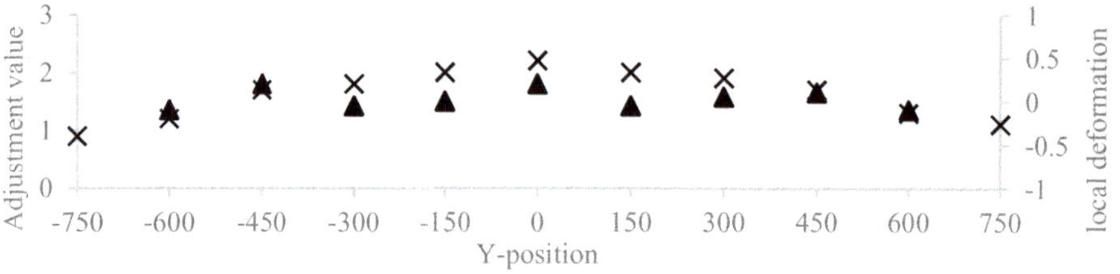

Fig. 10. Local deformation of the hemming crown along the grille

3 Construction of an Adjustable Hemming Die

The adjustable hemming die enables an adaptation of the inserting and clamping situation of the parts before and during the roller hemming process so that the dimensional accuracy of the respective assembly – especially with regard to any transition offset defects – can be significantly influenced and thus improved during the series launch as well as in series production.

In the following a hemming die for the assembly of an engine bonnet is constructed which principally enables the adjustment of the active surface according to the requirements of Sect. 2. Basically, the construction consists of three components: a base frame, an adjustable crown and an adjustable fixing frame (Fig. 11). In addition, 28 standardized setting units are provided to adjust the crown. The distance between these setting units is chosen to be 150 mm, so that a local, regional or global adjustment of the crown is possible in a reversible way (Fig. 12). The direction of action of the setting units is orthogonal to the active surface.

The connection between the 3d-printed fixing segments and the fixing frame is standardized. Adjustable angle sheets can be used to react to different settings of the crown and to keep the inner and outer part in place during the hemming process.

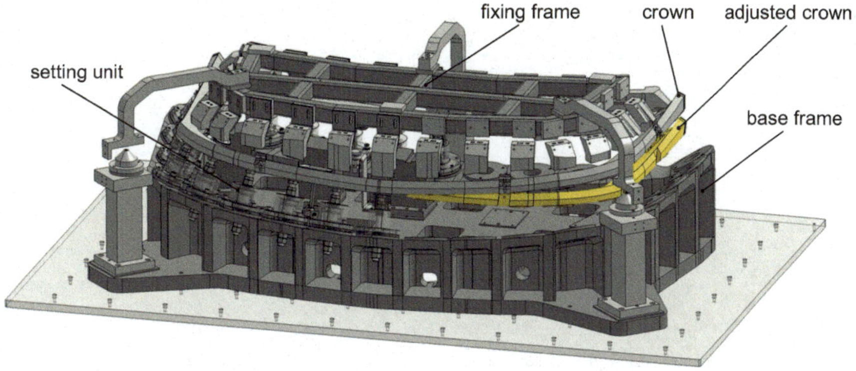

Fig. 11. 3D-model of the adjustable hemming die

When deforming the hemming crown the setting units apply tension or pressure forces in orthogonal direction. As the geometric non-linearity of the parts have to be taken into account during the adjustment of the hemming crown, the setting units are equipped with the degrees of freedom of a bearing: Translational degrees of freedom in the plane which is parallel to the active surface and rotational degrees of freedom in order to change the angle between these planes (Fig. 12). By this means the risk of a critical stress occurrence, which would deform the hemming crown permanently, is significantly reduced.

Due to manufacturing restrictions, the hemming crown (made of EN-GJS-500-7) is constructed with a thickness t of 50 mm. Calculations show that a deformation of the crown by 0.4 mm, corresponding to the average adjustment in (Fig. 7), requires a force of nearly 80 kN and leads to bending stresses close to the yield limit. However, for practical applications it would be desirable to achieve an adjustment range as wide as possible, at best even more than 0.4 mm, and involving stresses *below* the yield limit only. A decrease of the induced bending moment could, for instance, be achieved by modification of the thickness t of the crown, the distance between the setting units l or by changing the material to one with more favorable material parameters; possible modifications of the construction according to these a considerations will be part of further research.

Now, the hemming die crown has been simulated[2] according to the above construction in order to check the resulting stresses induced when the crown is deformed as shown in (Fig. 7). The von Mises stress resulting from the simulated adjustment is depicted in (Fig. 13 – left-hand side) and is calculated to be very close to the yield strength of the employed cast iron EN-GJS-500-7 of 320 MPa. Only in the critical area between the indicator and the A-pillar the yield strength is exceeded. Further calculations reveal that the deformation of the crown remains reversible only if the adjustment value at the fender is reduced from −1.0 mm to −0.8 mm (Fig. 13 – right-hand side).

[2] The deformation is simulated in a purely elastic way (E = 1,7e5 MPa, ν = 0,28).

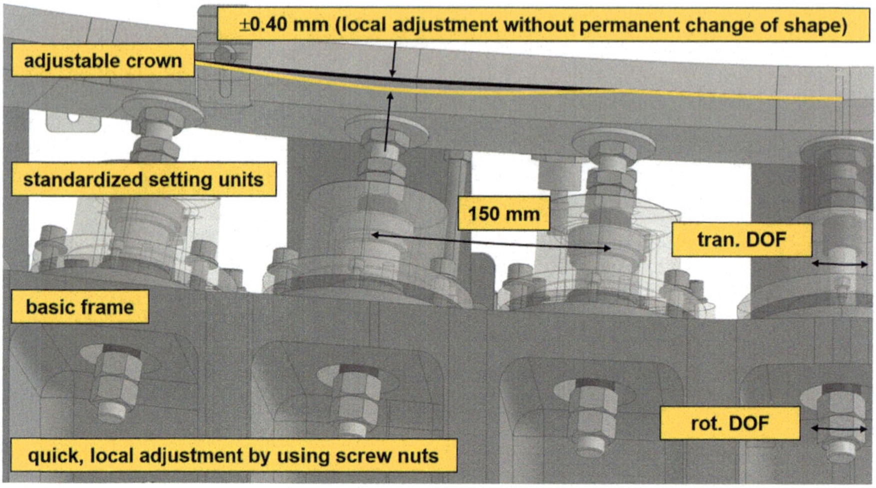

Fig. 12. Construction setting for the local adjustment

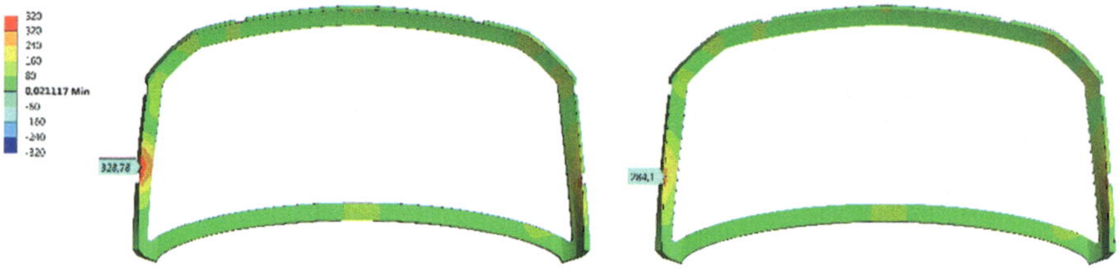

Fig. 13. Von Mises stress as a result of the adjustment of the hemming crown

4 Validation of an Adjustable Hemming Die Prototype for the Manufacturing of an Engine Bonnet

Following the mechanical construction of an adjustable hemming die in Sect. 3, a realized prototype of the same is validated for a hemming process of an engine bonnet. Certainly, a primary objective of experiments would be, first, an adjustment of the hemming die according to near-optimal rules obtained by the iterative simulation procedure described in Sect. 2 and, second, a comparison of the really produced and optically measured assembly with the result predicted by numerical simulation. However, the focus of this *very first* experimental investigation is rather to explore the *limits* of the adjustment range of the constructed hemming die while the detailed investigation of optimal settings will be part of forthcoming research.

The reason to explore the adjustment limits is twofold. In practice, it is often necessary to produce hemmed parts deviating from the target geometry; for instance, in order to compensate for gaps to neighbored hang-on parts that already exhibit dimensional inaccuracies. Thus, for a flexible application of the presented hemming die it would be of interest to achieve compensation ranges as wide as possible. Furthermore, it is highly relevant to check whether extreme adjustment settings, which might expose the hemming crown to plastic deformation effects, lead to undesirable shortcomings. In order to explore the limits of the adjustment range, the values obtained by numerical simulation (Fig. 7) are multiplied by a factor of 1.5 which, according to the considerations of Sect. 3 and additional calculations, leads to a predicted plastification of the hemming crown, being accepted, but not yet to a failure of the crown.

In the conducted experiments, two nearly identical pairs of inner and outer parts are roller hemmed. The first pair is roller hemmed with the default settings, which correspond to the classical hemming die with an active surface identical to the geometry of the target assembly. The accuracy of the assembled engine bonnet is obtained as the distance between the optically measured parts and their ideal target geometry, measured

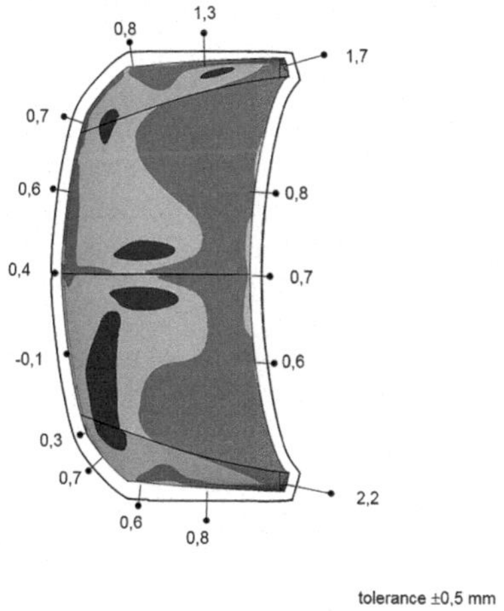

Fig. 14. Fitting accuracy of the assembly with a classical hemming die

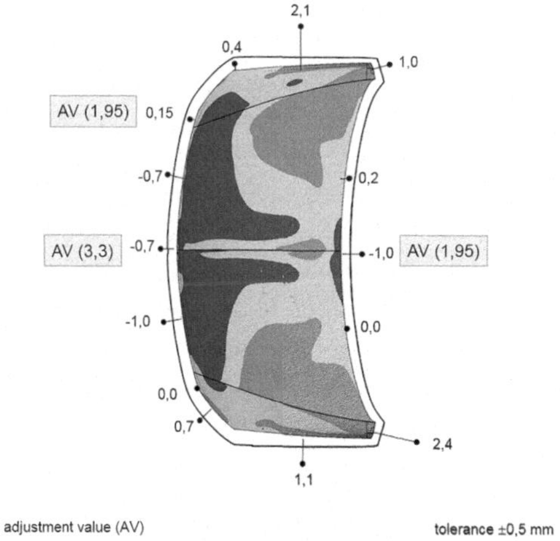

Fig. 15. Fitting accuracy of the assembly with an adjusted hemming die

in normal direction from the latter, and the corresponding results are provided in (Fig. 14). The second pair is roller hemmed with adjusted settings of the hemming die, the values in (Fig. 7) multiplied by a factor of 1.5, and the resulting fitting accuracy is depicted in (Fig. 15).

As predicted by simulation, the experiments do not lead to a failure of the hemming crown. Furthermore, a comparison of the results in (Fig. 14) and (Fig. 15) demonstrates that an adjustment of the hemming die parameters has a significant impact on the fitting accuracy of the assembly. The highest adjustment value of 3.3 mm at the grill influences the fitting accuracy by 1.1 mm. The area near the windscreen is adjusted by 1.95 mm and the fitting accuracy changes by 1.2 mm while, around the indicator, an adjustment of 1.95 mm leads to a change in fitting accuracy of 0.58 mm. The identification of a general relationship between variations of the adjustment parameters and their impact on the fitting accuracy is non-trivial and will be object of upcoming investigations.

The results are very promising at the present stage, however, for a realistic application of the hemming die even higher changes in the fitting accuracy would be of interest in order to close undesired transmission offsets between neighbored hang-on parts. For this purpose, forthcoming investigations will consider modifications of the hemming die construction, as proposed in Sect. 3, and even a systematic local weakening of the hemming crown, for instance by predefined material cuts, in order to allow for larger deformations.

5 Summary and Conclusion

In automotive industry, the dimensional accuracy of roll hemmed hang-on parts can suffer from both the undesired deformation of the single stampings, as a result of spring-back effects in the forming process, as well as from a deficient conception of the hemming die itself. In contrast to fixed hemming facilities, commonly used in industry, an adjustable hemming die allows for a flexible reaction to such uncertainties and thus potentially permits huge cost and time savings in the serial production.

As part of the present work such an adjustable hemming die has been simulated, mechanically constructed and a realized prototype has been experimentally explored. The results of the hemming simulation with an iteratively optimized active surface of the hemming die, Sect. 2, demonstrate that the fitting accuracy of the assembly can, in principle, be improved dramatically. Based on these results a possible design of the hemming die is presented in Sect. 3. Experiments in Sect. 4 reveal that by adjustment of the hemming crown, using extreme parameters, the fitting accuracy of the engine bonnet could be influenced by up to 1.7 mm. An improvement of the die design in order to allow for a possibly larger compensation range and an analysis of the correlation between the adjustment value of the active surface and the resulting influence on the fitting accuracy of an assembly will be part of further investigations.

References

1. Eckert, A.: Prognose der Maßhaltigkeit punktförmig mechanisch gefügter Karosseriean-bauteile. Dissertation, Fraunhofer-Institut für Werkzeugmaschinen und Umformtechnik IWU, Chemnitz (2012)
2. Birkert, A., Haage, S., Straub, M.: Umformtechnische Herstellung komplexer Karosserieteile. Auslegung von Ziehanlagen. Springer Vieweg Verlag, Berlin (2013)
3. Endres, M.: Entwicklung einer aktiven Steuerung für die geometrischen Qualitätsziele der Prozesskette Karosseriebau in der Vorserie. Dissertation, Chemnitz (2012)
4. Hönle, S.: Methode zur Bewertung und Prognose der Anmutungsqualität und der Her-stellbarkeit von Falzschlaufen an Karosserieteilen aus Aluminium. Dissertation, Stuttgart (2016)
5. Kästle, C.: Simulationsmethode zur Beurteilung der Maßhaltigkeit von rollgefalzten Karosseriebaugruppen im Zusammenbau. Dissertation, IFU, Universität Stuttgart (2016)
6. Hammett, P., Baron, J., Majeske, K.: Evaluating multi-dimensional sheet metal assemblies using a biased sample of stamped components, University of Michigan (1997)
7. Kaiser, C.: Effiziente Produkt- und Prozessabsicherung für gefalzte Karosseriebauteile durch ein metamodellbasiertes Assistenzsystem. Dissertation, München (2017)
8. Birkert, A., Hartmann B., Straub, M.: New method for springback compensation for the stamp-ing of sheet metal components. 36th IDDRG Conference. J. Phys. Conf. Ser. **896**(1), 012067 (2017)

Production of Thin Outer Skin Car Body Panels by Using Novel Short Cycle Stretch-Forming (SCS) Technology

Mathias Liewald(✉) and Adrian Schenek

University of Stuttgart, 70174 Stuttgart, Germany
mathias.liewald@ifu.uni-stuttgart.de

Abstract. Large sheet metal parts with low drawing depths usually reveal only low work hardening levels in the central area of the components. As a result, such parts often do not meet specified property requirements. Short-cycle stretch technology (SCS) represents a novel forming process for combined stretching and deep drawing, which allows high-quality production of such large sheet metal parts with improved mechanical properties. Due to a consistently simple tool design with changeable die inserts, SCS technology additionally offers a cost-effective and relatively flexible alternative to currently known processes for combined stretch forming and deep drawing. This paper defines the basic principle and presents close-to-production tool concepts of the newly developed SCS process. Furthermore, the high energy saving potential of the novel forming technology is illustrated in comparison to conventional deep drawing.

1 Introduction and State of the Art

The reduction of fuel consumption by applying lightweight construction concepts constitutes a substantial development focus in the automotive industry. Therefore, lightweight materials and optimized sheet metal components are increasingly used in modern vehicle body design. In addition, modern sheet metal parts must fulfil complex technological and design aesthetic requirements. Automotive outer skin components, for example, do not only have to meet highest standards of appearance but also require high resistance to dynamic buckling (e.g. hail or stone impact). This me-chanical quality aspect can be controlled by work hardening sheet metal materials during the forming process [1–3], which usually consists of combined stretching and deep drawing operations [3, 4]. Main advantages of pre-stretching prior to the deep drawing process have already been examined by [5, 6]. A defined pre-stretching resulted in a stepwise increase of buckling resistance of sheet metal components [7, 8]. However, a disadvantage of conventional stretch forming methods is given by the complex tool design according to valid standards in industry, which leads to comparatively high investment costs and increased energy consumptions to run such established tool concepts in production. In the view of today's high degree of diversification of outer skin car design in automotive industry related to reductions in production quantities, new stretch forming and deep drawing processes are necessary to optimize existing production lines.

P. Weißgraeber et al. (Eds.): *Advances in Automotive Production Technology – Theory and Application,* ARENA2036, pp. 286–292, 2021. https://doi.org/10.1007/978-3-662-62962-8_33

In this respect, the SCS technology discussed in this contribution offers a cost-effective and flexible alternative to meet today's challenges in car body manufacturing in terms of flexibility of car body making equipment and reduced budget for such equipment.

2 Short Cycle Stretch-Forming (SCS)

2.1 Basic Function of SCS-Technology

The SCS-Technology has been developed and patented at the Institute for Metal Forming Technology (IFU) at the University of Stuttgart and continuously improved in recent years. It is primarily intended for the production of flat outer skin panels such as car doors and car roofs in medium-sized volume production. The SCS process is based on a simple tool design including two bead sets surrounding the outline of deep drawing punch. While the SCS-tool closes, the bead sets interlock and bend the blank alternately in the fringe region. Due to that tension impact, material becomes stretched laterally caused by the friction conditions prevailing between blank and upper/lower bead set. Once the punch contacts the pre-stretched blank, the last interfering bead element acts as a drawing ring and provides required restraining forces. Due to this tool setup, no blank holder is needed. Fig. 1 shows the basic function of the SCS tool design.

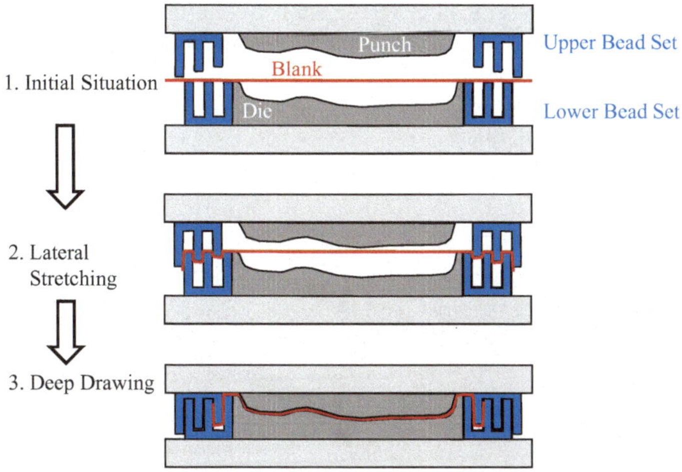

Fig. 1. Process stages of SCS-Forming Technology within same ram movement

2.2 Experimental Research and Validation of Simulation

In order to optimize SCS bead set geometries and for validation objectives of simula-tion parameters, a comparatively simple test device was developed at IFU. In this device, a small blank strip is clamped on one side to measure induced tensile force. On the other side the upper and lower bead set is mounted to simple press frame to interlock both bead sets and to induce tensile force leading to a certain amount of uniaxial plastic strain in the sheet stripe. In combination with a tactile distance measuring system attached to the press ram, the induced tensile force can be recorded as a function of press stroke.

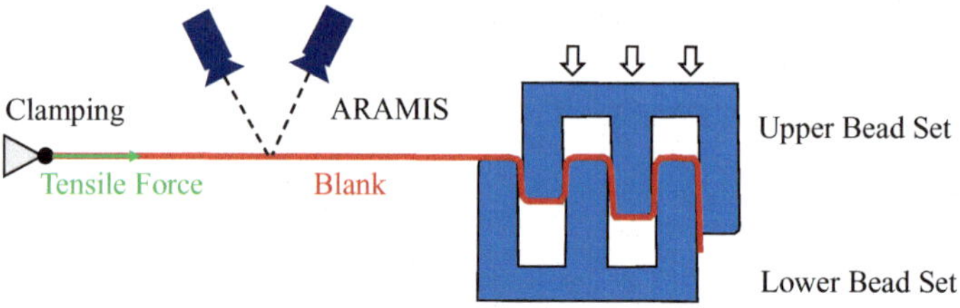

Fig. 2. Schematic of the test device for the validation of simulation parameters and bead shape

The validation of assumed friction parameters in numerical simulation and mathematical optimization of the upper/lower bead shape radii as well as height of beads in research was conducted with respect to the sheet metal material properties and to comparison of numerically calculated and experimentally determined tensile force curves. Fig. 3 shows such a comparison between measured and simulated tensile forces for the sheet metal material DC04 having a sheet thickness of 0.77 mm and a stripe width of 50 mm. The diagram shows the high agreement between FE-model and experiment for this example after optimisation of Coulomb's friction modelling. For performing the numerical investigations, the software codes LS-Dyna and PAM-STAMP were used. The type of finite elements were defined as shell elements, as it is commonly used in sheet metal forming simulations.

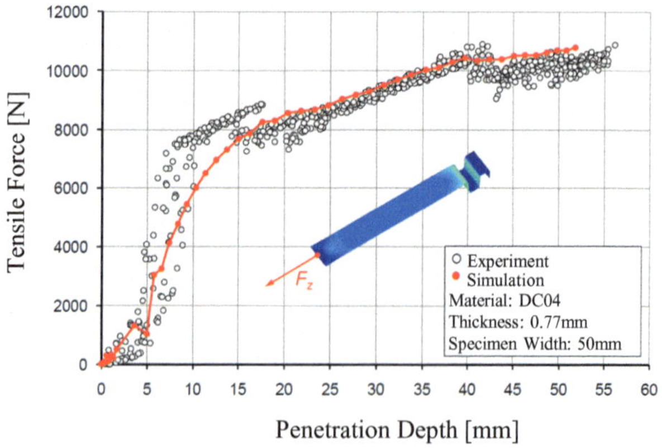

Fig. 3. Validation of simulation parameters – tensile force [7]

Besides comparing the simulated and experimentally measured tensile force curves, the plastic strain value physically induced in a sheet metal strip can be determined by using the GOM-ARAMIS measuring system. As shown in Fig. 4, comparatively high values in effective strain of φ_v = 0,08 can be achieved using sheet metal material DC04 with a sheet metal thickness of 0.77 mm assuming an optimised bead design.

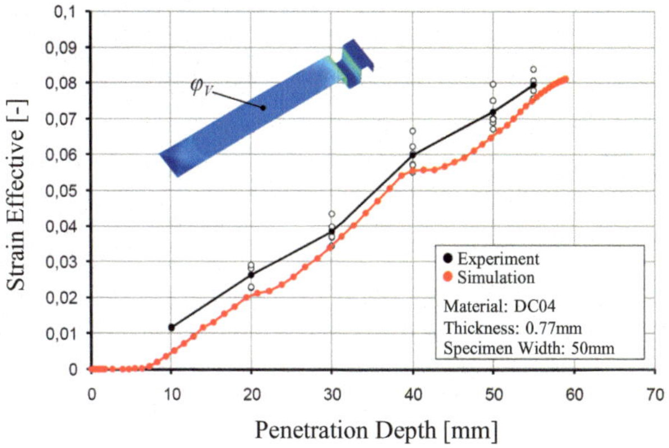

Fig. 4. Validation of simulation parameters – strain effective [7]

In the case of conventionally deep drawn, flat sheet metal components (e.g. automotive car doors or roofs), only comparative small values in effective strain of about φ_v = 0,01 to φ_v = 0,02 can be achieved (also see Fig. 7) in today's production. This leads to lower work hardening especially in central areas of deep drawn components. However, high levels of work hardening are particularly required for outer skin components, as these significantly increase the denting resistance to withstand any hail and stone impact.

2.3 Close-To-Production Tool Concept and Improved Sheet Metal Part Properties

Vlahovic [7] and Papaioanu [8] have already successfully used the SCS technology for new tool designs close to production in order to manufacture scaled car body outer skin panels and to improve robustness of SCS technology. In this context, forming tools for the production of size-scaled automotive outer skin components such as a front door, a rear door and a roof were manufactured. The automotive car door, which is shown in in Fig. 5, was made of the mild steel DC04 having a sheet metal thickness of 0.8 mm.

For the production of a simple car roof, sheet metal aluminium material AC170PX having a thickness of 1.04 mm was used. As shown in Fig. 6, a major strain value of φ_1 = 0.041 could be experimentally measured in the central area of the car roof produced.

In order to compare the SCS process with conventional deep drawing, a conventional deep drawing simulation was carried out using the vehicle roof. Fig. 7 shows the comparison of numerically determined major strain values using SCS-Technology and conventional deep drawing. Gained results of the forming simulations do illustrate, that the SCS process can achieve significantly higher amounts of plastic strain especially in the central area of the roof compared to conventional deep drawing tools consisting of a blank holder, punch and matrice.

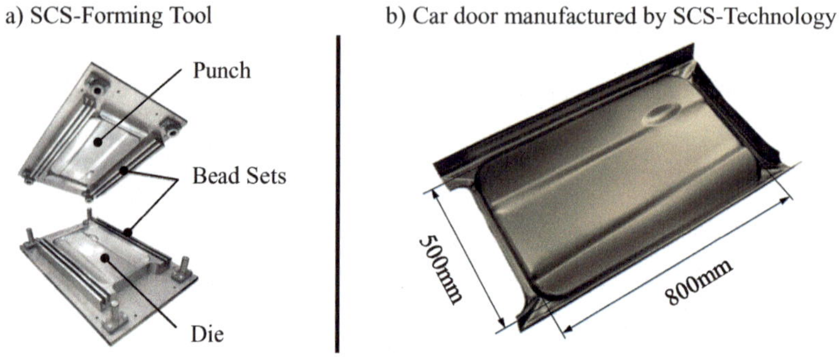

Fig. 5. SCS-Forming tool for an outer skin of a front car door [7]

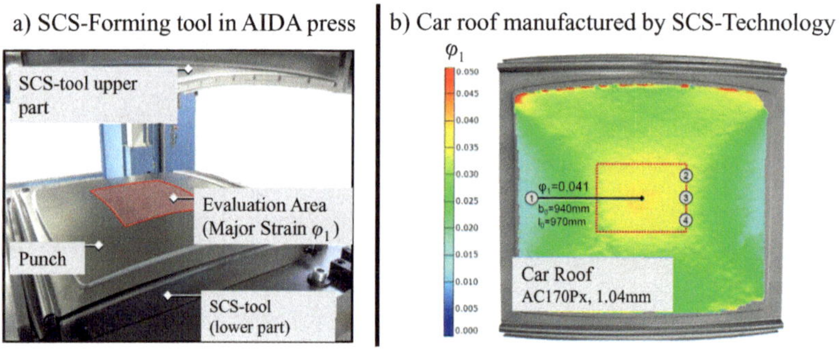

Fig. 6. SCS-Forming Tool and car roof manufactured by SCS-Technology [8]

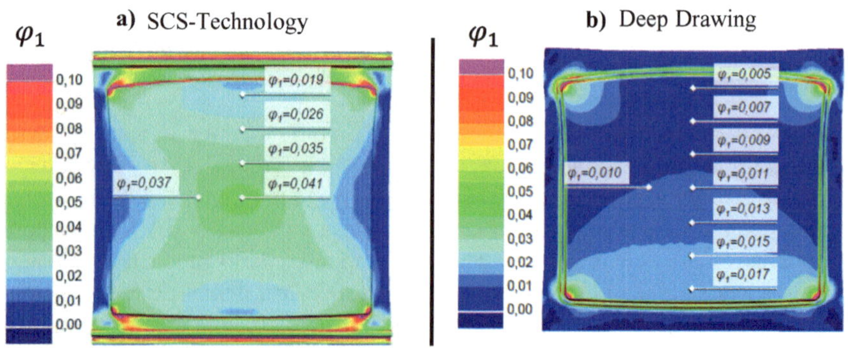

Fig. 7. Numerical comparative study of SCS-Technology and conventional deep drawing [8]

The expected increase in work hardening of the sheet metal material was experimentally verified in [8] and resulted into a significantly increased denting resistance of the sheet metal component produced that way. In addition to performed properties of sheet metal component, a further advantage of the SCS technology was found in a remarkable energy saving potential compared to the conventional deep drawing process, as described in the following chapter.

2.4 Energy Saving Potential of the SCS-Process

The material draw-in of part flange in conventional deep drawing processes is controlled by hydraulic cushion forces acting onto the lower side of the blank holder, draw beads and precisely adjusted tribological conditions. A large amount of ram force is therefore required to displace the hydraulic cylinders in the cushion versus the ram movement. In contrast, the SCS process controls material flow solely by means of the bending and friction effects prevailing between the blank and bead sets. Since no cushion is required for the SCS process, the press ram does not have to operate against a (constant) force applied by hydraulic cushion cylinders. Fig. 8 illustrates the much higher energy requirement of the conventional deep drawing process compared to the new SCS process, when manufacturing above mentioned car roof. For the SCS-based forming process, a total energy amount of 19.54 kJ is required. In contrast, conventional deep drawing of the same sheet metal part requires a total energy demand of 84.18 kJ. Thus, SCS technology not only enables the production of sheet metal components with optimized properties in terms of buckling resistance. The use of SCS technology also enables a significant improvement in the energy balance of the forming process.

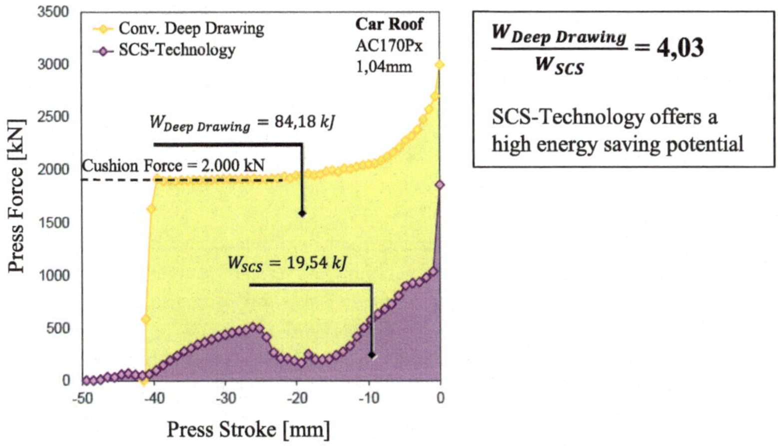

Fig. 8. *Energy to be provided by the press for deep drawing a car roof by conventional deep drawing (yellow) and by SCS-Technology (purple)* [8]

3 Conclusions

Short Cycle Stretch-Forming (SCS) is an innovative forming process to produce flat shaped outer skin panels for car body applications. The SCS technology combines process sequences of lateral pre-stretching of blank and subsequent deep drawing with same stroke. Newly invented technology for car makers and suppliers offers the following advantages:

- Cost-effective tooling concept, since no blank holder and no cushion system in the press bolster is required
- Combined stretch-forming and deep drawing process sequence without using complex clamping systems
- Improved sheet metal part properties in terms of denting resistance
- In comparison to conventional deep drawing, the SCS process requires only 25% of the energy consumption.

Future research at IFU will focus on the transfer of SCS-technology into industrial applications, presumably in terms of further bilateral projects together with car makers and their suppliers. Next projects in this field are intended to develop SCS process strategies ensuring sound hardening effects of component under 'close to production' conditions. This includes investigations on the usability of medium and high strength sheet metal materials and the implementation of corresponding automation concepts.

References

1. Bath, C.: Streckziehen von Karosserieteilen. Werkstatt Betrieb **98**(3), 143–144 (1965)
2. Siegert, K.: Blechumformung. Verfahren, Werkzeuge und Maschinen. Springer Vieweg, Heidelberg (2015)
3. Liewald, M., Siegert, K., Vlahovic, D.: Blank part improvement by defined pre-stretching. In: New Developments in Sheet Metal Forming, pp. 467–490. Verlag MAT INFO, Frankfurt a. M. (2006)
4. Schleich, R., Vlahovic, D., Liewald, M., Sindel, M.: Effect of strain dependent material properties on formability of sheet metals. Materialwissenschaft und Werkstofftechnik. Wiley-VCH (2007)
5. Kiesewetter, T.: Entwicklung und Erprobung einer flexiblen segmentierten Streckziehanlage. Dr.-Ing. Dissertation, Institut für Umformtechnik, Universität Stuttgart (1994)
6. Fann, K.-J.: Streckziehen mit vorgebbaren finalen Bauteileigenschaften. Dr.-Ing. Dissertation, Institut für Umformtechnik, Universität Stuttgart (1996)
7. Vlahovic, D.: Neue technologische Ansätze zum kombinierten Rechen und Tiefziehen von Außenhautbeplankungen aus Feinblech. Dr.-Ing. Dissertation, Institut für Umformtechnik, Universität Stuttgart (2009)
8. Papaioanu, A.: Einsatz eines neuartigen Verfahrens zum kombinierten Recken und Tiefziehen von Außenhauptbeplankungen aus Feinblech. Dr.-Ing. Dissertation, Institut für Umformtechnik, Universität Stuttgart (2016)

Automated Generation of Clamping Concepts and Assembly Cells for Car Body Parts for the Digitalization of Automobile Production

Andreas Zech[1], Ralf Stetter[1], Markus Till[1(✉)], and Stephan Rudolph[2]

[1] Ravensburg Weingarten University (RWU), 88250 Doggenriedstraße, Germany
zechan@rwu.de, andreas.zech@eks-intec.de

[2] Institute of Aircraft Design, University of Stuttgart, Pfaffenwaldring 31, 70569 Stuttgart, Germany

Abstract. A central success factor for the digitalization of production processes is the provision of a consistent database across domains and life cycles. The current situation in the automotive industry is characterized by a multitude of engineering tools with diverse, proprietary data formats, which require complex conversion processes and show dramatic deficits in the consistency and accessibility of the data. In recent years, graph-based design languages have been refined and their range of application expanded to an extent that they represent an interesting approach to addressing these problems. The focus of this paper is on the automated generation of assembly processes and assembly resources (e.g. type-related production equipment such as clamping devices) using the example of automotive body parts (front flap and B-pillar). Within the presented engineering framework of automated production planning, clamping concepts and assembly cells (with automated wiring) are generated using graph-based design languages. Based on the manufacturing concept, a joining sequence and fastener planning is carried out, from which a product-specific clamping and fixing concept is derived. The central advantage of this procedure is, in addition to the high degree of automation, the possibility of providing a consistent database. This database allows the automatic derivation for the various specialized engineering tools.

1 Introduction

The main objective of the presented research is to enable an automated generation of assembly processes and assembly resources using the example of automotive body parts which are use cases of the Center for Applied Research (ZAFH) "Digital Product Lifecycle (DiP)". The aim of the ZAFH research project is the complete digital representation of the product life cycle [1, 2]. The digital modelling of the product life cycle is implemented by means of a language-based engineering framework consisting of graph-based design languages. Extensive research work has made it possible to map such languages in UML (Unified Modeling Language) and led to the development of a suitable design compiler (DC43, IILS GmbH).

P. Weißgraeber et al. (Eds.): *Advances in Automotive Production Technology – Theory and Application*, ARENA2036, pp. 293–301, 2021. https://doi.org/10.1007/978-3-662-62962-8_34

Current processes in industry frequently exhibit a low level of efficiency, intelligence and sustainability in product design, manufacturing and service phases [1]. A close look at the current industrial reality leads to the insight that currently applied product development processes (PDPs), especially in the early phase of order initiation and the subsequent phases of product development are inadequate for automation. Currently, only few suitable, holistic approaches can be identified in industrial practice. In industry, numerous domain-specific programs (e.g. to perform calculations, create geometry or execute simulations) are applied. Although, the sequence of development tools is accelerated as much as possible and the mutual integration of design functions is steadily improved, holistic approaches and truly integrated process chains in development tools are currently not yet applied. The unique contribution of the presented research is the combination of processes, methods and tools which allow the complete digital representation and machine execution and the seamless integration in the digital factory. The main focus of this paper lies on an automated generation of assembly resources (e.g. type-related production equipment such as clamping devices, an assembly cell and wiring) using the example of automotive body parts. Within the framework of production planning, clamping and centering units with parameterized components are supposed to be produced. That leads to a defined setup and fixing concept which is, in combination with layout respectively arrangement planning processes, the basis for an automated geo-station planning process (the term geo-station (GS) describes stations in which the individual parts are brought into the correct position and are finally spatially fixed e.g. before a welding process). In addition to this, a successful geo-station planning process requires further analysis steps such as availability analysis, collision detection, material flow analysis and ergonomics analysis.

2 State of the Art

2.1 Graph-Based Design Languages

A central success factor for the digitization of production processes is the cross-domain and lifecycle-wide provision of a consistent database. An innovative approach is to use a common language to describe the product lifecycle by means of the unified modeling language (UML) with vocabularies (UML classes) and rules (UML activities). Rudolph [4] firstly proposed to use graph-based design languages (GBDL) for engineering design. GBDLs can be understood as a rule-based extension of building-block-oriented configuration tools. At the center are graphs, which are a manifold networked representation of a set of objects. Extensive research has enabled the mapping of such languages into UML and led to the development of a design compiler [5], DC43 of IILS GmbH [6]. A design-graph (product model) is created using a program (in UML: activity diagram) built from vocabulary (in UML: classes) and rules (in UML: model transformations). The program is executed in an engineering framework (the above-mentioned DC43) and a model (design graph) is generated, from which further domain-specific models (for example geometry (e.g. computer aided design (CAD))) and simulation (e.g. finite element method (FEM) models) can be generated automatically. The required abstract representation of a concrete CAD product geometry in the form of a so-called "abstract geometry" (in UML—compare [7]) and for further multi-physical product properties (e.g. in the form of multi-body system, FEM and continuous flow dynamics) models for simulations in the form of a so-called abstract physics (also in UML) in GBDLs, allows the information representation in UML independent of a specific data format in a special domain or from a specific vendor-specific or proprietary data formats. The final transformation from UML into one or more target languages (Domain-Specific Languages) takes place later and through a further machine processing step in the language compiler or its plug-ins. Obviously, other design automation methods or engineering frameworks such as developed in the areas of computational design synthesis [8] or cyber-physical system development [9] allow similar realizations; the unique advantage of GBDLs is the possibility to compile geometry.

2.2 Product Development Process and Geometric Modeling

The product development process proceeds iteratively form early stages such as the definition of requirements to the synthesis of the final detailed design. The research described in this paper builds upon the well-known design methodologies (compare e.g. [9, 10]); it turned out that there is a need for an appropriate design methodology (applied by using GBDLs) which enables automation [11]. Common CAD programs offer a variety of different complex surface modeling approaches with large differences in robustness and result quality. Based on these findings and on some insights concerning appropriate modeling techniques (compare [12]), a procedure for the creation of robust surfaces was developed [11]. The described design procedure (based on design parameter) equals one variant of the general solid model representation - the Boundary Representation (BRep). In addition to the general BRep approach, the already implemented GBDL for abstract geometry of Schmidt [7] offers another variant for modeling geometric shapes—the Constructive Solid Geometry (CSG).

3 Methodology

In this section, the product development process of geo-stations is illustrated in general terms. In this case, the design process is implemented by using a GBDL; the design process is described and performed by UML classes and UML activity diagrams. The aim of this section is to give an overview of an automated design process for a GS.

In the automotive industry, most production plants are highly automated; the technologies include, for instance, welding, gluing, riveting, clinching and folding. The production systems consist of various stations, in which the individual components of the body are interconnected by different joining technologies. The handling and moving, i.e. the transport of the components, as well as the joining processes are taken over by robots, which are equipped with different tools (e.g. grippers, welding guns, gluing tools, etc.). The product components which need to be joined are mostly placed in geo-stations where they are aligned and spatially fixed. The sequence in which the components have to be connected to each other is specified in the joining sequence. The centering and clamping plan specifies the points for aligning and fixing the components in space and in the joining plan (e.g. welding spot plan), the positions of the joints are specified. In general, a geo-station is needed to ensure the position and orientation of several product components which need to be assembled. This is realized by centering and clamping units whose centering and clamping positions and directions are specified locally in the product design (compare to centering and clamping plans in industry). Therefore, geometric CAD models of geo-stations include both the product components themselves and the individual centering and clamping units. These units consist of several parts, such as purchase, standard and/or self-made parts. Figure 1 shows the proposed design process for the automatic generation of requirement-driven geo-stations.

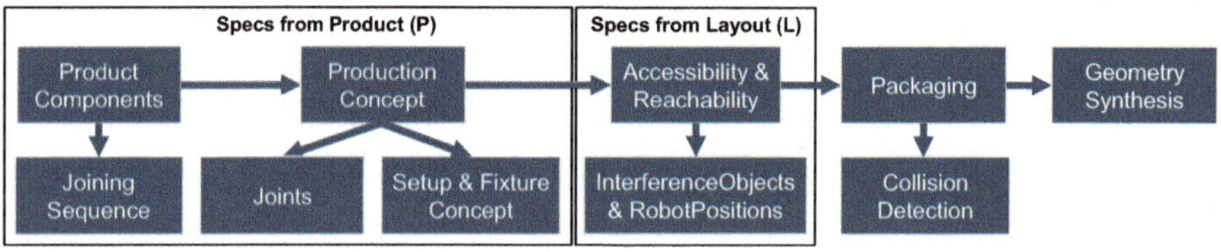

Fig. 1. Overview of individual process steps of a holistic and completely digital represented product development process for geo-station design by using graph-based design languages.

The design process includes the mapping of requirements that are derived from both the product design and the underlying layout design. Based on this, the geometry models of the centering and clamping units are synthesized and finally placed in the space around the product without any collision. Figure 2 shows some derived and abstracted classes and the corresponding class diagram for the geometry synthesis of centering and clamping units. Centering and clamping units consist of a console (UML class *Console*) which is screwed orthogonally onto a base plate. On the consoles, pneumatic components (*PneumaticComponent*) are screwed, which in turn serve as the basis of the clamping arms (*Arm*) respectively centering pins (*Pin*). The actual fixation on the surface of the product component takes places via common and standardized angles (*NCBlock*), which can be spatially adjusted via shim packages (*Shim*). The shim packages serve as intermediate pieces between the clamping arms and NCBlocks.

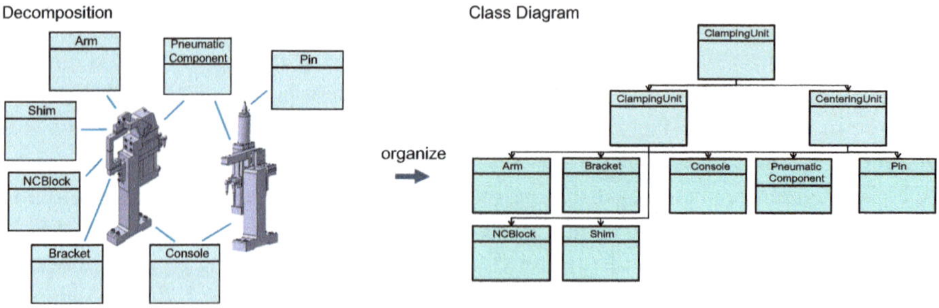

Fig. 2. Decomposition and abstraction of the components of a physical centering and clamping unit (compare [13]—highly simplified representation).

4 Results and Discussion

4.1 Central Data Model

By decomposition of existing, industrial geo-stations, an UML class diagram can be modeled in an abstract manner (compare Fig. 3). As a result of rule-based instantiation processes within a UML activity diagram, Fig. 3 shows a highly simplified representation of the design graph of one specific design variant for geo-station design and its geometric expression in CAD. As shown, the graph includes instances of product components, corresponding production concepts and, of course, geometry instances. Furthermore, the graph already includes corresponding cost objects.

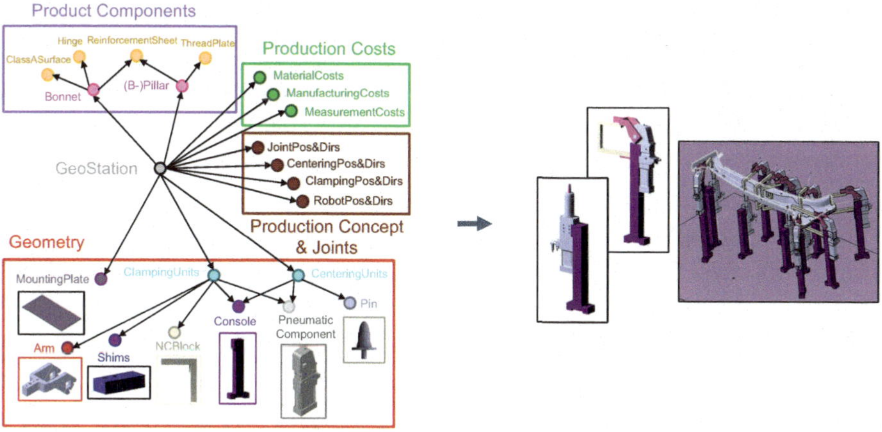

Fig. 3. Simplified central data model (design graph) of a design variant for geo-station design based on a welding process for B-pillars (compare [13]).

4.2 Design Variants

An essential process step within the GS design process is, besides the geometry generation, an initial and fully automatic arrangement of centering and clamping units based on product- and layout-specific requirements. One crucial requirement is the collision-free positioning of the units alongside the related product contour based on specific robot positions in the superordinate layout model as well as the setup and fixture concept of product design. For this, a packaging algorithm has been developed. Figure 4 shows an application example of the algorithm which leads to the corresponding packaging respectively physical model of the geo-station. The packaging model delivers the resulting positions where units may be placed without collision to any interfering object. Interfering objects are for instance robot tools such as welding guns, other centering and clamping units and individual part geometries of the product. The algorithm iteratively investigates possible positions for centering and clamping units by performing collision detections. For a quicker and simplified performance of the collision detection, the interfering objects and thus the resulting positions are initially represented by simple, geometric shapes such as cones. As an example, a turquoise-colored cone in the packaging model represents the corresponding reference volume which will be replaced in the subsequent process step of geometry synthesis by an appropriate geometry model of a clamping unit (Fig. 4).

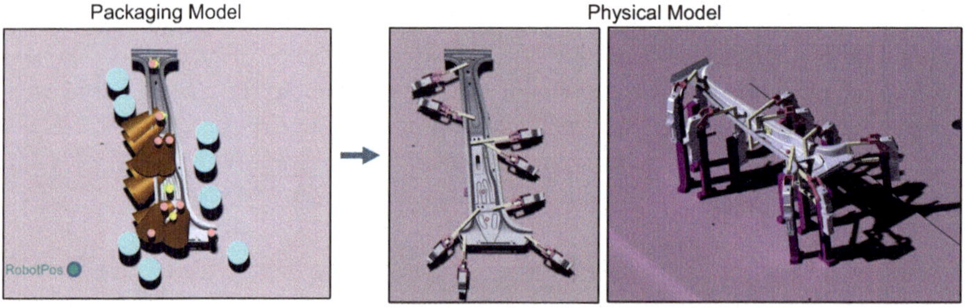

Fig. 4. Application example of the packaging algorithm which lead to the corresponding physical model of the geo-station.

Analogous to the geo-station design process related graph-based design languages were developed for the geometry synthesis of other production resources (such as fences, handling racks, conveyor systems, etc.); these cannot be discussed in detail within this paper. Figure 5 shows one specific geometric design variant of a predefined layout arrangement which was generated in a fully automatic manner.

4.3 Automated Wiring

A routing algorithm was developed for the automated laying of cables. This initially allows solution paths to be specified, i.e. specify a routing network, on which the cables are then laid using the shortest path. The purpose of the cables is initially irrelevant, i.e. it does not matter which medium (e.g. electricity, pneumatics, water, etc.) the cables will later be used for. Figure 6 shows an application example of the routing algorithm for geo-station design. The source of the routing process is a valve terminal attached to a table. Cable ducts are mounted on the side of the table, along which the cables (in this case pneumatic hoses) must be routed to the individual clamping and centering units (pneumatic components).

5 Summary and Outlook

In this research contribution, a new and innovative approach to develop assembly resources, such as geo-stations, is shown. The main objective of this new approach is to automate the underlying design processes. In the developed procedure based on graph-based design languages, the individual product architecture of geo-stations is modeled by means of the unified modeling language in an eclipse-based software tool (DesignCockpit43). The focus of the described product development processes primarily

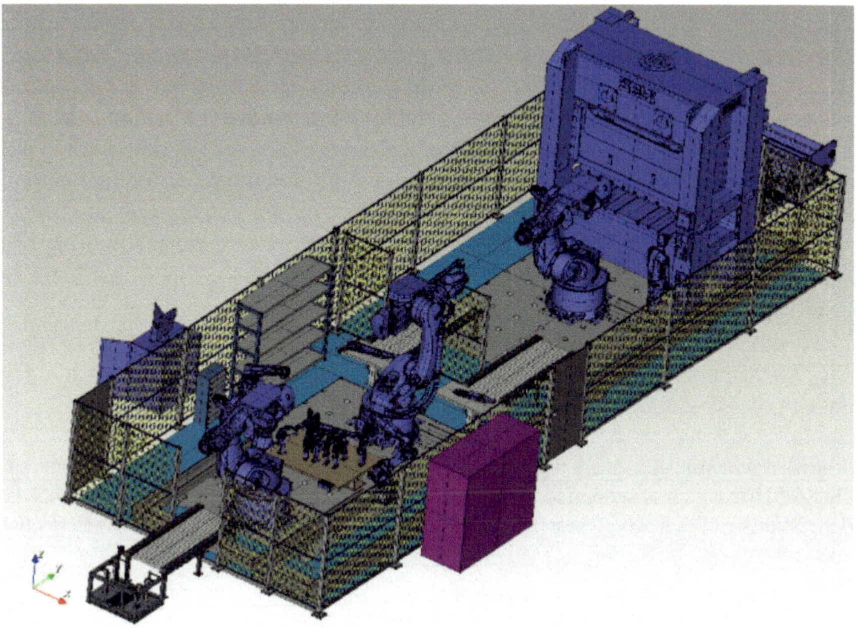

Fig. 5. Example of a design variant of a production cell with corresponding production resources based on a predefined layout configuration (robots and press are imported STEP-files).

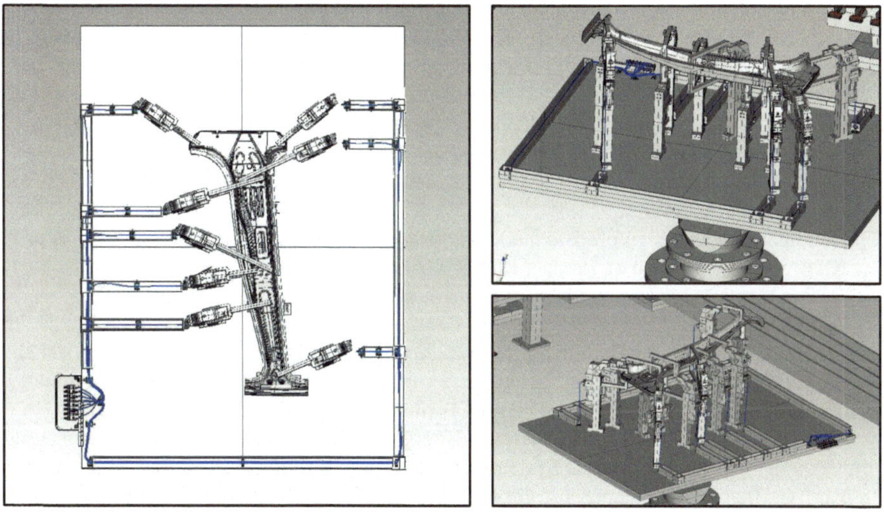

Fig. 6. Application example of the routing algorithm which leads to the wiring of geo-stations.

is on the automated generation of specific geometry models of geo-stations. Geo-stations belong to the group of type-specific equipment in assembly processes. Clamping and centering units with parameterized components are created based on the production concept defined in the product design. Further important aspects for the geo-station design is accessibility and collision testing (e.g. collision detection between the product components and robotic tools such as welding guns). For example, the positions of joining robots are necessary for the collision-free positioning of the centering and clamping units alongside the product contour. The geometry of positioned, individual clamping and centering units has been created automatically within a specific synthesis process. In addition to this, a routing algorithm was developed for an automated wiring of the whole geo-station design. At the present time, work is also being carried out in the field of evaluation of geo-stations. Evaluation processes for geo-stations are e.g. efficiency calculations, package and mass inspections or comprehensive cost and energy balancing. Basically, the goal is to evaluate multi-domain product property designs in an automated iterative process to find the best product designs.

Acknowledgements. The work was supported in the scope of the project „digital product lifecycle (ZaFH)", which is supported by a grant from the European Regional Development Fund and the Ministry of Science, Research and the Arts of Baden-Württemberg, Germany (https://efre-bw.de/).

References

1. Till, M., Stetter, R., Rudolph, S.: Multi-disziplinäre digitale Repräsentation des Produktlebenszyklus auf der Basis graphenbasierter Entwurfssprachen. Forschungsreport für den Maschinenbau in Baden-Württemberg; Bingen, ISSN 2196–8659, pp. 3–6 (2016)
2. Kiefer, J., Stetter, R., Rudolph, S., Till, M.: Digitale Konzeptplanung von Montageanlagen auf Basis multi-disziplinärer Repräsentationen des Produktlebenszyklus. Ingenieurspiegel 3; Bingen, pp. 47–49 (2018)
3. Tao, F.: Digital twin-driven product design, manufacturing and service with big data. Int. J. Adv. Manufac. Technol. 2010 **94**, 3563–3567 (2018)
4. Rudolph, S.: Übertragung von Ähnlichkeitsbegriffen. Habilitation thesis, University of Stuttgart, Stuttgart (2002)
5. Rudolph, S.: On design process modelling aspects in complex systems. Proceedings of the NASA-ESA Workshop on Product Data Exchange (PDE 2011)
6. Design Compiler 43 is a registered trademark of IILS GmbH. www.iils.de (2020)
7. Schmidt, J.: Eine Entwurfssprache für Geometrie; Studienarbeit; Institut für Statik und Dynamik der Luft- und Raumfahrtkonstruktionen; Universität Stuttgart; Germany (2012)
8. Rigger, E.; Lutz, A.; Shea, K.; Stankovic, T.: Estimating the Impact of Design Automation: The Influence of Knowledge on Potential Estimation. In: Proceedings of ICED19 (2019)
9. Zheng, C., Hehenberger, P., Le Duigou, J., Bricogne, M., Eynard, B.: Multidisciplinary Design Methodology for Mechatronic Systems Based on Interface Model. Res. Eng. Des. (2017)
10. Pahl, G., Beitz, W., Feldhusen, J., Grote, K.H.: Engineering Design: A Systematic Approach. Berlin: Springer (2007). https://doi.org/10.1007/978-1-84628-319-2
11. Ponn, J., Lindemann, U.: Konzeptentwicklung und Gestaltung technischer Produkte: Systematisch von Anforderungen zu Konzepten und Gestaltlösungen. Heidelberg: Springer Vieweg. https://doi.org/10.1007/978-3-642-20580-4 (2011)
12. Zech, A., Stetter, R., Holder, K., Rudolph, S., Till, M.: Novel approach for a holistic and completely digital represented product development process by using graph-based design languages. Procedia CIRP **79**, 568–573 (2018)

13. Dungs, S. Wissensbasierte Geometriemodelle zur Strukturanalyse; Dissertation; Institut für Produkt Engineering; Universität Duisburg-Essen; Deutschland (2008)
14. EKS InTec GmbH. Danziger Straße 3, 88250 Weingarten, Germany. https://www.eks-intec.de/about.html. Accessed 8. June 2020

A self-programming painting cell »SelfPaint«: Simulation-based path generation with automized quality control for painting in small lot sizes

Nico Guettler[1(✉)], Niklas Sandgren[2], Stefan Weber[3], Philipp Knee[1], Raad Salman[2], Jens Klier[3], Fredrik Edelvik[2], and Oliver Tiedje[1]

[1] Fraunhofer Institute for Manufacturing Engineering and Automation IPA, Nobelstrasse 12, 70569 Stuttgart, Germany
nico.guettler@ipa.fraunhofer.de

[2] Fraunhofer-Chalmers Centre, Chalmers Science Park, SE-412 88 Gothenburg, Sweden
niklas.sandgren@fcc.chalmers.se

[3] Fraunhofer Institute for Industrial Mathematics ITWM, Fraunhofer-Platz 1, 67663 Kaiserslautern, Germany
stefan.weber@itwm.fraunhofer.de

Abstract. The increasing diversity of products and variants requires a flexible and fast path generation in robot-based painting processes. In the state of the art, path generation in the painting industry is a time consuming and cost-intensive iteration process in which the generated paths are evaluated and optimized via painting trials. In this paper, we present a novel concept for a self-programming painting cell, which is based on the key technologies 3D-scanning, multi-physics painting simulations, and a contactless film thickness measurement using terahertz technology. The core element of this cyber-physical painting system is a unique combination of numerical painting simulations with a gradient-based multi-objective optimization method, to virtually compute painting paths that produce a homogeneous thickness on the painted object. In order to drastically reduce the time and computationally intensive numerical fluid dynamic simulations, a step-by-step coupling of an offline and online simulation was implemented. In a final step, a collision free robot motion without singularities is generated automatically from the computed painting path. The concept was validated under pilot plant conditions by the painting of a fender using an electrostatically assisted high-speed rotary bell atomizer. The paint film thickness, measured with terahertz technology was used as the target and validation criterion, as it shows a strong correlation to other quality values. The results show that the achieved film thickness was within the process specification, although deviations between simulated and measured film thicknesses were found in the edge zones of the workpiece.

P. Weißgraeber et al. (Eds.): *Advances in Automotive Production Technology – Theory and Application,* ARENA2036, pp. 302–310, 2021. https://doi.org/10.1007/978-3-662-62962-8_35

1 Introduction

Spray painting is a widely used process in the automotive industry for fully automated production of high-quality coatings at high-efficiency levels. An essential basis of every paint application is a homogeneous paint film thickness within the process window of the specified target film thickness.

To achieve this goal, in the state of the art, the painting robot paths are usually generated by a high amount of manual effort in offline programming, whereby homogeneity of the film thickness is optimized in an iterative process of painting prototypes and optimization of the painting paths. This results in long ramp-up times and high material and personnel costs. To enable fast reactions to changing customer requirements up to entirely personalized products in painting processes, versatile systems need to be created, where the painting paths will be automatically generated taking the required target film thickness and other specification into account.

To tackle this problem, we have developed a cyber-physical system of a painting cell, which is referred to as »SelfPaint« in the following. In our view, a cyber-physical system consists of sensors (e.g., optical measurement devices), actors (e.g., painting robot) and software services for intelligent data analysis, which are connected and communicating using state of the art network technology (e.g., internet). The modular structure of »SelfPaint« and the connection of real-world and software services is illustrated in the following Figure 1. The workflow of »SelfPaint« is divided into four modules, which are used in the actual painting process. In the first module *3D-Scanning*, a virtual image of the workpiece is generated as a CAD model. This CAD model is then transferred to the second module *Path Planning and Optimization*, which contains a multi-stage simulation-based path generation process. In order to achieve high accuracy of the film thickness prediction, numerical multi-physics simulations are utilized to evaluate and optimize the painting path. For these painting simulations, experimental as well as simulated input data must be provided and stored in a database. The calculated painting robot program is then embedded in the higher-level painting system program after which the painting application is carried out in the third module *Painting*. In the final module *Quality Control* the paint film thickness is measured by a contactless and robot-based measuring system using terahertz technology. For the state of the art, the reader is referred to the work of Edelvik et al. [1], which was published about the self-programming painting cell at an earlier stage of development [1]. This paper aims to extend their paper on a more technical side as well as presenting and discussing the results of entire framework of »SelfPaint«. The results presented in this paper were produced on a fender painting of an agricultural machine in pilot plant scale. As an applicator, an electrostatically assisted high-speed rotary bell-cup atomizer was used, with which a two-component green topcoat was applied.

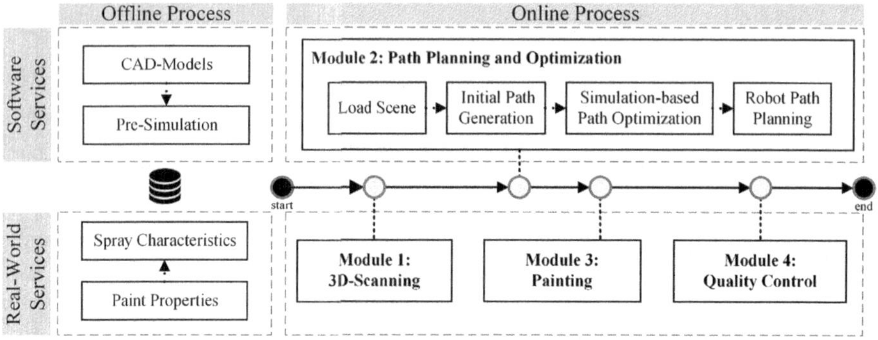

Fig. 1. Schematic illustration of the modular concept and the connection between real-world and software services. The black dots visualize the start and end point of the workpiece and the grey dots marking process milestones within the »SelfPaint« process chain.

2 Input Database

To perform numerical multi-phase flow simulations with electrostatics for prediction of film thickness, experimental and digital input data need to be provided. The most comprehensive core of the database is a pre-simulation of the shaping air near the high-speed rotary atomizer (so-called »Near-Bell« simulation) [2]. Due to high physical gradients and compressible flow in the »Near-Bell« region, very fine computational grids are required, which results in computationally intensive simulations on high performance computers. As an input for the »Near-Bell« simulation, the 3D-model (CAD) of the atomizer as well as atomizer-relevant process parameters are required. The results of the »Near-Bell« simulation (cell-coordinates, velocity, and turbulence) are stored in the database in a proprietary format for online »Laydown« simulations. The pre-simulations are performed with a semi-automatic workflow for every process setting used with the atomizer. Figure 2 shows the cylindrical interface between »Near-Bell« and »Laydown« simulation on the high-speed rotary bell-cup atomizer used in this paper. For the »Laydown« simulation, paint material properties such as density and non-volatile content are required to compute the film thickness of the cured paint. Spray related distribution values of the droplet size, the film thickness distribution as well as the integral discharge current of a spray pattern measured according to DIN EN 13966-1, are used to calibrate model parameters of the simulation [2]. Besides the simulation environment, a CAD model of the workpiece is required for the registration method used in the module *3D-Scanning*, which will be provided in most cases by virtual product development in automotive industry anyway. The provision of this data is currently the only manual effort of the self-programming painting cell, which needs to be done in preparation by qualified personnel.

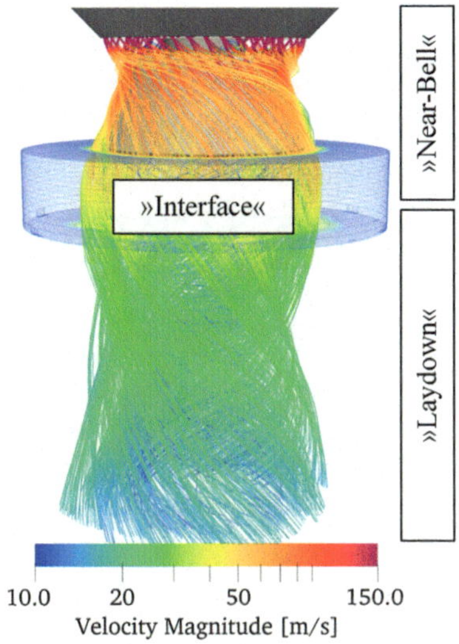

Fig. 2. Interface (blue cylinder) between »Near-Bell« and »Laydown« simulation. Colored streamlines are used to visualize the shaping airflow.

3 Module 1: 3D-Scanning

The specifications for object detection in »SelfPaint« are mainly based on the requirements of the software services for numerical simulations and path planning. Usually for numerical simulations, a so-called waterproof CAD model is required to generate the computational grids. A special feature of the used »Laydown« solver IPS Virtual Paint is that CAD surface information (e.g. triangulated formats such as STL) is the only required input and the solver is not sensitive to the quality of the CAD as long as it captures the curvature of the object.

To fulfill the requirements cost-efficiently, a 3D-scanning system was developed for »SelfPaint«, which is based on Microsoft's Kinect 2.0 using a point cloud registration method to fit an existing CAD model of the workpiece into the scanned point cloud.

The acquisition accuracy of the scanning system was experimentally determined by scanning an identifiable marked plate which was moved parallel to the camera at fixed perpendicular distanced between 1 m to 3 m. The maximum standard deviation in the measurement data along this distance study is approx. 2 mm. No trend between increasing standard deviation and distance between Kinect sensor and plate was found in the selected measurement range. The core of this method is an Iterative Closest Point (ICP) Algorithm, whereby a mean deviation between CAD model and the scanned point cloud of 6 mm in depth and 4 mm in the horizontal was achieved (see Fig. 3). After completion of the registration, the CAD model position is transformed from the cameras coordinate system into the coordinate system of the painting robot, which was previously

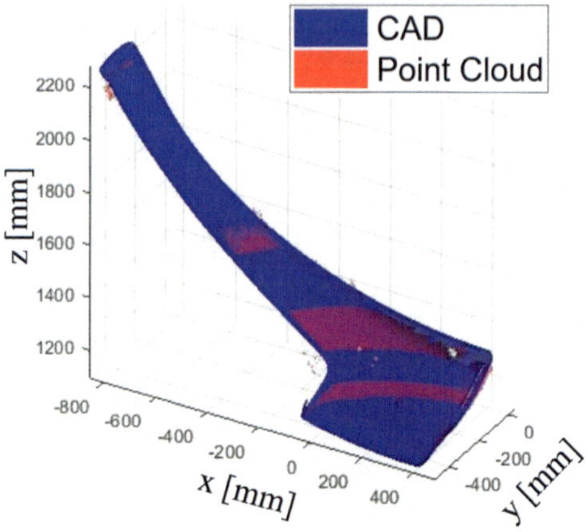

Fig. 3. ICP-aligned CAD model of the fender

calibrated. A drawback of the method is that is has problems when bending and twisting are present in the real workpiece. On the fender demonstrated here, larger deviations up to 20 mm to the CAD model could occur, especially at the thin end of the workpiece. These deviations are not caused by the manufacturing tolerances, but by slight elastic deformability in an unmounted state. Under certain circumstances, the suspension on the painting skids is carried out under tension, so that additional deviations could occur.

4 Modules 2 and 3: Path Generation and Painting

Automatic path planning and optimization is the software-based core of »SelfPaint«, which iteratively generates and optimizes a painting robot path taking paint film thickness homogeneity, paint transfer efficiency, and cycle time into account. The path generation is based on a workflow consisting of the steps *Load Scene*, *Create Initial Path*, *Optimize Path* and *Robot Path Planning* (see Fig. 1), which are fully implemented in the software IPS Virtual Paint developed and maintained by Fraunhofer-Chalmers Centre.[3,4]

The simulations require a virtual representation of the painting cell, which is why the geometry of the painting booth with relevant system components, as well as the CAD model of the workpiece are loaded first. To generate an initial path a so-called raster plane method is utilized, which is based on a meander (correct word?) structure with a sweep width typically given by the width of the spray pattern. The sweep width is typically based on the pattern width of the spray pattern, as this results in the most homogeneous overlap using parallel painting paths. In the raster plane method, a viewing direction is given to the workpiece, whereupon all points on the target surface of the workpiece are projected onto a plane orthogonal to the viewing direction.

At first, a fast simulation is performed using the so-called projection method, where the experimental or simulated spray pattern is used as a footprint of the atomizer to integrate it above the workpiece (see Fig. 4). The limited accuracy for strongly curved surfaces, and the lack of electrostatic and air turbulence effects is compensated by the second simulation step using the physics-based »Laydown« simulations (see Fig. 4d). This method is based on a multi-physics simulation that couples the fluid dynamics, electrostatics and motion of charged paint droplets [3]. To drastically reduce the computational effort and computing time for this simulation, the »Near-Bell« results stored in the database are utilized as input. In the optimization process both simulation methods are utilized in a stepwise procedure that balance calculation time and film thickness prediction accuracy.

A gradient-based optimization method, in which the path points and their orientations are used as variables, is utilized to achieve a homogeneous film thickness. In the final step of the path optimization, the physics-based simulation is performed and utilized to adjust the painting speed along the path. (see Fig. 4). The obtained optimized painting path can then be used to simulate the robot motion and obtain a realistic robot trajectory. The optimization of the painting path is a very complex mathematical problem that can be simplified by introducing knowledge-based constraints, e.g. constant painting distance, perpendicular application angle, turning points outside the part. The painting of the fender was carried out under controlled climatic conditions and constant sinking air velocity of $0.3\,\mathrm{m\,s^{-1}}$ in the paint booth at Fraunhofer IPA. The applied paint on the workpiece was flashed off and baked according to the given specification. Figure 5 compares the simulated and measured film thicknesses. The experimental film thicknesses are as far as possible within the given window (green area). Still, more significant deviations between simulation and experiment can be seen in the strongly curved fender areas. Since the painting was carried out with electrostatic support, the projection method's weaknesses can also be recognized in this figure. Especially on lines L1 and L5, significant deviations from the measured film thickness are visible. In the central areas, however, a good agreement was achieved, which is as good as the physics-based simulation. The physics-based simulation shows a good agreement in the trend of the film thickness, but some deviations are recognizable at the outer curved sections. In these areas, deviations from the 3D-scanned CAD model position are particularly significant, since stronger changes in the flow field and electric field occur there. It is also known that curved surfaces cause uncertainties in the film thickness measurement, which are difficult to quantify.

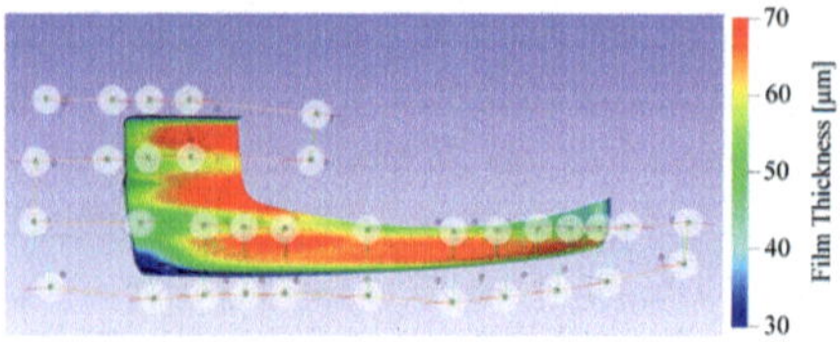

(a) Initial path

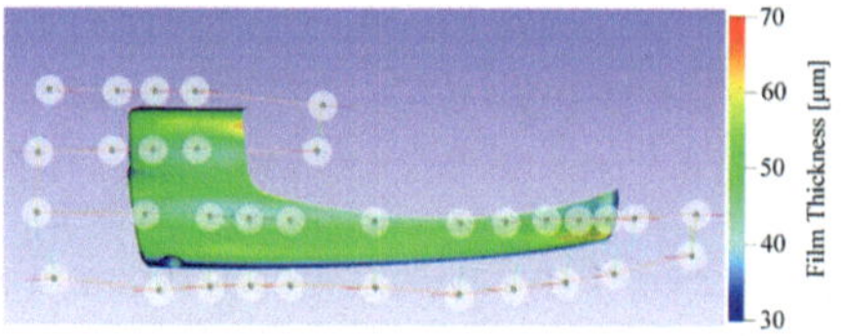

(b) Velocity-tuned path (projection)

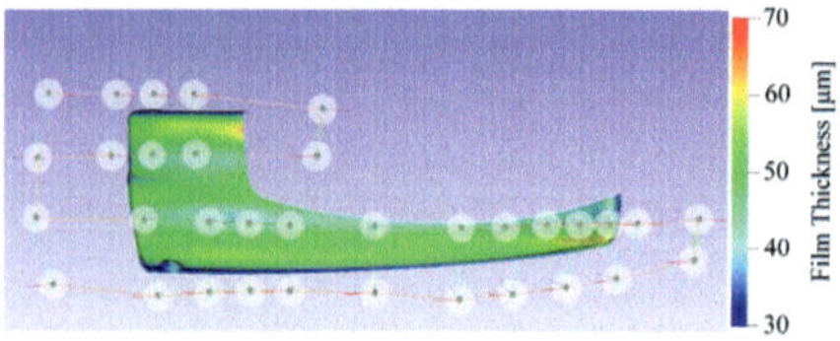

(c) Gradient-based optimized path (projection)

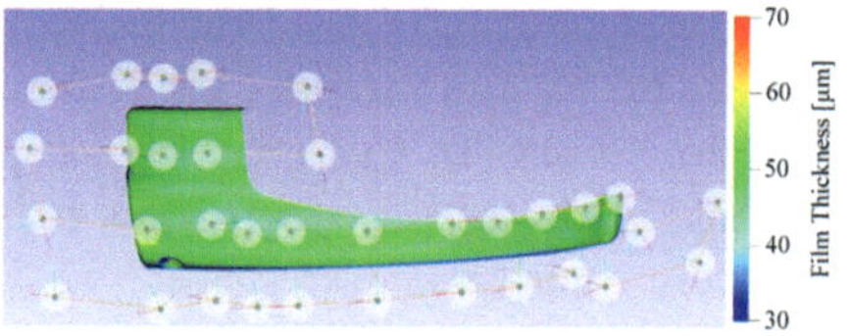

(d) Velocity-tuned path (physics-based)

Fig. 4. Simulated film thickness at different stages of the path generation workflow for a target film thickness of 50 μm

5 Module 4: Contactless Quality Control of Paint Film Thickness

Within this work scope, a multi-channel terahertz system by Fraunhofer ITWM was enhanced, which allows a contactless film thickness measurement up to seven individual layers in a multilayer paint film with micrometer resolution. The functional principle is based on a short terahertz pulse, which is reflected and transmitted at each painted layer. A sensor detects the reflecting parts and the film thickness of the individual layers is calculated with known material parameters. For a detailed description and comparison to conventional film thickness measurement methods like magnetiv-inductive the reader is referred to Pfeiffer et al. [5] and Weber et al. [6,7].

In this work, a robot-guided measuring system, which aligns itself automatically via a triangulation system, was used to measure the film thickness on vertical and horizontal lines of the painted fender (see Fig. 5). Figures 5b to 5c show the quantitative comparison of the measured film thicknesses. The results were already discussed in the previous Section 4.

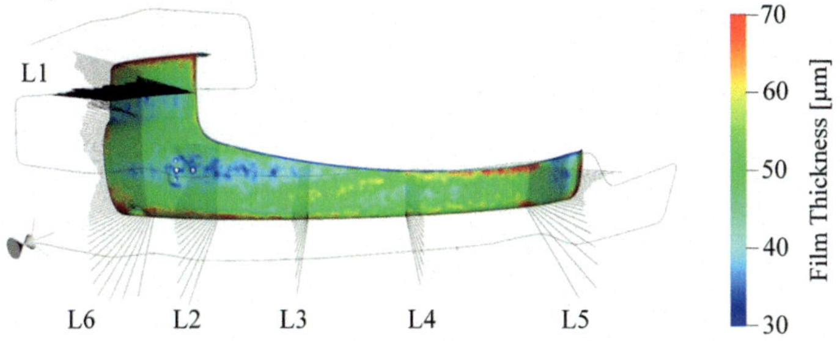

(a) Measurement lines on the physics-based simulated fender

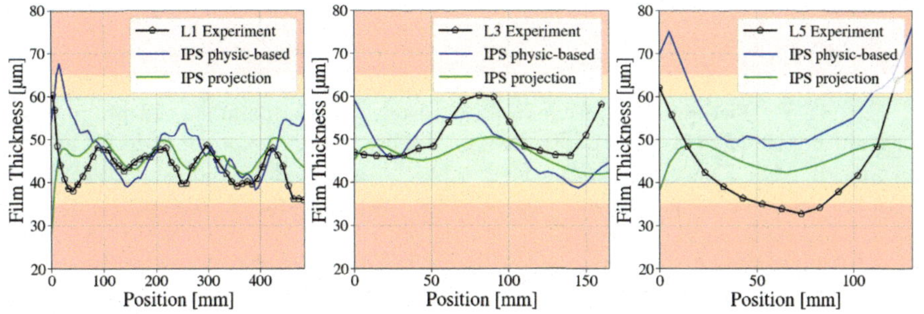

(b) Line L1 (c) Line L3 (d) Line L5

Fig. 5. Comparison of simulated and experimental determined film thickness on the painted fender using the self-programming painting workflow. The background color in the plots shows the risk of a surface defect due to the film thickness: red (high), orange (moderate), green (low).

6 Conclusion

In this paper, a concept for a self-programming painting cell was presented, which generates painting paths automatically using a simulation-based path planning and optimization algorithm. The proof of concept was demonstrated on a fender painting using electrostatic high-speed rotary bell paint application in the pilot plant scale. The modular structure of the cyber-physical system allows for a customized integration into existing as well as new painting lines.

The fully automated simulation-based workflow for the generation of a painting robot path is mostly capable of generating a painting path within the given specifications. However, minor deviations between simulation and experiment can be detected. As already mentioned in the beginning, the entire simulation-based workflow strongly depends on the accuracy of the input data. This requires a more accurate scanning technique to reduce the position and orientation uncertainties as well as precise rotary atomizer model and reliability of paint and spray characteristics.

The modularity of »SelfPaint« also allows the modules to be used individually or in combination for other industrial applications. For example, numerical simulations are already being used to accelerate offline painting robot programming through virtual operational tests. The robot-assisted contactless film thickness measurement, in combination with 3D-scanning, can also be used to measure specific positions on the workpiece.

Acknowledgments. The authors are very grateful to all colleagues who have been involved in the Fraunhofer joint project »SelfPaint« but are not listed as authors. This research was supported by Fraunhofer-Gesellschaft zur Foerderung der angewandten Forschung e.V. within the project »SelfPaint«.

References

1. Edelvik, F., Tiedje, O., Jonuscheit, J., Carlson, J.S.: Selfpaint - a self-programming paint booth **72**, 474–479 (2018). https://doi.org/10.1016/j.procir.2018.03.167
2. Guettler, N., Knee, P., Ye, Q., Tiedje, O.: Initial droplet conditions in numerical spray painting by electrostatic rotary bell sprayers. Journal of Coatings Technology and Research **24**(11), 1647 (2020). https://doi.org/10.1007/s11998-020-00352-1
3. Edelvik, F., Mark, A., Karlsson, N., Johnson, T., Carlson, J.S.: Math-based algorithms and software for virtual product realization implemented in automotive paint shops. In: L. Ghezzi, D. Hömberg, C. Landry (eds.) Math for the digital factory, *Mathematics in industry. The European Consortium for Mathematics in Industry*, vol. 27, pp. 231–251. Springer International Publishing (2017). https://doi.org/10.1007/978-3-319-63957-4_11
4. Mark, A., Bohlin, R., Segerdahl, D., Edelvik, F., Carlson, J.S.: Optimisation of robotised sealing stations in paint shops by process simulation and automatic path planning **9**(1), 4 (2014). https://doi.org/10.1504/IJMR.2014.059597
5. Pfeiffer, T., Weber, S., Klier, J., Bachtler, S., Molter, D., Jonuscheit, J., von Freymann, G.: Terahertz thickness determination with interferometric vibration correction for industrial applications **26**(10), 12558–12568 (2018). https://doi.org/10.1364/OE.26.012558
6. Weber, S., Klier, J., Ellrich, F., Paustian, S., Guettler, N., Tiedje, O., Jonuscheit, J., von Freymann, G.: Thickness determination of wet coatings using self-calibration method. In: 2017 42nd International Conference on Infrared Millimeter and Terahertz Waves, pp. 1–2. IEEE (2017). https://doi.org/10.1109/IRMMW-THz.2017.8067107
7. Weber, S., Liebelt, L., Klier, J., Pfeiffer, T., Molter, D., Ellrich, F., Jonuscheit, J., von Freymann, G.: Influence of system performance on layer thickness determination using terahertz time-domain spectroscopy **41**(4), 438–449 (2020). https://doi.org/10.1007/s10762-020-00669-3

Less Chemicals and More Power: Pulsed Electric Field-Treatment for Reduction of Microorganisms

A biocide-free bath maintenance method in pre-treatment of dip coating plants for high-volume car body painting plants

Philipp Preiß[1(✉)], Monika Eva Bohem[2], Christian Gusbeth[3], Martin Sack[3], Dennis Herzog[3], Thomas Schwartz[4], Stefan Dekold[2], Norman Poboss[2], Claus Lang-Koetz[1], and Wolfgang Frey[3]

[1] Institute for Industrial Ecology (INEC), Pforzheim University, Tiefenbronner Str. 65, D-75175 Pforzheim, Germany
philipp.preiss@hs-pforzheim.de

[2] Eisenmann Anlagenbau GmbH & Co. KG, Daimlerstraße 5, D-71088 Holzgerlingen, Germany

[3] Institute for Pulsed Power and Microwave Technology (IHM), Karlsruhe Institute of Technology (KIT), Hermann-von-Helmholtz-Platz 1, D-76344 Eggenstein-Leopoldshafen, Germany

[4] Institute of Functional Interfaces (IFG), Karlsruhe Institute of Technology (KIT), Hermann-von-Helmholtz-Platz 1, D-76344 Eggenstein-Leopoldshafen, Germany

Abstracts. Decisions on the implementation of innovative concepts and technologies into automotive pre-treatment lines are regularly marked by uncertainty regarding trade-offs between economic, ecological and technical aspects. Large amounts of water are consumed during the production in car body painting plants. It is hardly possible to avoid that certain microorganisms (MOs) proliferate in process water and pre-treatment bath tanks. If the bacterial load increases too much, the quality of the paint finish is likely to be impaired. Therefore, chemical biocides are regularly used in pre-treatment and dip coating plants. However, repeated use of the same biocides can lead to resistance of some MOs strains. In addition, stricter legislation is gradually withdrawing certain biocides from the market, making it more difficult to obtain approval for newly designed active substances. Hence, conventional decontamination methods might no longer work in the future. The Pulsed Electric Field (PEF)-Treatment is an innovative technology within the field of pre-treatment lines. By applying high voltage pulses (kV range, μs duration), a high field strength is generated in the process fluid, across the cell membrane of the MOs, which permeabilises the cell membrane. As a result, the MOs lose their cell interior (cytoplasm), and most likely die.

The results of the project show that PEF-treatment has a high market potential, several advantages but currently also higher cost compared to conventional biocide treatment.

P. Weißgraeber et al. (Eds.): *Advances in Automotive Production Technology – Theory and Application,* ARENA2036, pp. 311–318, 2021. https://doi.org/10.1007/978-3-662-62962-8_36

1 Introduction

Automotive series painting is performed in several layers. Paint is applied in dip-coating and spray-coating processes. For this, the car body passes through a pre-treatment and painting line with various tanks and stations. These include dip tanks, spray booths and dryer stations. The waste water generated during the painting of cars is usually more than 600 dm^3 per car body. With a share of more than 90%, this is the area in automotive production with the highest water consumption [1]. A large share of water and energy is used, especially in pre-treatment and cathodic dip coating processes (more general, electrophoretic deposition coating, EDC).

A process flow chart of a typical pre-treatment plant is shown in Fig. 1. The pre-treatment line can be subdivided into three main sections: cleaning, zinc phosphating and rinsing. The subsequent EDC is followed by several rinsing steps and eventually curing of the component parts in the EDC-oven.

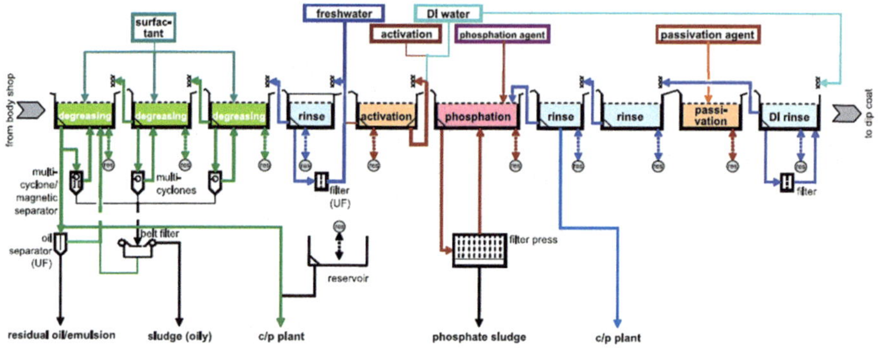

Fig. 1. Simplified process flow chart of a pre-treatment plant, (VDI 3445, 2013).

A car body coming from the body shop is contaminated with oils, welding dirt, mechanical particles and traces of glue. In order to provide corrosion protection, it is therefore crucial to thoroughly clean the component parts during pre-treatment. This is necessary, since high-quality EDC depends on a well prepared surface. In order to reduce the freshwater consumption to a minimum, different water treatment and bath maintenance measures, like oil separators, cyclones as well as different separator and filter systems are installed [2].

The intensification of environmental legislation has more and more led to omitting environmentally hazardous substances and to reduce waste water as much as possible [3]. Therefore, the number and amount of biocides and other environmentally hazardous substances in process fluids have been reduced. This in turn established a more advantageous culture substrate for diverse microorganisms (MOs) [4]. Because of lower process temperatures, the wide use of water-based paint systems and the inevitable entry of contaminants, the tanks are prone to microbial growth. An increased bacterial load can lead to a reduced quality of the paint finish. Moreover, the water requirement may be increased if the service life of the EDC tank is reduced, because of higher cleaning frequency and the need to dump the tank content. In order to reduce the risk of MOs proliferation, up to now, certain chemical biocides are used in pre-treatment and dip coating plants in regular intervals and additionally on demand. However, repeated use of the same biocidal active substances can lead to resistance of some MOs [3] and [5]. Stricter legislation is gradually removing certain biocides from the market and making it more difficult to obtain approval for new active substances [6]. Therefore, there is a risk that these conventional method of MOs control will no longer work in the future. On the other hand, legislative restrictions have already initialized a search for alternative and innovative techniques during the past decades, especially in the automotive coating industry. Examples are the possibilities to operate at ambient temperatures or the replacement of conventional zinc phosphate process by thin-layer deposition [7].

1.1 The DiWaL Project of BMBF

The aim of the R&D project DiWaL (Decontamination of industrial WAter in Lacquering processes) is the development of resource-efficient water management and plant concepts for pre-treatment and dip coating plants using pulsed electric field (PEF) technology for the reduction of microbial contamination [8]. At KIT/IHM for more than 15 years, the effect of the PEF-treatment has been applied on a large scale for the effective extraction of cell contents in wine production [9], for the extraction of sugar from sugar beets [10], for the drying of green biomass, for growth stimulation and, since 2010 for the extraction of valuable substances from microalgae [11–16]. Moreover recently, for production of organic fertilizer from microalgae and for phycocyanine extraction from spirulina microalgae [17, 18]. In DiWaL, PEF-treatment is applied to paints and water from pre-treatment and the dip coating processes. The aim is to reduce and prevent MOs contamination of various process liquids. For the PEF-technology, novel semiconductor switched pulse generators were developed and applied. Within an integrated sustainability assessment, economic and ecologic models were developed and potentials for optimization based on stakeholder-involvement were derived [19–21].

The project industry and research partners are: BMW Group Leipzig; Eisenmann Anlagenbau GmbH & Co. KG; Emil Frei GmbH & Co. KG; PPG Deutschland Business Support GmbH; Pforzheim University, Institute for industrial Ecology (INEC); KIT—Institute for Pulsed Power and Microwave Technology (IHM) and the Institute of Functional Interfaces (IFG). The project was funded by the German Federal Ministry of Education and Research—BMBF from 2016–2020. Grant Number: 02WAV1405C.

2 Pulsed Electric Field-Treatment

An innovative option for the reduction of MOs contamination is the Pulsed Electric Field (PEF) treatment. The concept of PEF-treatment is depicted in Fig. 2. By applying a high voltage (kV range) for the fraction of a second, a high electric field strength is generated across the cell membrane of the MOs, which irreversibly permeabilises the cell membrane. As a result, the MOs lose their cell interior (cytoplasm) and die [22]. The formation of resistant MOs after PEF-treatment has not been observed [23]. A further important advantage of PEF-treatment is the possibility to work in turbid liquid media, unlike other physical decontamination methods that require clear fluids.

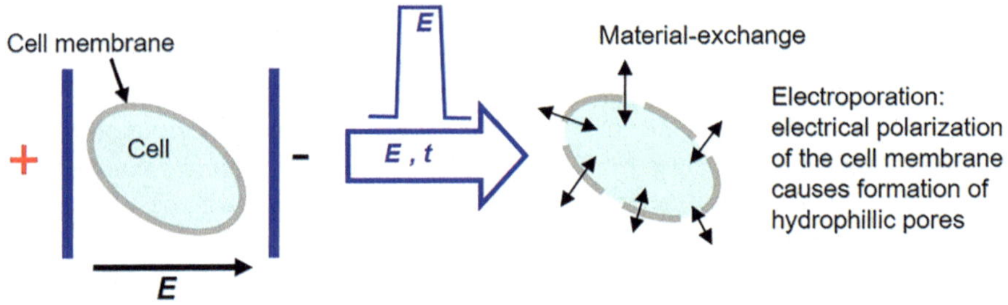

Fig. 2. PEF-treatment concept.

2.1 Technical PEF-Concepts and New Developments

In order to realize a high energetic efficiency of bacterial inactivation, the electric field strength E has to be selected at 40 kV/cm and higher [22]. Consequently, at an electrode distance of several centimeters, the required pulse voltage amplitude has to be on the order of 100 kV. Such pulse voltages usually are generated by voltage adding circuits, e.g. Marx generators. Here, capacitors are charged in parallel to a comparatively low voltage level. For pulse release, the charged capacitors are switched into series connection. This multiplies the output voltage of the generator according to the number of capacitors. Conventional concepts of Marx-type PEF generators use gas-insulated spark gaps for switching. Spark gaps including gas pressurizing and gas cleaning auxiliaries are expensive and relatively complex to operate and maintain. Moreover, another disadvantage of this kind of pulse generators is its inflexibility in the choice of pulse parameters. The design values, in particular pulse shape and duration cannot easily be matched to altered process requirements. In contrast to spark gaps, semiconductor switches, e.g. IGBTs, can be switched off under full load conditions. This allows simple control of the pulse duration. Moreover, implementing single Marx stages in full-bridge configuration [24] additionally allows simple polarity change of the output pulse and the generation of bipolar output pulses [25] as required for PEF-treatment of EDC.

The aim of the DiWaL project was to be able to sufficiently reduce the microbial load of the various process fluids during plant operation. For this purpose, the PEF system is operated in a bypass, coming from the tank to be treated. Model calculations for bypass operation have shown that a comparatively low inactivation rate of 2 log steps, i.e. a germ reduction of at least 99% is sufficient to achieve growth stagnation in the affected basin. Inactivation experiments on bacterial species isolated from the EDC show that the required inactivation rate of 2 log levels can already be obtained with an electrical energy input of 40 kJ/dm3. Regarding pulse technique, IGBT-switched pulse generators consisting of stacked 1 kV/0.6 kA full bridge modules have been developed for this application. A demonstration plant has been set up and currently being set up and tested at paint manufacturers.

2.2 Results and Evaluation

PEF-treatment was demonstrated at pilot scale, to efficiently inactivate bacteria in electrocoating process media. In particular, for PEF-treatment of electrodeposition paints, pulse parameters have to be chosen appropriately, so that paint does not precipitate at the surface of electrodes of the treatment chamber, which might potentially result in blocking of the chamber.

Since paint coagulation and deposition is initiated by electrochemically generated H + and OH- ions, EDC deposition at the treatment chamber electrodes can be prevented by application of bipolar pulse protocols for PEF-treatment.

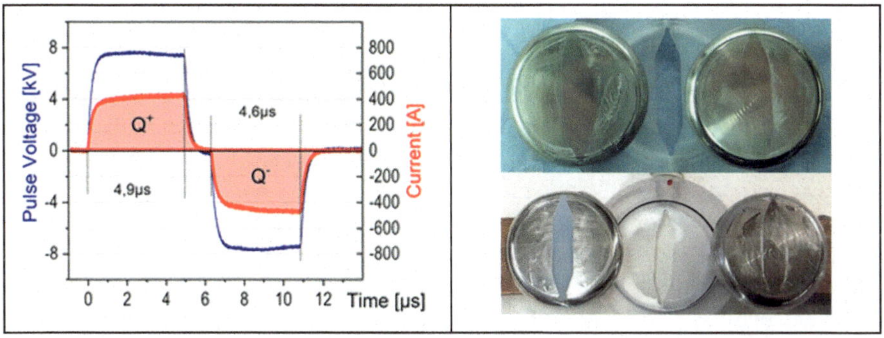

Fig. 3. Current and voltage waveform for PEF-treatment of EDC without precipitation of paint at the surface of the electrodes of the treatment chamber (upper right). In comparison, paint precipitation is visible at the surface of the cathode of the treatment chamber, (lower right, electrode to the left), when unipolar pulses were applied

H^+ ions occurring during the first half-wave of the bipolar pulse will be neutralized by OH^- ions generated during the second half-wave. As a requirement for complete neutralization, the amount of charge during the first half-wave Q^+ has to equal Q (Fig. 3). Since ohmic heating of the EDC during the first half-wave increases paint conductivity, and in consequence also the pulse current during the second half-wave, the pulse duration of the second negative pulse has to be shortened appropriately for fulfilling the requirement $IQ^+I = IQ^-I$. This procedure for PEF-treatment of electrodepositing media was disclosed in November 2019. This approach was successfully tested at pilot scale at KIT-IHM. For pulse generation a 8 kV IGBT-switched prototype bipolar pulse generator was applied.

2.3 Integrated Sustainability Assessment of PEF-Treatment

The integrated assessment attempts to take full account of the relevant sustainability aspects. The assessment aims to consider the whole life cycle of both alternatives, e.g. also the transport of the biocides. Moreover, it takes into account the viewpoints of the relevant stakeholders. The analysis is based on the assumption that PEF-treatment substitutes biocide treatment, whereas it also has additional benefits but also some drawbacks.

Economic analysis. The range of costs for treatment with chemical biocides is based on certain product costs and a lower and upper assumption of applications per year. The so called Total Costs of Ownership (TCO) approach has been applied [26]. The range of costs of PEF-treatment includes the costs for building the PEF module and the costs of energy for the operation including also circulation pumps and re-cooling.

The results show that the total costs of a sufficient reduction of MOs by biocides is currently at least five times less expensive than PEF-treatment if a price for electricity of approx. 10€-ct. per kWh_{el} is assumed [20].

Ecological analysis. The approach is based on the concept of environmental Life Cycle Assessment (LCA) [27]. The evaluation of ecological aspects of biocides includes their production and transportation. The corresponding impacts are therefore mainly related to the energy consumption. The application of biocides takes place in a closed system and therefore the impacts due to unintended emission of the substance to the environment have not been quantified. The PEF-treatment has a relatively high electricity demand and therefore, if based on the current German electricity mix, the quantified environmental impacts are higher than those of production and transport of the biocides. With an electricity mix consisting mainly of renewable technologies, the environmental impact will be much lower (e.g. current Norwegian electricity mix or a future German electricity mix). Under such conditions, the quantifiable environmental impacts of the application of biocides are still lower. Evaluated by the endpoint impact categories human health and eco-toxicity, cf. [28], the results for both alternatives are easier to compare. The uncertainty of this impact assessment is relatively high but the chemical biocides seem, as long as they are working as foreseen, to have much lower ecological impacts as the application of the PEF-treatment. However, the potentially negative impacts of resistance of some MOs strains after repeated use of the same biocides could not be taken into account due to lack of basic data so far.

Stakeholder Analysis. Requirements, obstacles and drivers were identified during a stakeholder workshop [21]. Among the most important technical obstacles were the lack of empirical values regarding effectiveness, the electricity consumption, investment and transaction costs. Besides, the conventional process currently works well and delivers high-quality products. As PEF-treatment is new in this field of application, there are no references for the integration of PEF-treatment in existing or new plants. Hence, some stakeholders saw the problem of liability as a further obstacle. On the other hand, several potential advantages have been identified. For some stakeholders, workers' health is particularly important and they would accept smaller disadvantages or higher costs of a new technology capable of reducing the occupational risks associated with the use of chemical biocides. Due to the sophistication and its adaptability the PEF-treatment could be used at various points in the process. Another identified advantage is the reduction of the unpredictability of MOs load through continuous MOs suppression. Therefore, a reduction of MOs carry-over from the pre-treatment is possible. Eventually, a reduction in potential environmental impact of chemical biocides or possible negative interactions with the paint system can be achieved.

3 Conclusion and Outlook

In technical developments of the PEF-treatment within the DiWaL-project, the technology was lifted to a user-friendly, semiconductor-based pulse generator version in modular design. The feasibility of PEF-treatment of EDC was successfully demonstrated for the first time. It was demonstrated that no adverse effects on paint and deposition quality take place. A reduction of the bacterial concentration in EDC by 2 log consumes an electric energy 40 kJ/dm^3. A 30 kV semiconductor based PEF-facility, with a mass flow of 800 dm^3/h is still in development undergoing operational testing.

The quantifiable results of the economic and ecological analysis do not explicitly advocate the immediate switch from biocides to PEF-treatment. However, for a future innovation process, important parameters have been identified and discussed with stakeholders. The results of the stakeholder involvement show that the PEF-treatment has several advantages.

It is an established routine to use biocides in a preventative manner, even if MOs have not even been properly detected. Legislative regulations of biocides may further restrict the use of this wide pool of agents. Supplying the process fluids with new supplements might eventually have an impact on the coating quality. In contrast, the PEF-treatment has neither a negative influence on the paint itself nor on the coating. Therefore, the PEF-treatment can be a functional alternative technique in the field of automotive pre-treatment lines with EDC. Other branches of industry could be, e.g. electroplating technology, cooling towers or even car washes in which the process water is circulated.

References

1. Schumacher, J., Gebicke, W.: Neue Konzepte zum Wassermanagement im Lackierbetrieb von PKW-Karossen, in: Tagungsband Industrietage Wassertechnik, ISBN: 978–3–944328–29–4 (2013)
2. Fröhlich, G., Halbartschlager, J.: Ressourceneffiziente Nutzung von Wasser in Anlagen zur Serienlackierung von Automobilkarosserien. Chem. Ing. Tec. **91**(10), 1417–1433 (2019)
3. Gühring, I.K.: Mikrobieller Befall von Elektrotauchlack in der Automobilindustrie. Universität Stuttgart. https://doi.org/10.18419/opus-1501 (2000)
4. Zwanger: Anwendungsgebiete für eine Anlage zur Dekontamination industrieller Abwässer mittels Elektroporation im Tauchlackierprozess. Studienarbeit (2019)
5. Rushton, L., Sass, A., Baldwin, A., Dowson, C.G., Donoghue, D., Mahenthiralingam, E.: Key role for efflux in the preservative susceptibility and adaptive resistance of Burkholderia cepacia complex bacteria. Antimicrob Agents Chemother **57**(7), 2972–2980 (2013). https://doi.org/10.1128/AAC.00140-13
6. EU: Regulation (Eu) No 528/2012 Of The European Parliament and of the Council of 22 May 2012 concerning the making available on the market and use of biocidal products (2012)
7. Streitberger, H.-J. (ed.): Automotive paints and coatings, 2. Aufl, WILEY-VCH (2008)
8. DiWaL: Development of resource-efficient water management and plant concepts for electrocoating lines: application of pulsed electric field technology for bacterial decontamination of industrial water and electro-deposition paints, BMBF R&D Project FKZ 02WAV1405C (2020)
9. Sack, M., Sigler, J., Frenzel, S., Eing, C., Arnold, J., Michelberger, T., Frey, W., Attmann, F., Stukenbrock, L., Müller, G.: Research on industrial-scale electroporation devices fostering the extraction of substances from biological tissue. Food Eng. Rev. **2**(2), 147–156 (2010). https://doi.org/10.1007/s12393-010-9017-1
10. Sack, M., Bluhm, H.: New measurement methods for an industrial-scale electroporation facility for sugar beets. IEEE Trans. on Ind. Applicat. **44**(4), 1074–1083 (2008). https://doi.org/10.1109/TIA.2008.926222

11. Eing, C., Goettel, M., Straessner, R., Gusbeth, C., Frey, W.: Pulsed electric field treatment of microalgae—benefits for microalgae bio-mass processing. IEEE Trans. Plasma Sci. **41**(10), 2901–2907 (2013). https://doi.org/10.1109/TPS.2013.2274805
12. Goettel, M., Eing, C., Gusbeth, C., Straessner, R., Frey, W.: Pulsed electric field assisted extraction of intracellular valuables from microal-gae. Algal Res. **2**(4), 401–408 (2013). https://doi.org/10.1016/j.algal.2013.07.004
13. Sack, M., Sigler, J., Eing, C., Stukenbrock, L., Stangle, R., Wolf, A., Muller, G.: Operation of an Electroporation Device for Grape Mash. IEEE Trans. Plasma Sci. **38**(8), 1928–1934 (2010). https://doi.org/10.1109/TPS.2010.2050073
14. Haberl, S., Miklavcic, D., Sersa, G., Frey, W., Rubinsky, B.: Cell membrane electroporation-Part 2: The applications. IEEE Electr. Insul. Mag. **29**(1), 29–37 (2013). https://doi.org/10.1109/MEI.2013.6410537
15. Kotnik, T., Frey, W., Sack, M., Haberl, S., Peterka, M., Miklavčič, D.: Electroporation-based applications in biotechnology. Trends Biotechnol **33**(8), 480–488 (2015). https://doi.org/10.1016/j.tibtech.2015.06.002
16. Silve, A., Papachristou, I., Wüstner, R., Sträßner, R., Schirmer, M., Leber, K., Guo, B., Interrante, L., Posten, C., Frey, W.: Extraction of lipids from wet microalga Auxenochlorella protothecoides using PEF-treatment and ethanol-hexane blends. Algal Res. **29**, 212–222 (2018). https://doi.org/10.1016/j.algal.2017.11.016
17. Akaberi, S., Gusbeth, C., Silve, A., Senthilnathan, D.S., Navarro-, E., Mo-, E., Müller, G., Frey, W.: Effect of pulsed electric field treatment on enzymatic hydrolysis of proteins of Scenedesmus almeriensis. Algal Res. **43**, (2019). https://doi.org/10.1016/j.algal.2019.101656
18. Akaberi, S., Krust, D., Müller, G., Frey, W., Gusbeth, C.: Impact of incubation conditions on protein and C-Phycocyanin recovery from Arthrospira platensis post-pulsed electric field treatment. Bioresour Technol **306**, (2020). https://doi.org/10.1016/j.biortech.2020.123099
19. Zielbauer, A.: Master Thesis: Technologierecherche, Szenario-Vergleich und Ökobilanz im Bereich der Automobilserienlackierung – Identifikation und Bewertung von Potenzialen der Badpflege-Technologien im Vorbehandlungs- und Tauchlackierprozess (2018)
20. Vogt, A. Master Thesis: Analyse und Quantifizierung von ökonomischen und ökologischen Vor- und Nachteilen der Elektroimpulsbehandlung in der Automobilserienlackierung (2018)
21. Gasde, J., Preiss, P., Lang-, C.: Integrated Innovation and Sustainability Analysis for New Technologies. TIM Review **10**(2), 37–50 (2020). https://doi.org/10.22215/timreview/1328
22. Frey, W., Gusbeth, C., Schwartz, T. Inactivation of Pseudomonas putida by PEF-Treatment: A Study on the Correlation of Treatment Parameters and Inactivation Efficiency in the Short-Pulse Range. J. Membrane Biol. **246**, 769–781 (2013)
23. Gusbeth, C., Frey, W., Volkmann, H., Schwartz, T., Bluhm, H.: PEF-treatment for bacteria reduction and its impact on hospital wastewater. Chemosphere **75**(2), 228–233 (2009). https://doi.org/10.1016/j.chemosphere.2008.11.066
24. Sack, M., Hochberg, M., Mueller, G. Design of a Semiconductor-based Bipolar Marx Generator. IPMHVC Conference. 978–1–4673–7323–4 (2014)
25. Sack, M., Herzog, D., Hochberg, M., Frey, W., Mueller, G. 8-Stage Pulse Generator for Bipolar Ground-Symmetric Operation. PCIM conference. 978–3–8007–4938–6 (2019)
26. Krämer, S.: Total Cost of Ownership (TCO). VDM-Verl, Saarbrücken (2007)
27. ISO 14044: Env. Management - life cycle assessment - requirements & guidelines (2018)
28. Goedkoop, M., Heijungs, R., Huijbregts, M., de Schryver, A., Struijs, J., van Zelm, R. ReCiPe 2008 - A life cycle impact assessment method which comprises harmonised category indicators at the midpoint and the endpoint level First edition (version 1.08) Report I: Characterisation (2013)

Safety in Electromobility – Technical Cleanliness Between the Poles of Design Requirements and Efficient Production

Patrick Brag(✉) and Markus Rochowicz

Fraunhofer IPA, Nobelstraße 12, 70569 Stuttgart, Germany
patrick.brag@ipa.fraunhofer.de

Abstract. Reliability and trust in the new technologies are essential if the autonomous, connected and electrified automobile is to become established in society. The more the performance of an electrical/ electronic (E/E) system increases, the more sensitive it becomes to particulate contamination. The risk of a short circuit increases exponentially rather than linearly. Integrating technical cleanliness into product design and production is essential to reduce the risk of a short circuit. But what is the right level of cleanliness? This question can soon be answered with the newly introduced *fault injection test* procedure. The method can contribute to increasing road safety in areas where normative or legislative requirements have been lacking up to now.

1 Introduction

Tomorrow's vehicles will be equipped with dozens of smart sensors and control units that will eventually supersede the cognitive capabilities of the driver. In the future, traffic could move smoothly on highways [1].

But the presence of a tiny particle inside a smart sensor or electronic control unit represent a potential safety risk for passengers. Electrically conductive particles could cause faulty signals or failures in domain control units or reduce the performance of traction batteries, culminating in explosions. Non-conductive particles could also cause pixel errors or faulty information [2].

In order to prevent failures caused by particulate contamination, design engineers specify cleanliness requirements for the components. But how much contamination a component can tolerate until it really fails is widely unknown so that cleanliness requirements are specified stricter than eventually necessary. As a consequence, the required cleanliness cannot be produced any more or only with excessive efforts, while alternative approaches in product design are neglected. The right level of cleanliness is the basis to derive appropriate measures in product design and production with respect to reliable sensors and control units.

P. Weißgraeber et al. (Eds.): *Advances in Automotive Production Technology – Theory and Application*, ARENA2036, pp. 319–326, 2021. https://doi.org/10.1007/978-3-662-62962-8_37

A new holistic concept of technical cleanliness is presented in this paper, based on a novel test procedure to validate the robustness of an electrical/electronic (E/E) component against particulate contamination. Current approaches in product design and production are summarized to complete the holistic concept, because a cooperative strategy is still missing in a lot of companies.

2 Status Quo

Technical cleanliness is defined as a condition in which the function of a product or component is not impaired by particulate or fibrous contamination. But keeping the component clean has become a major challenge in the era of electromobility.

A simple electronic control unit commonly consists of a printed circuit board (PCB) and two housing covers. In comparison, power electronics, such as a power control unit (PCU), consists of a multitude of single components, see Fig. 1. Each additional component and each additional assembly step poses a particle risk. Single items, such as busbars, cover plates or housings are delivered with surplus material from casting processes, have burrs or other imperfections which can be detached during handling or assembly. Joining processes, such as screwing or press-fit could generate a lot of particles.

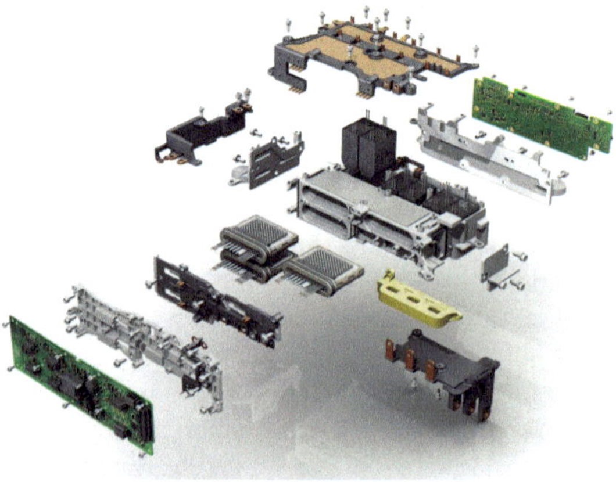

Fig. 1. Audi e-tron power control unit without housing, © Audi AG

Starting point for the cleanliness specification of any E/E system is the component with the lowest particle tolerance, i.e. the PCB with the smallest trace spacing or pitch (cf. Fig. 2). Today, a maximum permissible particle size of 0.1 mm[1] is common—regardless of the component complexity.

[1] In terms of technical cleanliness, particle size is defined as the longest dimension of a particle ($feret_{max}$).

Fig. 2. Connected pins of an integrated circuit

For battery cells, cleanliness requirements are even higher. Conductive particles with a size upwards of 30 μm can damage the separator and lead to a thermal runaway. Figure 3 shows the deviation between required and produced cleanliness. So far, however, there is no standard procedure or recognized method for reliably preventing particulate contamination [3].

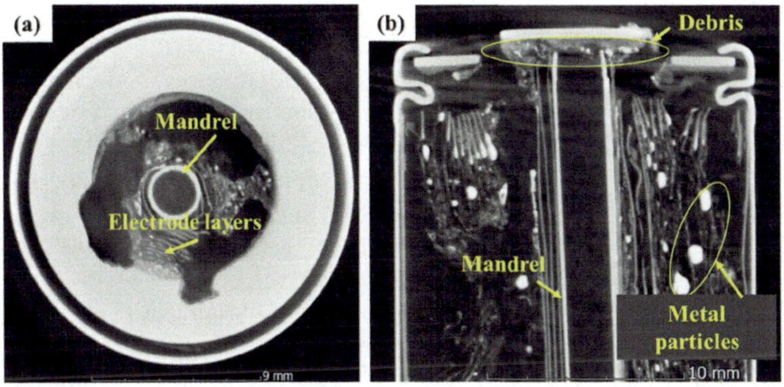

Fig. 3. CT scan of an exploded 18650 battery cell, (**a**) top view, (**b**) side view [4]

3 Current Approaches: Product Design

The achievable cleanliness of a product is primarily determined by its design. An improvement can be attained through the following approaches.

Material selection. Porous or brittle materials have a higher particle emission when joined or handled, e.g. sintered parts or duromers. Materials with a high resistance to mechanical abrasion are preferable. Coatings poses a particle risk when flakes are detached during production. Although uncoated substrates are preferable, coatings can be used as protective layers on opto-electronic parts or flat PCBs.

Particle sinks. Particle sinks are places where particles (permanently) disappear or adhere. Examples of geometric sinks (so-called particle trap geometries) include stagnant areas in flowing fluids (dead zones), or cavities in which particles can settle. Particles can also adhere to filters or adhesive and have a finite absorption capacity.

Mounting position. By making use of gravity, the position in which a part is mounted to the vehicle can help to ensure that particles settle in areas with a lower risk of short circuits. As far as PCBs are concerned, a vertical mounting position is preferable.

Easy-to-clean geometries and surfaces. The parts of E/E systems are increasingly being cleaned, and sometimes several times over. The fewer cavities and undercuts a component has and the more accessible the areas to be cleaned are, the better the cleaning result will be. A surface with low degree of roughness also has a positive effect on the cleaning result. A sintered or cast surface, for example, is harder to clean than a mechanically-processed surface.

4 Current Approaches: Production

Manufacturing processes (e.g. machining) emit a lot of chips and other contaminations. Near-contour casting can reduce the amount of waste during machining and contribute to resource-efficient production.

Burrs can detach when parts are handled and thus represent a particle risk. The formation of burrs in casting is influenced by the position of edges or parting lines, tool wear, form fit of dies and uniform mold filling [5]. The formation of burrs in machining is influenced by the angle and position of the tool on insertion into and withdrawal from the workpiece, cutting geometry or tool wear, cutting parameters (tooth feed, direction of rotation, etc.). Burrs can be controlled during machining processes by using chamfers and fillets [6].

Rework is another source of particle emission. Should rework prove unavoidable, it is important to find out which contaminants are introduced or generated by the process. If wet chemical processes are planned for cleaning parts anyway, sandblasting, grinding or polishing can be used to remove burrs. However, if these methods are not appropriate, deburring with a high-pressure water jet may be a suitable alternative [7].

The various parts making up E/E systems can generally only be deburred using a dry process without additional auxiliary media—the only exception to this is aluminum housings. One of the few suitable processes is the carbon dioxide (CO_2) snow-jet method. High-purity CO_2 liquid is used, which forms fine snow crystals due to its expansion on exiting the nozzle and the associated rapid drop in temperature. A jacketed jet of high-purity compressed air bundles the snow crystals and directs them at the surface to be cleaned. Besides deburring components, the CO_2 snow-jet technology can also be used to remove solder beads and flux material (cf. Fig. 4).

Fig. 4. Cleaning of an embedded PCB with CO_2

After manufacturing comes assembly. The objective for manufacturing is to minimize particle emission. Cleaning is a pro-active step to reduce the contamination level at the end although the cleanability of most E/E single parts is rather poor. In addition, the chance to clean components during assembly decreases with each added value. The objective for assembly is thus to keep the components as clean as possible. The VDA 19.2 guideline describes measures how to avoid additional contamination related to the environment, logistics, personnel and assembly facilities.

Companies are pushed by high cost pressure and high safety demands to revise their cleanliness concepts. A successful concept combines effective measures from both approaches, based on the expertise which cleanliness level is really required.

5 New Approach: *Fault Injection Test* procedure

With the *risk assessment tool for determining the short circuit risk from particles*, the Zentralverband der Elektronikindustrie e.V. (ZVEI) presented a mathematical model in 2018 that enables risks to be quantitatively assessed.

To refine this approach, the Fraunhofer IPA is currently working on an experimental risk assessment method. This is similar to robustness validation testing and could be used in reliability engineering in the future to ensure the properties of E/E systems according to the V-model [8]. The aim of the development is to improve the accuracy of forecasts by verifying calculation results on the basis of empirical data.

Between the theoretical and real failure rates of E/E systems, there is a divergence that cannot yet be fully explained. However, not every particle on a printed circuit board causes an immediate short circuit. Rather, there are probabilities of a short circuit occurring which are reflected in field failures. The *fault injection test* investigates the impact of various factors on the short circuit probability, see Fig. 5. Test particles with a spectrum of different characteristics are dispensed onto a test PCB, which is then exposed to realistic driving conditions on a test bench. Up to 7,000 short circuits per second and PCB can be measured with the hardware and analysing software.

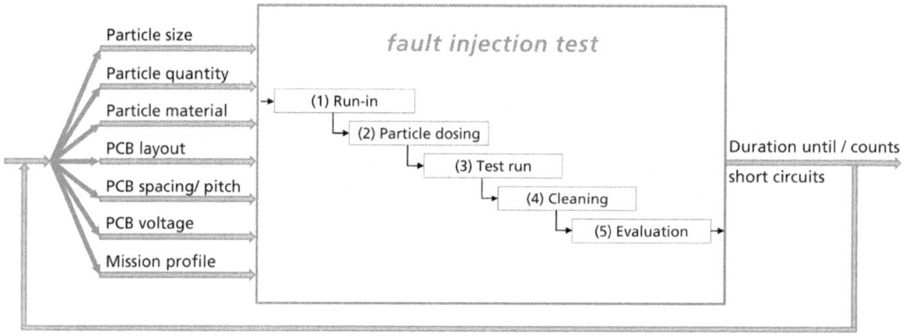

Fig. 5. Fault injection test procedure

Particle characteristics (size, quantity and material) are variables forming an essential part of the *fault injection test* method. Although particle shape also has an impact on the probability of a short circuit, this is not taken into account. Attempting to cluster particles by shape groups would not be helpful in this regard. Instead, the particles have a cuboid shape, allowing the current to flow ideally between two conductor tracks. With particles in Size Group $\mathrm{L} = \{100, 200, 300, 400, 500\}$ μm and in the quantities $\mathrm{n} = \{1, 2, 4, 8, 16, 32, 64, 128\}$, particle size ranges are generated which are representative for the contamination typically found on today's electronic parts according to ISO 16232 and VDA 19.1.

To analyze the impacts of trace spacing and voltage, test PCBs consisting solely of conductor tracks are used. To generate the highest possible density of events, the tracks are distributed as meander arrays over the entire surface of the PCB—in style of the IPC-9253 test board design. The maximum number of meander arrays is prescribed by the signal controller while the PCB dimension were prescribed by the overall weight of maximum 1 kg for the PCB including aluminium housing and connector.

Figure 6 shows the basic design with a fine pitch of 300 µm. Further design variants with a fine pitch of 200 µm and 400 µm are planned to follow next year. Via the measurement technology, the voltage can be adjusted from 5 V to 48 V.

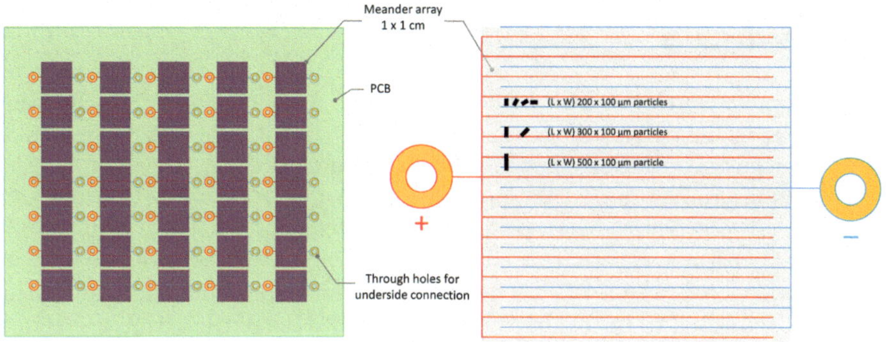

Fig. 6. 2D layout of the test PCB

The mission profile is also a variable that will be studied in the course of the development of the *fault injection test* method. However, this will be entered as a fixed variable in future in order to create standard test conditions. The mission profile is designed to simulate normal operating conditions in an automobile through vibration and shock excitation. The impulse acting on the particles must be strong enough for them to detach from the circuit board and redistribute or realign themselves several times per second. ISO 16750-3 defines various mission profiles, depending on the installation site in the automobile.

To perform the *fault injection test*, a test bench is needed which mechanically transmits the vibrations and shocks to the printed circuit board so that short circuit events occur.

A further requirement is a housing which is capable of absorbing the energy and transferring it to the PCB or particles. It also keeps the particles within the relevant test area, thus guaranteeing the availability of a constant quantity of particles over the entire test period. The short circuits are recorded and cumulated discretely over the entire length of the test; this allows the time until the first short circuit occurs, as well as the number of all short circuits within a given test period, to be specified. If the recorded values are below a defined limit, the particle size, count or types of material can be increased in a new test. If 5 different particle sizes and 8 different particle quantities are used, this results in up to 40 test cycles or iterations. However, before a new test can be started, the previously-dosed particles must first be removed without leaving any residues. Only then can a new/ different particle size range be applied to the PCB. As these are not tests to determine service life, a test period of a few hours up to a maximum of one day is sufficient. Particle-related failures usually occur at the beginning of a product's service life [9].

The *fault injection test* can finally support companies to determine the robustness of their E/E components against a particulate load. Recurring failures of the component indicate the maximum tolerable particulate contamination. This limit can be specified as minimalistic cleanliness requirement.

6 Conclusions

Electromobility is related with higher safety and cleanliness demands. Existing concepts in technical cleanliness need to keep pace with the higher demands. This paper has introduced a new holistic cleanliness concept, based on current approaches in product design and production in conjunction with the novel *fault injection test* procedure. Now, cleanliness requirements can be experimentally determined with a synthetic test contamination which can be individually created and dispensed onto a PCB. The test rig is under construction and first test results are expected to be available in March 2021. One of the future research is to transfer the test procedure from the test PCBs without any surface mounted device towards serial assembled PCBs.

References

1. Attias, D. (ed.): The Automobile Revolution: Towards a New Electro-Mobility Paradigm. Springer International Publishing, Cham and s.l. (2017)
2. Fahlbusch, E. (ed.): Batterien als Energiespeicher: Beispiele, Strategien, Lösungen, 1st edn. Beuth Wissen, Berlin (2015)
3. Beard, K.W., Reddy, T.B., Linden, D.: Linden's handbook of batteries, fifth edition edn. McGraw-Hill Education (2019)
4. Kong, L., Hu, X., Gui, G., Su, Y., Pecht, M.: Fire Technology **54**(8), 359 (2020). https://doi.org/10.1007/s10694-020-00996-w
5. Brown, J.R.: Foseco non-ferrous foundryman's handbook, 11th edn. Butterworth-Heinemann, Oxford England, Boston (1999)
6. Dornfeld, D.: Burrs – analysis, control and removal: Proceedings of the CIRP International Conference on Burrs, 2nd–3rd April, 2009, University of Kaiserslautern, Germany. Springer, Berlin (2010)
7. Gillespie, L.K.: Deburring and edge finishing handbook (Society of Manufacturing Engineers and American Society of Mechanical Engineers, Dearborn. Mich. and Fairfield, NJ (1999)
8. Handbook for robustness validation of automotive electrical/electronic modules (2012)
9. Portolan, M., Savino, A., Leveugle, R., Di Carlo, S., Bosio, A., Di Natale, G.: Proceedings of the European Test Workshop, vol. 2019-May (IEEE Xplore, 2019). https://doi.org/10.1109/ETS.2019.8791555

Highly Integrative Rear End Concept of Battery Electric Vehicles

Dominik Klaiber[1](✉), Dr. Philipp Kellner[1], Marc Meyer[1], Matthias Biegerl[2], Dr. Gabriele Gorbach[2], Thomas Goetz[3], and Dr. Marco Schneider[3]

[1] Porsche AG, Porschestr. 911, 71287 Weissach, Germany
dominik.klaiber@porsche.de
[2] ElringKlinger AG, Max-Eyth-Str. 2, 72581 Dettingen/Erms, Germany
[3] Fraunhofer IPA, Nobelstr. 12, 70569 Stuttgart, Germany

Abstract. In most current battery electrical vehicles the battery modules are positioned in the underfloor area. For sporty-emotional cars a lower seating position might make it necessary to use the package in the rear end. In addition, lightweight design solutions are necessary. This paper deals with the question of how a new innovative and integrative rear end for a battery electric vehicle can be designed. Therefore, different solutions are designed and assessed, whereby a stack of five battery modules and a specially devolved axle are designed. With regard to separate topics, parts are manufactured and different tests are applied. With this approach it is possible to reach ambitious goals by means of functional integration and close interdepartmental collaboration.

1 Introduction

Developments in vehicle design are increasingly concerned with the electrification of the powertrain. Lightweight design solutions are coming into focus as a key technology for meeting legal requirements e.g. CO_2-emission or energy consumption and customer demands e.g. range [1, 2]. However, in the automotive industry, approaches to weight reduction are hampered by the generally low acceptance of additional costs involved in lightweight design. Hence, for new battery electric vehicles (BEVs), it is imperative to develop highly innovative concepts in close collaboration between different departments in order to reduce costs and increase range via weight reduction.

Fig. 1. Area under consideration and goals of the project

The HigHKo project takes account of these challenges by developing a new lightweight and highly integrated rear end concept for a BEV (see Fig. 1). The project

P. Weißgraeber et al. (Eds.): *Advances in Automotive Production Technology – Theory and Application*, ARENA2036, pp. 327–334, 2021. https://doi.org/10.1007/978-3-662-62962-8_38

partners from Porsche AG, ElringKlinger AG and Fraunhofer IPA are working at key system components such as the battery system, chassis and body-in-white at an early stage of the development process in order to realize different goals. This involves material innovations, the integration of functions as well as new process and joining technologies.

2 State of the Art

2.1 Battery Design

Most current battery designs are either positioned at the underbody as in the Porsche Taycan (see Fig. 2) [3–5]. A few concepts in the center tunnel [6, 7] or in the rear end exist, however, with a much lower capacity than the HigHKo-Concept [8]. The battery package comes along with a separate battery frame which fulfills the functions of protecting the battery modules and tighten the area inside against dust and especially media.

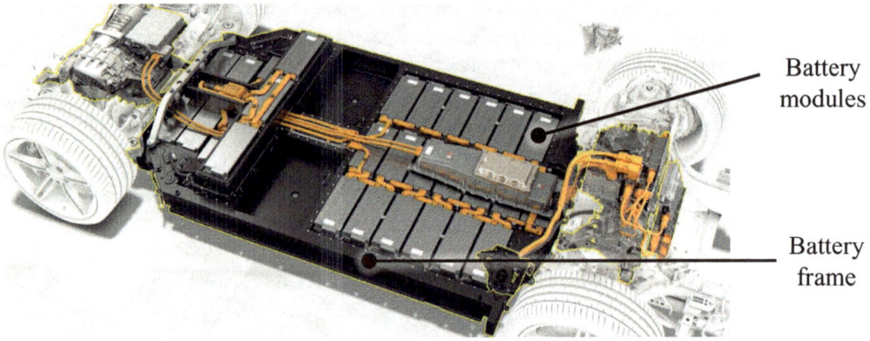

Fig. 2. Current battery design of Porsche Taycan [3]

2.2 Multi-material-Design

In lightweight automotive engineering a mere substitution of traditional metallic structures by modern materials such as carbon fiber reinforced plastics does not appear to be expedient in future due to the high material costs. An alternative to reduce the high costs is the use of multi-material designs, which enable the load and functionally appropriate interaction of different materials within a hybrid component [9, 10]. With regard to economic and production requirements, those materials are selected and combined which best meet the demands placed upon the respective component at minimum weight [10]. However, weight reduction on the material side is not the only advantage of the mixed design method. Thanks to the combination of different materials, maximum functional integration can also be achieved, comprising additional functions such as structural, thermal, sensory or electrical functions on the component level [11].

3 Design

In order to reach the project goals, lightweight, efficiency of installation space and costs are important factors. In this project the ambitious goals can be realized through a combined design and development of chassis, electric components and body-in-white. The different components which are under consideration in this project are shown in Fig. 3.

The concept has an integrated battery frame in the body-in-white structure. The battery assembly is composed of multiple battery modules and an electronic control unit. Compared to other concepts, the stack design of the battery modules in the rear end is an innovative arrangement. Each battery module consists of several battery cells protected by a load-carrying and media-tightened extruded aluminium profile and an integrated battery cooling. The plate at the underfloor protects the battery assembly and seals the body-in-white against media, dust and electromagnetic waves. Therewith, a separate battery housing is not required.

During the manufacturing process, first the chassis is mounted. Second, the battery assembly is built separately and later inserted into the body-in-white where it is fixed with several screws. Finally, the underfloor protection is mounted. [12].

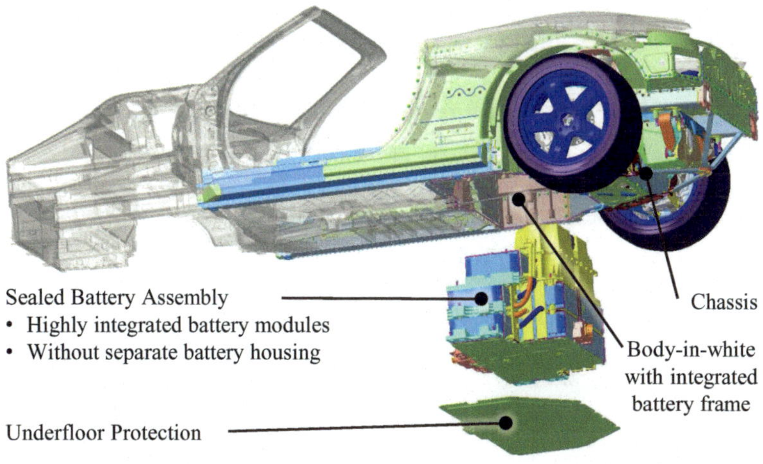

Fig. 3. Design with considered parts, according to [12]

The car is equipped with a McPherson axle for maximal weight gain (see Fig. 4). The wheelhub, the coupling rod and one connection point have been structure-optimized for different load cases like acceleration, braking, bumps and cornering. Additional weight savings have been achieved by

- connecting the axle to already existing hardpoints of the chassis,
- the use of a highly innovative leafspring made of fiber-reinforced plastic which substitutes the conventional metal helicoil springs and also the antirollbar

- and the use of an upside-down damper unit.

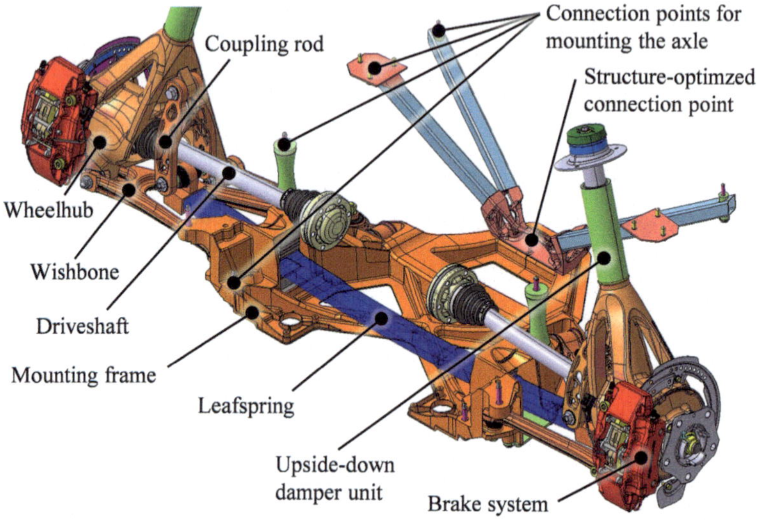

Fig. 4. McPherson axle of the HigHKo-Concept with different components

Different examples regarding functional integration are part of the project and important elements for the realization of the project goals. For stabilizing the temperature during driving the extruded profiles are cooled on both sides with a cooling media [13]. Therefore, a plate is glued directly on the extruded profile (see Fig. 5 left). By integrating the function, the used glue has to fulfill both a structural and a sealing function. Separate specimens are designed and leakage and ageing tests are performed. Fig. 5 (right) shows specimens for single-lap shear tests.

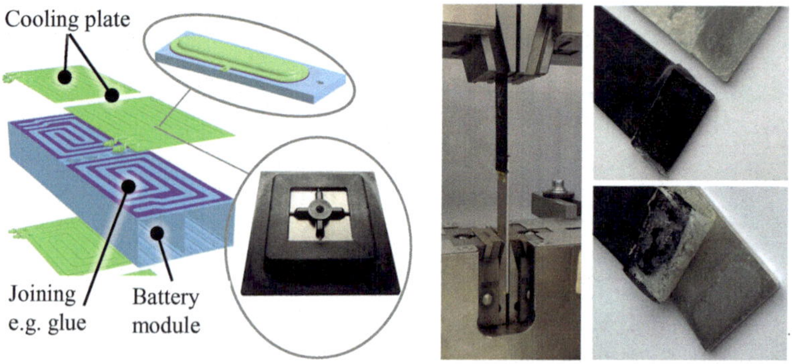

Fig. 5. Mounted cooling plate according to [13] with derived test specimens (left) and Lapped shear-strength specimens (right)

Another part with functional integration is the HigHKo technology demonstrator (see Fig. 6). It is a highly integral structural component, which has to meet strict requirements for overall stiffness, sealing to the single battery modules and carrying loads in case of a vehicle rear crash.

The production technology implementation using a One-Shot-Process allows part production with hardly any subsequent assembly despite sealing mounting and this in automotive usual cycle times to keep the production costs low.

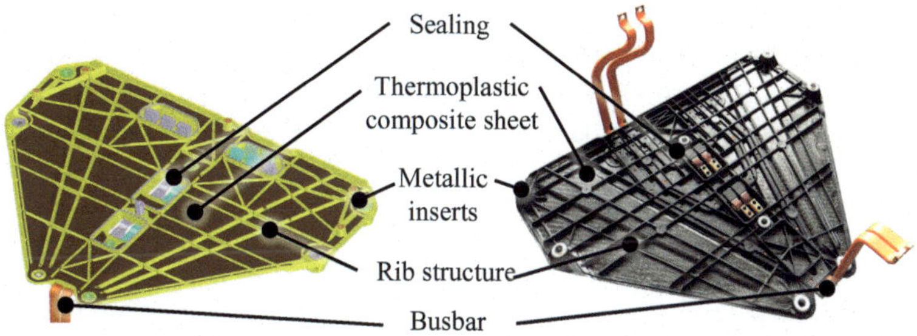

Fig. 6. Multi component thermoplastic composite part (HigHKo technology demonstrator) according to [14] as CAD Data (top view, left) and real part (bottom view, right)

4 Design and Manufacturing Process

4.1 Highly Integral Part Design

One main challenge in the project is the realization of the technology demonstrator including the integration of the busbars in the injection molding tool.

Figure 7 shows the section of the technology demonstrator with the multi material components. The base structure is made of glass fiber (GF) reinforced Polyamide (PA6) composites (organic-sheet). The component is additionally stiffened by the injection molded rib structure. Furthermore, functional elements such as sealing groves and screw-on domes are manufactured by the injection molding process. Moreover, the requirement of media tight interfaces between the injection molding material (PA6 GF45), the organic sheet and the single busbars must be fulfilled.

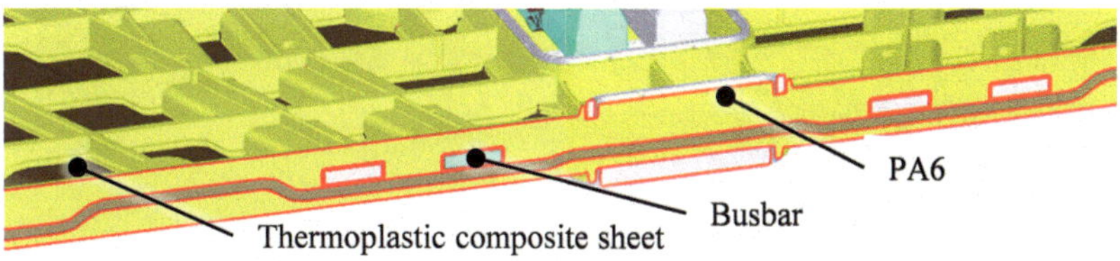

Fig. 7. Interfaces between single busbars and thermoplastic material

This is achieved by providing the busbars with additional insulation material for example high temperature silicone rubber, so that the injection molding material shrinks onto the insulation to provide a media-tight connection between all components. By overmolding the busbars, a media-tight connection between the injection molding material and the organic-sheet is also possible, whereby the organic-sheet is also media-tight due to an additional inserted barrier layer.

4.2 Manufacturing Process

The busbars are inserted as inserts in the injection molding tool and then overmolded. For this purpose, the busbars must be supported against the injection pressure to avoid bending, whereby active supporting elements are integrated in the molding tool with the ability to release the cavity before the holding pressure phase and the subsequent complete overmolding of the busbars (see Fig. 8). This enables complete media tightness.

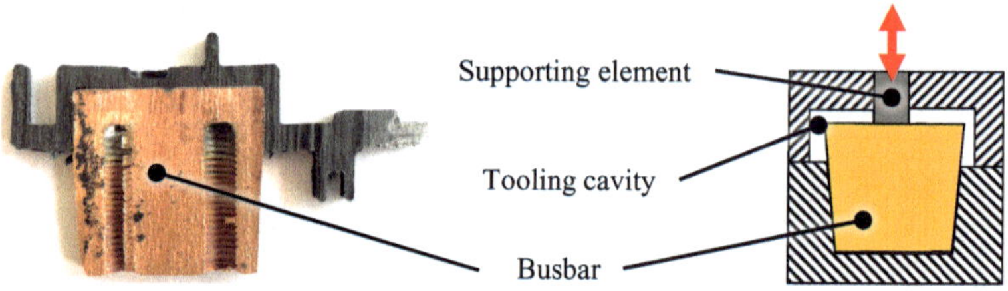

Fig. 8. Overmolded busbar with moveable tooling element to support inserts in the tool

During the manufacturing process in a One-Shot-Process the tailored organic-sheet is heated above melting temperature of the matrix in a convection oven. Afterwards the now deformable organic sheet is transferred and placed into the cavity of the molding tool. As the tool closes, the organic-sheet is thermoformed and immediately overmolded with PA6 GF45 to create the rib structure and all functional elements. The tool and the manufacturing facility are shown in Fig. 9.

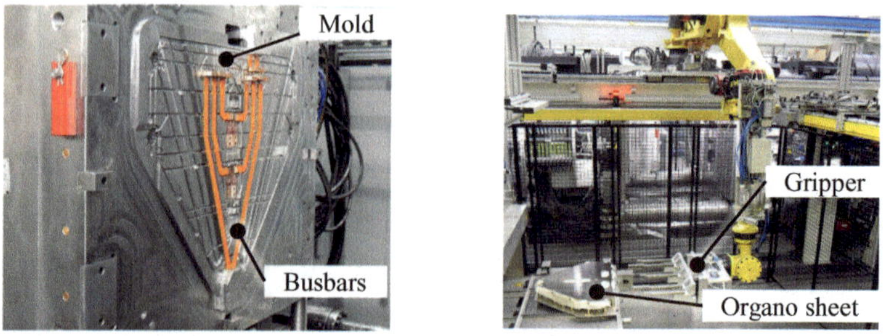

Fig. 9. Injection molding tool with inserted busbars (left), handling process of pre-heated organic-sheet (right)

5 Evaluation of the Highly Integrative Rear End Concept

The four main functional assemblies were finally evaluated with regard to possible weight savings. The rear end of the e-generation [15] was used as a reference. The reference weight already included surcharges due to increased crash safety requirements for battery electric vehicles (higher crash mass, structural connection of the high-voltage battery, high-voltage safety) as well as further necessary adjustments to the new vehicle concept (higher energy content, cooling on both sides). Figure 10 shows the overall assessment of the design and integration of all assemblies.

With the weight reduction of 55 kg as compared to the initial state, the battery assembly contributes significantly. For a lower percentage of weight reduction e. g. for the body-in-white, it should be noted that further functions e. g. from the battery assembly are integrated. As described in chapter 1, the total weight reduction was focused and with 74 kg total weight reduction the goal (46 kg) can be achieved.

Furthermore, with the integrative design compared to a solution in differential design benefits regarding the overall costs are expected.

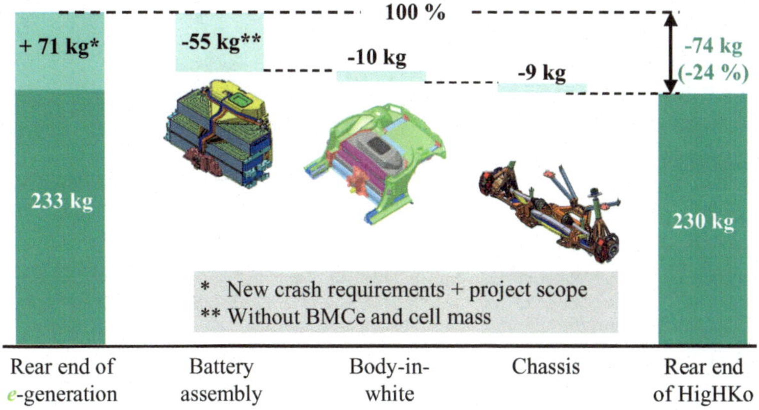

Fig. 10. Functional mass analysis of the new rear end concept, according to [12]

6 Conclusion and Outlook

Lightweight design is a key technology for future efficient and low-emission vehicle concepts. The aim of the project was the prototypical design of a highly integrated rear end of a BEV, consisting of the components battery system, chassis, body-in-white and drive. The analysis shows that weight reductions can be achieved through functional integration and new conceptual approaches.

Based on this concept, further developments for the specific car design are necessary. Additional tests of the single components and the building and testing of entire assemblies will be mandatory if the concept shall be used in future.

Acknowlegdements. The HigHKo project was funded by the Ministry of Economic Affairs and Energy (BMWi) within the framework of the funding program "Neue Fahrzeug- und Systemtechnologien" and supervised by the project sponsor TÜV Rheinland. The authors thank the BMWi for the funding provided and the project sponsor for its support.

References

1. Köth, C.-P.: Volkswagen: „Wir investieren in bezahlbaren Leichtbau statt in teure Batteriezellen", https://www.automobil-industrie.vogel.de/volkswagen-wir-investieren-in-bezahlbaren-leichtbau-statt-in-teure-batteriezellen-a-904626/. Accessed 8. May 2020
2. Strathmann, T.: Elektromobilität als disruptive Innovation. Herausforderungen und Implikationen für etablierte Automobilhersteller. Gabler, Mainz (2019)
3. Wienkötter, M.: Die Batterie: Ausgeklügeltes Thermomanagement, 800 Volt Systemspannung. https://newsroom.porsche.com/pdf/print/aeba095e-639f-457a-8e95-5c8f61b4abdc. Accessed 5. June 2020
4. Gomoll, W.: Batterie-Geheimnisse des Audi e-tron: Raus aus der Tesla-Falle. https://www.automobil-produktion.de/technik-produktion/fahrzeugtechnik/batterie-geheimnisse-des-audi-e-tron-raus-aus-der-tesla-falle-128.html. Accessed 6. June 2020
5. Schaal, S.: Mercedes EQC: Die Batterie im Detail. https://edison.media/erklaeren/mercedes-eqc-die-batterie-im-detail/23004130.html. Accessed 6. June 2020
6. Audi: Forschungsprojekt e-performance. https://www.audi-technology-portal.de/de/mobilitaet-der-zukunft/audi-future-lab-mobility/audi-future-engines/forschungsprojekt-e-perfomance. Accessed 6. June 2020
7. Slavnich, D.: Volvo cars to go "all electric" by 2019. https://www.enginetechnologyinternational.com/news/electric-powertrain-technologies/volvo-cars-to-go-all-electric-by-2019.html. Accessed 6. June 2020
8. Lambert, F.: 2017 Ford Focus Electric coming with new 33.5 kWh battery pack for 115 miles of range. https://electrek.co/2016/11/23/2017-ford-focus-electric-battery-pack-range/. Accessed 4. June 2020
9. Prüß, H., Stechert, C., Vietor, T.: Methodik zur Auswahl von Fügetechnologien in Multimaterialsystemen. In: 21st Symposium on Design for X, p. 13 (2010)
10. Sahr, C., Berger, L., Lesemann, M.: Systematische Werkstoff-Auswahl für die Karosserie des Superlight-Car. ATZ Automobiltechnische Zeitschrift **112**(5), 341–347 (2010)
11. Hoßfeld, M., Ackermann, C.: Leichtbau durch Funktionsintegration. Springer Vieweg, Wiesbaden (2020)
12. Dimitrov, M., Kellner, P., Mostofi, S., Meyer, M., Klaiber, D., Biegerl, M., Gorbach, G., Götz, T., Schneider, M.: Abschlussbericht Forschungsprojekt HigHKo (2020) – currently unpublished
13. Kellner, P., Bohlien, J., Klaiber, D.: Batteriemodul zur Verwendung bei einem Hochvolt-Energiespeicher, DE 10 2017 104 710.7 (2018)
14. Kellner, P., Klaiber, D., Kollmannthaler, M.: Lasttragelement, DE 10 2018 126 319 (2020)
15. Dimitrov, M.: E-generation/Key technologies for electric vehicles. In: Bargende, M., Reuss, H.-C., Wiedemann, J. (eds.) 14. Internationales Stuttgarter Symposium, pp. 19–29. Springer Vieweg, Wiesbaden (2014)

Modelling Defects of Unhardened Adhesives Resulting from Handling and Warpage

Silvio Facciotto[1(✉)], Daniel Sommer[1], Martin Helbig[2], André Haufe[2], and Peter Middendorf[1]

[1] Institute of Aircraft Design, University of Stuttgart, Stuttgart, Germany
facciotto@ifb.uni-stuttgart.de
[2] DYNAmore GmbH, Stuttgart, Germany

Abstract. The focus of this work is capturing the generation of instabilities in bondlines of uncured adhesives using the finite element method. Two geometries have been studied, a cylindrical probe and a specimen representative of a possible adhesive line used in the automotive production. Such geometries have been analyzed under tensile loading considering different width/thickness aspect ratios. A Neo-Hookian material model was chosen to describe the behavior of the material and the formation of instabilities during the process. A discretization and a parametric study have been performed to show the validity of the model and the results have been found to agree with results found in literature.

1 Introduction & State of the Art

There are several cases of elastic layers used in technological applications, such as the use of adhesives to bond hybrid and multifunctional parts in the automotive industry.

Often mechanical joints are additionally required to guarantee the integrity of the body in white in case of failure of the adhesive bonds; e.g. to stop catastrophic zipper-like failure of a bonded seam. However not all of these joints are placed in the early stages of the process and whilst the adhesive is still in an uncured state, the parts undergo several mechanical and thermal loadings which might induce deformations.

The most critical stages during automobile production are robotic handling of the body in white and the oven-curing of electrophoretic deposition coating. The metal parts are exposed to high accelerations and temperatures respectively, causing them to move from the original position. This leads to relative displacements of joint partners, which are also passed to the adhesive layer and might result in mechanical instabilities. Such instabilities affect the mechanical strength of the structure generating weak points where several failure mechanisms might occur jeopardizing the final product [1]. Several types of instability can be recognized, for example partial de-bonding creating undulations [2, 3]. In case of perfect adhesion of the adhesive layer to both joining partners, three types of elastic instability might occur when critical values of displacement are reached.

These instabilities are related to the geometry of the adhesive layer and can be divided into three categories: cavitation, fingering and fringe instabilities. Cavitation is created

P. Weißgraeber et al. (Eds.): *Advances in Automotive Production Technology – Theory and Application*, ARENA2036, pp. 335–342, 2021. https://doi.org/10.1007/978-3-662-62962-8_39

when the hydrostatic stress reaches a critical value in all the regions of the elastic layer and appears as a nucleation effect in the middle of the cross section [3–5]. Finger (see Fig. 1) and fringe instabilities both present themselves as protrusions in the external surface of the elastic layer which is in contact with the air. These phenomena result from nearly incompressible behavior and thus a volume constancy when being stretched. The main differences are the location of the instability and the stress-stretch response: while fingers present themselves in the middle of the meniscus and have a non-monotonic stress-stretch behavior, fringes appear at the ends near the rigid bodies and result in a monotonic curve [3, 6]. Apart from imperfections and problems related to the application of the adhesive, the main parameter affecting the onset of one instead of the other is of geometrical nature and is related to the width of the specimen compared with its height [3, 4, 7]. It was demonstrated that cavitation can coexist with both fingering and fringing instabilities, i.e. there is no aspect ratio where effects switch instantaneously, it is more likely a transition between effects [3, 4].

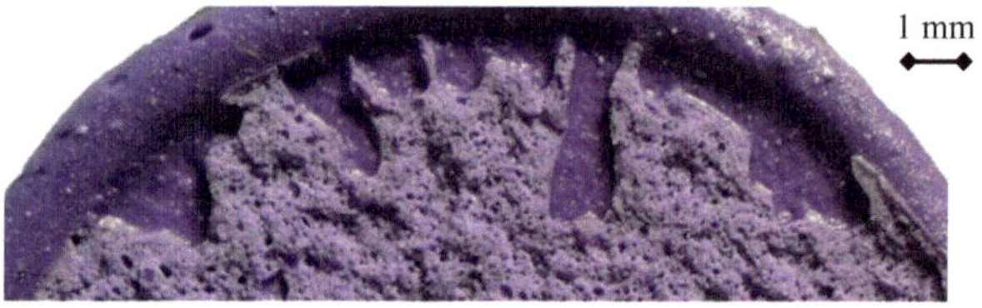

Fig. 1. Fingering in a round specimen subjected to relative displacement of joining partners obtained in a multiaxial test-rig in an oven simulating the hardening in the electrophoretic process

Often the analogy between such instabilities and the Taylor-Saffman (or so-called viscous fingering) instability is made. Although the resemblance, it must be understood that while the latter is generated by a dynamic process (two fluids of different viscosities interacting with each other), the onset of elastic instabilities occur under quasi-static conditions. It was observed in various experiments that while the first is a rate-dependent, supercritical and irreversible process, elastic instabilities are sub-critical, rate-independent and fully reversible [7, 8]. This is the relevant case for adhesives.

Computational simulations based on iterative solving of combinations of Laplace's, Poisson's and Darcy's equations yield a rough estimate of finger geometry and formation for fluids and may be eligible for refinement and adaption to adhesives [9]. The onset and evolution of elastic instabilities was studied by many authors from both an analytical perspective [3, 6, 8, 10] and from a numerical one [3, 4, 6, 7] by considering the elastic layer as a Neo-Hookian material in finite element (FE) simulations. Lin et al. [3, 4, 6] presented an analytical solution for the onset of the instabilities. Assuming axial symmetry and incompressibility they were able to calculate the analytical stress-stretch response prior the onset of fringe and fingering instabilities. Moreover the onset of bifurcation was obtained by linear perturbation analysis of the deformation.

The same authors presented an explicit finite element method to predict the evolutions of the instabilities. The method was validated with experimental results performed with hydrogel solutions and compared with the analytical results mentioned above.

Giovanardi et al. [7] showed that despite the robustness of explicit methods, the computational effort which is required from the onset of the instabilities grows drastically with the increase of the deformation and that techniques to maintain the computation time acceptable, such as mass scaling, may introduce unwanted inertia forces. In order to prevent this and also consider the quasi-static nature of the instabilities, Giovanardi et al. [7] proposed the use of the arc length method to solve the problem using an equilibrium method. Such methods allows for complex snap-back and snap-through effects to occur during the deformation of the specimen, which would be otherwise be difficult to capture with explicit simulations.

These methods have not yet been applied specifically for adhesives. The aim of this work is to apply the above mentioned techniques to typical material and shapes relevant to the automotive industry.

2 Physical Parameters Affecting the Instabilities

The main parameter defining the onset of finger over fringes is the geometry of the specimen, which is represented by its aspect ratio, defined as:

$$\alpha_O = \frac{D}{H} \text{ or } \alpha = \frac{W}{H} \tag{1}$$

where D is diameter and H is thickness of a circular specimen [3, 6]. Depending on this ratio the type of instability which will occur can be predicted. Lin et al. [3] showed the formation of fringe instabilities for values of $\alpha_O < 4$, while for values above the phenomenon transitions to fingering instabilities, which become visible for $\alpha_O > 6$.

In this work two studies are presented: a circular geometry, similar to Lin et al. [3] and Giovanardi et al. [7], allowing the validation by comparison with results found in literature, and a thin strip of adhesive which constitutes a more interesting and representative geometry for the automotive industry. In order to analyze this the aspect ratio parameter is modified here to α, where W is the width of the rectangular geometry considered. As the length of the line is significantly larger it is hence not accounted for. To introduce loading and to evaluate results, stress and stretch are defined by

$$\sigma_c = \frac{F}{A_0} \text{ and } \lambda = \frac{h}{H_0} \tag{2}$$

σ_c is the critical stress leading to instability as the fraction of force F and initial area A_0 and λ represents the stretch determined by current thickness h and initial thickness H_0.

Adhesive bonds in automobile applications typically have thicknesses below 1 mm, and are often as low as 0.15 mm. Resulting in large aspect ratios, within the range where fingering is to be expected and can observed in production environments.

3 Virtual Prognosis of Instabilities in Uncured Adhesives

3.1 Choosing the Modelling Strategy

The software chosen and set for this study is the multi-physics solver LS-Dyna R11.1 capable of different methods of solutions, from which the original finite element analysis (FEA) is the most-used and best-implemented. However, the migration of material when fingering occurs from the outer surface of the adhesive far to the inside is large compared to the thickness of a bond line, resulting in large deformations of the mesh required for FEA-analysis.

Finite element simulation with remeshing or using the Element-Free-Galerkin method, could lead to better discretized fingers, but seems to require more dedicated work. However, it was not necessary to implement this strategy, as it was possible to analyze the onset of the instabilities using a classical FE analysis setup with the simulation method derived within this paper as follows.

3.2 Selection of Suitable Material Models

In order to describe the incompressible, reversible and highly strained behavior of uncured adhesives, a robust material model is required which must be capable of handling Poisson´s ratios close to 0.5 and offer a hyperelastic constitutive formulation along with means of calibration such as lifting Hele-shaw cells [11].

Four material models were found eligible in LS-Dyna to fulfill these criteria: *MAT_MOONEY-RIVLIN_RUBBER, *MAT_OGDEN_RUBBER, *MAT_ARRUDA_BOYCE_RUBBER and *MAT_HYPERELASTIC_RUBBER [12]. These implement different hyperelastic material models which can be divided into phenomenological models, e.g. the Mooney-Rivlin or Ogden model and mechanistic models such as Arruda-Boyce or the Neo-Hookean method. Each of the models provides different parameters to be calibrated to represent the physical behavior of the material. For the scope of providing a method for simulation of the phenomena, the Neo-Hookian model as implemented in *MAT_HYPERELASTIC_RUBBER is chosen here. This is the model with least parameters able to describe a hyperelastic behavior [13], with only poisson's ratio, bulk modulus and one hyperelastic constant for the strain energy functional required:

$$W = C_{10}(I_1 - 3), \text{ with } C_{10} = \frac{G}{2} \tag{3}$$

where W is the strain energy density, C_{10} the first hyperelastic constant and I_1 the first deformation invariant, reflecting the principal stretches. C_{10} is directly related to half the shear modulus G for the Neo-Hookian model.

As baseline material model parameters, the card is filled with a density $\rho = 1.2 * 10^3 \text{kg/m}^3$ for the epoxy-based structural adhesive under consideration, a shear modulus $G = 5 * 10^{-6}\text{GPa}$ and thus $C_{10} = 2.5 * 10^{-6}\text{GPa}$ from literature [3, 6, 7].

3.3 Simulation Setup and a Study of Convergence and Aspect-Ratio

To study the feasibility of process simulation of these phenomena, a series of round glue points of diameter 20 mm and varying thickness is setup. Ideal adhesion is assumed representing complete wetting of the joint partners with the adhesive. Thus, nodes on the bottom are fixed in all directions, whereas the upper nodal layer is moved upwards with a prescribed motion, while being fixed in-plane in order to keep the moving plane parallel to the plane where the boundary conditions were applied. The chosen element formulation are 1 point nodal pressure tetrahedrons (type 13 within LS-Dyna), which alleviate volumetric locking and are suited for incompressible applications [12].

Results for different aspect-ratios

The ability of this modelling method to predict fingering and fringing is shown in Fig. 2. The specimens of different initial height H_0 from 0.5 to 50 mm show the aforementioned dependence of the phenomena on the geometry of the adhesive layer. Element length is 0.25 mm for all simulations shown in Fig. 2. This gives finer discretization over the height as specimen grow higher and shows, that a minimum of four elements are required to discretize the thickness. Two elements can then form the bottom/top of the finger and two are highly strained to form the wall of the finger. An under-discretized specimen (e.g. 0.5 mm, top left, Fig. 2) cannot represent correctly the onset of such instabilities.

The simulation results shown in grey in Fig. 2 are strained to same stretch $\lambda = 2$ and show a delayed onset of fingering with rising initial height, i.e. the onset of fingering will occur at higher λ with larger H_0. Modelling of fringing is possible as shown in blue in Fig. 2, right where a specimen of low aspect ratio of 0.4 is strained to $\lambda = 3.5$, past the onset of fingering at $\lambda = 2.6$. This shows the capability of the modeling method presented here.

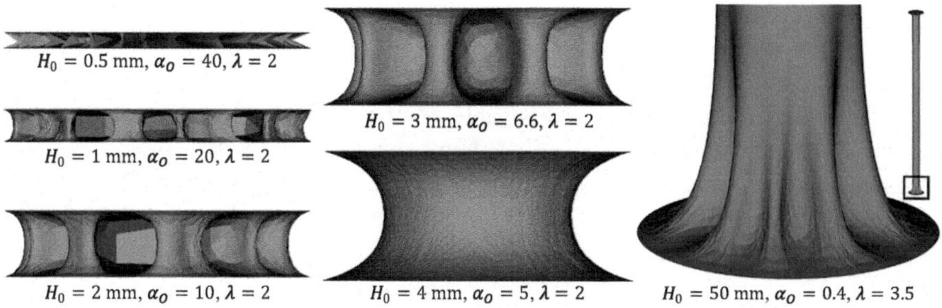

Fig. 2. Round specimen of $D = 20$ mm and element size of 0.25 mm with different H_0 and α

Influence of Discretization

This onset of the instabilities is not generated by initial mesh-disturbances or introduced by imperfections in the model. This was tested using a completely symmetrical discretization. The shape of fingers can, however, be initialized at instability points on a non-regular mesh, just as they would be at real-world imperfections on the adhesive line. This provides means of control, mapping and calibration to a physical reference geometry.

Another study performed is shown in Fig. 3, where the same specimen as before, only with identical initial height H_0 of 2 mm and varying characteristic element length l_E. Fingering occurs at the same stretch and the same force-displacement curve is obtained when thickness is discretized with at least 4 elements. Resolution of the fingers increases with smaller element size. Exact geometrical representation of physical fingering has to be evaluated on tests with scanned and mapped initial geometry to make a statement on correctness of finger reproduction. Figure 3 also depicts the extreme stretch elements on the finger's external surface, whereas elements on the inside of the fingers are highly compressed.

However, the robustness of the method was proven by varying the length of the elements and using different irregular and regular meshes resulting in the onset of the phenomena regardless of the changes applied.

Fig. 3. Round specimen of $D = 20$ mm, $H_0 = 2$ mm at $\lambda = 2.0$ with different element lengths

Force reaction and influence of material model parameters

To validate and calibrate the model, not only the geometry must be predicted, but also the correct strain-resistance response must be reproduced by the FE-analysis. Figure 4, left depicts the force-stretch response of two specimen already shown in Fig. 2. The specimen with an initial ratio of $\alpha = 20$ shows fingering and yields a non-monotonic force-strain response. After onset of the fingers, the slope changes and after a peak load the force needed for further straining declines with displacement. Here, fingers can merge and split, which this FE-Method is also able to reproduce. The fringing-case exhibited by a specimen with an aspect ratio of 0.4 shows a monotonically increasing force-strain reaction. Onset of the perturbation results in a slight change of slope, but force needed for further straining continues to rise constantly. This is in good agreement with findings in the literature along with the values obtained for the force magnitude in both types of instability [7].

Regarding the influence of parameters of the Neo-Hookian material model, the plot on the right of Fig. 4 shows five combinations of C_{10} and G. As introduced, the model does not depend on the given shear modulus, as it can be directly deduced from C_{10}. Thus, as shown in the figure, the internal energy, or accumulated force, is only increased when C_{10} is increased and vice-versa. Geometry and shape of fingers at same stretch are not affected. The same fingers are predicted. At same the force however, a different geometry results from different values of C_{10}. This provides a means of calibration with tests, as outlined in Chap. 3, which is enhanced even more when choosing a multi-parameter hyperelastic model. Changes in C_{10} and therefore in G also reflect changes in viscosity, as both are deeply connected [14]. In this paper low viscosity, fluid-like glues are not taken into account as these cannot be modelled with this method.

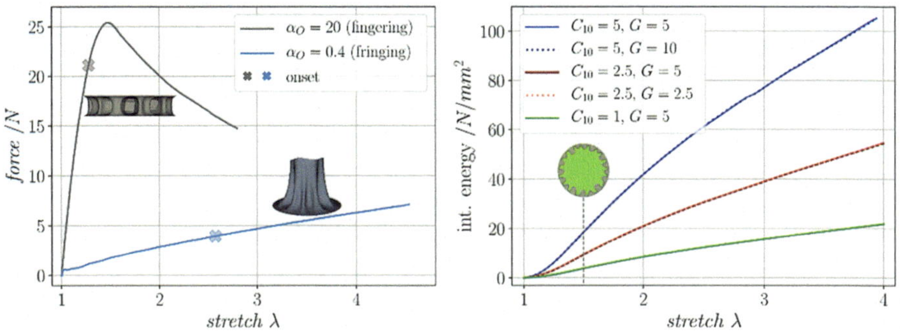

Fig. 4. Left: force-stretch response of two specimens of high and low aspect ratio; Right: internal energy of specimen with $\alpha_O = 20$ with varying material parameters in GPa

4 Loaded Uncured Adhesive Bondlines in Automobiles

Fig. 5. Long adhesive bondline with tetra 13 elements

In automobile manufacturing, adhesives are not used as point-connections as studied before. For this purpose, a strip with rounded edges was simulated in this work. The geometry considered has a length of 100 mm, a width of 5 mm and initial thickness of 1 mm, resulting in $\alpha = 5$. To validate the results obtained with Tetra 13 elements, Hexa 1, 2 or 3 were also investigated. Although the onset of fingering instability was initiated in some cases, these elements cannot reproduce stretches above 1.25, regardless of hourglass control, and were therefore discarded. The stable, symmetrical result is shown in Fig. 5 at $\lambda = 1.5$, which are unprecedented for adhesives and this geometry.

5 Conclusions and Outlook

This work presents an *FEA* model able to predict the onset of fingering and fringing instabilities for circular specimens and geometries representative of the adhesive lines used in the automotive industry. A discretization and parametric studies have been performed showing the effect of the aspect ratio of the specimens on the generation and type of instability. The model was validated using findings and results present in literature. A more detailed examination of long bondlines of different geometries and their real-world imperfections shall follow this work. Calibration with tests is currently undertaken and will result in a robust material card for two structural adhesives. A representative volume approach for long bondlines will be considered to reduce calculation effort. Additionally, to overcome the extreme stretch of elements a multi-stage simulation is suggested where the disturbed mesh is re-generated and the simulation is continued from that point. Using the results of this process simulation to generate the deformed geometry of the adhesive layer and assess the influence on mechanical parameters on automobile adhesive bondlines is an ongoing next step.

This work is funded by the *Federal Ministry for Economic Affairs and Energy* of Germany for project *DigiBody* under identification 19I19003C.

References

1. Creton, C., Ciccotti, M.: Fracture and adhesion of soft materials: A review. Rep Prog Phys **79**, 46601 (2016). https://doi.org/10.1088/0034-4885/79/4/046601
2. Chakrabarti, A., Chaudhury, M.K.: Direct measurement of the surface tension of a soft elastic hydrogel: Exploration of elastocapillary instability in adhesion. Langmuir **29**, 6926–6935 (2013). https://doi.org/10.1021/la401115j
3. Lin, S., Mao, Y., Radovitzky, R., et al.: Instabilities in confined elastic layers under tension: Fringe, fingering and cavitation. J. Mech. Phys. Solids **106**, 229–256 (2017). https://doi.org/10.1016/j.jmps.2017.05.011
4. Lin, S., Mao, Y., Yuk, H., et al.: Material-stiffening suppresses elastic fingering and fringe instabilities. Int. J. Solids Struct. **139–140**, 96–104 (2018). https://doi.org/10.1016/j.ijsolstr.2018.01.022
5. Lopez-Pamies, O., Idiart, M.I., Nakamura, T.: Cavitation in elastomeric solids: I—A defect-growth theory. J. Mech. Phys. Solids **59**, 1464–1487 (2011). https://doi.org/10.1016/j.jmps.2011.04.015
6. Lin, S., Cohen, T., Zhang, T., et al.: Fringe instability in constrained soft elastic layers. Soft Matter **12**, 8899–8906 (2016). https://doi.org/10.1039/c6sm01672c
7. Giovanardi, B., Śliwiak, A.A., Koshakji, A., et al.: A path-following simulation-based study of elastic instabilities in nearly-incompressible confined cylinders under tension. J. Mech. Phys. Solids **131**, 252–275 (2019). https://doi.org/10.1016/j.jmps.2019.06.020
8. Biggins, J.S., Mahadevan, L.: Meniscus instabilities in thin elastic layers. Soft Matter **14**, 7680–7689 (2018). https://doi.org/10.1039/c8sm01033a
9. Tarafdar, S., Nag, S., Dutta, T., et al.: Computer simulation of viscous fingering in a lifting Hele-Shaw cell with grooved plates. Pramana - J Phys **73**, 743–754 (2009). https://doi.org/10.1007/s12043-009-0142-4
10. Biggins, J.S., Saintyves, B., Wei, Z., et al.: Digital instability of a confined elastic meniscus. Proc. Natl. Acad. Sci. USA **110**, 12545–12548 (2013). https://doi.org/10.1073/pnas.1302269110
11. Hele-Shaw, H.: The flow of water. Nature **58**, 34–36 (1898). https://doi.org/10.1038/058034a0
12. Livermore Software Technology Corporation: LS-Dyna Keyword User's Manual: Volume 1&2 R11 (r:10572) (2018)
13. Ogden, R.W.: Non-Linear Elastic Deformations, Dover Civil and Mechanical Engineering. Dover Publications, Newburyport (2013)
14. Macosko, C.W.: Rheology: Principles, measurements, and applications. Advances in interfacial engineering series. VCH, New York (1994)

Experimental Study on Depth of Cure During UV-Post-Curing of Photopolymers Used for Additive Manufacturing

Jan Nitsche[1(✉)], Tristan Schlotthauer[2], Florian Hermann[1], and Peter Middendorf[2]

[1] Mercedes-Benz AG, Leibnizstraße 2, 71032 Böblingen, Germany
jan.j.nitsche@daimler.com

[2] IFB, Institute of Aircraft Design, University of Stuttgart, Pfaffenwaldring 31, 70569 Stuttgart, Germany

Abstract. For the production of automotive interior parts, there is a high demand regarding the surface quality combined with large production rates. These requirements can potentially be achieved by using Stereolithography (SL).

This work concerns the dimensional limitations of depth of cure during UV-post-curing of parts made of urethane acrylates by Digital Light Processing (DLP). Due to absorption of the UV-light in the photopolymer, the penetration depth is limited. A better understanding of the depth of cure is necessary to define optimal post-curing parameters and potential limits of the wall thickness. To investigate the depth of cure, green-state specimens are irradiated unidirectionally by UV-light with a wavelength of 365 nm and 405 nm. Subsequent hardness measurements according to Shore D are used to determine the corresponding depth profiles. Thereby the influence of photoinitiator concentration, irradiation dose and wavelength on depth of cure is analysed. Additional transmission measurements during post-curing contribute to a better understanding of the penetration depth of the UV-light for optimal production times.

For a non-pigmented photopolymer, the measured depth of cure is between 0.59 mm and 8.58 mm. The highest depth of cure is obtained by a low photoinitiator concentration of 0.75%wt, long wavelength of 405 nm and high irradiation dose of 70 J/cm^2.

1 Introduction

Additive manufacturing (AM) offers high potential for the low volume production of automotive parts by a reduced lead-time and high design freedom.

P. Weißgraeber et al. (Eds.): *Advances in Automotive Production Technology – Theory and Application,* ARENA2036, pp. 343–350, 2021. https://doi.org/10.1007/978-3-662-62962-8_40

Different processes like Fused Filament Fabrication (FFF) and Selective Laser Melting (SLM) are available to produce parts by polymers and metals [1]. Stereolithography (SL) was the first AM process established in 1986 [2]. During the SL process, a photosensitive resin is cured layer-by-layer via UV-light to form three-dimensional objects (Fig. 1a). Parts produced by SL or Digital Light Processing (DLP) are showing a high accuracy and surface quality that can meet the requirements for automotive interior parts. This opens the opportunity to produce personalized interior parts without expensive and time intensive tooling.

During processing in a stereolithography apparatus a degree of cure between 80% and 90% can be achieved [3,4]. A subsequent post-curing by UV-radiation and temperature further increases the degree of cure leading to a significant raise in strength, stiffness and hardness [5–7]. The UV-light penetration depth into the printed part is limited by the absorptivity of the photopolymer [8]. This affects the required post-curing time and potentially limits the wall thickness that can be post-cured.

This study investigates the effect of the photoinitiator concentration, which is a material parameter and the process parameters wavelength and radiation dose on the depth of cure during UV-post-curing. This contributes to a better understanding on depth of cure evolution and limitations of wall thicknesses for feasible cross-linking by UV-post-curing. To assess the depth of cure, specimens are post-cured by unidirectional UV-light followed by Shore D hardness testing. Additionally, the transmission of the photopolymer during post-curing is investigated.

2 Experimental

2.1 Design of Experiment

A design of experiment (DoE) approach with a full factorial design is applied to systematically investigate the effects of the material and process on the curing depth. For the technical application, the most relevant information is the part thickness that can be fully post-cured during the process. For that reason, the chosen output variable for the DoE is the depth where the measured hardness drops below a threshold value. Here this value is defined to 95% of the average of the maximum obtained hardness for all specimen (see Sect. 3.1). Hardness testing shows a good correlation to degree of cure for photopolymers and a high sensitivity at high levels of degree of cure [9].

The input variables for the DoE are the concentration of photoinitiator in the material, the wavelength of the radiation source and the post-curing light dose. The photoinitiator concentration and light dose are varied in three levels while the wavelength is varied in two levels: 365 nm and 405 nm. These are the wavelengths typically used for the curing of SL materials. The chosen values are listed in Table 1.

Table 1. Material and process parameters for the three factors of DoE

Factor	Level −1	Level 0	Level 1
Photoinitiator concentration (%wt)	0.75	1	1.25
Wavelength (nm)	365	n/a	405
Light dose (J/cm^2)	10	40	70

2.2 Materials

For the study an urethane acrylate is mixed by the following base components (Rahn AG, Swiss):

- Monomer: ACMO
- Oligomer: GENOMER 4277
- Photoinitiator: BAPO

The monomer and oligomer are always mixed by a ratio of 1:1 with three different levels of photoinitiator concentration (Table 2).

Table 2. Material compositions for the three levels of photoinitiator concentration

Level	ACMO	GENOMER 4277	BAPO
Level -1	49.625%	49.625%	0.75%
Level 0	49.5%	49.5%	1%
Level 1	49.375%	49.375%	1.25%

2.3 Test Specimen Production and Post-Curing

The specimen are printed on the DLP system Rapid Shape S60 (Rapid Shape, Germany) equipped with a 405 nm LED light source. The layer height is set to 50 μm and the light dose per layer is 3.2 mJ/cm^2 with a light intensity of 0.64 mW/cm^2 and exposure time of 5 s (Fig. 1a). For hardness testing 18 blocks of $12 \times 20 \times 40$ mm are produced, one for each factor combination. For post-curing the blocks are placed in an opaque housing that only allows exposure to one side of the block. This ensures unidirectional post-curing and a gradient in degree of cure in the material. The UV-chamber LED Cube 100 IC (Hönle AG, Germany) is used to post-cure the specimens. The chamber can be equipped with a 365 nm or 405 nm LED light source. The light intensity on the surface of the specimen is set to 16.1 mW/cm^2 (Fig. 1b). For transmission measurements two specimen for each initiator concentration are produced. The transmission specimens have a wall thickness of 0.75 mm. The specimens are not post-cured before application i.e. they remain in the green state (Fig. 2a)

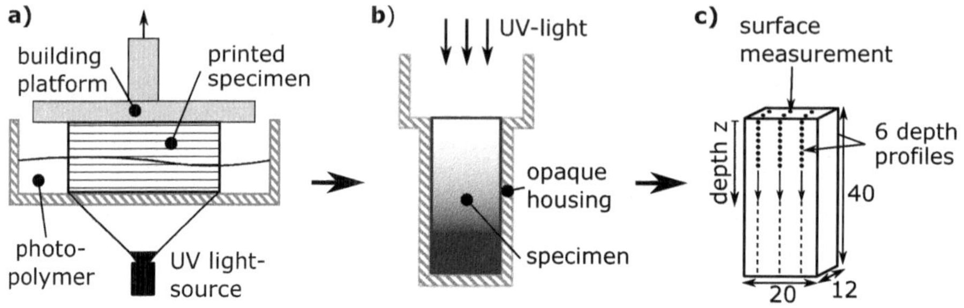

Fig. 1. Schematic of hardness specimen production and testing. **a)** Printing. **b)** Unidirectional post-curing. **c)** Hardness testing

2.4 Hardness Testing

Hardness testing is performed according to ASTM D 2240 Shore D by a Bareiss digi test II (Bareiss, Germany) hardness tester. To obtain a depth profile the hardness is measured at several points along three lines in the direction of the unidirectional post-curing (compare Fig. 1c). The minimal distance between two hardness measurements is set to 0.5 mm to obtain a sufficient resolution of the depth profile. For each specimen six depth profiles on different locations are generated. To generate a reference of the maximum obtained hardness six points on the top surface directly exposed to the UV-light are measured (Fig. 1c).

2.5 Transmission Measurements

For the transmission measurements the specimens are placed directly on a radiation meter (UV-meter, Hönle AG, Germany). The used sensor is suitable for the measurement of LEDs of wavelengths between 265 nm and 485 nm according to the manufacturer. The sensor and specimen are then placed in the UV-chamber LED Cube 100 IC (Hönle AG, Germany) and irradiated from above (see Fig. 2b). The transmission measurements are carried out with a light intensity of 60 mW/cm^2 on the surface of the specimen. This higher intensity compared to the unidirectional post-curing is chosen to obtain a sufficient transmission signal. By continuous measurement of the light intensity underneath the specimen, i.e. the light intensity that has been transmitted by the specimen, any change in transmission can be measured in-situ during post-curing. To calculate the transmission coefficient additional reference measurements without specimen are conduced.

3 Results and Discussion

3.1 Determination of the Depth of Cure

For the analysis of the DoE, the depth of cure is extracted from the measured hardness profiles. Figure 3 shows an exemplary hardness profile of a specimen with a photoinitiator concentration of 1.25% post-cured with 365 nm and

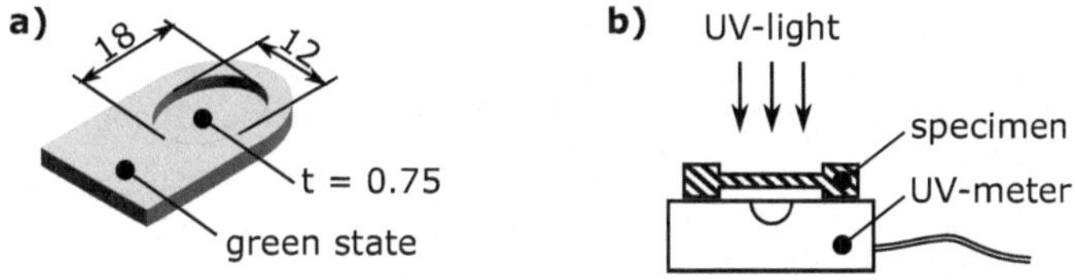

Fig. 2. Specimen for transmission measurements. **a)** Specimen geometry. **b)** Experimental setting for transmission measurements

70 J/cm^2. For this specimen hardness values between 67.0 and 79.5 Shore D are measured. For the first millimetres, the hardness profile shows a plateau close to the maximum hardness. This plateau represents the maximum achievable hardness and degree of cure for the given material and process parameters. A sharp decrease in hardness is followed by a second plateau close to the minimum hardness. The second plateau represents the hardness of a non-post-cured green specimen. Therefore, no post-curing took place in this area of the specimen. To define a methodology to determine a depth of cure value, the maximum hardness on the surface (depth of 0 mm) of all measured specimen is compared. The mean surface hardness is 79.75 Shore D with a standard deviation of 0.95 Shore D. This proofs the maximum achievable hardness to be a practical reference value. The depth of cure is then defined as the depth where 95% of the maximum hardness is achieved. A similar definition of curing depth was proposed by Alrahlah et al. [8] where 80% of the surface hardness is used as reference. By this definition the specimen can be divided in a fully cured area and an area with insufficient degree of cure, see Fig. 3. The depth of cure for all six hardness profiles per specimen is calculated and averaged by this methodology.

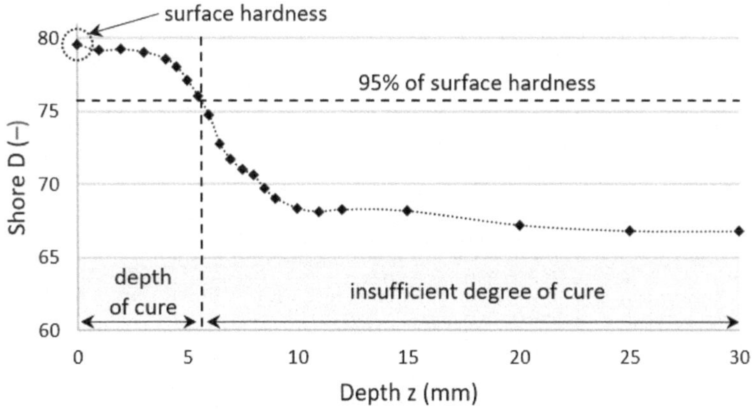

Fig. 3. Exemplary hardness profile of a specimen with a photoinitiator concentration of 1.25% post-cured with 365 nm and 70 J/cm^2 and definition of the depth of cure

3.2 Results DoE

Figure 4 shows the main effects plot of the depth of cure with the input variables photoinitiator concentration, wavelength and post-curing light dose. An increasing photoinitiator concentration shows a negative effect on the depth of cure. The increase from 0.75% to 1% shows a bigger effect than the increase from 1% to 1.25%. This result shows the effect described in Lee et al. [10], that the increased absorption due to a higher photoinitiator concentration exceeds any increase in reactivity due to more photoinitiator and therefore radicals. By changing the wavelength from 365 nm to 405 nm the mean curing depth increases. This effect is larger than the effect of the photoinitiator concentration. The post-curing light dose shows the biggest effect on the depth of cure. Increasing the dose significantly increases the depth of cure. While the increase from 10 J/cm^2 to 40 J/cm^2 shows a bigger effect than the increase from 40 J/cm^2 to 70 J/cm^2.

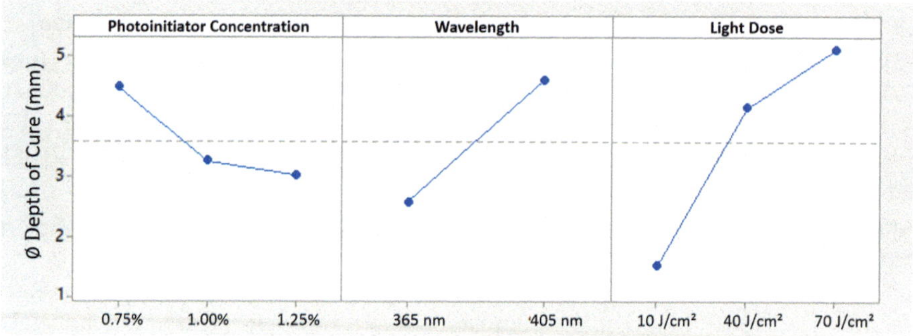

Fig. 4. Main effects plot for depth of cure of the DoE

The factor combinations with the lowest and highest depth of cure and the corresponding depth values are shown in Table 3. The minimum and maximum measured curing depth are 0.59 mm and 8.58 mm respectively. These values are representing the limitations of curing depth for the given parameters.

Table 3. Factor combinations with the lowest and highest depth of cure

Photoinitiator concentration	Wavelength	Light dose	Depth of cure
1.25%	365 nm	10 J/cm^2	0.59 mm
0.75%	405 nm	70 J/cm^2	8.58 mm

3.3 Results Transmission Measurements

Figure 5 shows the transmission of the specimens during post-curing with 365 nm and 405 nm. Since the specimens were not post-cured after printing the degree of cure changes during the measurement. In the beginning, i.e. the green parts show low transmission values between 0% and 18%. With increasing dose, the transmission increases significantly. For the measurements with 365 nm wavelength, the transmissivity is lower than for measurements with 405 nm. For a photoinitiator concentration of 1% and 1.25% and a wavelength of 365 nm no transmission at all is measured in the beginning. Apart from the wavelength, the photoinitiator concentration has a big effect on transmissivity. Increasing the concentration yields to a reduction in transmission.

The results show that for the studied wavelengths most of the absorption of the green part occurs due to the photoinitiator. With increasing degree of cure and decay of photoinitiator all specimen tend to a similar transmission of 70–80%, independently of the initial photoinitiator concentration and wavelength.

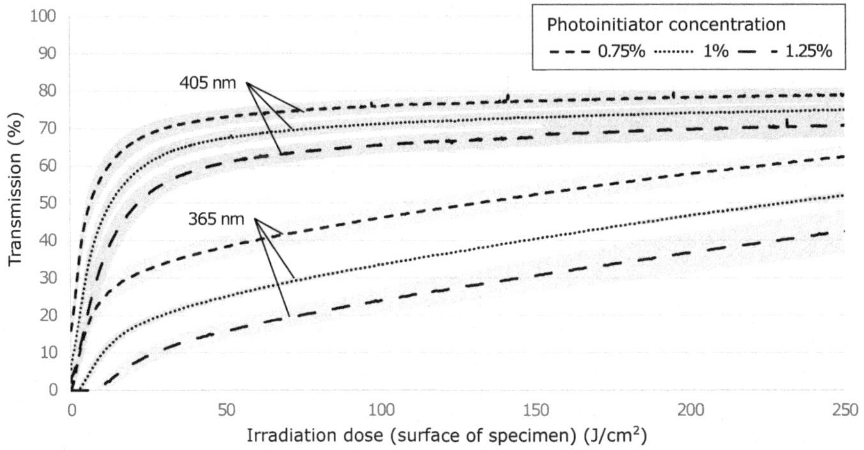

Fig. 5. Results of the transmission measurements, the grey areas are representing the 95% confidence interval

4 Conclusion

The UV-post-curing depth of photopolymers used for additive manufacturing is significantly influenced by the material and process parameters. Increasing the concentration of photoinitiator in the material decreases the curing depth. Increasing the wavelength of the UV-light source from 365 nm to 405 nm increases the curing depth. The post-curing radiation dose has the biggest influence on depth of cure. A higher dose increases the depth of cure. For the studied parameters the measured curing depths ranged from 0.59 mm to 8.58 mm. For technical applications with wall thicknesses between 1 mm and 5 mm the choice of material and process parameters has to be concerned to ensure a sufficient cured material. While the curing depth varied in a wide range the maximum hardness and therefore degree of cure on the surface remained constant for all specimen.

By transmission measurements the effect of initiator concentration, wavelength and post-curing dose on the transmissivity can be investigated. The photoinitiator has a big effect on light absorption in the material. Increasing the photoinitiator concentration decreases the transmission. During post-curing the photoinitiator decays which leads to increased transmissivity of the polymer with increasing post-curing dose. At a wavelength of 405 nm the transmissivity of the polymer is higher than at 365 nm. For time optimal post-curing the transmission must be maximised. This can be achieved by a long wavelength of 405 nm and a small photoinitiator concentration of 0.75%.

In further work a mathematical model to describe the post-curing process shall be elaborated and compared with the experimental results. With this model optimal post-curing process parameters can be defined for a given material and geometry. This will ensure a fully cured material and optimal process times.

References

1. Gebhardt, A.: Additive Fertigungsverfahren: Additive Manufacturing und 3D-Drucken für Prototyping-Tooling-Produktion. Carl Hanser Verl, GmbH Co KG (2017)
2. Jacobs, P. F.: Rapid prototyping & manufacturing: Fundamentals of stereolithography. Soc. of Manufacturing Engineers (1992)
3. Salmoria, G.V., Ahrens, C.H., Beal, V.E., Pires, A.T.N., Soldi, V.: Evaluation of post-curing and laser manufacturing parameters on the properties of SOMOS 7110 photosensitive resin used in stereolithography. Mat. & Des. **30**(3), 758–763 (2009)
4. Wiedemann, B.: Verzugsursachen stereolithorgaphisch hergestellter photopolymerer Bauteile und die Auswirkungen der Prozessfürung auf ihr Eigenschaftsprofil, Dissertation, Universität Stuttgart (1997)
5. Watters, M. P., Bernhardt, M. L.: Curing parameters to improve the mechanical properties of stereolithographic printed specimens. Rapid Prototyp. J. (2018)
6. Jayanthi, S., Hokuf, B., Lawton, J.: Influence of post curing conditions on the mechanical properties of stereolithographic photopolymers. Int. Solid Freeform Fabr, Symp (1995)
7. Zguris, Z.: How mechanical properties of stereolithography 3D prints are affected by UV curing. Formlabs Inc. (2017)
8. Alrahlah, A., Silikas, N., Watts, D.C.: Post-cure depth of cure of bulk fill dental resin-composites. Dent. Mater. **30**(2), 149–154 (2014)
9. Rueggeberg, F., Craig, R.G.: Correlation of parameters used to estimate monomer conversion in a light-cured composite. J. of Dent. Res. **67**(6), 932–937 (1988)
10. Lee, J.H., Prud'Homme, R.K., Aksay, I.A.: Cure depth in photopolymerization: Experiments and theory. J. of Mat. Res. **16**(12), 3536–3544 (2001)

Simulation Supported Manufacturing of Profiled Composite Parts Using the Braiding Technique

Jörg Dittmann[1,3](✉), Matthieu Vinot[2,3], Peter Middendorf[1], Nathalie Toso[2], and Heinz Voggenreiter[2]

[1] Institute of Aircraft Design (IFB), University of Stuttgart, 70569 Stuttgart, Germany
dittmann@ifb.uni-stuttgart.de

[2] Institute of Structures and Design, German Aerospace Center (DLR), 70569 Stuttgart, Germany

[3] ARENA2036 Research Campus, 70569 Stuttgart, Germany

Abstract. Composite materials have brought new development and sizing possibilities for structural components in transportation systems. Their high specific material properties are enabling weight reduction while increasing structural performance. On the downside, composite materials are generally related to high material and manufacturing costs and increased characterization efforts. Through the braiding technique, profiled structures can be manufactured in a highly automated and reproducible process. Moreover, braided composites can absorb more energy compared to their unidirectional or woven counterparts (Falzon P. J., Herszberg I., Bannister M. K., Leong K. H.: Compression and Compression-after-impact Properties of 2-D Braided Carbon/Epoxy Composites. Proceedings of the First Australasian Congress on Applied Mechanics: ACAM-96, pp. 297 (1996).).

In this paper, we describe the development and validation of a simulation framework as sustainable alternative to material- and cost-intensive experimental testing. Our work aims at considering the influence of manufacturing effects and textile architecture on the material properties and therefore at increasing the reliability of structure sizing. As validation basis, flat specimens of biaxial and triaxial braided composites are first manufactured and tested under quasi-static loading. We then develop a digital twin of the braiding process and its material characterisation. Within this framework, the braid's textile architecture is predicted with multiple finite-element simulations at the mesoscopic scale.

The numerical predictions show the strong influence of braiding angle and braiding core diameter on the textile architecture and consequently on the material properties. More particularly, crucial effects with negative impact on the mechanical properties (presence of gaps or yarn locking) are highlighted. On a pure numerical basis, we finally calculate the process window for braided structures, which links the process parameters to the resulting material properties. The present approach is a crucial step toward the reduction of experimental investigations in early development.

1 Introduction

Urbanisation of cities and electrification of transport systems like automobiles lead to more complex and heavier subsystems. These subsystems need to be lighter so that

P. Weißgraeber et al. (Eds.): *Advances in Automotive Production Technology – Theory and Application,* ARENA2036, pp. 351–360, 2021. https://doi.org/10.1007/978-3-662-62962-8_41

additional functions do not increase CO_2 emissions. To achieve this goal, lightweight components such as composite structures have to be used. One promising manufacturing technique is the braiding process, where a two-yarn or three-yarn system (cf. Fig. 1) automatically forms the structures geometry by braiding onto a near-net-shape braiding core [2]. Triaxial braids are reinforced with a axial yarns in the 0°-direction and are therefore suited for axial loaded structures like crash absorbers. The braiding process could be 100% automated, which is promising for industry 4.0 applications. By integrating sensors into the braiding process, every component could be a sensing, logging and communicating smart subsystem [3]. The braiding process also enables a maximum flexibility and reproducibility. In addition, braided components have a very high-energy absorption compared to woven or non-crimped composites, which makes them ideal for impact and crash loaded structures [1]. Furthermore, the braiding process enables the use of natural fibre material like flax or hemp, which stores CO_2 during growth and makes structures therefore more sustainable [4].

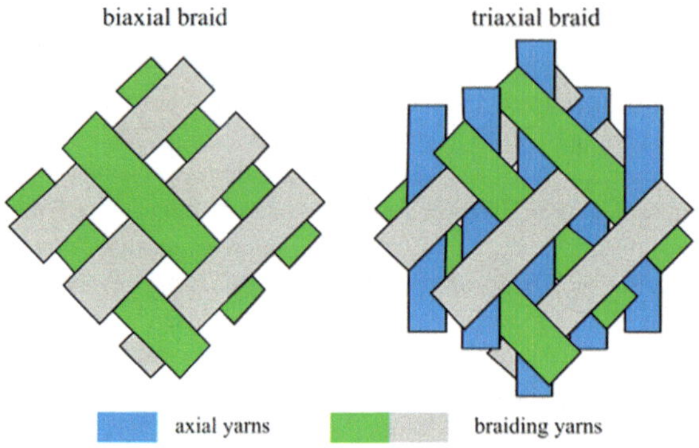

Fig. 1. Braid architecture for biaxial and triaxial braided composites

Existing approaches in the braiding simulation are based on kinematic modelling or Finite-Element Methods with different element types like bar, shell, membrane or solid elements [2]. Every approach intends to reach the most realistic fibre bundle behaviour while reducing simultaneously the computing time. In P. Böhler [5] it was shown that realistic material architectures could be modelled with shell elements but lacks the transfer to virtual mechanical testing. The transfer from solid descriptions to virtual tensile testing promises to improve the prediction of material properties but requires more computational time [6]. This work aims at creating a flexible numerical framework for industrial applications, linking different simulation approaches and modelling levels with a strong focus on accessibility and computational efficiency.

In this work, we first focus on the experimental investigation of braided composites. Flat specimens with different braiding angles are manufactured and tested under various conditions to access the mechanical properties in axial and transverse direction. A generic structure is then manufactured, from which in-situ specimens are extracted as a comparison to the previously measured properties. Then, we develop and validate a simulation chain based on the finite-element method at different levels of detail to predict braided architectures and their respective mechanical properties. The numerical results are finally combined to create a virtual process window, which links the process parameters to the mechanical properties of the material.

2 Experimental Characterization

To validate the predicted simulation results in Sect. 3 and measure the effects attributed to the braids architecture, we manufactured biaxial and triaxial braided flat specimens with a fibre orientation of 30° and 45°. As a comparison, specimens were directly cut out of a structural component with potential defects and discrepancies. The specimen preforms were manufactured on a Herzog RF1–176-100 and a Herzog RF 1–64-100 radial braiding machine at the Institute of Aircraft Design (IFB). We used a STS-40 24k carbon fibre (Toho Tenax) and Baxxodur EC5440/Baxxores ER5400 (BASF) as resin system. The specimen were infiltrated with the VARI process (vacuum assisted resin infusion), where an applied pressure gradient forces the resin to infuse the fibre layup. The VARI process uses a solid bottom mold and a flexible vacuum membrane as top mold, which leads to an uneven contour at the top surface of the specimen. Results related to this effect are discussed in Sect. 3.2.

2.1 Investigation on Flat Specimens

To characterize the material we performed tensile tests (DIN EN ISO 527–4) and compaction tests (DIN EN ISO 14126). As the generic structures braid architecture is made of a 45° triaxial braid, only the results for 45° triaxial braid are shown here in detail. In order to statistically verify the results, the specimens for each test type were randomly distributed over eight plates. For every specimen, we determined the fibre volume content (DIN EN 3783) and the void content (DIN EN 2564). We measured a mean fibre volume content of $FVC_{\text{Triax45}} = 57.9\%$ for the 45° triaxial specimens. Due to the lower fibre angle and thus to the higher number of fibre bundles within the same specimen dimension, the mean fibre volume content for 30° triaxial braided specimens increased to $FVC_{\text{Triax30}} = 62.5\%$. Fibre volume contents for the biaxial braided specimens were $FVC_{\text{Biax45}} = 65.2\%$ and $FVC_{\text{Biax30}} = 65.2\%$.

Every material test type was executed in axial tension (0°) and transverse tension (90°) direction. 45° biaxial braid possesses similar fibre orientation in 0°- and 90°-orientation, so that only axial tension was tested. The stress-strain diagrams showed a low scatter in material behaviour (cf. Fig. 2). Obtained values for young's modulus and strength are detailed in Fig. 6. Homogeneous material behaviour could also be investigated for biaxial braided specimen and for compression curves, which are not shown here. As a reference, in-situ properties of the axial tension tests were plotted to all curves. Indices as KP, KO, ZO etc. are referring to the different plates where specimens were cut out. Indices with the prefix Z represent the load direction and the last indices are the specimen's serial number.

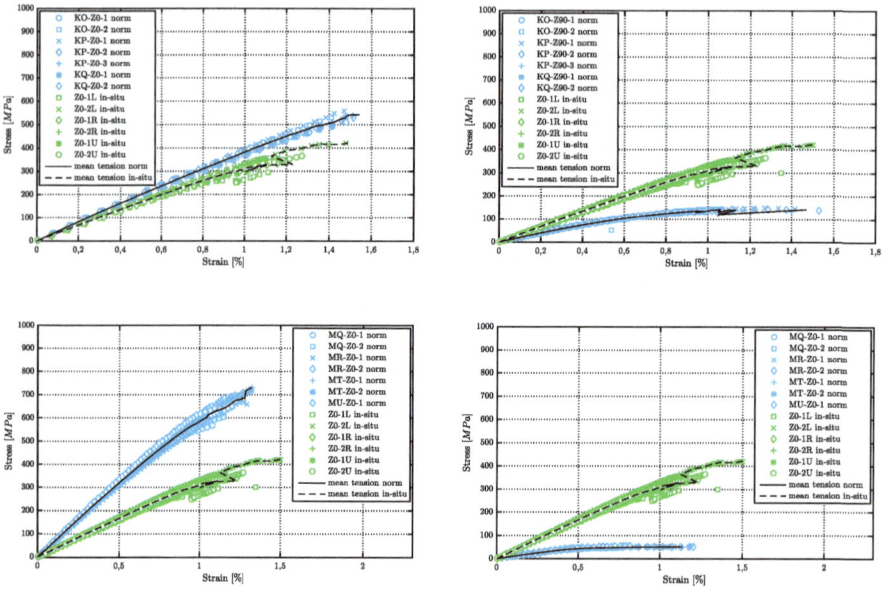

Fig. 2. Stress-Strain curves of a) Triaxial 45° axial tension, b) Triaxial 45° transverse tension, c) Triaxial 30° axial tension, d) Triaxial 30° transverse tension

2.2 Manufacturing and Investigation of a Generic Structure

To compare results on flat specimens with results closer to serial production, we designed a generic structure based on an automotive application. The structure is braided on a s-curved mandrel with triangular cross-section and round edges (cf. Fig. 3). It was manufactured with the same fibre and resin system as for the flat specimen with averaged fibre volume content of 56.1%. We investigated effects like fibre angle deviations, curvature of braiding yarns, detachments and thickness variations. Finally, we cut flat specimen out of the flanks of the generic structure.

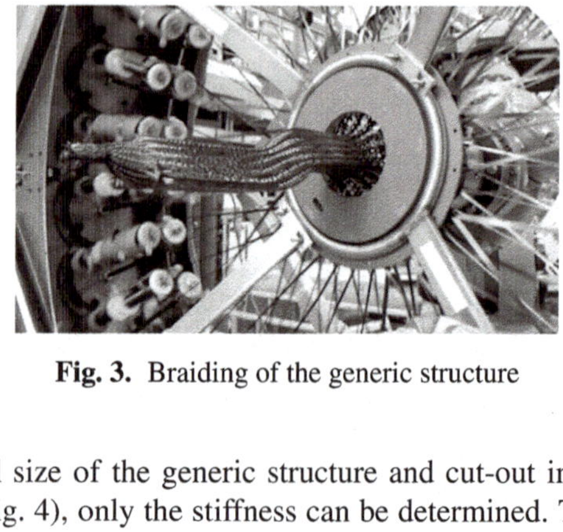

Fig. 3. Braiding of the generic structure

Due to the small size of the generic structure and cut-out in-situ specimens with 40 mm length (cf. Fig. 4), only the stiffness can be determined. The Young's modulus of the material has a small variation of 4.5% and is in average 33.1 GPa, which is comparable to values of the standard tensile specimens. Therefore, specimens from flat plates are representative to the in-situ material.

3 Predictive Simulations Within a Closed Process Chain

In this chapter, we illustrate the potential of the simulation chain developed within the BMBF project DigitPro for the prediction of braided composites' properties. We describe the numerical solutions used in this work and compare the simulated results with previously documented experimental tests. Simulation work has been performed within LS-DYNA explicit with the solver version R10.2.

3.1 Digital Twin

In this section, we create a digital twin of the braiding process. We first import the CAD file of the generic structure into a FEM simulation software. Afterwards, the fibre yarns, their bobbin positions and their braiding paths are modelled with triangular shell elements according to the setting of the braiding machine. Finally, fibre yarn tensioning, braiding mandrel movement, braiding speed as well as contact conditions are defined. In simulation, the fibre bundles are first dragged onto the braiding mandrel. Through the interlocking of bundles, the textile architecture is then progressively built (cf. Fig. 4). At this point, the accurate prediction of the textile architecture is strongly dependent of the numerical behaviour of dry yarns. Compared to classical approaches which rely on analytical solutions, the present approach predicts the braid architecture locally dependently on the core geometry. It is therefore possible to predict effects like yarn interlocking, fibre detachment or low fibre volume contents. Information about the resulting fibre architecture can finally be transferred to the virtual material testing using the mapping tool envyo [7]. There, material and structures properties will be predicted and compared to experimental measurements. The present digital twin is a flexible container which links different steps of the product lifecycle and can be extended with other simulations (infiltration, structure testing etc.).

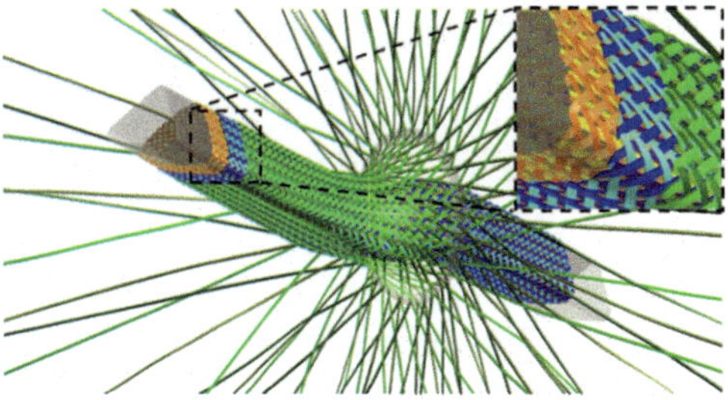

Fig. 4. Braiding process simulation of the generic structure

3.2 Prediction of Material Properties with a Mesoscopic Approach

The material properties of the consolidated textile are virtually generated with numerical simulations on the mesoscopic scale based on the principle of representative unit cell [8]. Compared to other work [9, 10], the present approach does not require any previous measurements on manufactured textiles and the architecture of the textile (yarn width, yarn spacing, size and thickness of the unit cell etc.) can be obtained from the previously described braiding simulation. In addition, only manufacturers' data for fibres and resin are necessary to reach a complete characterization of the braided material. This represents a clear benefit for the reduction of development costs, time and experimental efforts.

First, we generate and mesh a 3D model of the textile architecture with the open-source software TexGen [11]. Through this, we successfully recreated various architectures at different braiding angles and mandrel diameters (cf. Fig. 5). The presence of gaps or the too strong locking of yarns can reduce drastically the material properties. Predicting accurately these effects is therefore crucial for simulation tools. Increasing the braiding mandrel diameter and decreasing the braiding angle leads to a decrease of the cover factor. On the contrary, at a braiding angle of 55°, a locking effect appears, especially for small mandrel diameters. This locking effect is characterised in experiment by an overlapping of the yarns through severe lateral deformations (marked in blue in Fig. 5). In simulation it can be detected by the presence of penetrations between the yarns (marked in green in Fig. 5). This locking effect influences negatively the mechanical properties by introducing local defects in the material.

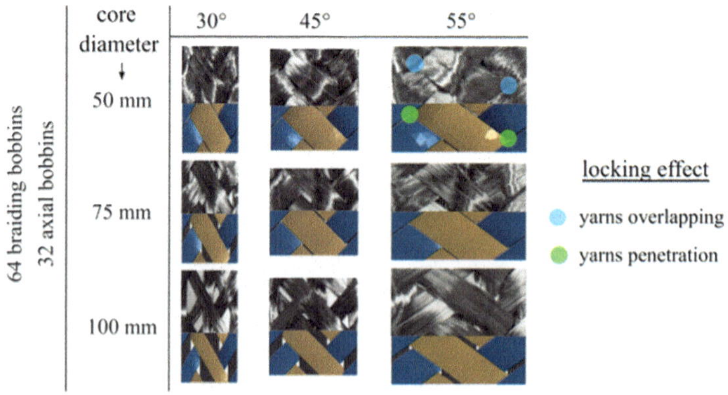

Fig. 5. Correlation between the virtual generated unit cells and the manufactured braid

After generation of the resin mesh, the complete unit cell is virtually loaded in tension, compression and shear in both axial and transverse directions via in-plane periodic boundary conditions applied on matching opposite faces of the unit cell. Surfaces in thickness direction are not constrained to simulate the finite thickness of the composite. Tetrahedron elements with an average length of 0.1 mm are used due to the meshing advantage they bring. The use of element formulation type 13 prevents volumetric locking inherent to tetrahedron elements, which often leads to a stiffer behaviour. For the mechanical simulation, tows and resin materials are simulated with use of *MAT_261 and *MAT_187 in LS-DYNA respectively [12, 13]. The material properties of the yarns are calculated with the rules of mixtures for a fibre volume content of 70% (cf. Table 1). Debonding mechanisms between yarns and resin are simulated using a bilinear traction-separation law of the tiebreak-contact in LS-DYNA, based on literature values (cf. Table 2) [14].

Table 1. Material properties for the simulation of impregnated carbon yarn bundles with a fibre content of 70%

E_{11}	$E_{22} = E_{33}$	$G_{12} = G_{13}$	G_{23}	$\nu_{21} = \nu_{31}$	ν_{32}
168 GPa	12 GPa	4.5 GPa	3.7 GPa	0.022	0.606
S_{11t}	**S_{11c}**	**S_{22t}**	**S_{22c}**	**S_{12}**	
3.23 GPa	2.47 GPa	0.062 GPa	0.233 GPa	0.20 GPa	

Table 2. Parameters for the bilinear traction-separation law

S_{Normal}	S_{Shear}	G_{IC}	G_{IIC}	K
0.050 GPa	0.050 GPa	1.0×10^{-4} GPa mm	1.0×10^{-4} GPa mm	100 GPa/mm

Without any optimisation loop in the simulation, we obtain a good correlation with an accuracy of at least 80% between simulated and experimental material properties for different loading cases (cf. Fig. 6). Increasing the braiding angle leads to a strong reduction of the axial material properties and a strengthening of the transverse direction. This is due to the previously mentioned introduction of defects or waviness by increasing the braiding angle. Moreover, 30° braids possess a higher axial yarn content than for high braiding angle. We observe a stronger discrepancy between predicted and measured values for the compression load cases (which are not shown here for purpose of clarity). The compression loading is strongly dependent on geometrical effects and induces stability issues, which do not occur on unit cell level.

3.3 Virtual Process Window

A concept of process window has been presented by K. Birkefeld [15] to qualify the manufacturing of braided composites and represents the optimal design space for the manufacturing of braided textiles in regard to manufacturing parameters (number of braiding bobbins used, core diameter, braiding angle). The generation required experimental observation of fibre patterns on braided preforms.

Combining the results from the braiding and from the mechanical simulations allows us to generate a process window on a purely virtual basis and therefore to reduce consequently the amount of experimental work required. Here, we not only bring in the

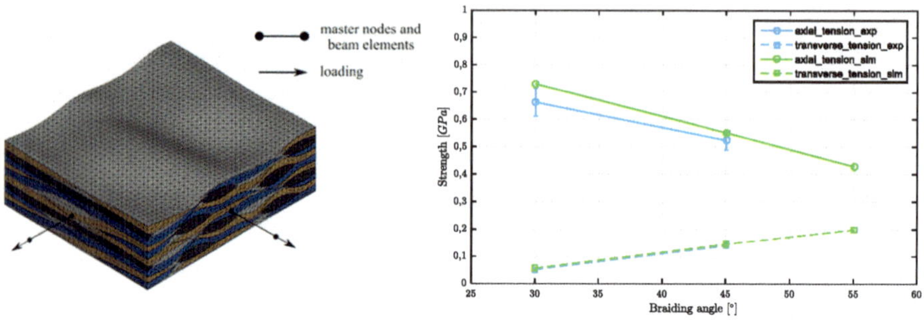

Fig. 6. Unitcell model for a 30° triaxial braided composite (left) and correlation between experimentally and numerically assessed strengths (right) for triaxial braided composites at different braiding angles

correlation of textile quality to the process parameters, but to the resulting material properties as well. Using this three-dimensional process window (cf. Fig. 7), it is possible to optimise the process parameters and to achieve the highest structure potential as possible. The process window shows in particular that braiding angles around 30° have optimal axial properties and are therefore adapted to crash loaded structures with small diameter. On the opposite, high braiding angles or higher diameters are more suited for structures under multiaxial loading.

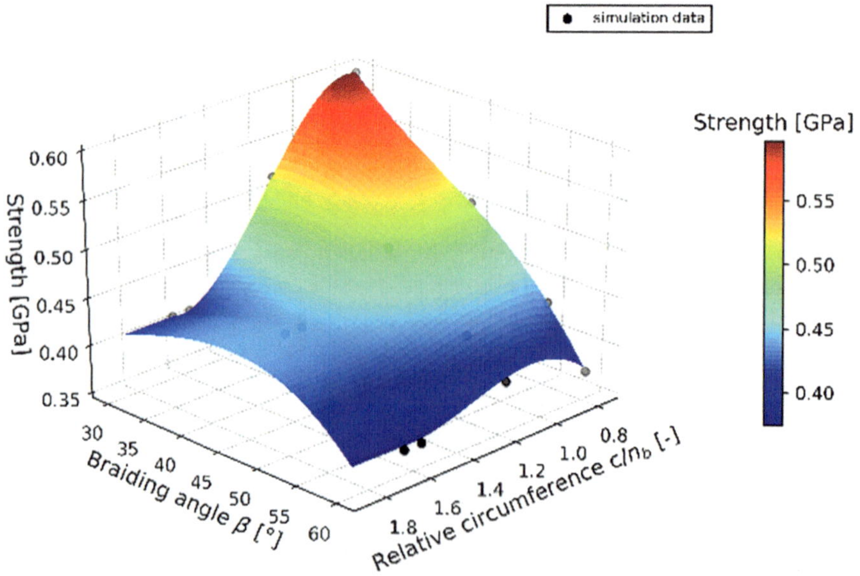

Fig. 7. Virtual process window of triaxial braided composites (Herzog RF 1–64-100)

4 Conclusion and Outlook

In this paper, we used a simulation approach to reach a better understanding of composite materials by investigating the architecture and material properties of braided composites on a pure numerical scale. In a first step, we developed a finite-element tool in which the manufacturing braiding process can be accurately simulated. To overcome the classical issue of high computational costs, we simplified the yarns in the form of 2D shell elements. The digital twin of the braiding process delivered crucial information about the local textile architecture at every position of the structure. In a second step, we abstracted the simulation results of the process simulation and used them as input for the generation of unit cells for virtual testing. At the cost of increased computational time, the unit cells are meshed with 3D-solid elements coupled to state-of-the-art material models. The modelling of delamination is made possible by traction-separation laws in cohesive contacts between yarns and resin material.

With the combination of process simulation and virtual testing, we reproduced accurately the influence of braiding parameters on the textile architecture and material properties. Moreover, the prediction capability of the suggested approach has been validated with experimental tests on flat specimens under tension and compression.

In this paper, we illustrated the potential of numerical tools for the virtual investigation of composites using the example of the braiding process. To reach an industrial 4.0 application, a framework was created to link the different simulation steps and more important to create a connection to real manufacturing processes. This enables a cost-efficient development and manufacturing of composite parts and the reduction of experimental tests.

Acknowledgement. The authors wish to acknowledge the funding provided by the Federal Ministry of Education and Research Germany within the Research campus ARENA2036—DigitPro and Digitaler Fingerabdruck.

References

1. Falzon, P. J., Herszberg, I., Bannister, M. K., Leong, K. H.: Compression and Compression-after-impact Properties of 2-D Braided Carbon/Epoxy Composites. Proceedings of the First Australasian Congress on Applied Mechanics: ACAM-96, p. 297 (1996)
2. Dittmann, J., Middendorf, P., et al.: Der digitale Prototyp - Ganzheitlicher digitaler Prototyp im Leichtbau für die Großserienproduktion. ARENA2036 Reihe-DigitPro, Springer-Vieweg (2019)
3. Raichle, A., Ritter, F., Vinot, M., Dittmann, J. et al.: Weiterentwicklung des Digitalen Prototyps zum Digitalen Fingerabdruck, ATZ-Magazin 03/2019, Springer-Vieweg (2019)
4. Blonk, H., Kool, A., Luske, B., Ponsioen, T., Scholten, J.: Methodology for assessing carbon footprints of horticultural products. Blonk Milieu Advies, S. 36–40 (2010)
5. Böhler, P.: Einzelfadenbasierte Modellierung von textilen Preform-Prozessen. Dissertation Universität Stuttgart (2019)
6. Czichos, R., Bareiro, O., Pickett, A. K., Middendorf, P., Gries, T.: Experimental and numerical studies of process variabilities in biaxial carbon fiber braids. International Journal of Material Forming (2020)

7. DYNAmore GmbH; Envyo® User's Manual-DRAFT, Germany (2020)
8. Whitcomb, J.D., Chapman, C.D., Tang, X.: Derivation of Boundary Conditions for Micromechanics Analyses of Plain and Satin Weave Composites. J. Compos. Mater. **34**, 724–747 (2000)
9. Wehrkamp-Richter, T., Pinho, S.T., Hinterhölzl, R.: Failure behaviour of triaxial braided composites, 17th European Conference on Composite Materials (2016)
10. Kier, Z.T., Salvi, A., Theis, G., Waas, AM., Shahwan, K.: Estimating mechanical properties of 2D triaxially braided textile composites based on microstructure properties, Composites: Part B **68**, 288–299 (2014)
11. Lin, H., Brown, L.P., Long, A.C.: Modelling and Simulating Textile Structures using TexGen. Adv. Tex. Eng. **331**, 44–47 (2011)
12. Pinho, S., Iannucci, L., Robinson, P.: Physically based failure models and criteria for laminated fibre-reinforced composites with emphasis on fibre kinking. Part II: FE implementation, Composites Part A: Applied Science and Manufacturing **37**, 766–777 (2005)
13. Kolling S., Haufe A., Feucht M., Du Bois P.A.: SAMP-1: A Semi-Analytical Model for the Simulation of Polymers, 4. LS-DYNA Anwenderforum (2005)
14. Pinho, S., Robinson, P., Iannucci, L.: Fracture toughness of the tensile and compressive fibre failure modes in laminated composites. Compos. Sci. Technol. **66**, 2069–2079 (2006)
15. Birkefeld, K.: Virtuelle Optimierung von Geflecht-Preforms unter Berücksichtigung von Fertigungsaspekten. Universität Stuttgart, Institut für Flugzeugsbau (2013)

A New Concept for Producing High Strength Aluminum Line-Joints in Car Body Assembly by a Robot Guided Friction Stir Welding Gun

Dominik Walz(✉), Martin Werz, and Stefan Weihe

Materials Testing Institute, University of Stuttgart, 70569 Stuttgart, Germany

Abstract. Materials with higher specific and buckling strength are increasingly used in order to reduce the weight of car bodies while maintaining or increasing passenger safety. Examples for such materials are ultra-high-strength aluminum alloys and press-hardened steels. A major challenge now is to combine the ultra-high-strength aluminum alloys with each other and, if necessary, with other metals without losing their outstanding strength properties.

High-strength aluminum alloys with virtually no loss of strength properties can be welded with the so-called friction stir welding process since the 1990s. In this process, a rotating tool is pressed into the joint gap of the joining partners and moved along the joint gap to produce the weld seam. An anvil adapted to the component is required on the opposite side of the weld seam in order to absorb the high process forces.

A joining process for the body-in-white assembly of aluminum and hybrid car bodies should have the following specifications: Joining high-strength aluminum alloys without hot cracks, high-strength joints, sufficient service life of the joining device, robot-supported execution, no complex and expensive component-adapted anvil and an accessibility similar to resistance spot welding guns.

Based on these requirements, a process concept for a new type of robot-guided friction stir welding gun for the production of short, merging weld seams was developed at the Materials Testing Institute (MPA, University of Stuttgart) and a patent application was filed. A first functional model of the welding gun was built and the underlying kinematics were successfully tested in initial tests. The concept differs from other known solutions because it compensates process forces through the weld gun completely. Moreover, the concept allows curved weld seams which are common in the automotive industry. These curved weld seams are realized as polygon courses in a robust manner and without component-specific anvils.

P. Weißgraeber et al. (Eds.): *Advances in Automotive Production Technology – Theory and Application,* ARENA2036, pp. 361–368, 2021. https://doi.org/10.1007/978-3-662-62962-8_42

1 Introduction

The trend of lightweight design for increasing resource efficiency and CO_2-saving leads to a growing usage of high-strength aluminum alloys and multi-material parts in the automotive sector.

Conventional fusion welding technologies such as resistance spot welding struggle with joining those high-strength aluminum alloys and multi-material parts. This is based on a loss of strength due to the high temperatures in the heat affected and fusion zone [1]. Another problem occurring with fusion welding are hot cracks, especially on high-strength aluminum alloys with a wide solidification range. According to the currently valid standard Eurocode 9, there is a loss in strength of fusion welded high-strength aluminum joints of up to 70% [2]. Further drawbacks for resistance spot welding are high costs for the equipment and operating costs like high electrode wear, a high power consumption and complex process control.

Friction stir welding (FSW) is a so-called solid-state pressure welding technology, invented at The Welding Institute (TWI) in 1991 [3]. During the welding process the materials temperature is always below its melting temperature, which qualifies it especially for joining high-strength aluminum materials. The three steps visualized in .

Figure 1 characterize the process: (1) a rotating tool is plunged into the joining partners with a high force and held in position so that the work pieces have time to heat up though friction. The tool usually consists of a pin, which locally stirs the work piece materials, and a shoulder for generating frictional heat and preventing the plasticized material from leaving the stir zone. (2) The tool then moves translational along the joining gap, stirring and joining the work pieces together. (3) The tool is retracted at the end of the joint leaving the characteristic pinhole.

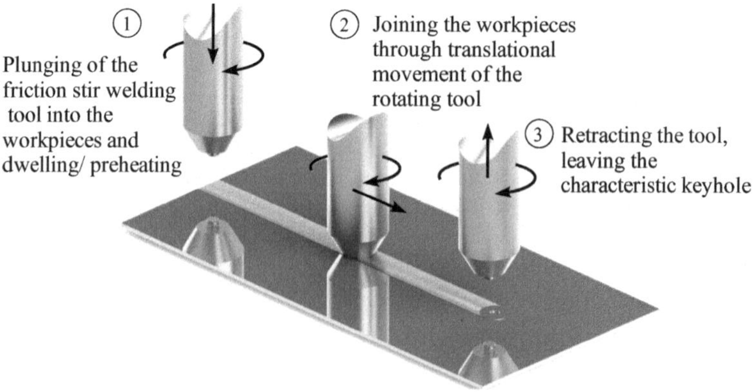

Fig. 1. The three process steps of friction stir welding

This paper presents a new concept for a robot guided friction stir welding gun based on a novel mechanical kinematic for short-line joints. The concept aims for high accessibility and can be used to replace long weld seams with short, interlocking seams, or low strength spot-welds with high strength short-line seams. The spindle uses a cost-effective bearing concept as well as an integrated water-cooling and is especially designed for the FSW process forces. The motorized anvil allows the fabrication of short line joints in variable length up to 45 mm. The C-gun is used to weld Al 6014-T4 aluminum sheets with 2 mm thickness in an overlap configuration. Lap-shear specimens with three different joint lengths were welded, tested and compared to show the potential of the joining technology and the FSW gun.

2 Literature Review

Conventional machines for FSW are usually built for long weld seams and to deal with work pieces with large dimensions. Therefore, the FSW machines are always larger than the work pieces, which results in high costs and inflexibility. Additionally, the high process forces require elaborate clamping tools. Those machines are often used in the maritime and train constructing sector, or even for aeronautics and space due to the large work pieces in those sectors.

A modified version of the FSW process especially used in the automotive sector is the friction stir spot welding (FSSW). Here the conventional FSW process is simplified by omitting the translational movement of the tool in order to realize the process with robot positioned welding guns. Major drawbacks of FSSW are lower weld seam strengths due to the high ratio of the leftover pinhole area to the joining area. This led to further process optimizations such as refill friction stir spot welding (RFSSW), "swing FSSW" and "stitch FSSW". The RFSSW is based on a multi-part tool, where pin, shoulder and a clamping ring can be moved independently. Either the pin or the shoulder are plunged into the material while the other part retracts to allow the plasticized material to flow into the gap. Then the movement of pin and shoulder is inverted, filling up the pin or shoulder hole and leaving a spot-weld. The tool is vulnerable to wear [4] and has to be cleaned frequently [5] because of the moveable tool parts and plasticized material in between them. Swing and stitch FSSW are both based on a gyrate movement of the tool. FSSW welds either in a centric (so-called swing FSSW) or oscillating movement in one direction (so-called stitch FSSW). Both variants produce higher strength welds than conventional FSSW. However, the weld length of stitch and swing FSSW is limited. Also they need an additional swing motor, which adds additional weight. Furthermore, the tool tilt angle varies depending on the gyrate movement, which makes the process more sensitive for flaws. In summary, there is a need for a FSW process variant for the automotive sector to weld robustly and which can be adapted to a robot.

FSW welding guns. The structure of most FSW/FSSW welding guns is similar. They consist of a C-shaped frame, a spindle, which holds the tool and is driven by a servomotor or hydraulic cylinder, and an additional motor for opening and closing the gun. Opposite of the spindle is an anvil, which absorbs the process forces, so that the force flux is a closed circuit within the C-frame. Those guns are used to spot weld back doors and tailgates at Mazda [6] and Kawasaki Heavy Industries [7]. Some FSSW welding guns have a mechanism allowing the shoulder and the pin of the tool to move independently. This makes it possible to fill up the pinhole, resulting in higher strength of the spot welds [8]. Stirtec patented a concept for visible seams in automobile production, swapping the position of the spindle and the anvil [9]. The characteristic weld seam and the pinhole are located on the backside of the car body with this assembly.

An additional servomotor is added to the welding gun for the swing and stitch FSSW processes. The spindle is flexibly for both of the processes. The motor moves the spindle via a lever and circles it around a fix point to enlarge the weld spot. Hitachi built and tested one such prototype [10]. However, up until now industrial use has not been demonstrated or is not documented in literature.

Lap-shear tests on FSW joints. Uematsu et al. [11] compared FSSW and resistance spot welded (RSW) lap-shear specimen made from EN AW-5005 aluminum alloy for tensile shear strength. FSSW welded joints showed significantly higher tensile shear strength with up to 295% higher values than RSW specimens. The tool pin height was identified as a main parameter affecting the tensile shear strength with FSSW. The shear strength was calculated without taking the weld nucleus diameter into account.

Okamoto et al. [12] carried out lap-shear tests on Al 6022-T4 aluminum alloy specimens welded on a Hitachi prototype swing-FSW gun. The lap-shear strength increases almost linear with the length of the stitch FSW joints and exceeds conventional FSSW and RSW joints due to the enlarged weld nugget. Further investigations on 1 mm thick Al 6111 aluminum lap-shear specimens showed a 38% higher shear strength of a 3 mm long stitch weld compared to conventional FSSW joints [13]. The linear increase of shear strength could be observed depending on the weld direction up to a weld length of 5 mm. Longitudinal welds showed lower shear strength than transversal welds. Zhang et al. [14] and Tweedy et al. [15] confirmed an increasing shear strength with increasing stitch weld length for Al 5052 aluminum alloy and swing stirred Al 2024-T3 aluminum alloy with an increasing swept spot diameter.

3 Solution Approach

A novel weld gun was developed and built in order to manufacture short line joints, called "step welds". It is possible to produce curved welds in every spatial direction when the welding gun is adapted to a robot. In this case, the curved welds of short step welds arranged in a polygonal chain. The robot repositions the welding gun after every step weld, so that the tool plunges into the pinhole of the prior weld. The tool tilt angle is realized with a tilted adapter piece. The production sequence for a step weld is shown in Fig. 2 and defined by four steps:

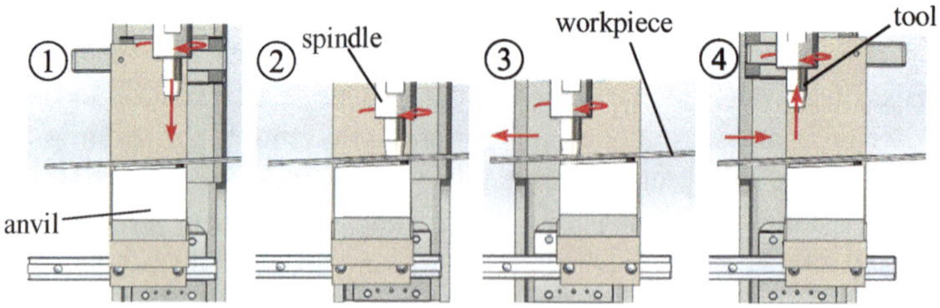

Fig. 2. Production sequence for producing a step weld

- The spindle starts rotating from its start position and the welding gun head moves down towards the work pieces.
- The tool plunges into the work pieces and holds its position until the work piece material is heated to process temperature.
- Then the anvil starts moving horizontally. The spindle head is adjusting the height due to the tilted adapter piece on the anvil.
- Lastly, the spindle and the anvil move back to their respective start position.

The components of the welding gun and the adaption to an industrial robot are shown in Fig. 3. Three weld configurations are possible with this weld gun. Those are i) single step welds, for example to substitute RSW or FSSW, ii) merging step welds to form a line joint or iii) a polygon course for curved structures. Currently, the welding gun is mounted on a test rig without a robot and work pieces are placed and moved by hand at this early stage of development.

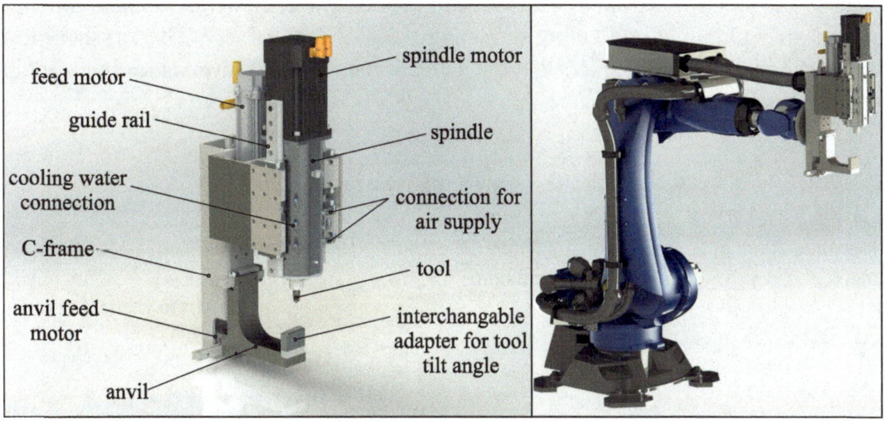

Fig. 3. Labelled CAD-model of the welding gun (left) and adaption to a robot (right)

The joint strength is estimated to be higher compared to RSW and FSSW due to the larger weld seams. The here considered important factors for an industrial use are good accessibility to, e.g. car bodies, the overall costs and the flexibility to use the weld gun with any robot system. Therefore, the design goal of such a machine is a minimized weld gun building volume and maximized accessibility to the work pieces.

A main advantage of this system is the production of friction stir welds without the need for component-adapted anvils and elaborate clamping. As a result this concept is usable for a variety of applications.

4 Experimental Studies

Lap-shear tests were carried out for specimens with different weld lengths according to DIN EN ISO 14273 [16] to validate the operability of the weld gun. The welded sheets were both out of 2 mm thick Al 6014-T4 with dimensions shown in Fig. 4.

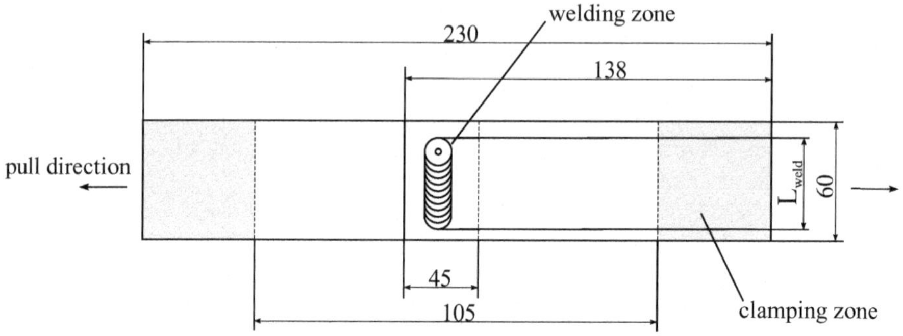

Fig. 4. Dimensions of the welded lap-shear specimens

The weld length LWeld of the specimens varied between 6 mm, 10 mm, 20 mm and 30 mm respectively. Five specimens were welded with the same process parameters (see Table 1) and a path-controlled process control for each weld length. The hold time was applied after plunging and before retracting the tool respectively. The lap-shear tests were performed on a Zwick Roell 100 kN electromechanical universal testing machine without compensation sheets.

Table 1. Process parameters used for the step welds.

Tool tilt angle [°]	Plunge depth [mm]	Rpm [1/min]	Direction of rotation	Hold time [s]	Feed rate [mm/min]	Plunge rate [mm/s]
2,2	0,1	2000 (plunge) 700 (weld)	Counter-clockwise	2	240	0,6

Additional experiments were carried out with the welding gun mounted to a test rig to show two different welding configurations. Figure 5a shows two aluminum sheets joined by a polygonal weld course. The polygonal weld course consists out of five 40 mm long step welds. After each step weld the sheet orientation was altered manually to simulate a swinging or rotating movement of the robot and the pin was centered to the pinhole of the prior step weld. This allows filling up the pinhole and creating a consistent seam. Fig. 5b shows an example of an edge bended sheet joint configuration with single step welds.

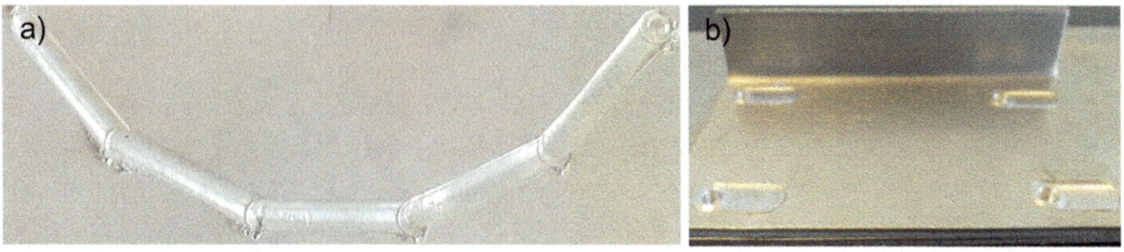

Fig. 5. **a** Polygonal weld course consisting of merged step welds, **b** step weld demonstrator

5 Results and Discussion

The results of the lap-shear tests can be seen in Fig. 6. The maximum shear force is an average value of lap-shear tests. As expected the smallest shear force of 2.17 kN was observed at the specimen with a 6 mm weld length, whereas the largest shear force of 11.33 kN was observed at the specimen with a 30 mm weld length. The relation between the weld length and shear force can be described as linear, which corresponds to prior investigations described in the literature [13]. The maximum shear forces could be improved by using a tool with larger pin and shoulder diameters, or by a specially designed tool for the overlap sheet configuration.

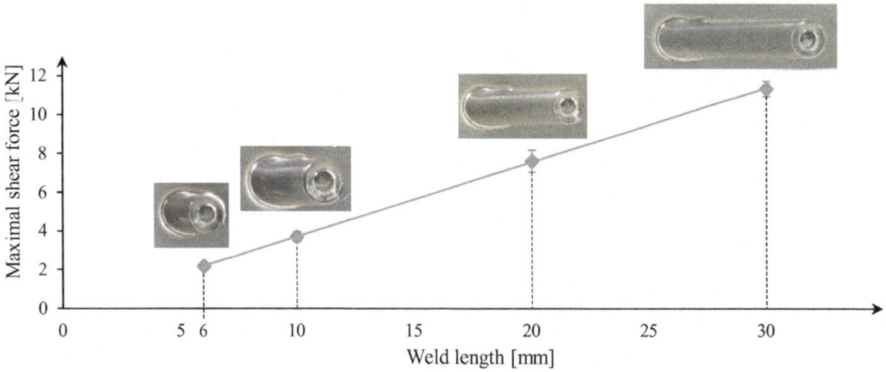

Fig. 6. Maximal shear force of lap-shear specimen as a function of weld length

6 Conclusion and Outlook

A novel welding gun for producing short line friction stir welds, so-called step welds, was developed and built. The welding gun is designed to be adapted to an industrial robot for automated welding applications, for example in automotive body assembly.

EN AW 6014-T4 aluminum sheets with a thickness of 2 mm were step welded. The sheets were welded in an overlap configuration according to DIN EN ISO 14273 with transversal weld direction for lap-shear testing. As expected the maximum shear force increases linearly with the weld length. Thus, validating the operability of the novel welding gun.

Further investigations have to be made to identify ideal process parameters in combination with different tool geometries. Additionally, the effects of weld orientation (0°/90°) and the quality of merged step welds on the lap-shear strength will be investigated. On-going development of the welding gun focusses on reducing weight, increasing stiffness, improving accessibility and increasing maximum possible welding speed.

References

1. Mishra, R., Ma, Z.: Friction stir welding and processing. Materials Science and Engineering: R: Reports **50**(1–2), 1–78 (2005)
2. DIN 1999–1–1: Eurocode 9: Bemessung und Konstruktion von Aluminiumtragwerken (2014)
3. Thomas W., Nicholas J. et al.: Friction stir butt welding (GB9125978.8)
4. Montag, T., Wulfsberg, J.-P., et al.: Influence of Tool Wear on Quality Criteria for Refill Friction Stir Spot Welding (RFSSW) Process. Procedia CIRP **24**, 108–113 (2014). https://doi.org/10.1016/j.procir.2014.08.015
5. Kluz, R., Kubit, A., et al.: A Weighting Grade-Based Optimization Method for Determining Refill Friction Stir Spot Welding Process Parameters. J. Mater. Eng. Perform **28**(10), 6471–6482 (2019). https://doi.org/10.1007/s11665-019-04378-4
6. Kallee S.: Industrial applications of friction stir welding. In: Friction Stir Welding, pp 118–163. Elsevier (2010)
7. Connolly, C.: Friction spot joining in aluminium car bodies. Ind. Robot: Int. J. **34**(1), 17–20 (2007)
8. Schilling C., Dos Santos J.: Method and device for joining at least two adjoining work pieces by friction welding (US6722556B2) (2004)
9. Stirtec GmbH: Gebrauchsmusterschrift: Punktschweißzange (AT 14153 U1 2015–05–15) (2015)
10. Hirano S., Aota K.: Method of Friction Stir Spot Welding and Welding Apparatus (US007635075B2) (2009)
11. Uematsu, Y., Tokaji, K.: Comparison of fatigue behaviour between resistance spot and friction stir spot welded aluminium alloy sheets. Sci. Technol. Weld. Joining **14**(1), 62–71 (2009)
12. Okamoto K., Hunt F. et al.: Development of Friction Stir Welding Technique and Machine for Aluminum Sheet Metal Assembly- Friction Stir Welding of Aluminum for Automotive Applications (2) -. In: SAE Technical Paper Series. SAE International400 Commonwealth Drive, Warrendale, PA, United States (2005)
13. Hunt F., Badarinarayan H. et al.: Design of Experiments for Friction Stir Stitch Welding of Aluminum Alloy 6022-T4 - Friction Stir Welding of Aluminum for Automotive Applications (3) 2034 (SAE/SP-2034) (2006)
14. Zhang, Z., Yang, X.: Effect of welding parameters on microstructure and mechanical properties of friction stir spot welded 5052 aluminum alloy. Mater. Des. **32**(8–9), 4461–4470 (2011)
15. Tweedy B., Widener C.: Factors Affecting the Properties of Swept Friction Stir Spot Welds. In: SAE Technical Paper Series. SAE International400 Commonwealth Drive, Warrendale, PA, United States (2008)
16. DIN EN ISO 14273: Widerstandsschweißen - Zerstörende Prüfung von Schweißverbindungen (2016)

Multi-robotic Composite Production of Complex and Large-Scaled Components for the Automotive Industry

Florian Helber(✉), Stefan Carosella, and Peter Middendorf

Institute of Aircraft Design, University of Stuttgart, Pfaffenwaldring 31, 70569 Stuttgart, Germany
helber@ifb.uni-stuttgart.de

Abstract. For the automotive industry, automated composite manufacturing via state-of-the-art Automated Fibre Placement (AFP) or Automated Tape Laying (ATL) is of limited interest due to moderate cycle times and limitations concerning applicable reinforcement materials. The novel Advanced Ply Placement (APP) on the other hand offers a high degree of freedom concerning fabric utilization. A unique multi-robotic handling process allows the APP to process dry fibers as well as pre-impregnated fibers with a large variety of fiber areal weights and fabric types. With the APP approach, automated, wrinkle free placement of wide, unidirectional fabrics onto complex, double curved geometries is possible.

Offline robot path planning of the APP process is realized via FlexiCAM, a CAE interface for geometry based trajectory generation that allows databased parameter adaption and process simulation.

The present paper will explain the working principles of the multi-robotic Advanced Ply Placement and distinguish between established automated manufacturing processes. Results on geometry based path generation via FlexiCAM and their applicability in the automotive industry will be discussed and the benefits of knowledge-based offline path planning will be highlighted.

1 Introduction

The Automated Tape Laying and the Automated Fibre Placement are two manufacturing approaches for the production of fibre-reinforced components. Both technologies are mainly utilized in the aerospace industry, due to their high financial invest and limited production rates. Automated Tape Laying allows layup of unidirectional (UD) fabrics with material widths of up to 300 mm but is therefore restricted to structures with low complexity and limited curvatures (e.g. wing skins, etc.).

The Automated Fibre Placement is the evolution of the ATL technology. Instead of deploying only one broad unidirectional fabric, the AFP approach allows layup of up to 32 single tows, with a tow width of 1/8″ (3.175 mm), 1/4″ (6.35 mm) or 1/2″ (12.7 mm) [1]. The number of single tows that can be placed simultaneously next to each other depends on the layup head and has a significant influence on the systems productivity. Most AFP systems utilize a separate (refrigerated) bobbin creel for material storage.

P. Weißgraeber et al. (Eds.): *Advances in Automotive Production Technology – Theory and Application,* ARENA2036, pp. 369–376, 2021. https://doi.org/10.1007/978-3-662-62962-8_43

Refrigeration is necessary, if prepreg materials shall be used during material layup. A complex hose system guides the single tows to the layup head. Flexible compaction rollers, integrated in the layup head, consolidate the reinforcement tape onto the tooling geometry. The layup head has a cutting system integrated, which is able to cut single tows separately and conveyor rollers for each tow enables separate refeeding. Tensile stresses during layup are minimized in order to produce concave structures and non-geodetic geometries. The segmentation of the fabric also allows feeding, stopping and cutting of the single tows at different speeds, which enables steering within one plane, as well as production of complex geometries with double-curved features. During automated deposition processes the pre-bindered fabrics need to be thermally activated in order to achieve fabric adhesion. Sufficient tack is an important factor, since it avoids shifting of the fabric on the laminate or the mould during layup. Several binder activation technologies for AFP and ATL processes are available [2].

2 Advanced Ply Placement

Szcesny et al. [3] introduced a novel layup system for the automated manufacturing of complex and large-scaled fibre-reinforced plastic (FRP) components. The Advanced Ply Placement (APP) was developed to apply unidirectional, pre-impregnated fabrics [4] onto double-curved surfaces, without causing wrinkles. The APP was especially developed for cost-efficient and flexible out-of-autoclave composite processes. The APP process utilizes three cooperating robots to imitate the manual layup process of reinforcement fabrics, as it can be seen in Fig. 1a. Two industrial robots, that are equipped with pneumatic gripper elements [2, 3], transport the pre-cut fabrics from a material supply station over the tooling mould in the center of the unit. These gripper robots are mounted on linear axes in order to extend the applicable range. A third robot, equipped with a shape-adaptive application roller, drapes the fabric onto the mould surface, as illustrated in Fig. 1b. Therefore, the roller presses the fabric against the mould and moves in direction of the unidirectional fibers to one of the grippers. The gripper robots adjust their position according to pre-defined trajectories. In order to avoid wrinkling of the fabric, tension is applied to the ply at all times [5]. As soon as the compaction roller has reached its dedicated end position, it moves back to the initial starting point and repeats this process towards the second gripper unit. Plies are placed next and on top of each other, acc. to the respective ply book. The tooling mould is mounted on a rotary axis, so that different fiber orientations can be applied.

Fabrics can either be pre-cut on a separate material supply station or fed by an automated fabric creel. However, the material storage and supply is separated from the robotic devices in order to reduce the system complexity to a minimum and reduce maintenance susceptibility. The single plies are clamped by segmented gripper units. Each gripper unit consists of 15 single gripper elements with a clamping width of 20 mm and can be moved relatively to each other. To compensate length differences during layup, axial shearing of the unidirectional fabrics is used and at any time during the entire process a defined ply tension is ensured due to a pneumatic control system.

The shape adaptive compaction roller is the second key component to guarantee wrinkle free layup of the fabrics. The compaction roller of the APP process consists of a flexible silicone membrane which is mounted on pneumatic tandem cylinders in order to adapt to concave and convex geometries. This enables the compaction roller to adapt to complex surfaces and a defined pressure across the entire tape width can be set and guaranteed throughout the entire layup process.

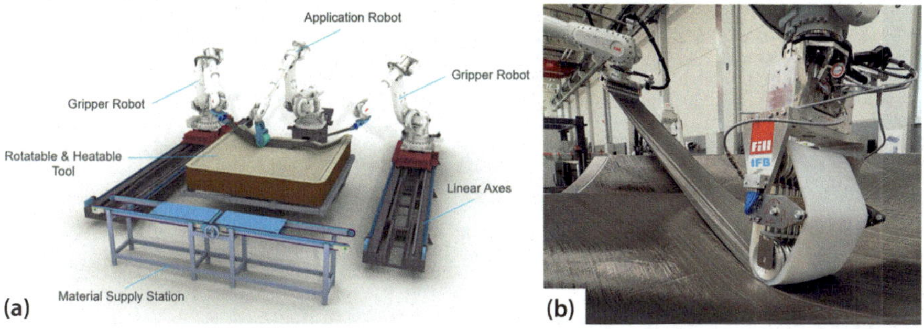

Fig. 1. **a** APP Process Scheme; **b** APP Layup with shape adaptive compaction roller

Offline process programming is realized via FlexiCAM. FlexiCAM is a software, which was developed at the Institute of Aircraft Design. It is based on the open-source platform Open Cascade and can be used for offline programming of different robotic processes, such as AFP, pick and place, robotic milling, etc. For the APP offline programming two different subroutines have been developed. A kinematic draping simulation and a subsequent path planning tool generate robot trajectories, which can then be exported to the respective robotic systems. For time-efficient calculation and processing of the robotic trajectories a kinematic draping approach, based on simple shear [6], was chosen. A MATLAB optimization routine calculates the deformation of the fabric during the APP layup process and adjusts the position of adjacent plies according to set parameters. Based on a set discretization, a point cloud with coordinates for the tool center point (TCP) of the application robot is generated. These vectors and the corresponding orientation of the tool surface are fundamental for the generation of the respective gripper trajectories. Relevant parameters such as ply width, ply orientation, gap width, etc. can be set and adjusted via a graphical user interface (GUI). The stacking sequence and number of plies with possible offsets are given in the according ply book. FlexiCAM carries out a deformation analysis, in which the maximum shearing angle of the unidirectional ply is compared with a material data base [7]. The shearing angle is dependent on the UD fabric as well as the given gripper kinematics. Furthermore, an accessibility study and collision control is carried out in order to assess the layup feasibility and can be visualized via process simulation (Fig. 2).

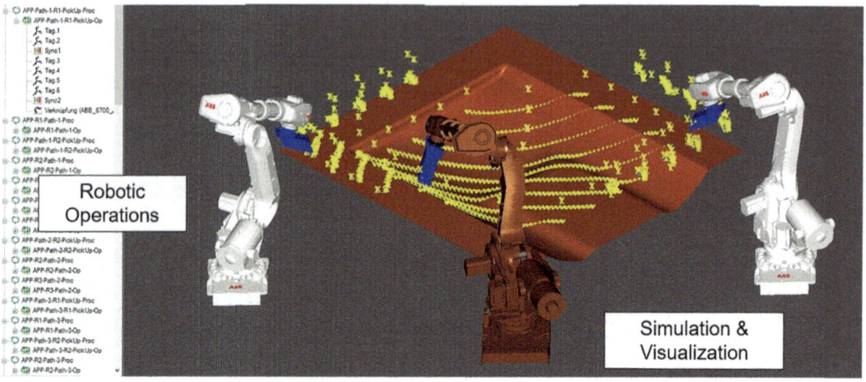

Fig. 2. APP Offline Programming (FlexiCAM GUI)

3 Analytical Layup Comparison

For assessing the productivity of different automated fibre deposition processes, an analytical comparison has been carried out. Various parameters such as robotic layup speeds (ν_{LU}), ply/course width (w_p), fibre areal weight (FAW), downtime, time for material feed and ply take up as well as inspection can be considered. The layup geometry (front wall for truck semi-trailers) is split into sections with different complexities (Fig. 3), since at regions with no or little single curvature higher layup rates can be applied, while robotic speeds at double curved surfaces need to be reduced in order to minimize the risk of layup defects, like wrinkles, gaps or overlaps [8].

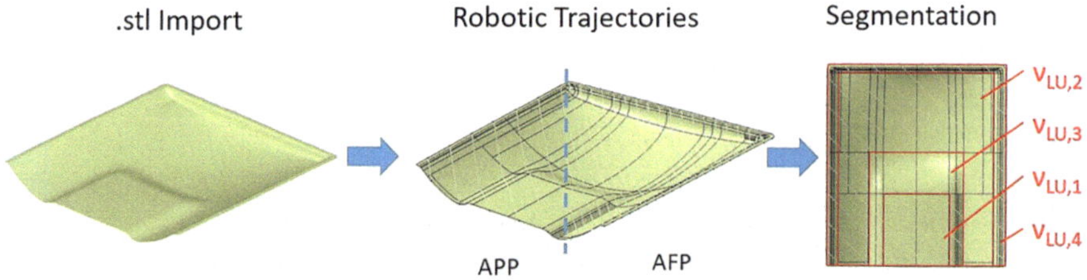

Fig. 3. Processing of Benchmark Geometry

Layup rates (R_{LU}) of the AFP process were taken from the literature [1]. Values range from 45.4 kg/h to 8.5 kg/h. For the present study, a value of 19.60 kg/h is assumed, resulting from robotic layup speed of 400 mm/s for flat sections, a course width of 101.6 mm (16 tows; ¼" tow width) and a fibre areal weight of 134 gsm (HEXCEL HiTape; [9]). In sections with higher complexity, layup rates have been adjusted accordingly. 75% of ν_{Lu1} has been applied for sections with low complexity and slight double curvature (Sect. 2), 50% for sections with high double curvature (Sect. 3) and 10% for border regions (Sect. 4). Layup rates for the APP process have been experimentally obtained during production of the respective demonstrator geometry.

Table 1. Deposition rate comparison (AFP vs. APP for different geometry sections)

Section	Geometry Complexity	v_{AFP}	$R_{LU,AFP}$	v_{APP}	$R_{LU,APP}$
1	Flat Section; Single Curvature	500 mm/s	24,51 kg/h	6 mm/s	19 kg/h
2	Low Complexity	250 mm/s	12,25 kg/h	5 mm/s	15 kg/h
3	High Complexity	100 mm/s	4,90 kg/h	3 mm/s	10 kg/h
4	Tool Borders	50 mm/s	2,45 kg/h	3 mm/s	10 kg/h

Table 1 shows the robotic layup speeds of the AFP and APP process. It is obvious that the layup speed of the AFP process is several times faster than the APP process, while the layup rate is nearly equal. This is due to the fact that during APP a UD fabric with 300 mm tape width and a FAW of 300 gsm has been used. While AFP processes are limited to low FAWs the APP shows it's potential when it comes to higher FAWs Fig. 2.

Direct comparison of the layup process with the given geometry can be seen in Table 2 and is illustrated in Fig. 4. Two different AFP setups (8 tows & 16 tows) have been taken into account using the same tape material (134 gsm, ¼" tape width) and are directly compared to the APP process (300 gsm; 300 mm ply width). Time for acceleration and deceleration of the AFP end effector have not been considered, but time for starting a new course, tow cutting, turning of the layup head and repositioning has been taken into account (7 s) [1]. Time for inspection, rework, maintenance and material change has been considered according to Anderton et. al [10]. One advantage of the APP process is the external material storage and feeding operations. Material preparation and supply is carried out parallel to the actual layup process and therefor has no impact on machine downtime or overall layup time. Furthermore, the low technical complexity of the layup head and gripper systems results in low susceptibility for maintenance and down time. 3% machine downtime is an estimation that is considered, since during the experimental testing no maintenance was necessary. An inspection duration of ~120 s per ply and a pick-up duration of ~60 s is anticipated.

Table 2 shows that the AFP process with 16 tows is calculated with an overall layup duration of 13.70 h (3.28 h layup; 8.63 h idle time). The APP process is ~24% slower with an overall layup duration of 17.09 h (14.42 h layup; 2.67 h idle time). It is obvious that the AFP process with only 8 tows and 50.8 mm layup width respectively is less efficient due to the high number of courses and cut & turn processes. An overall layup duration of 26.33 h is to be expected.

4 Cost-Benefit Analysis

A cost-benefit analysis evaluates the different layup technologies (ATL, AFP & APP) and incorporates economic aspects. An initial benefit analysis considers essential factors for the different deposition processes. Acquisition cost of the layup technology is an

Table 2. Layup time comparison (AFP vs. APP)

	AFP		APP
	8 Tows	16 Tows	1 Ply
w	50.8 mm	101.6 mm	300 mm
FAW	134 g/m²	134 g/m²	300 g/m²
t_{total} [h]	26.33	13.70	17.09
$t_{Program}$ [h]	6.32	3.28	14.42
$t_{Inspection}$ [h]	16.59	8.63	2.67

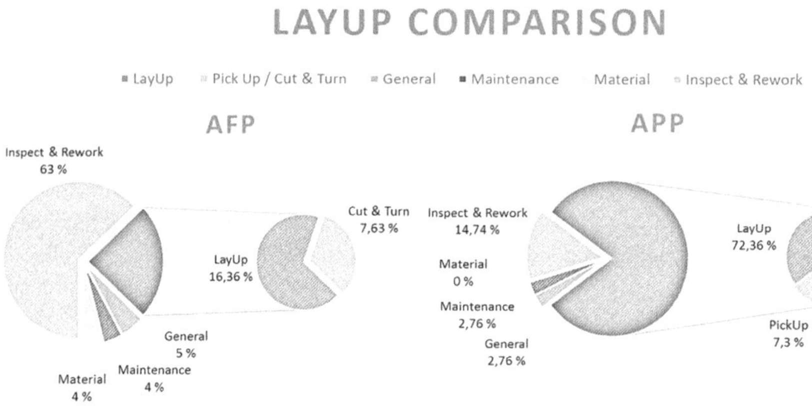

Fig. 4. AFP vs. APP layup comparison (Process Times)

important economic factor for the customer. Automating processes is key for quality insurance and economic manufacturing, however many customers cannot afford the initial invest of cost-intensive equipment. Furthermore, high TRL will result in higher customer confidence, when investing in automated solutions. Addressing the producible parts, tool orientation, part complexity and part size flexibility are aspects that need to be assessed during the decision making process. Conclusively, the layup process itself has to be investigated. Aspects such as layup quality, technical complexity and machine down time vary between the different manufacturing technologies. Furthermore, material flexibility, scrap rates, ability for thermoplastic layup and fibre steering are important factors that need to match the component portfolio of the customer.

To indicate the importance of each criterion, a benefit analysis compares the different criteria with each other and a ranking with respective weighting factors is created, acc. VDI 2225-3 [11]. The assessment reveals, that down time and layup quality are highly ranked. Each layup concept (ATL, AFP & APP) is then given a score (fulfilment factor, f_{full}) ranging from 0 (unsatisfactory) to 4 (very good) that depends on the level of fulfilment for each criteria and is subsequently weighted acc. to the aforementioned factors. These point scores allow quantitative measurement and direct comparison with other layup concepts. According to this assessment the AFP process achieves the highest score ($f_{full,AFP} = 3.06$), followed by APP ($f_{full,APP} = 2.54$) and ATL ($f_{full,ATL} = 1.7$).

A strength-diagram acc. VDI2225-3 [10] is created in order to get a better understanding of the correlation between the economic and technical value of the different layup technologies. The economic value (V_e), is the ratio of the cost of an ideal concept (c_{ideal}) and the cost of the respective layup concept ($c_{concept}$). The ideal concept cost is defined as 90% of the cost of the cheapest technology in question. The technical Value (V_t), describes the ratio of the respective fulfillment factor ($f_{concept}$) and the fulfillment factor of the ideal concept (f_{ideal}). The single values for V_e and V_t are calculated for each concept and are plotted in Fig. 5. The results for both factors are between 0 and 1, while the ideal concept has a value of 1 for V_e and V_t. The closer a concept is to the diagonal line, the more balanced it is. The strength of each concept can be calculated and is a measure to express the suitability of a concept to its requirements and cost. The strength s is defined as the squared root of the product of V_e and V_t for each concept.

From the cost-benefit analysis it can be seen that the APP process reaches a strength score of 0.76, which is higher than AFP (s = 0.64) and ATL (s = 0.24). The high strength value of the APP process mainly results from the high economic value, due to low APP acquisition costs. The AFP process on the other hand convinces with a high technical value and high TRL levels. The disproportionate invest for ATL equipment and low part complexity and flexibility results in the low strength value.

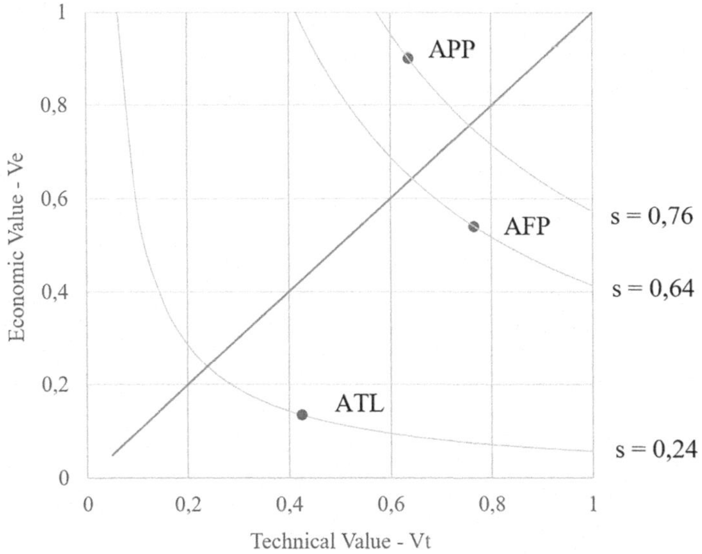

Fig. 5. Strength Diagram for Automated Layup Concepts

5 Results and Summary

The present paper explains the working principles of the Advanced Ply Placement. Gripper segmentation and a shape-adaptive application roller allows manufacturing of double-curved geometries while using UD materials. Offline programming of this multi-robotic process is realized via FlexiCAM.

An analytical comparison has been carried out to assess and compare different fibre deposition processes, considering relevant parameters. For a given geometry with several complex double-curved features, the AFP process (8 tows & 16 tows; 1/4" tow width) and the APP process have been compared. The state-of-the-art 16 tow AFP process turns out to be most efficient, directly followed by the APP process. However, it needs to be highlighted, that the parameters for APP layup have been determined at an early TRL level. Due to the optimization of the path planning via FlexiCAM the actual layup rates have to be reevaluated and higher values are to be expected.

A subsequent cost-benefit analysis of the different processes reveals the potential of the APP. The high economic value of the APP process makes it attractive for industrial users, if a highly flexible process for complex, large-scaled components is required.

References

1. Lukaszewicz, D., Ward, C. Potter, K.: The engineering aspects of automated prepreg layup: History, present and future. Composites: Part B 43 997–1009 (2012). https://doi.org/10.1016/j.compositesb.2011.12.003
2. Helber, F., Amann, A., Carossela, S., Middendorf, P.: Intrinsic Fibre Heating: A Novel Approach for Automated Dry Fibre Placement. IOP Conference Series: Materials Science and Engineering **406**(1), (2018). https://doi.org/10.1088/1757-899X/406/1/012064
3. Szcesny, M., Heieck, F., Carosella, S., Middendorf, P., Sehrschön, H., Schneiderbauer, M.: The advanced ply placement process – an innovative direct 3D placement technology for plies and tapes. Advanced Manufacturing: Polymer & Composites Science. **3**(1), 2–9 (2017). https://doi.org/10.1080/20550340.2017.1291398
4. Meinhardt, T., Harismendy, I., Heieck, F.: LOWFLIP – Tailored Snap-Cure Prepregs for Novel Composite Production Processes. In ECCM 17 – 17th European Conference on Composites Materials
5. Engelfried, M., Nosek, T., Heieck, F., Böhler, P., Middendorf, P.: *LOWFLIP – Draping Simulation of Prepregs at Mesoscopic and Macroscopic Level*. In: ECCM 17 – 17th European Conference on Composites Materials
6. Zouari, B., Daniel, J.-L., Boisse, P.: A woven reinforcement forming simulation method. Influence of the shear stiffness. Comput. & Struct. **84**(5–6), 351–363 (2006). https://doi.org/10.1016/j.compstruc.2005.09.031
7. Heudorfer, K., Engelfried, M., Fial, J., Carosella, S., Middendorf, P.: Characterisation of the Forming properties of wide unidirectional prepreg tapes using the advanced ply placement (APP) process. In: 8th EASN – CEAS International Conference on Manufacturing for Growth & Innovation
8. Rudberg, T., Nielson, J., Henscheid, M., Cemenska, J.: Improving AFP Cell Performance. SAE Int. J. Aerospc. **7**(2), 317–321. https://doi.org/10.4271/2014-01-2272
9. HiTape Brochure (HEXCEL Website). https://www.hexcel.com/user_area/content_media/raw/HiTape.pdf. Accessed 20 June 2020
10. Anderton, C., Colvin, A.: Roadmap to Automated Composites. https://www.compositesworld.com/cdn/cms/uploadedfiles/roadmap-to-automated-composites.pdf. Accessed 20 June 2020
11. VDI – The Association of German Engineers: VDI 2225–3: Design Engineering Methodics – Engineering Design at Optimum Cost – Valuation of Costs. (1998)

Integrated Machining, Quality Inspection and Sealing for CFRP Components

Philipp Esch(✉), Andreas Gebhardt, Oliver Tiedje, and Andreas Frommknecht

Fraunhofer Institute for Manufacturing Engineering and Automation IPA, Nobelstr. 12, 70569 Stuttgart, Germany
oliver.tiedje@ipa.fraunhofer.de

Abstract. Manufacturing processes of carbon fiber reinforced plastics (CFRP) components are often characterized by a high fraction of manual activities especially in terms of handling, cleaning and edge sealing. The innovative machining process introduced in the project *CFKComplete* comprises a fully integrated and automated workflow reducing unnecessary handling and transportation activities. The concept has been realized on a 5-axis machine tool, providing enough workspace for large structural workpieces and the developed technology units. The units themselves are defined by their modularity, scalability and independency of the guiding machine. Key aspect of the innovative concept are four purpose built and developed interchangeable technology units, which are driven by the base machinery. Thus, the machine's working scope is being enhanced from machining to a fully integrated workflow including quality inspection, cleaning, sealing and curing.

1 Introduction

Since the usage of carbon fiber reinforced plastics has grown significantly in the last years due to its lightweight potentials, the processing has to be optimized in order to generate an efficient workflow [1–3]. There are several interacting process steps as cutting, cleaning of the edges and sealing which are demanding for an automated and integrated process to enable process stability. It is still state of the art to realize these steps separately with a huge amount of manual steps for handling, quality control etc. Corrosion protection is usually either applied manually or performed after assembly e.g. by seam sealing with PVC plastisols. This paper will present an integrated process consisting of machining (cutting), brushing and cleaning, inspection and sealing (painting and curing) of the edges of CFRP in one machine with one clamping of the workpiece. To do so, it is necessary to avoid media like water in the cutting process, to generate a paintable quality in the edge rounding which is assured by an optical inspection. Furthermore, contamination

P. Weißgraeber et al. (Eds.): *Advances in Automotive Production Technology – Theory and Application,* ARENA2036, pp. 377–385, 2021. https://doi.org/10.1007/978-3-662-62962-8_44

and solvent evaporation has to be avoided in sealing process and only low temperatures can be applied in the curing process. Due to the hazardous dusts and to the interaction of the cutting and painting process innovative air flow concepts and exhaustion systems were developed [4, 5]. The cleaning and brushing was performed by a multipurpose tool constructed for the application. A new overspray free application process using a solvent free coatings material with 100% solid content prevents the machine from contamination and risk of explosion. UV curing with LED technology is a new approach to yield a fast curing without significant heat transfer.

2 Fiber Dust Protection and Collection

Machining of fiber reinforced plastics is often performed by milling under dry cutting conditions. The compound material fails with brittle characteristics resulting in a high amount of machining volume of dust and particles. The amount of dust and the scale distribution depend on process parameters and tool design [6]. The number of particles increases with increasing cutting speed, which is relevant for modern production processes. Small particles, which fulfill WHO-criteria can stay airborne and harm the operator when inhaling [7]. The hazardous effect of small particles has been studied intensively. Without precaution or technical measures, fiber dust exposure can potentially affect cancer [8]. An efficient dust distraction system is therefore required for machining CFRP.

In order to counteract dust exposure, a holistic concept was developed and implemented. First, the workpiece clamping itself features dust collection cavities, realizing a nearby acquisition to the cutting process. Second, a dust distraction hood covers the spindle from above capturing raised up dust. The third measure comprises a direct air stream which overflows the workpiece thus collecting the remaining airborne particles and carrying them via air stream to an opposing suction place. The air stream is filtered coarsely and air volume is returned to the ventilation nozzles. Only a small ratio of about 30% of air volume is fine filtered and emitted to the surroundings. This so-called Air-Return-System (ARS) optimizes energy efficiency in conjunction with maximized dust collection rate. Figure 1 illustrates the principle of ARS. System integration was realized in the moving gantry providing a surrounding air stream around the tool without relative motion in transverse direction. In order to overflow the workpiece in optimum relation to the workpiece height, the jet nozzles can be vertically adjusted.

The direct measures for dust collection are combined with a dust concentration detector, which monitors the concentration and regulates access permissions. For highest safety, the entire machine is fully enclosed, too. The otherwise half-encapsulated machine is now safe for operation.

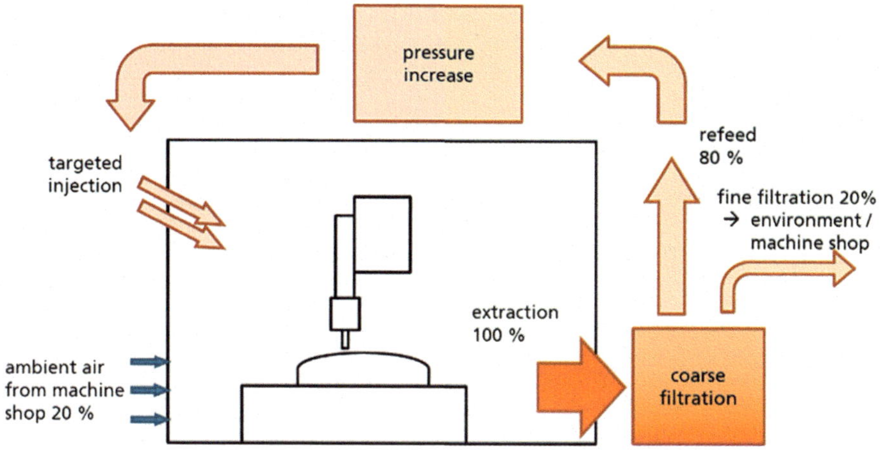

Fig. 1. Working principle of Air-Return-System

3 Process Steps

3.1 Machining

For machining CFRP, dry cutting is preferential to wet conditions in terms of quality and wear [9]. In return, the lack of temperature dissipation due to missing cutting fluids must be addressed by careful selection of process parameters and process guidance. Research findings also indicate a positive effect of chilled air [10]. Tool wear mainly occurs at the cutting edge of the cutting tool due to abrasive characteristics of the carbon fibers. Increasing cutting edge radius leads to delamination and fiber protrusion [11]. Diamond or PCD coatings can reduce wear but also raise initial cutting edge rounding [12].

Modern cutting tools for CFRP comprise a complex cutting edge design. The resulting spatial position of the cutting edge in relation to the fiber cutting angle at contact is defined by superimposed tilt and lead angles [13]. The right engagement of the cutting edge is essential for cutting CFRP with reduced edge delamination or fiber protrusion.

Fiber orientation in relation to the cutting edge effects wear and quality output significantly. For the purpose of determining process parameters with low edge damage, the critical fiber cutting angle for five different cutting tools has been investigated. For the given material, the critical fiber cutting angle was established between 90° and 135°. For milling, the spread of information about recommended process parameters is quite vast and mainly attributed by different tooling concepts. The range of applicable cutting speeds ranges from $v_c = 80$m/min to $v_c = 500$ and in terms of feed rates per tooth from $f_z = 0.01$mm–$f_z = 0.08$mm. The investigated tools exhibit lowest edge damage in terms of delamination at the most critical fiber orientation angle for the experimentally defined process parameters given in Fig. 2.

Next to the condition and design of the cutting edge, the workpiece clamping is of significant importance. The investigation included an evaluation of different clamping situations. A vibration sensor mounted in the center of the workpiece recorded the resulting vibrations induced by milling for two different set ups. A highly flexible but less

Tool	Optimum feed per tooth [mm]	Number of main cutting teeth	Optimum cutting velocity [m/min]
WZ1	0.01	8	200
WZ2	0.04	8	500
WZ3	0.04	8	500
WZ4	0.04	4	400
WZ5	0.03	2	500

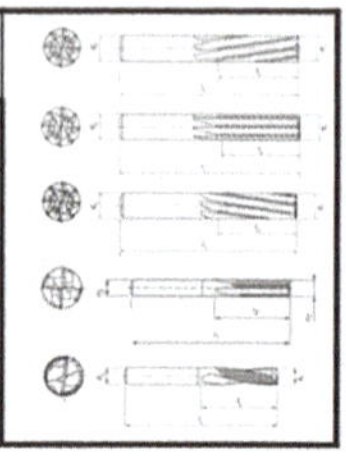

Fig. 2. Recommended process parameters for cutting CFRP

rigid clamping system exposes maximum accelerations of about 45 g in perpendicular direction to the workpiece surface plane. The rigid state-of-the-art workpiece clamping system reduces maximum acceleration by half. The effect of inadequate fixation is also evident in terms of quality (Fig. 3).

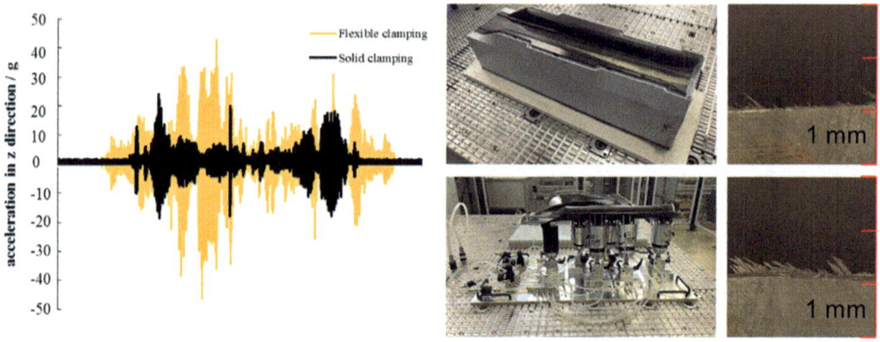

Fig. 3. Comparison of different clamping scenarios

3.2 Brushing and Cleaning

With increasing wear and excessive cutting edge rounding CFRP is rather crushed in the cutting zone instead of cut. Loose fibers can easily deflect a blunt cutting edge as their transverse rupture strain is not attained. Brushes with specialized fillings can remove remaining fiber protrusions in order to achieve quality requirements of the milling edge.

Treatment of high abrasive materials is a current subject of research and development in the domain of tooling material. Cutting edge preparation is performed with brushes, drag finishing or abrasive flow machining. Regarding CFRP post-treatment, brushing of the cut edge can also be considered appropriate. Fast rotating brushes clear the milling edge from fiber protrusions. The non-defined cutting edge of brushes minimizes the erosion of the material and no dimensional deviation is provoked. Only allowing fibers to be cut away, delamination phenomena cannot be treated with this technique. Non-abrasive brushes are further used to clean the workpiece surfaces.

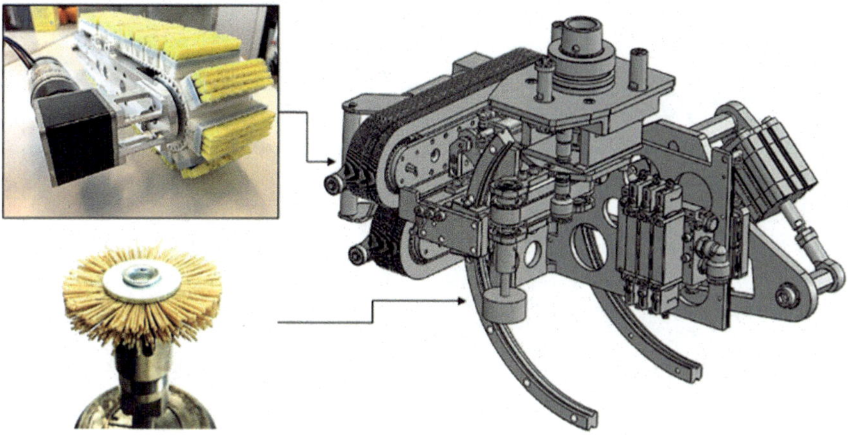

Fig. 4. Integrated cleaning unit (by courtesy of HOMAG)

In an integrated and exchangeable active unit, both brushing technologies for edge and surface cleaning are implemented for simultaneous processing. Figure 4 illustrates the developed cleaning unit for 5-axis treatment.

3.3 Inspection

After the cleaning of the part, it has to be inspected before the sealing. It is necessary to ensure that the quality for the next process step is sufficient (e.g. the coating can only cover a certain fiber protrusion) and no rejection or reworking is necessary. For a fast and at the same time comprehensive inspection, optical methods should be favored. So any direct contact with the part is avoided, which normally increases the measurement time compared to e.g. tactile sensors.

Typical defects during the machining process are fiber protrusions, delaminations and disruptions. Some delaminations do not affect the surface of the part and cannot be detected by optical methods. Therefore, other methods like ultrasonic, thermography or computer tomography have to be considered [14]. If the delaminations lead to bulges of the part, they can be inspected with similar data acquisition and processing procedures as the other defect types.

Clearly most of the defects influence the three dimensional shape of the parts. Therefore, it was necessary to use a 3D sensor. Due to speed and accuracy requirements, only two type of start of the art systems are adequate: fringe projection or laser line sensors. Both sensor types have been used in pretests. Finally the fringe projection system ZEISS AIMax cloud (working distance: 165 mm, measurement area: 80 mm × 80 mm × 40 mm, camera resolution: 2048 Pixel × 2048 Pixel) was integrated in the guiding

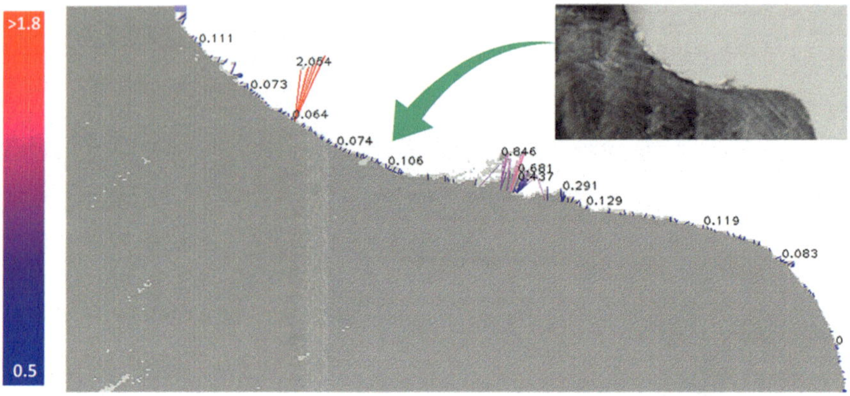

Fig. 5. Color coded fiber protrusions (values in mm)

machine. The advantages compared to laser line systems are that stationary measurements are possible and no eye safety measurements have to be considered. The most critical areas of the machined CFRP parts have been digitalized with this sensor. The digital workflow is managed by a measurement computer (MC), which is connected to the guiding machine and the sensor. For each measurement area the guiding machine indicates its position to the MC, then the MC triggers the sensor, then the sensor sends the acquired 2D pictures back to the MC. On the MC the final 3D data is reconstructed.

This 3D data has been analyzed with existing and also specifically developed and adapted software. The implemented algorithms are described in detail in [15]. For analyzing and visualization purposes, the 3D data is mapped in a 2D gray value image, where each gray value corresponds to a specific height, as shown in Fig. 5. In the above right corner of the figure, a photo of the original part is attached. The achieved analysis result allows to measure the lengths of the fiber protrusions and to visualize them. Similar results can be obtained for disruptions and delaminations (as long as they affect the surface of the part).

Considering again the digital workflow, the final analysis results can be stored as a table containing position and size of the different kind of defects. This information can be used for defining necessary rework or tool changes. Furthermore, the long-term process quality can be monitored. This is very useful for process optimization for example with artificial intelligence.

3.4 Sealing and Curing

Suitable coatings were evaluated with regard to the achievable edge coverage and layer thickness, their stability on the edge, their cross-linkability using suitable UV LED technology, as well as their coating adhesion and compatibility in cataphoretic painting. The varnish should also be able to reliably cover minimal fiber protrusions. Solvent-free UV coating systems are used that also meet the requirements for explosion protection, have an optimal ecological balance sheet and no flash-off times.

The varnish is applied with a piezo-actuated micro-dosing system [16], in which the application is carried out by means of contactless dosing of free-flying material drops

Fig. 6. Sealing process at the work piece inside the machine

(Fig. 6). These are deposited on the edge of the component without overspray and joined to form a continuous line (width 1 to 1.5 mm). Several painting lines are required to cover wider edges. The piezo control allows to compensate for the different material properties by means of an adapted opening and closing movement. Numerical simulation [17] and measurement of complex rheological data were used to find the application window, validated by experimental investigations also using a high-speed camera. In parallel, a unit for UV-LED curing of the coatings was developed by the project partner IST Metz. All results of the laboratory tests could finally be transferred to the demonstrator machine.

For edge sealing in the machine, it soon became apparent that free guidance of the sealing unit was not expedient due to the high demands on the component position. Even small deviations in clamping and distortion lead to the fact that a reproducible sealing is not possible. In order to compensate for position and shape tolerances, the design of the sealing unit has been modified in such a way that it is positively guided along the edge by a guide roller. The basic paths are offline programmed, the exact guiding along the edge is ensured by a touching and resilient guide roller. In the subsequent process step, the paint is cured by a "light finger". The LED Spot developed by IST Metz has an LED adapted to the wavelength of the coating and is guided along the edge by the machine.

4 Summary and Conclusions

The project shows that it is possible to integrate the complete CFRP tooling process in one machine. The interaction between the process steps were investigated and mastered to ensure a stable process. Regarding the individual process steps innovative solutions have been developed as multi-function cleaning tools, booth air concepts, overspray free painting and fast curing with UV-LED, monitored by new optical inspection algorithms.

There is a huge area of application of the project results from automotive industry over aircraft production to machine construction. The modular and flexible concept allows adaption to other demands and is scalable to production units. In automotive industry, there are several duroplastic parts due to the excellent compromise between weight and crash behavior that can be much more efficient produced by this new technology. In aerospace there is a necessity to automate manual processes as the edge sealing of the enormous amount of carbon fiber parts. Furthermore, ongoing developments will be done to transfer the results also to edges of glass fiber plastic parts and metallic parts.

Acknowledgement. The present investigations were implemented in the project CFKComplete (Grant no.:02P14A086), funded by Bundesministerium für Bildung und Forschung under the support of Projektträger Karlsruhe (PTKA). We like to thank our project partners from the companies Homag, Audi, Carl Zeiss Automated Inspection, IST Metz and Krautzberger.

References

1. Rummenhöller, S.: Werkstofforientierte Prozessauslegung für das Fräsen kohlenstofffaserverstärkter Kunststoffe. Dissertation. Rheinisch-Westfälische Technische Hochschule Aachen. Aachen: Shaker (1996)
2. Schneider, M., Birenbaum, C., Forbes, A., Mayer, T., Burkhardt, J.: Spanende Bearbeitung von Leichtbauwerkstoffen. Einführung und Überblick. Stuttgart: e-mobil BW GmbH – Landesagentur für Elektromobilität und Brennstoffzellentechnologie Baden-Württemberg (2012)
3. N. N.: Wertschöpfungspotenziale im Leichtbau und deren Bedeutung für Baden-Württemberg. Leichtbau BW GmbH (2014)
4. N. N.: Bearbeitung von CFK Materialien. Orientierungshilfe für Schutzmaßnahmen. Fachbereich Holz und Metall der DGUV (2014)
5. Pöhler, D., Thümmler, J.: CFK-Bearbeitung – Schutz der elektrischen Ausrüstung. In: CCeV-Thementag „Ertüchtigung von Werkzeugmaschinen für die CFK Bearbeitung". 15.02.2017, Fulda (2017)
6. Haddad, M., Zitoune, R., Eyma, F., Castanie, B.: Study of the surface defects and dust generated during trimming of CFRP: Influence of tool geometry, machining parameters and cutting speed range. Compos. A Appl. Sci. Manuf. **66**, 142–154 (2014). https://doi.org/10.1016/j.compositesa.2014.07.005
7. TRGS 900; Fassung 12.5.2020.
8. Liste der krebserzeugenden: keimzellmutagenen und reproduktionstoxischen Stoffe (KMR-Stoffe). IFA, Stand Februar (2020)
9. Kerrigan, K., Scaife, R.J.: Wet vs dry CFRP drilling: Influence of cutting fluid on tool performance. Procedia CIRP **77**, 315–319 (2018). https://doi.org/10.1016/j.procir.2018.09.024
10. Khairusshima, M.N., Sharifah, I.S.S.: Study on Tool Wear during Milling CFRP under Dry and Chilled Air Machining. Procedia Engineering **184**, 506–517 (2017). https://doi.org/10.1016/j.proeng.2017.04.121
11. Sauer, K., Witt, M., Putz, M.: Influence of cutting edge radius on process forces in orthogonal machining of carbon fibre reinforced plastics (CFRP). Procedia CIRP **85**, 218–223 (2019)
12. Hrechuk, A., Bushlya, V., M'Saoubi, R., Ståhl, J.-E.: Experimental investigations into tool wear of drilling CFRP. Procedia Manufacturing **25**, 294–301 (2018). https://doi.org/10.1016/j.promfg.2018.06.086

13. Hintze, W., Brügmann, F.: Influence of spatial tool inclination on delamination when milling CFRP. J. Mater. Process. Technol. **252**, 830–837 (2018) https://doi.org/10.1016/j.jmatprotec.2017.10.041
14. Effenberger, I., Felix, T., Frommknecht, A., Fulga, S., Springhoff, A.: Machine vision platform for non-destructive testing methods of fibre reinforced plastics. In: Proceedings 19th World Conference on Non-Destructive Testing (WCNDT), Munich (2016)
15. Frommknecht, A., Effenberger, I.: 3D Milling Edge Inspection of CFRP Parts by Optical Methods. In: International Symposium on Structural Health Monitoring and Nondestructive Testing, Saarbrücken (2018)
16. Hess, T., Tiedje, O. and Niemeier, W.: Overspray-free painting technology-From the lab to the painting object, 90th JSCM Anniversary conference, Japan (2017)
17. Ye, Q., Hess, T., Niemeier, W., Tiedje, O.: Numerical Investigation on viscous droplet generation for selective coating processes, ICLASS 2018, 14th Triennial International Conference on Liquid Atomization and Spray Systems, Chicago, IL, USA (2018)

A Universal Machine: Enabling Digital Manufacturing with Laser Technology

Thomas Graf[1,2,3](✉), Max Hoßfeld[1,3], and Volkher Onuseit[1,3]

[1] Institut Für Strahlwerkzeuge (IFSW), University of Stuttgart, Pfaffenwaldring 43, 70569 Stuttgart, Germany
thomas.graf@ifsw.uni-stuttgart.de
[2] Graduate School of Advanced Manufacturing Engineering (GSaME), University of Stuttgart, 70569 Stuttgart, Germany
[3] InnovationCampus Future Mobility, University of Stuttgart, Pfaffenwaldring 43, 70569 Stuttgart, Germany

Abstract. The laser is the only tool that can address all six main manufacturing groups of the German standard DIN 8580 simply by applying different processing parameters. Given the present state of technology, however, different dedicated machine concepts are still being used for the respective applications. To enable digital manufacturing in its full consistency, novel, fully reconfigurable machines need to be developed. We therefore outline the further developments that are required to combine all presently known laser-based manufacturing processes on one and the same machine. As both the laser devices and the knowledge about the fundamentals of laser materials processing are already very advanced, research must now be intensified on system engineering. The vision is an intelligent machine, which is fed with CAD data, semi-finished products, or sub-components and that is capable to autonomously produce the desired components with a 100% quality guarantee at a batch size of 1 – "first time right" – and at the costs of comparable mass-produced items. The machine independently selects the best production strategy, i.e. the best combination and sequence of different manufacturing processes. Such a fully flexible and autonomous laser machine will not only boost the implementation of digital manufacturing on a broad scale and provide an approach to resolve the "polylemma of production" but will also enable the relocalization of value creation and manufacturing back into high-wage countries and by this potentially disrupt today's globalized value creating networks.

1 Introduction

One of the central postulates of digital manufacturing in the era of Industry 4.0 is the implementation of fully flexible and highly integrated production systems which are dynamically reconfigurable to adapt themselves to product and market changes while simultaneously and efficiently mastering a high product variety. [1–3]. The aim of such systems has at least in part to be to resolve what can be summarized as the "polylemma

P. Weißgraeber et al. (Eds.): *Advances in Automotive Production Technology – Theory and Application*, ARENA2036, pp. 386–393, 2021. https://doi.org/10.1007/978-3-662-62962-8_45

of production" [4] which incorporates the central dichotomies of today's manufacturing: economy of scale versus economy of scope and value versus planning orientation.

From a production planning perspective, the present situation may be summarized as follows. Specialized, dedicated machines and production systems generally perform very well when producing standardized products in high quantities and a low variety. They offer great opportunities for automation and process integration but lack flexibility and versatility especially when facing different requirements, product changes or volatile market conditions like changes in demand or product mix, hence some of the machines might easily become underworked or even obsolete. In contrast to that, especially small and medium-sized enterprises and contract manufacturers which typically have to deal with smaller batch sizes, a higher product variety and a dynamically changing manufacturing program use production systems that combine and network different more or less specialized production machines and processes such as milling, turning or welding. This flexibility is typically gained at the cost of efficiency and overall equipment effectiveness (OEE) by increasing expenditures for handling and logistics, setting up machines and programming, longer cycle and processing times.

Closing the gap between flexibility, versatility and efficiency requires the development of machines that are as versatile and efficient as possible and can be reconfigured in real time. Despite ongoing research on different approaches, be it just the combination of different cutting tools on one machine [5], the simultaneous interaction of different process mechanisms and/or energy sources/tools [6], or the integration of additive and subtractive machine tools [7], a real breakthrough is still hampered due to the fact that in most of the investigated approaches a reconfiguration of the machine implies significant changes of the involved hardware. Even more, most process combinations result in an increased overall system complexity and often strongly conflicting hardware requirements such as mechanical and thermal load, precision or kinematic dynamics. This is where laser technology may now be able to herald a disruptive change. The laser is the only tool that is able to address all six main manufacturing groups (primary forming, forming, separating, joining, coating, change material properties) as defined by the standard DIN 8580 of the German Institute for Standardization [8], by essentially only adjusting the process parameters while using a consistent hardware framework for all these manufacturing applications.

2 Laser Materials Processing

The reason for the unparalleled versatility of the laser is that the vast majority of laser material processing is based on simply heating the processed material to a specific temperature [9]. For fusion cutting of metals for instance the laser beam is used to heat the material that is to be removed to above melting temperature and the melt is removed by a gas jet. Evaporation temperature needs to be exceeded for deep-penetration welding, where a deep vapor-filled keyhole is formed inside a pool of molten material that forms the welding seam after re-solidification at its trailing edge. For ablation or drilling, the evaporation is further intensified to completely remove the heated material from the workpiece. Hence, it is essentially the locally reached temperature and the heating time which define what process is induced by the incident laser beam. The parameters required to carry out a particular laser-based manufacturing process can therefore be determined by solving the heat conduction equation. For a laser beam with a Gaussian intensity profile moving with the velocity v across the surface of an extended workpiece one can derive the relation [10]

$$\frac{P_A}{d} = (T - T_0) \cdot \lambda_{th} \sqrt{\frac{\pi}{8}} \sqrt{\frac{F}{\tau} \frac{1}{\kappa} + 4}, \tag{1}$$

where P_A is the absorbed laser power, d the diameter (at $1/e^2$ of the maximum intensity) of the beam on the workpiece, T is the resulting maximum temperature on the surface of the workpiece, T_0 the ambient temperature, λ_{th} the heat conductivity, κ the thermal diffusivity, $F = \pi d^2/4$ the cross-sectional area of the beam on the surface of the workpiece, and $\tau = d/v$ the duration of the irradiation of the area F. In this form the relation analogously also applies to a resting but pulsed laser beam with a pulse duration of τ.

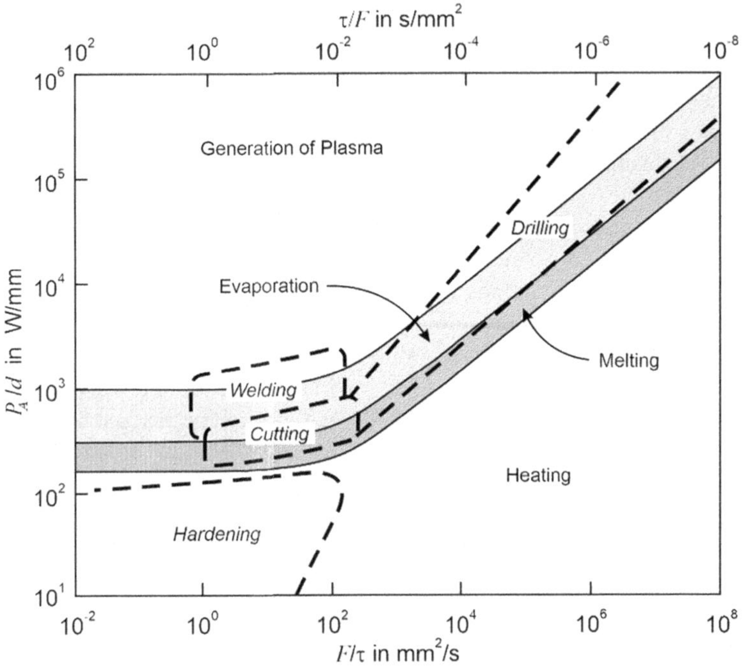

Fig. 1 Length-related absorbed power P_A/d and irradiated area per unit time (lower abscissa) – or irradiation time per unit area (upper abscissa) – required to reach certain values of the process temperature (e.g. melting or evaporation temperature) and typical parameter ranges (dashed lines) of some manufacturing process. The solid curves which define the phase changes between solid, liquid, vapor, and plasma (assumed at a temperature of 10^4 K) were calculated by (1) for the example of iron.

It is evident from equation (1), that in order to attain a desired temperature T at the surface of a given processed material, only the two processing parameters *length-related absorbed power* P_A/d and *irradiated area per time* F/τ need to be set accordingly. As illustrated by Fig. 1 for the example of iron, this means that all the different processes can be performed by just adapting these two parameters.

The parameter ranges shown by the dashed lines in Fig. 1 apply to the processing of an extended workpiece of solid iron. This is why the parameter fields for additive manufacturing processes are not shown in this same figure because the heat conduction in the additional material (e.g. powder) differs from that in the solid material. The same also applies to forming processes such as laser peening, which is typically based on the ablation of a layer of a different material applied to the surface of the processed workpiece. As an orientation one may state that the parameter fields for additive laser processes are comparable to those of welding and cutting and laser peening is typically located in the upper right area outside the section shown in the picture.

In summary one can conclude that with laser materials processing it is possible to address all the six main manufacturing groups specified by the standard DIN 8580 by only changing two processing parameters with no or only little need for a change of the involved hardware. This is illustrated by the examples given in Fig. 2.

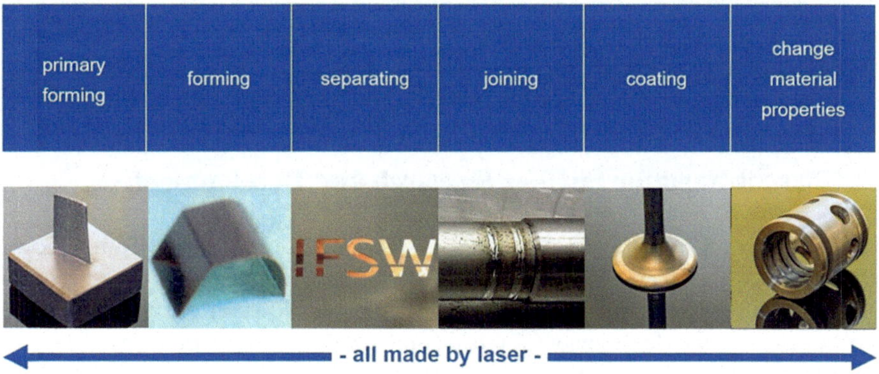

Fig. 2 The six main manufacturing groups as specified by the standard DIN 8580 and corresponding examples of laser-manufactured parts (from right to left: selective laser melting, bending, cutting, welding, coating, hardening).

2.1 Current State of Technology in Industrial Manufacturing

The conclusion of the previous section suggests that all laser-based manufacturing processes can be performed with one and the same laser device by only adapting the length-related power P_A/d and the processed area per unit time F/τ (or rather the interaction time per unit area τ/F). To date, however, quite different lasers and machines are used for different applications. The main reason for this is that so far the setup of ultrafast lasers delivering pulses with durations in the picosecond or femtosecond ranges differs significantly form the one of a free-running pulsed or continuous-wave (cw) laser. Additionally, the average powers provided by these different lasers were several orders of magnitude apart, with the consequence that high-precision micro processing with ultrafast lasers used to be much less productive than macro materials processing with high-power cw lasers.

With the most recent advances in the field of laser development this is however about to change. Since the newly developed ultrafast lasers providing several kilowatts of average output power [11, 12] are all based on a modular architecture with one or more amplification stages that work equally well for cw radiation, it is only a small further step to set up lasers that can be switched between different operation modes and that are therefore suitable to address all the common material processing applications. Hence, this kind of lasers will be available in the near future. Together with the significant and extensive progress on the fundamental understanding of the physics of laser materials processing during the last decades, whereby it is known to a large extent how an ideal process should be performed, this leads to the fact that the guiding of the laser beam to the workpiece and the physical implementation of the machine, its dynamics and precision, becomes the limiting factor of an efficient, versatile and flexible – i.e. universal – laser machine. Novel approaches to system engineering are now required to bridge this gap by providing completely new solutions for highly dynamic and precise kinematics, controls, and novel laser machine concepts as outlined in the next section.

In view of Industry 4.0 and digital manufacturing the manufacturing machines should also be able to flexibly and dynamically switch between the many different applications and be equipped with significant (artificial) intelligence to autonomously decide on how and with which processes the ordered parts should be manufactured.

3 Current Need for Further Research and Development

This calls for an intensified research on system engineering especially for the full exploitation of the huge potential and flexibility offered by laser-based manufacturing. Despite the fact that the specific solutions may look quite different, the current needs for further research and development can be summarized by three main fields of research and generally valid requirements:

- **Universal system hardware** for the adaptation and combination of different manufacturing processes using high average laser powers and even higher peak powers in the case of pulsed laser processing to increase the productivity with scalable processing spaces from dm^3 to m^3, constant process quality, and precision in the sub-μm range.
- **Universal control architecture** with increased scalability and configurability for handling high-frequency acquisition of measured variables, the derivation of corresponding states and universal process control instead of empirically set control parameters.
- **Universal autonomy**, hence the ability of the machine to autonomously identify an optimum production sequence, processing parameters and needed control parameters solely from the product description with AI-supported machine learning abilities.

In principle, the system hardware for laser materials processing only needs a comparatively low adaptation from process to process as in most of the cases the latter can mainly be reduced to an adaptation of the processing parameters such as laser power or focusing conditions and kinematic parameters like the feed rate, see section 2. This can be illustrated by the comparison of two completely different laser processes e.g. ablation cutting of carbon-fiber reinforced plastic by an ultrafast laser with a pulse duration of a few picoseconds [13] and laser welding of metals with a cw-laser [14], which can both be implemented as a remote process with the same processing head to just move the laser beam without any kind of additional process gas or process material such as filler wire or metal powder. The needed adjustment of the feed rate which in this case needs to cover more than one order of magnitude from about 1 m/s for laser welding to 30 m/s for ablation cutting is easily possible with state-of-the-art galvanometer scanners. However, further increasing the productivity of the ablation cutting by increasing the average laser power also requires a significant increase of the processing area in order to avoid heat accumulation [15, 16].

A first measure to reduce heat accumulation is to allow only one laser pulse to be deposited per irradiated area during one scan. Hence the feed rate has to be at least

$$v = d \cdot f, \tag{2}$$

where f is the repetition rate of the used laser. For typical applications such as ablation or ablation cutting with ultrafast lasers the beam diameter d is defined by the needed precision and the length-related power (see section 2) required for the application (for pulsed processing the other parameter F/τ is determined by the pulse duration τ).

Assuming a typical beam diameter d ranging between 10 μm and 100 μm and a repetition rate of $f = 10$ MHz, the feed rate v for processing according to Eq. (2) needs to range between 100 and 1000 m/s to avoid heat accumulation between consecutive pulses. As these velocities are far beyond of what can be achieved with present 2D scanners, new kinematics concepts and components are needed which can provide and reliably withstand extremely high accelerations while simultaneously fulfilling the high precision requirements of all laser-based processes. Since fast scanners are usually limited with respect to the processing areas, a commonly followed approach presently is to combine the scanner with slower kinematics to cover larger areas.

The high required feed rates v together with the small dimension d of the processing beam also pose a challenge for data processing. Typical control systems for laser scanning today work with a control rate of 100 kHz. For a feed rate of 1000 m/s this results in one data point every 10 mm along the processing path. To achieve a path accuracy in the order of typical beam diameters of less than 100 μm, a control rate of at least 10 MHz becomes necessary. To ensure this path accuracy with the combined kinematics (linear axes with scanners) mentioned above, it is necessary to develop new concepts for faster control architectures especially if a closed-loop control is to be included, which adds additional requirements in terms of computing power of the control system.

Although remote processing with scanning systems cover many possible laser applications, other processes require additional support material such as the process gas for conventional laser cutting, additional material for additive manufacturing, or filler wire for some welding applications which lead to additional constraints for the movement of the processing head along a defined trajectory. To really achieve a universal machine, new system concepts and methods have to be developed to integrate these processes.

In order to realize the full potential of a universal machine, a digital representation of all processes along the whole value chain is needed beyond the above-mentioned challenges for the hardware and the control systems. This not only includes a consistent and extensive description of the different integrated manufacturing processes, their features and interactions but also the subsequent and steadily updated integration and implementation of the growing knowledge as capabilities of the machines into application-independent software modules. This will allow for an automatic program generation, that autonomously selects the optimum manufacturing strategy and processing sequences and directly derives complementary control instances and feedback loops.

Such a seamless process integration will not only improve the overall quality by eliminating error sources such as handover points but can also improve performance and functionality of the manufactured products by actively considering and exploiting mutual influences of the different manufacturing processes and their sequences, e.g. the time-dependent change of the mechanical properties of the material during processing [17]. Supported by methods such as artificial intelligence, this may lead to a comprehensive or even a counter-intuitive reconsideration of today's standard manufacturing approaches which still are largely defined by human experts based on experience.

4 Conclusions and Outlook

The laser as a universal tool provides the opportunity to develop a versatile and efficient material processing system which can be reconfigured in real time without significant hardware changes. To fully exploit this potential, further research is mainly required with respect to faster kinematics, faster control technology, and machine learning in order to be prepared for the further expected power increase of available laser systems. For this, laser machines have to become about 100 times faster while retaining the same precision as today. With a successful advance in these three fields, the resulting laser-based universal machines have the potential to significantly affect if not even disrupt today's global value creating networks.

The aggregation and merging of manufacturing processes will shorten the value chain and simplify the supplier pyramid by decreasing the lead time, the need for material buffers and as a result the volume of both goods in process and of semi-finished goods. The system-inherent elimination of non-value-added activities such as handling and logistics or programming and setting up machines will strongly re-enhance the focus on value creation and – since those activities are mostly labor-intensive or require qualified personnel – will improve the cost structure disproportionately in high-wage countries. Summed up, this offers the chance to relocate manufacturing back closer to markets and customers, typically located in high-wage countries and by this exploit market potentials and competitive advantages such as a better time-to-market, a more dynamic response to volatile demand changes, a more resilient value creation chain.

References

1. Koren, Y., Shpitalni, M.: Design of reconfigurable manufacturing systems. J Manuf. Syst. **29**(4), 130–141 (2010)

2. Forschungsunion (forschungsunion.de) and acatech: Recommendations for implementing the strategic initiative INDUSTRIE 4.0. April 2013. https://forschungsunion.de/pdf/industrie_4_0_final_report.pdf
3. Cohen, Y. Faccio, M., Pilati, F., Yao, X.: Design and management of digital manufacturing and assembly system in the Industry 4.0 era. Int. J. Adv. Manuf. Technol. 105, 3565–3577 (2019)
4. Schuh, G., Klocke, F., Brecher, C., Schmitt, R.: Excellence in Production, p. 31–51, Apprimus, Aachen (2007)
5. Moriwaki, T.: Multi-functional machine tool. CIRP Ann. Manuf. Technol. 57, 736–749 (2008)
6. Lauwers, B., Klocke, F., Klink, A., Tekkaya, A.E., Neugebauer, R., Mcintosh, D.: Hybrid processes in manufacturing. CIRP Ann. Manuf. Technol. 63, 561–583 (2014)
7. Flynn, J.M., Shokrani, A., Newman, S.T., Dhokia, V: Hybrid additive and subtractive machine tools – Research and industrial developments. Int. J. Mach. Tools Manuf **101**, 79–101 (2016)
8. DIN 8580:2003–09, Fertigungsverfahren - Begriffe, Einteilung
9. Hügel, H., Graf, T.: Laser in der Fertigung, 3rd edn. Springer, Wiesbaden (2014)
10. Graf, T., Berger, P., Weber, R., Hügel, H., Heider, A., Stritt, P.: Analytical Expressions for the Threshold of Deep-Penetration Laser Welding. Laser Phys. Lett. **12**(5), 056002 (2015). https://doi.org/10.1088/1612-2011/12/5/056002
11. Negel, J.-P., Löscher, A., Voss, A., Bauer, D., Sutter, D., Killi, A., Abdou Ahmed, M., and Graf, T.: Ultrafast thin-disk multipass laser amplifier delivering 1.4 kW (4.7 mJ, 1030 nm) average power converted to 820 W at 515 nm and 234 W at 343 nm. Opt. Express 23 (16), 21064–21077 (August 2015). DOI:https://doi.org/10.1364/OE.23.021064.
12. Müller, M., Klenke, A., Steinkopff, A., Stark, H., Tünnermann, A., Limpert, J.: 3.5 kW coherently combined ultrafast fiber laser. Opt. Lett. 43 (24), 6037–6040 (2018). DOI: https://doi.org/10.1364/OL.43.006037.
13. Freitag, C., Wiedenmann, M., Negel, J. P., Loescher, A., Onuseit, V., Weber, R., Abdou Ahmed, M, and Graf, T.: High-quality processing of CFRP with a 1.1-kW picosecond laser. Appl. Phys. A, 119(4), 1237–1243 (2015). DOI: https://doi.org/10.1007/s00339-015-9159-3.
14. Hagenlocher, C., Fetzer, F., Weber, R., Graf, T.: Benefits of very high feed rates for laser beam welding of AlMgSi aluminum alloys. J. Laser Appl. **30**(1), 012015 (2018). https://doi.org/10.2351/1.5003795
15. Weber, R., Graf, T., Berger, P., Onuseit, V., Wiedenmann, M., Freitag, C., Feuer, A.: Heat accumulation during pulsed laser materials processing. Opt. Express **22**(9), 11312–11324 (2014). https://doi.org/10.1364/OE.22.011312
16. Weber, R., Graf, T., Freitag, C., Feuer, A., Kononenko, T., Konov, V.I.: Processing constraints resulting from heat accumulation during pulsed and repetitive laser materials processing. Opt. Express **25**(4), 3966–3979 (2017). https://doi.org/10.1364/OE.25.003966
17. Hossfeld, M.: Time-dependency of mechanical properties and component behavior after friction stir welding. Int J Adv Manuf Technol **102**, 2297–2305 (2019). https://doi.org/10.1007/s00170-019-03324-x

Advancing from Additive Manufacturing to Large-Scale Production of Face Shields During the COVID-19 Pandemic

Frieder Heieck[1(✉)], Fabian Muhs[1,2], Marlies Springmann[1,2], Nicolas Unger[1,3], and Philipp Weißgraeber[1]

[1] ARENA2036 Research Campus, Stuttgart, Germany
frieder.heieck@arena2036.de
[2] Institute of Aircraft Design, University of Stuttgart, Stuttgart, Germany
[3] German Aerospace Center (DLR), Institute of Vehicle Concepts, Stuttgart, Germany

Abstract. This work describes why additive manufacturing is a key technology for efficient design iterations and rapid production ramp-up with large-scale manufacturing technologies. Laser cutting, injection moulding and folding were used to increase the production capacity of face shields for health care workers during the COVID-19 pandemic. We applied systematic learnings from the iterative processes used for additive manufacturing to these large-scale manufacturing technologies and the respective face shield designs. In cooperation with medical experts, structural and functional design requirements of face shields were identified and are described in detail in this work. The regulatory design requirements according to EN 166 are introduced, which were considered to receive a CE certification for three of the presented designs. The employed manufacturing techniques are specified and the respective implications on the design solutions are discussed. The paper concludes with a summary of the production initiative at the research campus ARENA2036 with a total output of over 13 000 face shields from April to June 2020, which were distributed internationally.

1 Introduction

Face shields are part of the personal protective equipment (PPE) of medical staff. The World Health Organization (WHO) Guidelines for PPE [18] recommend face shields for health care workers providing direct care to COVID-19 patients. Typically, a face shield consists of a transparent front visor covering the complete length and sides of the face, a frame to fixate the visor and an elastic band as head fixation.

P. Weißgraeber et al. (Eds.): *Advances in Automotive Production Technology – Theory and Application,* ARENA2036, pp. 394–404, 2021. https://doi.org/10.1007/978-3-662-62962-8_46

Especially in case of airborne transmission via droplets, face shields can reduce the inhalational exposure of aerosol particles by 96% in the period immediately after a cough [7]. However, smaller aerosol-sized particles can remain in the air, flow around the face shield and get dispersed widely. Thus, face shields should not be used as a substitute for respiratory protection but only as additional equipment to reduce the risk of droplet infection [12].

With the outbreak of COVID-19, governments immediately reacted in an attempt to mitigate the spread of the disease, also interrupting global supply chains and halting exports [9]. This led to numerous decentralized so-called *community level* manufacturing efforts, especially in the field of PPE, which were accelerated by the use of additive manufacturing (AM) ([8], [16], [13]). AM, providing a low-invest method for rapidly producing parts and digitally sharing designs among a global and open source network, proved to be a valuable manufacturing technique at the beginning of the pandemic to boost decentralized production.

The findings presented in this paper are based on a face shield production initiative, which mainly took place from April to June 2020, by a team of researchers and students from ARENA2036, the University of Stuttgart, the German Aerospace Center and numerous industry partners. In total, over 13 000 face shields were produced and donated to hospitals, doctor's offices, emergency services and care facilities in Germany and abroad. Additive Manufacturing was complemented with large scale production equipment with the intensive support of industry partners such as TRUMPF GmbH & Co. KG (laser cutting) and Ensinger GmbH (injection moulding).

2 Manufacturing and Design Approach

The "COVID-19 Technical Specifications for Personal Protective Equipment" by the WHO state that face shields should be made of clear plastic, offer good visibility, have an adjustable band to attach firmly around the head and completely cover the sides and length of the face [17]. The applicable test standard for face shields is given in the European norm for personal eye protection EN166 EN2002 [1]. Based on the feedback of medical staff, the wearing comfort of face shields is another important feature, which can be improved by lowering the overall weight and avoiding pressure marks by ensuring a high bending flexibility of the frame component. In addition, a quickly adjustable fixation system as well as a precise adaptation to different head shapes are relevant requirements of medical staff. The possibility to disinfect all components guarantees a safe application. Due to the planning of a worldwide distribution, high durability and reparability are also required.

In order to meet these requirements, first prototypes were manufactured with additive manufacturing (AM) and handed over to medical staff to receive feedback and iteratively improve the design. As one of the major design adoptions,

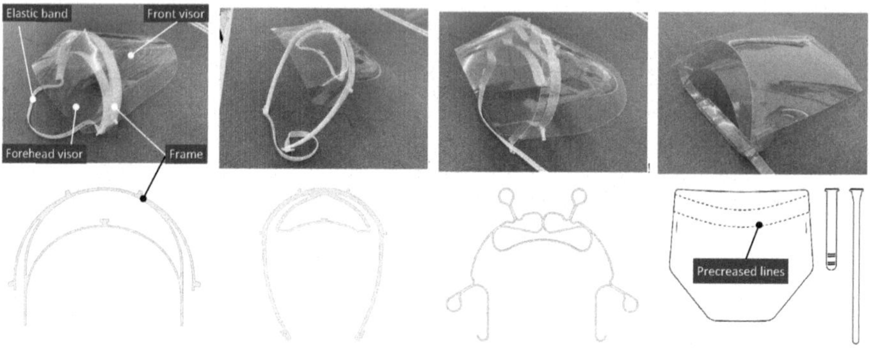

(a)FFF/Injection Moulding (b) Laser Cutting (c) SEAM (d)Precreasing (Foldable Design)

Fig. 1. Overview of applied face shield (frame) designs

an additional forehead visor was added to close the gap between forehead and front visor (see Fig. 1).

Starting with Fused Filament Fabrication (FFF) as a low-cost method, the optimized designs were adapted for the production of frame elements with more efficient manufacturing technologies, such as Screw Extrusion Additive Manufacturing (SEAM) and laser cutting (LC), significantly increasing the production volume to several hundred parts per day. While SEAM and LC do not require tools either, the optimized designs were transferred simultaneously to Injection Moulding (IM) experts in order to prepare moulds for the production of frame components with high production rates. Eventually, foldable face shield designs (FO) were developed, requiring only a single type of material (a transparent sheet) with precreased folding lines and providing a highly scalable production method. An overview of the realized designs is presented in Fig. 1.

3 Materials

In the presented work, PETG, PC, POM, PVC and PLA are used for both frames and visors as shown in Table 1 depending on their availability as well as their optical and mechanical properties. Due to the extreme demand for PETG and PC and subsequent market shortage at the time of production, a substitute front visor material had to be found. Thus, sheets from PLA were produced at the Institut für Kunststofftechnik (IKT, University of Stuttgart) throughout the initiative. In case of the elastic back fixation, a silicone rubber is used due to its high elongation at break and disinfectability. In case of front visors, a thickness of 0.3–0.6 mm has been found appropriate in terms of stiffness and weight, while in case of the forehead visor, a thinner material of around 0.15–0.2 mm was found to be sufficient due to its smaller size. It should be taken into account that a lightweight design with thin materials can increase wearing comfort and decrease

Table 1. Applied face shield materials for frames and front visors.

	FFF	SEAM	InjectionMoulding	LaserCutting
Frame	PETG	PC	PP	POM-C
Front Visor	PLA, 0.5 mm	PLA, 0.5 mm	PVC, 0.25 mm	PVC, 0.3 mm (DIN A4)

shipping costs, while thicker materials increase stiffness and robustness of visor elements. The forehead visor of all designs was made from 0.175 mm PET sheets and in case of the elastic head fixation, a silicone band with a thickness of 1.5 mm was used. Foldable designs were manufactured with precreasing technology from a PET sheet with a thickness of 0.5 mm as solitary material for the whole face shield, including the fixation band.

4 Manufacturing Technologies and Designs

In this section, five different manufacturing technologies for face shields as well as their respective designs are described.

In case of the FFF, IM, LC and SEAM design, front visors and elastic bands were produced on a plotting machine (Aristomat TL 1625-8) at the Institute of Aircraft Design (IFB, University of Stuttgart). Forehead visors were cut with a lab-scale laser cutter (Speedy400 flexx, trotec), which was possible due to the low material thickness and smaller size.

4.1 Fused Filament Fabrication (FFF)

Fused Filament Fabrication (FFF), also known under the trademark Fused Deposition Modelling (FDM), is one of the most widely used techniques in additive manufacturing [3], especially in small-scale applications. FFF is an extrusion technology which processes thermoplastic materials fed through a nozzle, thus generating line segments on a platform and building up parts layer-by-layer. Due to their process-induced microstructure, FFF parts demonstrate lower mechanical properties compared to injection moulded parts using the same type of material [2].

For the presented work, several Prusa i3 MK3 systems have been used with a max. build volume of $25 \times 21 \times 21\,cm^3$ for the production of frames. The overall production time per frame could be reduced to 1.5 hours by increasing the extrusion temperature and speed, with an overall part height of 20 mm and layer height of 0.3 mm. In order to reduce machine downtime, at least two headpieces were simultaneously manufactured on the build platform. The main advantage of using FFF is the ability to rapidly improve designs by several iterations per day and to accumulate distributed work force with available printers.

The FFF face shield design is based on the original Prusa RC3 Face Shield design [11] and has been improved based on feedback received from medical staff.

Therefore, prototypes were handed over to clinics and analyzed with regards to their wearing comfort, protection potential and disinfectability. The final FFF face shield design features an additional forehead visor and a front visor with smaller height and larger width to facilitate head movement and improve eye protection as well as additional fixation slots for different types of elastic bands. The FFF design was certified as PPE according to EN 166.

4.2 Screw Extrusion Additive Manufacturing (SEAM)

Screw Extrusion Additive Manufacturing (SEAM) or Fused Granular Fabrication (FGF), like FFF, is an extrusion-based AM process, which processes thermoplastic materials. The raw material is fed in pellet form and plasticized with an open single-screw extruder, which means the plastification is achieved with a combination of shearing and heat transfer as opposed to only heat transfer in FFF. This, as well as the vast selection of materials available in standard pellet form, allows the processing of a wide range of filled or unfilled thermoplastic materials, such as PA6, PP, TPE, PMMA or PC. In addition, the process allows significantly higher plastification rates (up to 28 times that of FFF), leading to increased productivity [14]. For the presented work, a Yizumi SpaceA-2P-2000-500 H2 hybrid manufacturing system operated by the DLR Institute of Vehicle Concepts, as shown in Fig. 2, was used. It provides a build chamber of 1500 x 2000 x 1500 mm^3 with a heatable build platform of 1000 x 1000 mm^2, max. extruder temperatures of 400 °C and extrusion rates of up to 20 g/min. The positioning of the extruder inside the build space is realized via a 6-axis industrial robot allowing additional degrees of freedom compared to gantry-based systems.

To benefit most from the technology and to maximize productivity, the original FFF face shield frame design was completely redesigned for the SEAM process (as shown in Fig. 1). Since the extruder is best utilized when it is continuously extruding, the design was changed to be one single continuous wall without any non-printing travel moves. For this, a specialized slicing/path-planning algorithm was developed inside the CAD Software Rhinoceros/Grasshopper. This allowed to design the part as an open shell and plan the paths in an oscillating fashion. The necessity of continuous extrusion led to four pin loops as shown in Fig. 1, which were cut off manually in order to leave only the required pins for fixating the front and forehead visors.

With this setup and an extrusion rate of around 6 g/min, a layer height of 0.5 mm, wall thickness of 1.7 mm and an overall part height of 18 mm, a production time of 5:30 min per part was achieved. To further increase productivity and decrease down time for part removal, the full build plate was utilized to build nine parts in one print job with a duration of 50 min. The used material was a medical grade Polycarbonate (Makrolon® 2458), generously provided to the cause by Covestro AG. The SEAM design was certified as PPE according to EN 166.

4.3 Laser Cutting (LC)

The main advantage of laser cutting is production speed without the need for a tool, compared to other mass production techniques such as injection moulding. The disadvantages are the restriction to 2D contours and high material scrap rates, as frames only cover a small part of the overall area of a plate, even when nested optimally. The scrap rate was calculated to around 85% for the applied frame design. However, the thermoplastic material can be recycled, which compensates the negative effect of low material utilization. Frames were produced on a 3D lasercutting machine (TruLaser Cell 7040) at TRUMPF GmbH in Ditzingen, Germany, from 6 mm thick POM-C plates with a production time of around 18 s per part. This allows for a daily production of more than 1 200 frames within an 8 h shift with one machine.

The face shields consist of laser-cut frames and standardized DIN A4 sheets for the front visor with the goal to significantly increase production capacity. First examples of lasercut frames for face shields produced in large quantities are known from Georgia Tech [4] with stated production capacities of ca. 2 000 frames per day. In order to increase wearing comfort, the Georgia Tech design has been iteratively adapted based on the previous user feedback and learnings to reduce stiffness and add a spring effect along the forehead section as well as along the visor fixation pins to facilitate assembly. In contrast to other frame designs, the laser cut version uses office supply DIN A4 sheets for the front visor which are perforated by a standard punch at the pin fixation points. This significantly reduces production efforts and enables health care workers to easily replace front visors themselves if required. However, care must be taken to choose material with sufficient transparency, in this case, highly transparent PVC films with a thickness of 0.3 mm were found to be satisfactory. The LC design was certified as PPE according to EN 166.

4.4 Injection Moulding (IM)

Face shield frames of commercial suppliers are often produced with injection moulding (IM) [6], as it allows for a high productivity and low costs when producing large amounts of parts. Injection moulding requires a closed tool, where the investment costs for the produced frame component with a single cavity including an extraction device were 26.200 EUR. Moulds for the production of face shield frames are also often designed with a multi-cavity to manufacture two frames at once. The cycle time for the produced frame was 30 seconds. With multi-cavity tools and fast aluminium moulds, typical production capacities can reach around 3 000 - 5 000 parts per day [8].

The IM frame design of this work is based on the FFF design and does not differ significantly except for minor changes to account e.g. for demouldability or suitable flow path length. The IM design was not considered for PPE approval at the time of this work.

(a) Prusa i3 MK3 (FFF)

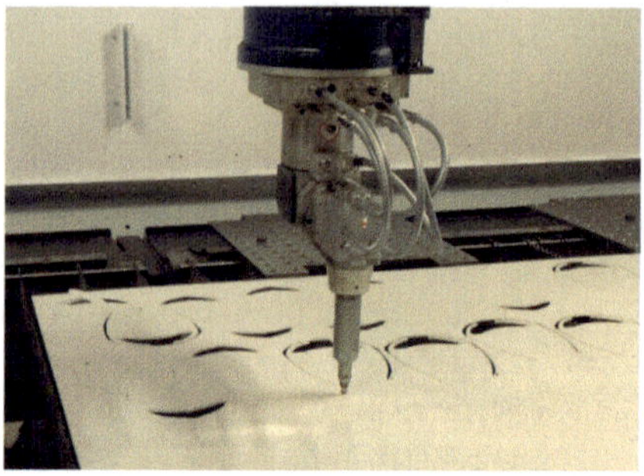

(b) TruLaser Cell 7040 (LC)

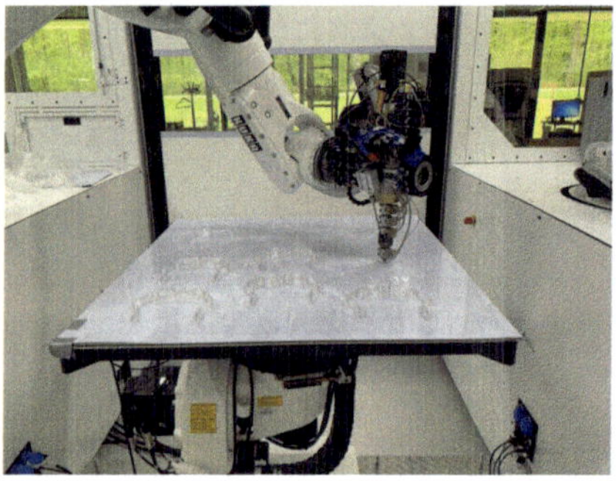

(c)Yizumi SpaceA-2P-2000-500H2 (SEAM)

(d) Aristomat TL 1625-8 (FO)

Fig. 2. Applied manufacturing technologies.

4.5 Folding (FO)

In order to further decrease production time and product costs, a foldable face shield made from a solitary material was developed for precreasing and punching technologies. Foldable face shield designs are known e.g. from the open source project "HappyShield" by the University of Cambridge [15]. This method is inspired by the so-called Origami technique, in which paper is folded to a three-dimensional structure [5]. In the first step of folding the face shield, the fold edges are applied on the flat sheet. This process is called precreasing or pressure creasing. In case of precreasing, the application of the fold edges can be done manually, by a mechanical process such as a CNC plotter or by using CNC controlled lasers. For pressure creasing, lines are embossed into the sheet with a press-driven tool. In the course of this work, a PET film was used and the fold edges were applied with a CNC plotter (Aristomat TL 1625-8). The time for precreasing a complete face shield contour including fixation bands as seen in Fig. 1 is around 20 seconds. In order to further reduce production times, a combination of punching and pressure creasing can be used [10]. Although not realized in the course of this work, an estimated production of 7 500 - 10 000 face shields per day with production times of ca. 3 s per face shield could be realized

Table 2. Comparison of manufacturing techniques for face shield frames and foldable designs. FO values are estimated based on punching and pressure creasing.

	Frames				Foldable
	FFF	SEAM	IM	LC	FO
Production Speed	ca. 1.5 h	ca. 330 s	ca. 30 s	ca. 20 s	ca. 3 s
Tool required	no	no	yes	no	yes
Tool costs [k€]	–	–	ca. 26	–	ca. 3
Material scrap rate	ca. 0%	ca. 0%	ca. 0%	ca. 84%	ca. 14%

in consultation with packaging manufacturers, where punching tool costs are in the range of 3 000 €.

When selecting the semi-finished product, care must be taken to ensure that the thickness of the material and thus its bending stiffness is not too high, otherwise folding is not possible or errors in the final design will occur. When creating the fold edges for the face shield, care must be taken to ensure that they are generated as conic sections. This is a basic requirement for the fold edges, otherwise the three-dimensional design cannot be folded flat and consequently cannot be produced by folding. Several fold edges on a single sheet have to be generated by cuts from one cone. After processing the flat semi-finished product, the folding is done manually.

The face shield design is limited by the folding kinematics, which only offer one degree of freedom. Due to the kinematic coupling, the folding process automatically transforms the 2D sheet into the three-dimensional face shield according to its designed geometry.

By choosing the same material for the fixation band, which is fed through cut out slots, only one type of material is required to produce the whole face shield. Another advantage is the design's low packaging volume, as foldable face shields can be delivered as flat, precreased products and folded manually on site, which also allows for significantly reducing shipping costs. The foldable design was not considered for PPE approval at the time of this work.

5 International Distribution of Face Shields

The pre-manufactured components were assembled and packed at the Research Campus ARENA2036 with the support of students and researchers on a voluntary basis. Special care has been taken to ensure a high level of hygiene standards during production, e.g. by maintaining a safety distance during work and wearing masks. Additionally, all face shields were quality-checked and disinfected prior to packing. In total over 13 000 face shield donations were delivered to clinics, doctor's offices, care facilities and emergency services. While around 60% went to German facilities and institutions, 40% were shipped to international partners in Italy, Greece, Egypt, Chile, Brazil and Bolivia. Regional distribution was organized by setting up online platform to allow for sending requests for a face shield donation.

6 Conclusions

The presented work shows how face shields can be rapidly improved and manufactured with additive manufacturing and how this allows for a production ramp-up with large-scale manufacturing in a matter of days. Tested by professional health care workers, the presented face shield designs were improved iteratively first. Small changes such as reducing the weight, improving visor fixation pins or adapting the shape of visor elements proved to have significant impact on being accepted by medical staff and fulfilling the official certification requirements. An additional forehead visor to close the gap on top of the face shield, which is hardly found in other designs, was set as requirement by medical staff and realized with an inclined, thin sheet element using the same fixation points as the front visor. Knowledge derived from additively manufactured face shields was transferred to designs suitable for mass-production technologies such as laser cutting, injection moulding or precreasing/punching (foldable design), which significantly reduce production time.

The presented work shows how additive manufacturing supports rapid design processes by implementing user feedback and transferring knowledge to mass-production techniques for rapid manufacturing ramp-up. This was key to quickly react and support health care workers during the COVID-19 pandemic on a large scale. By receiving approval for the presented FFF, LC and SEAM design according to EN 166, it is shown that both additive manufacturing as well as large-scale manufacturing, despite their significant implications on the design, are suitable for the production of face shields for medical use.

Acknowledgments. The face shield production initiative at the research campus ARENA2036 was only possible with the extraordinary support from researchers, students and industrial partners. The active engagement took place on a voluntary basis. We would like to thank the supporters from ARENA2036 (P. Froeschle, K. Haug), the German Aerospace Center (C. David, K. Frey, G. Kopp and D. Krüger), the Institute of Aircraft Design of Univ. of Stuttgart (J. Greiner, J. Schwingel, L. Münch and D. Sommer) as well as the Institut für Kunststofftechnik (J. Thurnherr). We would further like to thank the rectorate of the University of Stuttgart, especially Prof. W. Ressel, Prof. P. Middendorf and W. Holtkamp. On the side of the industry partners we would like to thank R. Bannmüller, R. Geissler, M. Gillert, T. Harrer and G. Ramirez (TRUMPF Laser- und Systemtechnik GmbH), A. Stehle (Ensinger GmbH), M. Wadlinger and C. Wilczek (BASF SE), K. Przybysz-Herz and R. Kusumoto (NAiSE GmbH) as well as Y. Klett (Foldcore GmbH). Finally, we would like to honor the commitment of our student team, which has made a decisive contribution to the initiative: J. Baireuther, M. Bilegee, S. Durek, J. Gremmelspacher, D. Hesse, S. M. Hönes, V. Hofmann, J. Hommel, P. Junger, I. Tien Khaw, A.-M. Kostelecky, V. Krein, S. Kurz, M. Lintzeri,

B. Maliczek, C. Osso, E. Salanitro, J. Seiffert, S. Strohmaier, D. Thon, S. Uecker, S. Wach, E. Wegener, N. Wölki, F. Wolff and F. Zeisberger.

References

1. Personal eye-protection – Specifications. Standard DIN EN 166-1:2002-04, Deutsches Institut für Normung e. V., Berlin, Germany, Apr. (2002)
2. Ahn, S.-H., Montero, M., Odell, D., Roundy, S., Wright, P.K.: Anisotropic material properties of fused deposition modeling abs. Rapid prototyping journal (2002)
3. Gebhardt, A., Kessler, J., Thurn, L.: 3D printing: understanding additive manufacturing. Hanser (2018)
4. Georgia Tech. Georgia tech community covid-19 response. https://sites.gatech.edu/rapid-response/face-shields (2020). Accessed 30. Nov 2020
5. Klett, Y.: Auslegung multifunktionaler isometrischer Faltstrukturen für den technischen Einsatz. Verlag Dr, Hut (2013)
6. Kunkel, M.E., Vasques, M.T., Perfeito, J.A.J., Zambrana, N.R.M., dos Santos Bina, T., de Melo Passoni, L.H., Ribeiro, T.V., Rodrigues, S.M.S., de Castro, R.O.M., Ota, N.H.: Mass-production and distribution of medical face shields using additive manufacturing and injection molding process for healthcare system support during covid-19 pandemic in brazil (2020)
7. Lindsley, W.G., Noti, J.D., Blachere, F.M., Szalajda, J.V., Beezhold, D.H.: Efficacy of face shields against cough aerosol droplets from a cough simulator. J. Occup. Environ. Hyg. **11**(8), 509–518 (2014). PMID: 24467190
8. Lück, T., Seifarth, C., Malauschek, P., Schendel, O., Nopper, H.: Ramping-up 3d-printed corona devices – additive manufacturing opposing injection moulding. Transactions on Additive Manufacturing Meets Medicine 2 (2020)
9. Mostaghimi, A., Antonini, M.-J., Plana, D., Anderson, P.D., Beller, B., Boyer, E.W., Fannin, A., Freake, J., Oakley, R., Sinha, M.S., Smith, L., Van, C., Yang, H., Sorger, P.K., LeBoeuf, N.R., Yu, S.H.: Regulatory and safety considerations in deploying a locally fabricated, reusable face shield in a hospital responding to the covid-19 pandemic. Med (2020)
10. Oh, W.-Y. (2004). Punch capable of punching and embossing simultaneously. https://patents.google.com/patent/US20050211041A1/en. Accessed: 30 Nov 2020.
11. Prusa Research. 3d printed face shields for medics and professionals. https://www.prusa3d.com/covid19/ (2020). Accessed 30 Nov 2020
12. Roberge, R.J.: Face shields for infection control: A review. J. Appl. Occup. Environ. Hyg. **13**(4):235–242 (2016)
13. Sinha, M.S., Bourgeois, F.T., Sorger, P.K.: Personal protective equipment for covid-19: Distributed fabrication and additive manufacturing. Am. J. Public Health **110**(8), 1162–1164 (2020)
14. Stier, T., Heinz, C., Lammert, N., Unger, N.: Additive manufacturing based on screw extrusion: machines, material and applications in the automotive industry. VDI-Congress Plastics in Automotive Engineering (PIAE) (2020)
15. University of Cambridge. Happyshield - curved crease origami face shield for infection control. https://happyshield.github.io/en/ (2020). Accessed 30 Nov 2020
16. Westphal, E., Maul, R., Dreier, T., Seitz, H.: 3d printing of frames for anti-coronavirus face shields using different processes and materials. Transactions on Additive Manufacturing Meets Medicine, 2 (2020)

17. World Health Organization. Covid-19 technical specifications for personal protective equipment and related ipc supplies. Technical report, 20 Avenue Appia, 1211 Geneva 27, Switzerland (2020)
18. World Health Organization. Rational use of personal protective equipment (ppe) for coronavirus disease (covid-19): interim guidance. Technical report, 20 Avenue Appia, 1211 Geneva 27, Switzerland (2020)

Part D New Concepts for Autonomous, Collaborative Intralogistics

Towards an Artificial Perception Framework for Autonomous Robots in Logistics

Christopher Mayershofer(✉) and Johannes Fottner

Technical University of Munich (TUM), Garching, Germany
christopher.mayershofer@tum.de, j.fottner@tum.de

Abstract. Autonomous robots in logistics are a promising approach towards a fully automated material flow. In order to use their full potential however, they must be able to extract semantic information from logistics environments. In contrast to other application areas of autonomous robots (e.g. autonomous driving, service robotics) the logistics domain lacks a common dataset and benchmark suite covering multiple sensor modalities and perception tasks. This paper conceptualizes a framework for artificial perception research in logistics that aims to close this gap in a sustainable, data-driven way. Our framework consists of three components: (1) A foundation, based on logistics-specific standards, concepts and requirements. (2) An open dataset, covering multiple sensor modalities and perception tasks and (3) a standardized benchmark suite. As shown in other research areas, a common and open platform for data-driven research facilitates novel developments and makes results comparable and traceable over time.

1 Introduction

Increasing demand, diversification of product variants and the lack of skilled workers in industrialized nations are reasons for a cross-sectoral change in logistics [1]. New technologies and innovations not only address these challenges, but can also generate a sustainable competitive advantage. Especially development in the field of robotics is increasingly applied in logistics and is thus considered a key technology towards a fully automated and flexible material flow [2]. These robots are gradually transforming from rigid automatons to intelligent machines. When robots found their way into production facilities of automotive manufacturers in the early 1960s, they were only capable of performing repetitive tasks by previously programming a fixed sequence of process steps. However, dynamic and less structured applications, as often found in logistics, require more sophisticated approaches in order to safely, robustly and

This work was supported by BMW Group

P. Weißgraeber et al. (Eds.): *Advances in Automotive Production Technology – Theory and Application,* ARENA2036, pp. 407–415, 2021. https://doi.org/10.1007/978-3-662-62962-8_47

autonomously execute complex process chains [1]. Hence, a shift towards partly intelligent machines is observable that mimic the human behavior with respect to adaptivity, flexibility and robustness. Perception is one enabling technology for these systems and describes the process of semantically understanding the outside world similar to how we humans do. In contrast to other application areas of autonomous systems (e.g. autonomous driving, service robotics) the logistics domain lacks a common dataset and benchmark suite covering multiple sensor modalities and perception tasks. We therefore propose a holistic, open, data-driven framework to support artificial perception research for autonomous robots in logistics. This paper describes our framework concept aiming to enable future research in the area of artificial perception within the application field of logistics. It focuses on delivering a solid foundation based on standards and concepts, multi-modal and multi-task datasets as well as standardized benchmarks.

2 Motivation

2.1 Robotics in Logistics

Logistics is a multi-trillion a year worldwide industry. As mentioned above, increasing demand, diversification of product variants and the lack of skilled workers are examples for cross-sectoral challenges which future logistics attempts to solve [1]. In e-commerce, for example, a worldwide increase in turnover of almost 20 percent was forecast for 2020.[1] In the manufacturing industry, on the other hand, the steady increase in product variants is also a contributing factor that poses challenges for logistics operations. A vivid example of this is the German automotive industry, where a model increase of almost 50% from 2007 to 2019 was predicted.[2] This contrasts with the shortage of skilled workers in logistics. In addition to demographic change (especially in industrialized countries), this is further supported by forecasts of an increase in the level of education [3]. In its labor market forecast, the German federal government reports an expected decline in the total number of employees by 1.4 million in 2030 compared to 2010.[3]

These challenges are countered by new technologies and innovations that not only solve them, but also generate a competitive advantage. Especially developments in the field of robotics are increasingly applied in logistics and are regarded a key technology for achieving the vision of a fully automated material flow. These automated systems stand out among other things due to scalability [1], independence from the labour market [4] and cost-efficiency [1]. However,

[1] https://www.statista.com/statistics/379046/worldwide-retail-e-commerce-sales/.

[2] https://www.pwc-wissen.de/pwc/de/shop/publikationen/How+to+stay+No+1!/\penalty-\@M?card=13052.

[3] https://www.bmas.de/SharedDocs/Downloads/DE/PDF-Publikationen/a756-arbeitsmarktprognose-2030.html.

today's systems are still far from the vision of flexible, fully automated material flow. The reasons for this include the lack of robustness and flexibility of the systems in the complex industrial environment as well as the high variance of logistics processes [2]. In practice, these problems are currently being countered with humans in key positions who have a general, non-task-specific flexibility [2].

Meanwhile, the logistics industry is perceiving mentioned problems as novel requirements for logistics systems of the future. In particular, there is a need to automate complex, difficult-to-automate processes in a sustainable manner using flexible, adaptive, cost-efficient and yet robust systems.

2.2 Towards Intelligent Machines

In the future, these systems are expected to achieve more and more human-like capabilities, including flexibility, adaptivity and rationality, through techniques of artificial intelligence and advanced robotics. Broadly speaking, the field of artificial intelligence is concerned with understanding, formulating, testing and ultimately artificially replicating the foundations of human intelligence. The term Intelligent Agent (IA) accumulates the ideas of a human-like, artificial intelligence into an abstract capsule, and hence formulates the ultimate goal of intelligent robotics. Similar to humans, an IA perceives the environment, draws conclusions and makes decisions before finally acting in order to optimally achieve a certain goal [5]. The underlying infrastructure for information acquisition, processing, reasoning and (re)action is called cognitive architecture [6]. In order to be able to reason about what action to take next, information about the outside world needs to be acquired first. Therefore, the ability to observe and understand the environment is a key technology for IAs [5].

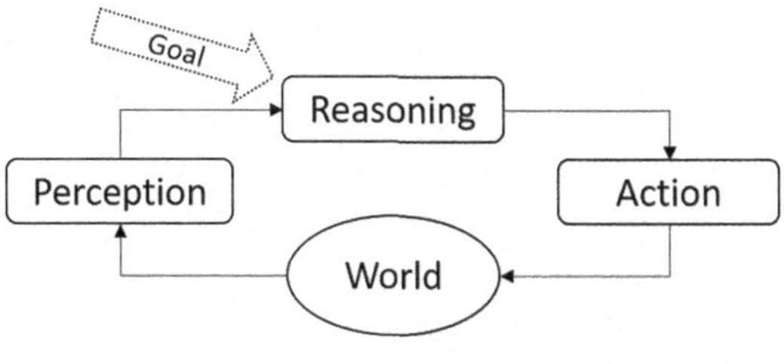

Fig. 1. Model of an intelligent agent from the perspective of *Grounded Cognition*' [7]

Based on the same human capability to autonomously acquire information from the environment, this process is referred to as *perception*. Psychology describes human perception as a means of forming an image of the physical outside world which is appropriate for our actions. This image is built upon our everyday intuitions and is acquired by our senses [8]. Perception thus enables a semantic understanding of the outside world and arranges information gained into an internal representation. Next, we briefly summarize state-of-the-art perception research related to this work.

3 Related Work

3.1 Perception for Autonomous Driving

Especially in the field of autonomous driving the necessity of artificial perception is confirmed in literature [9]. Several different sensor modalities are applied here. Besides low to high range radars, cameras, sonars as well as costly multi-line lidars are used. From the acquired multi-modal raw data semantic information is extracted next. Starting with classical object detection (e.g. vehicles and people [10]) over semantic segmentation (e.g. path or free-space[4]) up to the analysis of temporal relationships (e.g. tracking [11]) various methods are applied. The extracted information is subsequently used to derive situation-specific decisions [9]. These often data-driven methods are developed using publicly available datasets and evaluated against their benchmarks. Geiger et al. presented the KITTI Vision Benchmark Suite [12] in 2012, which has been continuously expanded over the years [13,14]. In addition, further datasets and benchmarks were published covering novel challenges, such as [15,16].

3.2 Perception for Service Robotics

Service robots are to provide services for humans in the future, such as care taking or the preparation of meals [17]. Correspondingly, service robots also need information from the environment in order to carry out autonomous actions to achieve their goal. Here, too, various exteroceptive sensor modalities are installed, including cameras, microphones and single-line laser scanners. Again, application-specific information is extracted from raw data using methods for object detection (e.g. kitchen utensils [18] or household items [19]) and recognition of object properties and states (e.g. cracked egg [20]). Again, data sets and benchmarks are available to track progress of the research community and make results comparable. Similar to KITTI in the field of autonomous driving, OpenLORIS comprises two tasks and various sensor modalities. In contrast to the comparatively limited object classes to be expected in the field of autonomous driving, service robots have to understand and handle a much higher object varience. Correspondingly, data sets were presented that cover a large number of the objects to be recognized, for example [18,19].

3.3 Perception for Industrial and Logistics Robotics

Industry is also striving for robots that perceive the environment in a human-like manner. Again, different sensor modalities are used to acquire raw data from the environment. Among others, (depth) cameras, lidar and tactile bumpers as well as sonars are used. Again, methods are used to locate and classify objects (e.g. people [21], load carriers [22]) or to predict information about an object (e.g. 6DoF gripping point [23], pallet pickup pose [24]). Building on this high-level information, robot systems are supposed to make smart decisions and act

[4] https://developer.nvidia.com/drive/drive-perception.

human-alike. In contrast to the application areas mentioned above, only a few, very specialized data sets are available in the logistics sector. So far, no dataset and benchmark suite is known in the literature. Reference [25] provides a lidar dataset for pallet recognition. Reference [23] provides data(on request) on 71 supermarket objects. Given the absence of necessary datasets, [22,26,27] describe the evaluation on custom recorded data. Thus a valid comparison is not possible.

4 Problem Formulation and Objective

We aim to enable high level robotic perception within logistics environments to overcome the above detailed challenges. To do so, we first formulate the key challenges, before introducing our framework concept.

Artificial perception is widely researched in the context of autonomous systems. These methods are applied in different types of systems such as autonomous vehicles, as well as service and logistics robots. In particular within the field of logistics, however, only isolated functions (e.g. object detection) on individual sensor modalities (e.g. camera, lidar) have been investigated so far. To the best of our knowledge, a multi-modal, multi-task perception system that mimics human perception has not yet been proposed.

Such a multi-modal, multi-task perception must be learned by the system from experience. While "hard-coded" perceptual skills are feasible for standardized and restricted scenarios under given developmental efforts, it is to be assumed that generic perception has to be (primarily) learned from experience. These experiences must be made available to the system in the form of data. In contrast to other fields of application, multi-modal, multi-task datasets dealing with object recognition and scene understanding in logistics are not available.

Finally, research results in the context of artificial perception in logistics must be comparable and make progress traceable. In order to achieve the goal of artificial perception, standardized and thus comparable tests are required. These tests enable the systematic analysis of research results and allow conclusions to be drawn about research progress over time. Although methods developed in logistics research have been evaluated, they rarely allow direct comparison due to different test setups (e.g. test design, sensor technology, environmental conditions).

5 Artificial Perception Framework for Autonomous Robots in Logistics

After formulating the problem, this section presents our concept towards a holistic artificial perception framework for autonomous robots in logistics. Our concept is a framework aiming to enable future research focusing on various artificial perception capabilities within the application field of logistics. It aims as a ground work, delivering three essential building blocks:

1) Foundation. Firstly, requirements for an Artificial Perception System (APS) need to be analyzed, defined and documented. The framework is based on a thorough consideration of all requirements, analyzed using information sources ranging from theoretical to practical ones. We differentiate between hardware, software and process requirements. Available information such as literature and international standards, logistics processes as well as market research data get processed in order to form a baseline of requirements which need to be considered when developing APS for logistics. This building block tries to find answers to questions such as, "Which semantic information is useful?", "Which sensors can be used to acquire it?", "How can this information be modeled?" and derive perception features suitable for autonomous systems in logistics.

2) Open dataset. Secondly, an open dataset is conceptualized, developed and compiled. The dataset allows the logistics community to get started with and develop novel, application-specific solutions towards artificial perception within logistics. The dataset will consist of data captured in different logistics environments (e.g. automotive industry, distribution centers, wholesale) using multiple sensor modalities (e.g. cameras, lidar) covering various perception capabilities. In addition, we encourage third parties to contribute high quality annotations and datasets. The dataset is composed of several subsets. Subsets offer the possibility to model various perception capabilities required for different use cases and to extend the framework with additional annotations as well as data in the future. We differentiate the subsets by sensor modality used, origin of the data and the tasks annotated. Examples for different sensor modalities used in logistics are color and depth cameras, 2d laser rangefinders and sonars. The data can originate from real logistics environments (natural), simulations and virtual environments (artificial) or from the Internet (online). Tasks to be annotated include object detection, depth estimation, semantic and panoptic segmentation or gripping point estimation. Table 1 lists possible subset candidates.

3) Benchmark suite. The third and final component of our framework is a standardized benchmark suite. This allows drawing comparable conclusions from defined benchmarks in order to make research progress traceable over time. The benchmark suite is envisioned to contain both, a decisive definition of performance indicators on the one hand and standardized test scenarios on the other hand. The benchmarks are based on the previously introduced dataset. They define the development process (e.g. definition of the training data to be used) as well as the evaluation process (e.g. definition of the test set) and thus enable a comparison of different models and methods. This allows, for instance, to test and analyze object detection models for the logistics domain in a standardized manner. Moreover, splitting the dataset into subsets with different properties (i.e. modality, data source and annotated tasks) allows designing benchmarks with increasing levels of complexity. This, for example, allows benchmarking the transfer of synthetically trained models into real applications. Evaluation metrics to be used are derived from relevant literature and extended if necessary.

The flexible design of individual benchmarks enables the research community to jointly and transparently push the state-of-the-art.

Table 1. Subset candidates for the open dataset. The proposed framework provides an open dataset consisting of different subsets which we categorize by sensor modality, data source and annotated task

Sensor modality	Data source	Annotated task
Color camera	Natural	Object detection
Color camera	Online	Depth estimation
Color camera	Artificial	Panoptic segmentation
Depth camera	Natural	Object detection
Depth camera	Artificial	Panoptic segmentation
2D laser rangefinder	Natural	Object detection
2D laser rangefinder	Artificial	Panoptic segmentation

6 Conclusion and Future Work

Autonomous robots in Logistics are a promising approach towards a fully automated material flow. In order to use their full potential, they must be able to extract semantic information from logistics environments. We motivate the necessity of artificial perception for logistics and formulate a key challenge. In contrast to other application areas, logistics research has so far not dealt with creating a data-driven foundation that allows developing and comparing novel perception capabilities. In this paper, we conceptualize an artificial perception framework for autonomous robot research in logistics. The framework lays a foundation by distilling logistics-specific standards, concepts and requirements, provides an open, multi-modal, multi-task dataset and defines standardized benchmarks.

Future work will focus on creating and releasing the frameworks initial version. Accordingly, we are working on defining the requirements as well as annotating and providing initial data. As a first version, we plan on releasing lidar, camera and synthetic data data for logistics-specific 2D object detection.

References

1. Molzow-Voit, F., Quandt, M., Freitag, M., Spöttl, G.: Robotik in der Logistik. Springer Fachmedien Wiesbaden, Wiesbaden (2016)
2. Rohde, A.-K.: Robotik in der Logistik – Einsatzpotenziale, Herausforderungen und Trends. Robotik in der Logistik, pp. 23–40 (2016)
3. Allmendinger, J., Ebner, C.: Arbeitsmarkt und demografischer wandel. Zeitschrift für Arbeits- und Organisationspsychologie, **50**(4): 227–239
4. Hahn-Woernle, C.: Neue Anforderungen für die Logistik des 21. Jahrhunderts. Internet der Dinge in der Intralogistik, pp. 9–13 (2010)
5. Russell, S.J., Norvig, P.: Artificial intelligence: A modern approach (2016)

6. Langley, P., Laird, J.E., Rogers, S.: Cognitive architectures: Research issues and challenges. Cognitive Systems Research **10**(2), 141–160 (2009)
7. Barsalou, L.W.: Grounded cognition. Annu. Rev. Psychol. **59**, 617–645 (2008)
8. Mausfeld, R.: Wahrnehmungspsychologie. Psychologie. Eine Einführung in ihre Grundlagen und Anwendungsfelder (2010)
9. van Brummelen, J., O'Brien, M., Gruyer, D., Najjaran, H.: Autonomous vehicle perception: The technology of today and tomorrow. Transp. Res. Part C. Emerg. Technol. **89**, 384–406 (2018)
10. Du, X., Ang, M.H., Karaman, S., Rus, D.: A general pipeline for 3d detection of vehicles. 2018 IEEE International Conference on Robotics and Automation (ICRA)
11. Voigtlaender, P., Krause, M., Osep, A., Luiten, J., Sekar, B.B.G., Geiger, A., Leibe, B.: MOTS: Multi-object tracking and segmentation. IEEE/CVF Conference on Computer Vision and Pattern Recognition (CVPR)
12. Geiger, A., Lenz, P., Urtasun, R.: Are we ready for autonomous driving? the kitti vision benchmark suite. Conference on Computer Vision and Pattern Recognition (CVPR) (2012)
13. Fritsch, J., Kuehnl, T., Geiger, A.: A new performance measure and evaluation benchmark for road detection algorithms: International Conference on Intelligent Transportation Systems (ITSC) (2013)
14. Menze, M., Geiger, A.: Object scene flow for autonomous vehicles. Conference on Computer Vision and Pattern Recognition (CVPR) (2015)
15. Sun, P., Kretzschmar, H. et al.: Scalability in perception for autonomous driving: Waymo open dataset. Proc. of the IEEE/CVF International Conf. on Computer Vision and Pattern Recognition (CVPR) (2019)
16. Patil, A., Malla, S., Gang, H., Chen, Y.: "The h3d dataset for full-surround 3d multi-object detection and tracking in crowded urban scenes," in. International Conference on Robotics and Automation (ICRA) **2019**, 9552–9557 (2019)
17. Schraft, R.D., Schmierer, G.: Service Robots. A K Peters/CRC Press 2000
18. Lin, T.-Y., Maire, M., Belongie, S., Bourdev, L., Girshick, R., Hays, J., Perona, P., Ramanan, D., Zitnick, C.L., Dollár, P.: Microsoft coco: Common objects in context. European Conference on Computer Vision (ECCV) Proceedings (2014)
19. Kasper, A., Xue, Z., Dillmann, R.: The kit object models database: An object model database for object recognition, localization and manipulation in service robotics. Int. J. Robot. Res. **31**(8), 927–934 (2012)
20. Jelodar, A.B., Sun, Y.: Joint object and state recognition using language knowledge. 2019 IEEE International Conference on Image Processing (ICIP)
21. Lang, A., Fottner, J.: Konzeption eines kamerabasierten kollisionswarnsystems zur prävention von arbeitsunfällen an gabelstaplern
22. Poss, C., Irrenhauser, T., Prueglmeier, M., Goehring, D., Zoghlami, F., Salehi, V.: Perceptionbased intelligent materialhandling in industrial logistics environments. 11th International Conference on Computer and Automation Engineering (2019)
23. Bormann, R., de Brito, B.F., Lindermayr, J., Omainska, M., Patel, M.: Towards automated order picking robots for warehouses and retail. ICVS (2019)
24. Molter, B., Fottner, J.: Real-time pallet localization with 3d camera technology for forklifts in logistic environments. 2018 IEEE International Conference on Service Operations and Logistics, and Informatics, pp. 297–302
25. Mohamed, I.S., Capitanelli, A., Mastrogiovanni, F., Rovetta, S., Zaccaria, R.: A 2d laser rangefinder scans dataset of standard eur pallets. Data in Brief (2019)

26. Hoang, D., Stoyanov, T., Lilienthal, A.J.: "Object-rpe: Dense 3d reconstruction and pose estimation with convolutional neural networks for warehouse robots," in. European Conference on Mobile Robots (ECMR) **2019**, 1–6 (2019)
27. Mayershofer, C., Ge, T., Fottner, J.: Towards Fully-Synthetic Training for Industrial Applications. IEEE International Conference on Logistics, Informatics and Service Sciences (LISS2020)

Concept of a Safety-Related Sensor System for Collaboration Between Human and Automated Guided Vehicles

David Korte(✉)

Institute of Mechanical Handling and Logistics, University of Stuttgart, Stuttgart, Germany
david.korte@ift.uni-stuttgart.de

Abstract. This paper presents a concept of a safety-related sensor system that is intended to ensure the detection of people in the vicinity of automated guided vehicles (AGV). This becomes necessary due to the increased usage of these vehicles in coexistence and in collaboration with people in industrial environment. The normative requirements will also change in a way that in the future people will have to be explicitly detected by safety sensors. Until now, only very simple test specimens have been used in the current standards to represent people. In this contribution, the efforts to expand these test specimens are shown. The experimental setup of the safety-related sensor system consists of three sensors from the sector of consumer electronics and is based on the open source framework Robot Operating System (ROS).

1 Introduction

1.1 Change in Automotive Production

Today, most of automobile production is based on the principles of Henry Ford, who revolutionized automobile production with the introduction of the assembly line in 1908. Apart from better assembly line systems, mechanized tools, more ergonomic workplaces and the partial mechanization of power-intensive processes, not much has changed in this principle for over 100 years [1]. The automotive industry in particular is subject to change: An extreme model offensive with a large number of additional derivatives within each model series, plus new drive concepts and material combinations.

This trend in product individualization is leading to an almost unmanageable product and process variance and thus to significantly smaller batch sizes. In combination with volatile demand, production must adapt more frequently, react to constantly changing product mixes in the production line and constantly manage new productions starts. Conventional production systems with fixed cycles and rigid processes can only meet this demand for flexibility and changeability to a limited extent.

Due to the rigid interlining of the individual production stations, the classic assembly line is very susceptible to faults. The failure of one station leads to an obstruction of the entire assembly system.

P. Weißgraeber et al. (Eds.): *Advances in Automotive Production Technology – Theory and Application*, ARENA2036, pp. 416–423, 2021. https://doi.org/10.1007/978-3-662-62962-8_48

Another disadvantage due to the rigid assembly line, but also due to the optimized production supply, is the very limited or not realizable resequencing of the vehicles to be produced. Damaged, incorrect or missing components, which are usually provided in a pre-defined sequence, would require the vehicle to be produced and the components assigned to this vehicle to be removed. However, since this is usually not possible, the vehicles remain on the assembly line and are further assembled. Reworking, which often leads to manual dismantling, is then carried out afterwards [2].

Both in the field of research and in practice, it has been shown that these challenges can be overcome with the help of new production concepts and the usage of AGVs.

Examples are the matrix production or the fluid production. In the matrix production, workstations are flexibly connected with each other, thus enabling a combination that meets the requirements. With high variance and small quantities, this production form is superior to the classic line production [3]. In fluid production, there is also a resolution of the fixed defined locations as well as the fixed allocation of the operating resources to the workstations. In this way, it is possible to shift the work contents of the work sequence and the distribution of work in the production system. The production can thus be changed permanently [4].

The new production facilities at Lamborghini, Porsche and Mercedes-Benz place particular emphasis on the use of AGVs. The classic assembly lines have been replaced by AGVs, thus achieving great flexibility and changeability. Since AGVs are also used in the area of material provision, the number of vehicles in production is increasing rapidly. Due to the more dynamic design of the environments, the zones in which people and AGVs used to move separately will merge. And despite a high degree of automation, people will further play an important role in future production.

Sensors previously used on AGVs to safeguard their environment will no longer meet these requirements. Usually, these sensors can only detect objects in one plane and are not able to differentiate between objects and people [5].

1.2 Normative Requirements

The requirements for the safety of AGVs are defined in the currently valid DIN EN 1525. The follow-up standard DIN EN ISO 3691–4 has so far only been published as a draft and is to replace the previous standard from 1997. For the North American market the ANSI/ITSDF B56.5 defines the safety requirements for AGVs [6, 7].

DIN EN 1525 defines that AGVs must be equipped with sensors that can detect objects that might be people in the guideway. The sensors must be able to monitor the entire width of the AGV, including the load, and must generate a stop signal so that the vehicle comes to a halt before it collides with a person followed by serious consequences. The standard defines two cylindrical horizontal and vertical test specimens (shown in Fig. 1) that represent parts of the human body and must be detected by the sensors. In addition, the standard defines that the architecture of the sensors used needs to comply with category 3 of the European DIN EN ISO 13849-1 [8].

DIN EN ISO 3691-4 The draft of the follow-up standard also prescribes the recognition of the two known test specimens. Instead of a category 3 architecture, a Performance Level d is now required [9].

ANSI/ITSDF B56.5. The American standard also requires the use of the two known test specimens. In addition, a third test specimen in the form of a square plate represents random obstacles [10].

However, all these standards do not take into account an adaptation of the behavior of the AGV depending on the presence of people or a differentiation in the detection of people and objects [11].

1.3 Approaches to the Further Development of the Standards

With the aim of improving the safety of AGVs by improving the test procedures itself, Hedenberg identified typical objects in the field of use of AGVs that could represent a source of danger and derived new test objects [12]. He identified obstacles, such as an object reaching into the guideway, hanging objects, barrier tapes or for example a ladder. The majority of these objects cannot be detected with the sensors used so far.

The authors Bostelman et al. aimed at further developments with regard to the detection of people. In the course of the revision of ANSI/ITSDF B56.5, the authors Bostelman et al. defined scenarios that may occur in the vicinity of people when using AGVs. In addition to the known test specimens and the scenario when people stepping into the guideway of the AGVs, the authors also wanted to enable the detection of people who are not in the close detection range of the sensor but could step into the guideway. The authors suggest that in order to be able to operate AGVs efficiently, in particular in future versatile production environments and in coexistence with people, speed of the AGVs should adapt to the environment and the respective situation [11].

Bostelman et al. also recommend in a publication from 2014 to expand the test specimens for representation of people. In their opinion, some body postures cannot be depicted with the two known test specimens alone.

Therefore, they suggest that the attitudes of people shown in Fig. 1 should also be taken into account when developing or testing the sensors for people detection.

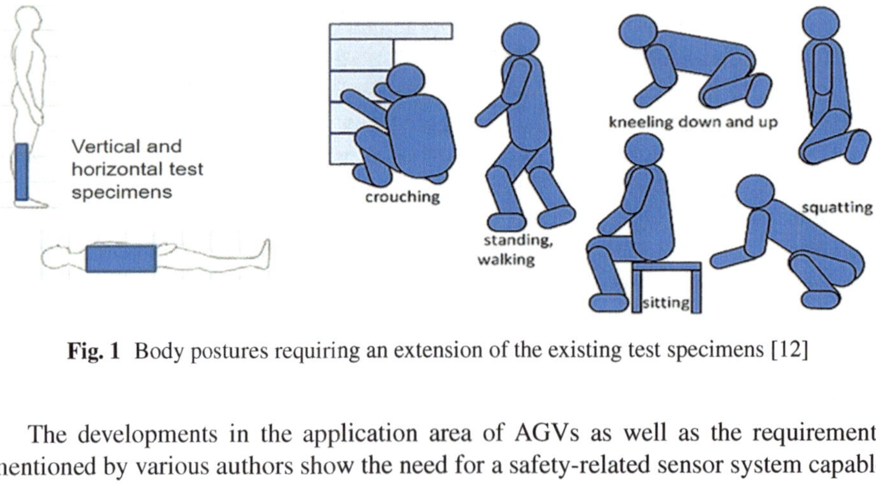

Fig. 1 Body postures requiring an extension of the existing test specimens [12]

The developments in the application area of AGVs as well as the requirements mentioned by various authors show the need for a safety-related sensor system capable of the explicit detection of people to enable a safe collaboration.

1.4 Approaches for the Detection of People

The authors Bell et al. presented in 2018 an approach for the safe detection of people based on 3D lidar sensors to be used on driverless transport vehicles in a kiwi plantation. With reference to the DIN EN ISO 13849-1, they chose a Category 3 architecture, combining two lidar sensors to be able to safely detect people. The people to be detected should wear reflective clothing, which is reliably detected by one of the sensors due to its high reflectivity. The second sensor measures the distance to obstacles and is supplemented by an algorithm that uses a trained neural network to detect people [13].

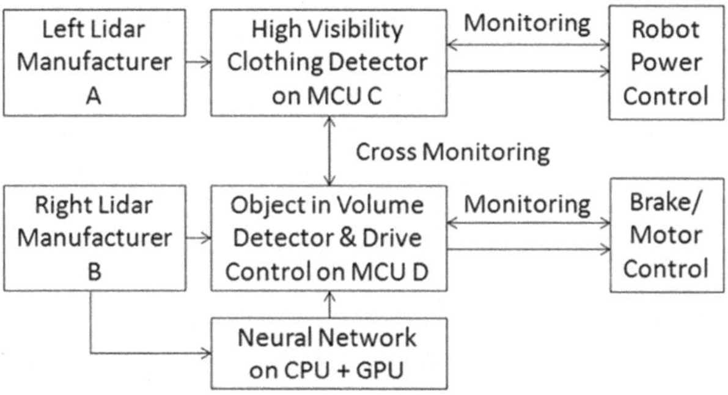

Fig. 2 Architecture of the people detection system [13]

The authors Indri et al. also presented in 2018 an approach combining a mono camera with a laser scanner to detect people in the environment of a mobile robot. They used simple and inexpensive hardware and a pre-trained neural network for people detection. With the help of the sensor combination, the respective weaknesses of the used sensors can be compensated. The demonstrator is realized with the help of the ROS framework. The authors point out that the application can be used as a supplement to existing safety sensors [14].

An approach for the algorithmic detection of people to control an AGV has been developed at the Karlsruhe Institute for Material Handling and Logistics. The AGV "FiFi" can be controlled by gestures and voice commands. The detection of people is also used, for example, for the follow mode in which the AGV follows the operator at a defined distance [15]. In this approach the safety-relevant aspects were not considered. Lang developed a system based on a time-of-flight camera and with the aided by machine learning methods for detecting people in the vicinity of forklift trucks. The system is intended to serve as an assistance system for the driver of the forklift truck and avoid collisions with people [16].

2 Concept of the Safety-Related Sensor System

In this paper the two main functions of the safety-related sensor system are the detection of people and the measurement of distances (see Fig. 3). With regard to Category 3 of the DIN EN ISO 13849-1 and in accordance with the state of the art, these two functions are to be implemented redundantly. In this approach, a combination of three sensors is used, in which one sensor can perform both functions (Time-of-Flight-camera (ToF)) and is supplemented in both functions by a further one (laser-scanner and RGB mono camera).

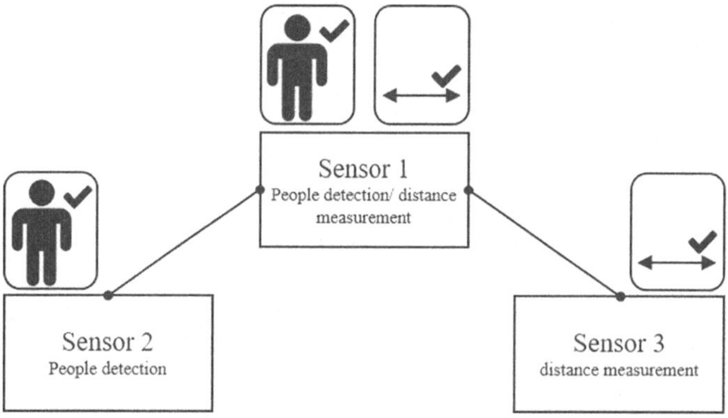

Fig. 3 Relation of the three sensors

2.1 Architecture and Sensor Types

Based on Category 3 of the DIN EN ISO 13849-1, the architecture shown in Fig. 4 was chosen to combine three sensors. The Category 3 architecture is characterized by its two-channel design, which increases reliability. For the present concept the architecture was transferred to the combination of three sensors. The sensors are chosen based on a detailed analysis of the state of the art and the results of a utility value analysis. Among other things, characteristics such as the ability to detect people, classify objects or measure distances in an area relevant to AGVs were taken into account.

The central sensor of the safety-related sensor system, a ToF-camera, can both detect people and measure distances. The distance measurement function is redundantly supplemented by a laser-scanner. An RGB mono camera was selected for redundant people detection.

The operating states of the sensors are monitored among each other, so that the overall status goes into an error state in case of the failure of one sensor.

For object and person detection the convolutional neural network MobileNet in combination with OpenCV is used in the demonstrator. The main advantage is that these methods can be used to detect people in different postures, like shown in Fig. 1, as well as people with a partial overlap.

2.2 Experimental Setup

The experimental setup for the validation of the safety-related sensor system was realized as a modular design (see Fig. 5). Standardized small load carriers were used for this

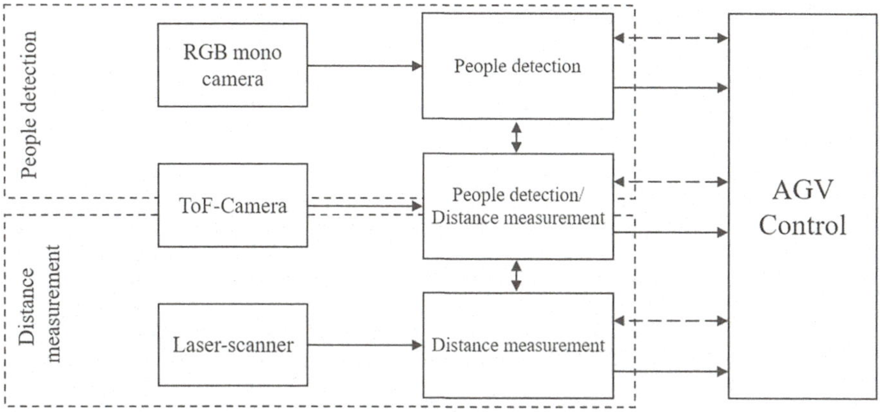

Fig. 4 Architecture of the safety-related sensor system

purpose, which are equipped with the sensors and the associated evaluation unit and networked with each other.

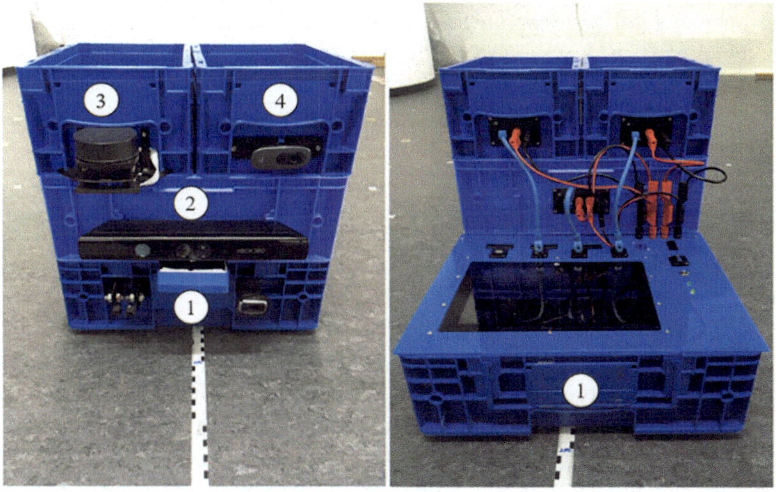

Fig. 5 Modular test setup of the safety-related sensor system

In addition to the sensors, a central basic module (1) was set up. It contains the self-sufficient power supply of the demonstrator, a central computer for recording and evaluating the data and a network switch for communication between the sensors and the central computer. Furthermore, there is an additional mono camera at the front of the basic module to document the experiments carried out. As a replacement for a professional ToF-camera, a Microsoft Kinect camera (2) was used, which has a comparable functionality. The function of the laser scanner was demonstrated using a simple RPLIDAR A1 360° 2D laser-scanner (3). The mono camera is a conventional Logitech webcam (4). The connections for the sensor units and an operating panel are located on the rear of the test setup. The framework ROS was chosen for linking the sensors and operating the demonstrator.

Due to the modular design, the defined interfaces and the use of ROS, additional or alternative sensors can easily be integrated into the experimental setup.

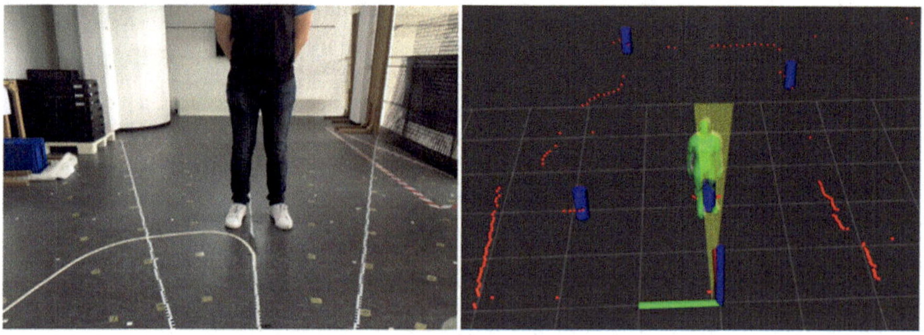

Fig. 6 Person in test field and visualization using ROS

The demonstrator was examined in stationary tests and it could be shown that the redundant and diverse architecture increased the detection probability in a defined protection field from 79.75% (ToF-camera) and 65.03% (RBG mono camera) to 93.77%.

3 Conclusion and Outlook

A concept for a safety-related sensor system was developed to adapt to the changing conditions in the use of AGVs and to enable the detection of people in their environment. By using Category 3 of the DIN EN ISO 13849-1 and the associated requirements, a redundant function of the sensors can be guaranteed. In the test setup for validation of the concept, sensors from the consumer electronic sector have been used so far. In the following experiments, the reliability of the detection of people will be investigated and evaluated. The developments in the use of artificial intelligence in the context of safety are of particular importance. Here progress is still needed to realize the concept. The selected sensor combination is well suited for integration into existing AGVs, as they are usually equipped with laser scanners and in a few cases even with ToF-cameras.

References

1. Lotter, B., Wiendahl, H.-P.: Montage in der industriellen Produktion. Springer, Berlin, Heidelberg (2012)
2. Herold, L.: Kundenorientierte Prozesssteuerung in der Automobilindustrie. Die Rolle von Logistik und Logistikcontrolling im Prozess „vom Kunden bis zum Kunden". Dt. Univ.-Verl, Wiesbaden (2005)

3. Foith-Förster, P.; Eising, J.-H.; Bauernhansl, T.: Effiziente Montagesysteme ohne Band und Takt. Sind modulare Produktionsstrukturen eine konkurrenzfähige Alternative zur abgetakteten Linie? In: wt Werkstattstechnk online 107 (2017) 3, S. 169–178. Düsseldorf: Springer-VDI-Verlag. Internet: https://www.werkstattstechnik.de
4. Fries, C., Al Assadi, A., Fechter, M., Grimm, S., Bauernhansl, T., Bozkurt, A., Schulz, R.: Zeitschriftenartikel; Fluide Fahrzeugproduktion; VDIz Ausgabe 12 (2019)
5. Korte, D., Neuschwander, B., Castro, E., Schweiker, M., et al.: Sichere 3D-Umgebungssensorik für mobile Roboter. MikroSystemTechnik Kongress 2019, 654–657 (2019)
6. Bell, J., MacDonald, B.A., Ahn, H.S., Scarfe, A.J.: An Analysis of Automated Guided Vehicle Standards to Inform the Development of Mobile Orchard Robots. IFAC-PapersOnLine 49(16), 475–480 (2016)
7. Hedenberg, K.: Obstacle detection for driverless trucks in industrial environments (Halmstad University Dissertations, 7). Halmstad University Press, Halmstad (2014)
8. DIN 1525, 12/1997: Sicherheit von Flurförderzeugen - Fahrerlose Flurförderzeuge und ihre Systeme. Beuth Verlag GmbH, Berlin
9. DIN EN ISO 3691–4 - Entwurf: Flurförderzeuge – Sicherheitstechnische Anforderungen und Verifizierung – Teil 4: Fahrerlose Flurförderzeuge und ihre Systeme. Beuth Verlag GmbH, Berlin (2018)
10. ANSI American National Standards Institute ANSI/ITSDF B56.5, 12.08.2019: Safety Standard for Driverless, Automatic Guided Industrial Vehicles and Automated Functions of Manned Industrial Vehicles. Washington: Industrial Truck Standards Development Foundation.
11. Bostelman, R., Norcross, R., Falco, J., Marvel, J.: Development of standard test methods for unmanned and manned industrial vehicles used near humans. In: Multisensor, Multisource Information Fusion: Architectures, Algorithms, and Applications 2013. SPIE Defense, Security, and Sensing. Baltimore, Maryland, USA, Monday 29 April 2013 (SPIE Proceedings). SPIE, 87560P (2013)
12. Bostelman, R., Hong, T., Eastman, R.: Safety and Performance Standard Developments for Automated Guided Vehicles. In: Gurvinder S. Virk, Krzysztof Kozlowski und M. O. Tokhi (Hg.): Mobile service robotics. 17th International Conference on Climbing and Walking Robots and the Support Technologies for Mobile Machines. Poznan, Poland, 21 – 23 July 2014. World Scientific (Firm), pp. 487–494. World Scientific Pub. Co, Hackensack (2014)
13. Bell, J., MacDonald, B.A., Seok Ahn, H.: Diversity in Pedestrian Safety for Industrial Environments Using 3D Lidar Sensors and Neural Networks. 2018 IEEE/RSJ International Conference on Intelligent Robots and Systems (IROS) 7743–7749 (2018)
14. Indri, M., Sibona, F., Cheng, P., David C.: Sensor data fusion for smart AMRs in human-shared industrial workspaces. IECON 2019 - 45th Annual Conference of the IEEE Industrial Electronics Society, Bd. 1, pp. 738–743 (2019)
15. Trenkle, A., Seibold, Z., Stoll, T., Furmans, K.: FiFi – Steuerung eines FTF durch Gesten- und Personenerkennung. Logistics Journal (2013)
16. Lang, A.: Evaluation of an Intelligent Collision Warning System for Forklift Truck Drivers in Industry. In: Vincent G. Duffy (Hg.): Digital Human Modeling. Applications in Health, Safety, Ergonomics, and Risk Management, pp. 610–622. Springer International Publishing, Cham (2018)

Novel Autonomous Guided Vehicle System for the Use in Logistics Applications

Javier Stillig(✉) and Nejila Parspour

Institute of Electrical Energy Conversion, University of Stuttgart, Stuttgart, Germany
info@iew.uni-stuttgart.de

Abstract. Providing a flexible logistics is an important contribution to retaining and expanding the convertibility of productions. It can be achieved by reducing storage capacities, whereby the goods movement through production is performed ideally in a continuous flow. In combination with the intelligent floor, the novel autonomous guided vehicle provides a cost-efficient and highly automated flow of goods for small-sized payloads up to 50 kg with short idle times. This solution is compared with a manually operated logistics and an in-plant milkrun system for a car interior parts production.

1 Introduction

The presented logistics concept consists of the two basic components: A light-weight *Autonomous Guided Vehicle* (AGV) and the *Intelligent Floor* (IF). Using a 70 m^2 large-scale demonstrator of the IF that interacts with a functional sample of the hereafter named *BoxAGV*, will highlight the expected gains for future logistics systems.

As shown in Fig. 1 the BoxAGV bases on a simple frame construction made of aluminum profiles, to which four meccanum wheels are attached. Two opposite drive motors are connected each to one inverter, which receives his commands directly from the controller unit in real-time. Actually, the BoxAGV has installed a camera for identifying the route, which is mounted centrally in the frame and connected to the controller via USB. By using the *Wireless Power Transfer* (WPT) system the battery of the vehicle can be charged by the battery charging system. The entire vehicle technique is embedded in a standard box [1] made of polypropylene with the outer dimension 600 mm x 400 mm x 213 mm (L x W x H) and offers after its integration still 50% transport volume in the box.

P. Weißgraeber et al. (Eds.): *Advances in Automotive Production Technology – Theory and Application*, ARENA2036, pp. 424–432, 2021. https://doi.org/10.1007/978-3-662-62962-8_49

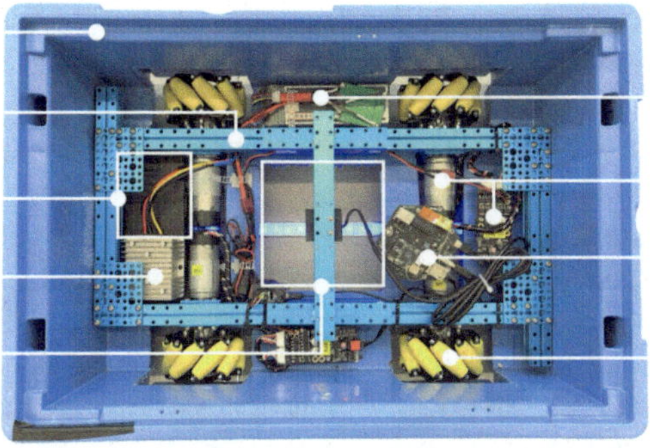

Fig. 1. The BoxAGV prototype design, where frame, driving and control components are built-in a standard VDA box. The first set-up was performed by ThingOS company.

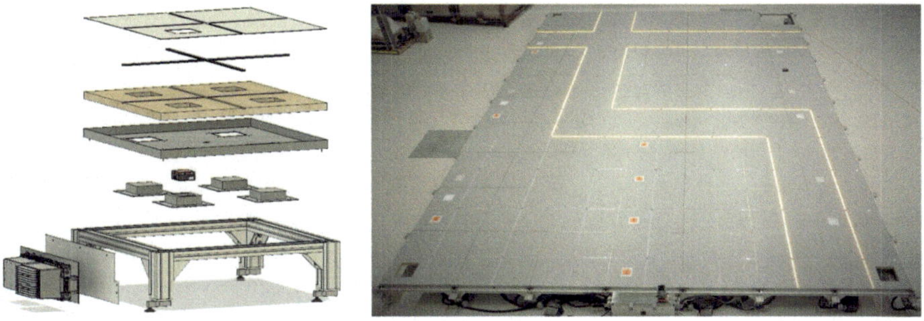

Fig. 2. Sketch of the BoschRexroth IF set-up showing the power and control unit, the substructure with load cells mounted on each stand, four functional modules, one of which is equipped with a WPT unit. The sheet steel tray contains the plywood panel together with the LED system and the floor covering. The IF demonstrator has a size of 10 m x 7 m (L x W). When the picture was taken, a walking path in the production was reproduced on the floor as an example.

Figure 2 shows on the left side a sketch from the IF set-up and on the right side the 70 m^2 sized prototype that is located in the ARENA 2036, a research platform for the mobility and production of the future that is located in Stuttgart (Germany). The IF concept is patent pending by the author [2]. It is designed as a modular raised floor system with a grid size of 1 m length and width, which is enriched with specific sensor / actuator functionalities. It poses as a universal infrastructure platform for the convertible production. Due to its modular design, the panel configuration can be continuously adapted to operational requirements. At its current stage of development, the prototype version V0.5 offers a wireless electrical power supply of production equipment. Further, the platform includes a display system based on four individually controllable Light Emitting Diodes (LED) stripes. The LED are located on the top of the panel in a cross-shaped form. In addition, each panel has four load cells that can measure the current surface load of the panel. Since the functionality of IF is described in detail in [3], existing AGV solutions are classified below and their differences to the BoxAGV highlighted.

[4, p.73] differentiates ten categories of AGVs, which differ mainly in their AGV design and payload capacity. The novel AGV described in this paper would therefore fall into the 8th category *Mini-AGV*, which typically transports standard boxes with a payload of usually less than 25 kg. Typical examples of the Mini-AGV category are the vehicles *KATE*, *Box Carrier*, *CompactMover* and *SALLY* [5–8]. Among other things, this AGV catagory is engineered for long operating times (→ large batteries on-board the AGV), autonomous navigation (→ precise localization sensor technology, powerful on-board computers) and high transportation speeds (→ complex safety functions and sensor technology).

Especially in the category of Mini-AGVs, the intelligent cooperation of many individual AGVs modelled on ants or bee swarms is advantageous as described in [4, p.83]: *"The vehicles should communicate with each other, develop strategies and solve tasks together – even without a separate AGV control system. The related fields of research are called multi-agent system and swarm intelligence but require a lot of sensors and sensor data as well as considerable computing power for their evaluation – this does not correspond to the vehicle characteristic "inexpensive" mentioned at the beginning."*

In contrast to the existing Mini-AGV solutions, the new BoxAGV uses the capabilities of the IF infrastructure platform to remove decentralized functions of AGVs, such as on-board power supply, navigation functionality and collision avoidance, to map them centrally on the IF. This significantly reduces the manufacturing cost of an AGV and thus also the barrier to investment in highly automated intralogistics system based on single standard boxes.

In particular, the supply of contactless transfered electrical power to the BoxAGVs is a key functionality to simplify and to reduce the costs of Mini-AGVs. For this purpose, the IF currently offers three WPT systems that are capable of transferring permanently 240 W respectively 1 kW to the vehicle. [9] calculates that the power requirement of an AGV with a 500 kg payload is 347 W at a steady ride and according to [3] a continuous power output density of 500 W/m^2 is adequate to start the supply of mobile production equipment. Today's WPT systems can transfer energy to AGVs along a fixed track – so-defined 1D systems, as shown in Fig. 3 – or to a fixed spot (0D) only. This is a first approach, but it does not match the future intralogistics requirements perfectly. At least a 1.5D or 2D WPT system is needed, to remove batteries from the AGVs completely.

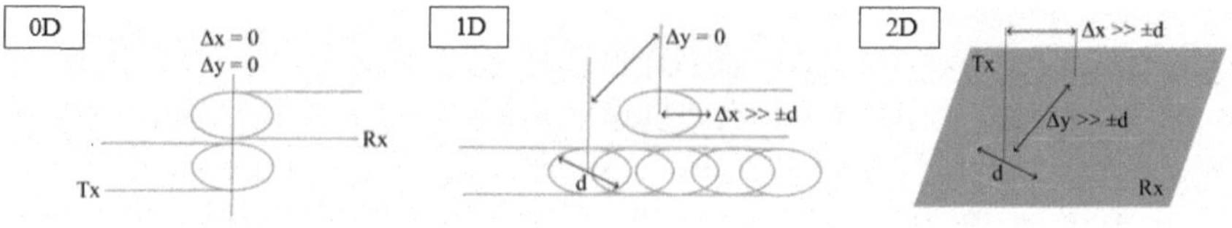

Fig. 3. Three categories of coupling between WPT's primary and secondary side units.

2 Mode of Operation

With the help of the novel AGV concept, goods movements can be organized along dynamically controllable logistics routes. The visualization function using LED is ideal for guiding an AGV as shown in Fig. 4. The inherently static process is made more flexible by the possibility of dynamic route modification, so that it fulfils the requirements of convertible production.

Fig. 4. The three pictures show a time sequence in which the BoxAGV follows the dynamically controllable blue line on the IF to a WPT charging unit.

In practice, it is not necessary to display the entire AGV route for track guidance. As shown in Fig. 5 it is sufficient to include a single LED segment, consisting of e.g. eight LED. The segment does not only determine the AGV driving direction, but also the driving speed through its dynamic visualization. Using only three colors can realize a speed control as follows:

- The AGV remains at its current speed, while detecting a green light only.
- If the track guidance sensor detects a yellow light, the AGV control will interpret this as an acceleration command.
- While the red light signal is detected, the AGV decelerates from its current speed to zero. The deceleration is the more delayed the more green light is received by the sensor.

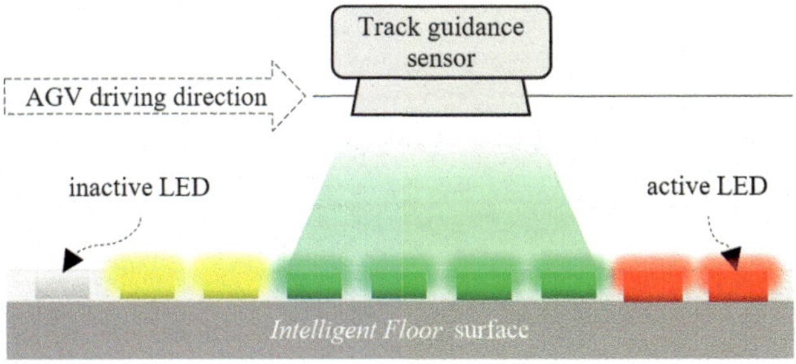

Fig. 5. Based on the author's patent [10] an easy to use optical guidance system for mobile robot applications in combination with the IF is proposed. The use of visible wavelengths only is not mandatory. Infrared or ultraviolet wavelengths can also be used to communicate contactless to the AGV.

3 Logistics Application

Using the simple designed lightweight BoxAGV, navigated centrally by LED stripes and powered by a WPT system, is a promising approach for material transport in the convertible production. To ensure high transport volumes many AGV systems must be provided. As today's AGV systems are complex and thus cost-intensive in terms of their technical equipment (battery, on-board computer, localization system and safety equipment), the investment costs for setting up automated goods traffic are high as well.

Based on the work of [11] three logistics concepts to operate a production logistics are adapted to the production layout shown in Fig. 6. The Table 1 provides a rough cost evaluation of the three logistics concepts *Manually Operated Logistics* (MOL), *In-plant Milkrun System* (IMS) and *Novel AGV System* (NAS). The table bases on a general operating lifetime of the transport equipment of eight years [12]. This period was also applied for the IF, so that the investment sum in the individual logistics concepts is depreciated on a linear method over eight years. Furthermore, the annual income of a logistics employee is set at 33,200 EUR for the comparison [13], to which an employer's contribution of 20% to social insurances is added to determine the staff costs.

Table 1. Investment and operating expenditures of the logistics concepts MOL, IMS and NAS in quick comparison.

Cost category	MOL	IMS	NAS
Building	44,388 EUR	44,388 EUR	39,604 EUR
– *Space costs /month /m²* [14]		*2.74 EUR*	
– *Total production area*	*1,350 m²*	*1,350 m²*	*1,205 m²*$^{(1)}$
Logistics equipment		– are not considered –	
Transport equipment	1,270 EUR	2,392 EUR	83,813 EUR
– *4 x Electric pallet truck* [15]	*10,156 EUR*		
– *2 x Platform truck* [16]		*5,580 EUR*	
– *6 x Trailer* [17]		*3,558 EUR*	
– *351 x IF panel*			*280,800 EUR*
– *854 x IF cover panel*			*170,703 EUR*
– *438 x BoxAGV* $^{(2)}$			*219,000 EUR*
Staff costs	159,360 EUR	79,680 EUR	0 EUR
– *Workers per shift*	*2*	*1*	*0*
Annual expenditures	205,018 EUR	126,460 EUR	123,417 EUR

$^{(1)}$ *Elimination of the supermarket and its traffic area* → 145 m².
$^{(2)}$ *Each BoxAGV can transport up to 2 PU.*

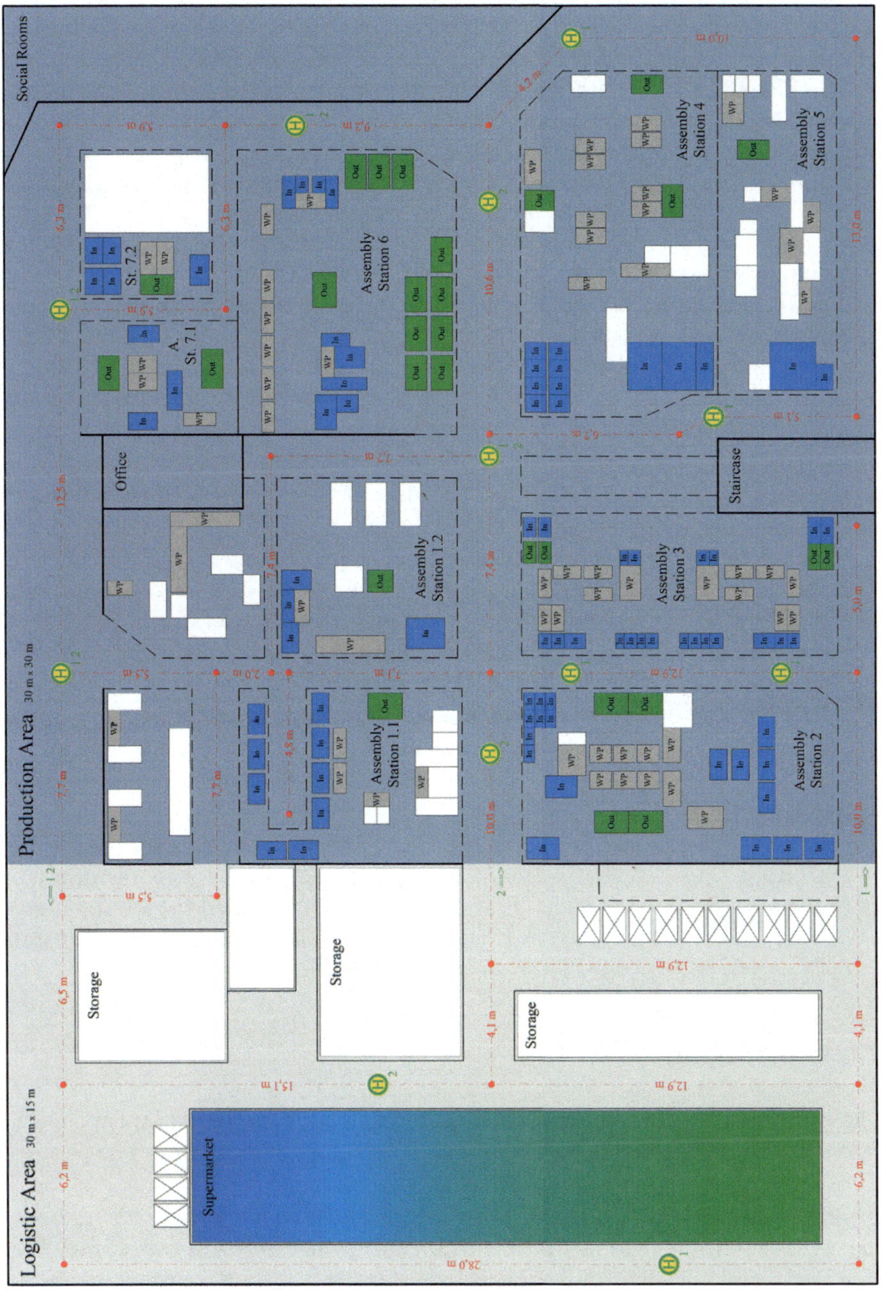

Fig. 6. Layout of the logistics and production area for car interior parts, consisting of seven assembly zones. The Ⓗ–symbols indicate the routes of two milkruns described in [11].

In addition to investment and operating expenditure considerations, the logistics concepts can also be evaluated according to other criteria. For example, the examination of adaptability to changing market and production conditions is an interesting evaluation criterion that should be considered before deciding on a specific logistics concept. Due to the focus of this paper on the NAS introduction, comparing and delimiting studies are described in subsequent papers. Therefore, a preliminary qualitative list of NAS logistic concept advantages is given here:

- Dynamic LED markings replace static adhesive tape markings, for example to indicate temporary storage locations for materials.
- The IF can detect obstacles along the traffic routes and redirect the flow of materials automatically by changing the route of the BoxAGV or by informing the responsible production employee.
- Batch tracking can be carried out via the visualization in the floor (display module and/or LED) in order to make required batches easier to find or to switch to a space-saving chaotic storage in the shop floor.
- Not all routes in production are suitable for the use of a tugger trains and thus limit the automation and efficiency gain possibilities of the IMS concept. As shown in the layout Fig. 6, tugger trains are only used on the main routes in the shop floor, because curves are not large enough or necessary safety distances [18] cannot be guaranteed. Logistics employees are therefore still forced to carry out many material transports by foot.
- The completely manually operated logistics is basically flexible, but sensitive to disruptions due to the low degree of interconnection between production and logistics. Bottlenecks that arise here affect the entire manufacturing process. Due to its high degree of interconnection, the NAS is more robust and less error prone.
- Compared to standard AGV solutions, the distances traveled by the BoxAGV are rather short, so that a BoxAGV can generally be built more simply and therefore more economically. Standard AGV are designed for a daily mileage of 15...20 km [19]. In contrast, the BoxAGV has an average travel distance of less than 200 m (warehouse → assembly station → warehouse).

4 Conclusion

The convertibility of production is becoming more and more essential and ensures the competitiveness and future viability of a production company. The use of the IF in an industrial environment opens a wide range of applications, making our current manufacturing processes more flexible and transparent. A contribution to the convertibility is made by novel AGV concept presented in this paper, which offers cost and / or space advantages over current logistics solutions. It reduces the material stock at the assembly stations by providing a demand-driven supply. Supermarkets can be downsized by chaotic warehousing or eliminated at all by shifting logistics to the traffic routes. This enables rapid changes in company internal logistics and thus a quick response to market-relevant developments.

The close integration of production and logistics also makes manufacturing processes faster and more efficient than today. It is not necessary to search for specific batches (boxes) because each BoxAGV can navigate to a defined point on command or the IF can show users the way to the BoxAGV via its LED display system. The BoxAGV can be used to save on personnel costs or to assign logistics staff to higher-value tasks too.

The promising possibilities of the presented logistics system will be put into practice step by step in the continuing research and development of the BoxAGV and IF. For example, questions regarding the optimal design of the IF's LED lines (AGV guiding lines) regarding their geometry and size – currently a square grid with an edge length of 1 m is used – are to be answered. Also feasible would be a guideline structures consisting of hexagons or octagons.

Furthermore, a detailed consideration of the investment and operating costs based on specific use cases in both greenfield and brownfield applications is planned. This calculatory consideration is particularly relevant for the economic success of the BoxAGV / IF solution. However, due to the versatility of the IF, it cannot be reduced to a logistics application alone, but should always be understood as an infrastructural solution. Thus, in addition to the management of goods flows through production, further functionalities such as the contactless power supply of production equipment can be realized with the IF.

In order to understand and develop the behavior of many BoxAGVs involved in the flow of production goods, a second BoxAGV is currently being built and tested on the prototype of the IF in the ARENA 2036. For example, dynamic traffic rules for the navigation of the BoxAGVs, which are influenced by individual demand situations of the production, are to be tested.

References

1. Verband der Automobilindustrie, *Kleinladungsträger (KLT)-System.* VDA 4500 Part 1 (2018)
2. Stillig, J.: Doppelbodenelement für einen Doppelboden. Patent application no. 10 2018 220 561.2, November 29 (2018)
3. Stillig, J., Parspour, N.: Advanced Manufacturing based on the Intelligent Floor. 20th IEEE Mediterranean Electrotechnical Conference 2020 (MELECON), Virtual Conference, pp.248–253, ISBN: 978-1-7281-5199-1 (2020)
4. Ullrich, G., Albrecht, T: Fahrerlose Transportsysteme. Springer, Berlin (2019). ISBN: 978-3-658-27471-9
5. Götting, KG: KATE – Kinetic Automat for Transport Enhancement. https://www.goetting-agv.com/kate. Accessed 19 Sept 2020
6. AGILOX Services GmbH: Intralogistics on a new level. https://www.agilox.net/en/agilox-igv. Accessed 19 Sept 2020
7. Oceaneering International, Inc.: AGV Solutions. https://www.oceaneering.com/automated-guided-vehicles/agv-solutions. Accessed 19 Sept 2020
8. DS Automation GmbH: SALLY – The Compact Class within our Product Range. https://www.ds-automotion.com/en/solutions/mini-load-agv. Accessed 19 Sept 2020

9. Kesselring, J.: Prozessbegleitende Planung und Konfiguration von Fördertechnikanlagen unter Zuhilfenahme von virtuellen Konfigurationsmustern und Konfigurationsmodellen. Dissertation, p. 160 (2017)
10. Stillig, J.: Verfahren zum Betrieb eines Spurführungssystems. Patent application no. 10 2019 108 353.2, March 29 (2019)
11. M. Knez and B. Gajšek, Implementation of in-plant milkrun system for material supply in lean automotive parts manufacturing. International Conference on Logistics and Sustainable Transport: ICLST. At Celje, Slovenia (2015)
12. Bundesfinanzministerium: AfA-Tabelle für die allgemein verwendbaren Anlagegüter. https://www.bundesfinanzminis-terium.de/Content/DE/Standardartikel/Themen/Steuern/Weitere_Steuerthemen/Betriebspruefung/AfA-Tabellen/Ergaenzende-AfA-Tabellen/AfA-Tabelle_AV.html. Accessed 18 July 2020
13. Stepstone: Logistik-Mitarbeiter/in Gehälter in Deutschland. https://www.stepstone.de/gehalt/Logistik-Mitarbeiter-in.html. Accessed 18 July 2020
14. Haufe: Logistik: Nebenkosten für Lager und Cross Docks leicht gestiegen. https://www.haufe.de/immobilien/entwicklung-vermark-tung/marktanalysen/jll-logistik-oscar-2018-nebenkosten-fuer-lager-steigen_84324_453104.html. Accessed 18 July 2020
15. Still: ECU 15 C. https://www.still.shop/ecu-15-c.html. Accessed 18 July 2020
16. Still: LTX 20. https://www.still.shop/ltx-20.html. Accessed 18 July 2020
17. Beutlhauser: Routenzug Katalog. https://stapler.beutlhauser.de/media/Media-Beutlhauser/05-Media/Intralogistik/191001_Broschuere-Routen-zuege_Beutlhauser.pdf. Accessed 18 July 2020
18. DGUV Fachbereich Handel und Logistik: Routenzüge – Einsatz von Schleppern und Anhängern als Routenzüge. Fachbereichsinformation FBHL-008, p. 19 (2018)
19. Robotize: Sichere Intralogistik mit den AMR von Robotize. https://www.mobile-robots.de/70/robotize. Accessed 18 July 2020

Increased Agility by Using Autonomous AGVs in Reconfigurable Factories

Daniel Strametz[1(✉)], Michael Reip[2], Rudolf Pichler[1], Christian Maasem[3], Martin Höffernig[4], and Michael Pichler[1]

[1] Graz University of Technology, Inffeldgasse 25f, 8010 Graz, Austria
daniel.strametz@tugraz.at
[2] Incubed IT GmbH, Pachern Hauptstraße 26, 8075 Hart bei Graz, Austria
[3] Center Connected Industry, Campus-Boulevard 55, 52074 Aachen, Germany
[4] Siemens AG Austria, Strassganger Straße 315, 8054 Graz, Austria

Abstract. Automated Guided Vehicles (AGVs) have come up to a quite familiar picture of nowadays manufacturing sites and in general they follow some rigidly given maps of the shopfloor. Future-oriented concepts require a much more agile approach that allows AGVs to continue its logistic tasks at the shopfloor securely and safely though being confronted with recently newly aligned facilities, new barriers and boundaries without necessary reprogramming the AGV's environment in front.

The actual paper deals with solutions which were worked out in the research and learning factory "smartfactory@tugraz" where all relevant facilities for these research goals have been available. Additional insights and results derive from the research facility of the Center Connected Industry and its industrial application eco system. So numerous Mobile Working Stations (MWS), the latest version of a Real-Time Locating System (RTLS) and a commercially available AGV belong to the core equipment of the presented research.

As a result there could be found, that the specified requirements and new applications of future production and logistic concepts can be achieved via dynamic targets. With this development, targets can be changed locally and the AGV automatically moves to the current target position, without necessary changes in the control software.

1 Introduction

There are still topics that can drive even European Manufacturing into a competitive future. Much more than the permanent issues of quality and price, it is a customer-oriented provision of a maximum in individualization and a nearly instant availability of products and services that will make the difference. Here the magic concept of agility [1] steps in.

Agility in a pure manufacturing context means the capability to adopt all available facilities and resources of a manufacturing site so quickly that even completely new products or highly individualized products can be realized at nearly the same level of costs, precision and time as known from standard products. Despite being aware

P. Weißgraeber et al. (Eds.): *Advances in Automotive Production Technology – Theory and Application,* ARENA2036, pp. 433–440, 2021. https://doi.org/10.1007/978-3-662-62962-8_50

that this is quite a challenge, agility will save market shares and an economic sustainability as mass customization or customer co-design improves the value perception of a product [7].

First of all an agile manufacturing concept needs an adaptive and multi-skilled equipment for value adding as machine tools, presses, assembly units and the like [13]. Second, there is the requirement of an adaptive layout [15] and a performant internal logistic system which connects all these mentioned physical resources [2]. And last but not least there is the need of a powerful communication infrastructure that interlinks the system and makes all the necessary information flow targeted, fast and secure. These properties are combined in a cyber-physical system (CPS) using the 5C-architecture proposed by Lee [3].

In the following there will be described an agile conception with an innovative focus on the logistic part of such an agile system which is represented by a new prototype of the control software for an AGV. When talking about "enhanced" agility there is the statement that the performance level of agility is even increased by the use of modern AGVs as well as the usage of the newest Real-Time Locating System to address highly complex challenges of future-oriented manufacturing and logistic layouts. Dynamic programming and/or dynamic mapping of AGVs and their fleets still is not represented sufficiently in research and its related literature [14].

2 Introduction of the Research Field and its Elements

2.1 Automated Guided Vehicle (AGV)

Agility in logistics offers multiple advantages for the actual and modern production processes. Production schemes such as "just-in-time" or "just-in-sequence" for example require flawless and efficient logistic processes [9]. AGVs connected to Enterprise Resource Planning (ERP) system or Warehouse Management Systems (WMS) meet these requirements by automatically picking and delivering raw materials or components to defined destinations or in this case the later introduced Mobile Working Stations.

AGVs have a long history of development regarding both the applied technologies and the field of applications. Whereas in the 60s and 70s AGVs were wire-guided, nowadays a variety of localization and communication technologies allows a free navigation of AGVs, even in unfamiliar environments [8]. Especially using multiple technologies and make use of sensor data fusion increases the accuracy and reliability of localization systems [6].

Dynamic positioning and localization are key-factors to implement intelligent routing algorithms in the AGVS that will replace pre-defined routes as they are still used today [5].

Nonetheless, the productive usability of agile AGVs does not only depend on the chosen localization or communication technology, but far more heavily on the adequate algorithms to cope with the gathered data by the AGV hardware. This refers to both the algorithms that provide the position information of the AGV and the intelligent routing avoiding deadlocks and increasing efficiency.

2.2 Mobile Working Stations (MWS)

The physical and visible agile infrastructure of the smartfactory@tugraz is realized via so-called Mobile Working Stations (see Fig. 1). It is a development of the Institute of Production Engineering at Graz University of Technology in order to demonstrate a promising solution of realized agility. One of their major features is their autarky concerning their supply (energy, information, compressed air, etc.). So they are free of cabling what makes these facilities not only agile as for their mobility but they can easily be accessed by AGVs and other self-driving units because of their barrier free ground all around.

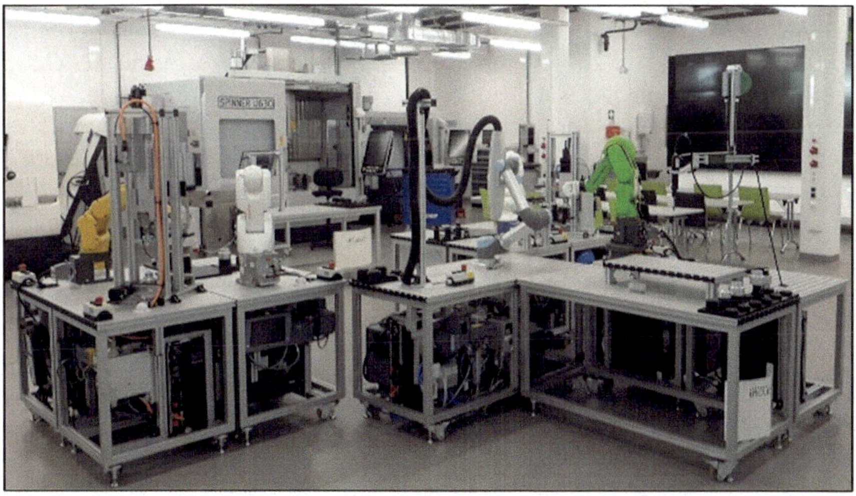

Fig. 1. Smartfactory@tugraz with freely accessible MWS

2.3 Real-Time Locating System (RTLS)

For an agile manufacturing infrastructure, its resources need to be equipped with preferably a Real-Time Locating System (RTLS), which provides the actual location data with an appropriate accuracy for the AGV. According to Siemens [10] a RTLS builds the basis for self-controlling processes in smart factories.

With the information about "what's where and when" from the RTLS, this system enables many valuable applications in the context of the fourth industrial revolution. For example the 3D model of the digital twin can be mapped with the real environment, new logistic concepts can be contrived and the RTLS paves the way to a dynamically changing production environment, where a reorganization and reconfiguration of the production is possible. [11].

The network-like infrastructure of the used Siemens RTLS consists of permanently installed reference components – called anchors and gateways – and transponders, which are attached to those objects whose position is of interest. The transponders send a radio signal based on the Ultra Wide Band (UWB) technology, and the reference components record those signals and transmit the captured data to the locating software, which is running on an Industrial PC (IPC). This software calculates the position of the assets and their movement, and provides the x, y, z coordinates via a standardized ISO interface for further actions [10].

On each MWS of the smartfactory@tugraz are two transponders of the type RTLS4030T installed, this in order to get calculated not only the position but also the orientation of the MWS. This transponder has an accuracy of localization of 200 mm and use the protocol type IEEE 802.15.4–2015 UWB HRP PHY [12].

3 Research Gap

Actual AGVs follow pre-defined routes and cannot cope with dynamically changing positions of their targets. As mentioned before, it is necessary to overcome this issue for realizing modern and reconfigurable production systems. Modern AGVs, which move autonomously and are connected with a Real-Time Locating System, allow to work in an environment of rapid changes and reconfigurations in a production. The relevant control software for AGVs can thus cope with changing environments and opens up completely new applications, which are exemplarily described as follows:

Maintenance. The AGV is able to bring the necessary or just recently ordered spare parts to a maintenance worker who is currently repairing any machine at a production site. In order to locate the worker, he must be equipped with a personal RTLS transponder.

Logistics. In chaotic storage systems, the use of modern AGVs and a Real-Time Locating System could decisively reduce the effort for the warehouse management software (WMS) and for the fleet management software of AGVs. Saving the positions where a certain object has been stored is no longer necessary. Even during the picking process in a logistics warehouse, the AGV could follow the worker, or it drives ahead and shows the worker the way to the next order.

Multi-Place Resource Allocation. Expensive facilities or instruments, which are required for certain production steps and some production lines only occasionally but also in different production areas could be transported to this lines with an AGV. This dramatically increases the use and cost-effectiveness of such special machines.

4 Approach of the Study

In this study a commercial available AGV of the company "incubed IT GmbH" is used. With this AGV it is possible to drive to any target position, because it moves completely autonomously in the environment and does not drive line-based, as widely known to be the state of the art. This flexibility is given by the software system it uses. It contains a central fleet management server that owns the view of the complete fleet, it acts as an interface to host systems like ERP or WMS, and it contains the configuration of the fleet. Additionally to this central control, each AGV is equipped with a software that allows localization, navigation and coordination with the rest of the fleet. By raising the capability level of each AGV as well as the fleet management server, installations differ by configuration only. In contrast to traditional AGV installations, there is no project specific programming, but configuration only.

The function of the RTLS in this concern is the steady tracking of any object that is equipped with a transponder. The always updated location information of an object either is written into a database or is communicated by a TCP/IP client.

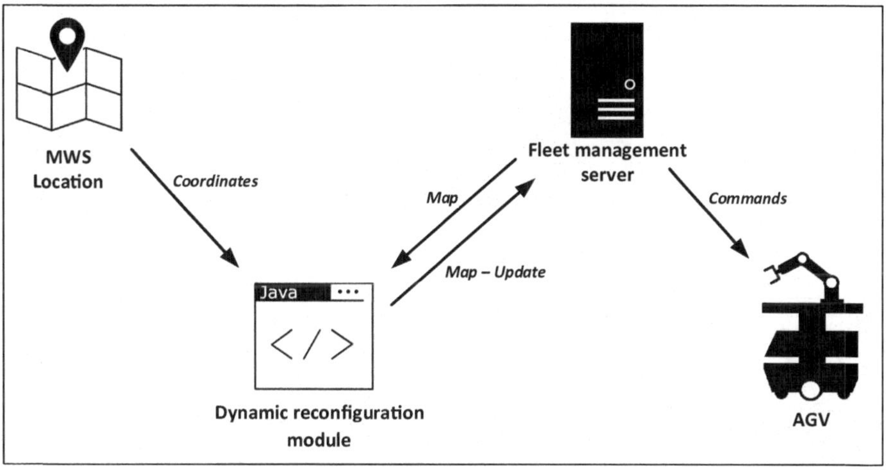

Fig. 2. Data processing of dynamic targets

As depicted in Fig. 2, the RTLS is connected to a dynamic reconfiguration module written in Java. Using a configuration file, the dynamic reconfiguration module distinguishes between relevant location information for MWS that are served by an AGV and non-relevant location information. If the location change of a relevant MWS exceeds a configured threshold, an automated update of the AGV's map is triggered. The map of the AGV contains contour information of the environment, targets (positions with specific actions like put, get or charge) and traffic regions (similar to car traffic rules). The dynamic reconfiguration module downloads the latest map by applying a RESTful (Representational State Transfer) web call, extracts the target that corresponds to the moved MWS, changes the pose of the target (x, y and orientation) and pushes the changed map back to the fleet management server. The fleet management server does a few sanity checks to make sure that AGV system is continuously operational and forwards the updated map to each AGV. Combining the updated map with the AGV's ability to navigate to arbitrary target positions results in a system that still can reach displaced MWS or any other object, because it is tracked with the RTLS.

5 Case Study for Validation

The networking of the RTLS with the newly developed control software of the AGV was tested in the research environment of the smartfactory@tugraz. A reconfigurable assembly line consisting of four Mobile Working Stations (MWS) formed the test setup.

In the first case, the production line was adapted for product A (see Fig. 3), where the MWS 3, which is also the target station for the AGV, was located at the right end of the line. In the second case, the production was reconfigured for assembling product B. In this layout, the target station MWS 3 was located on the left side of the line and has additionally changed its orientation.

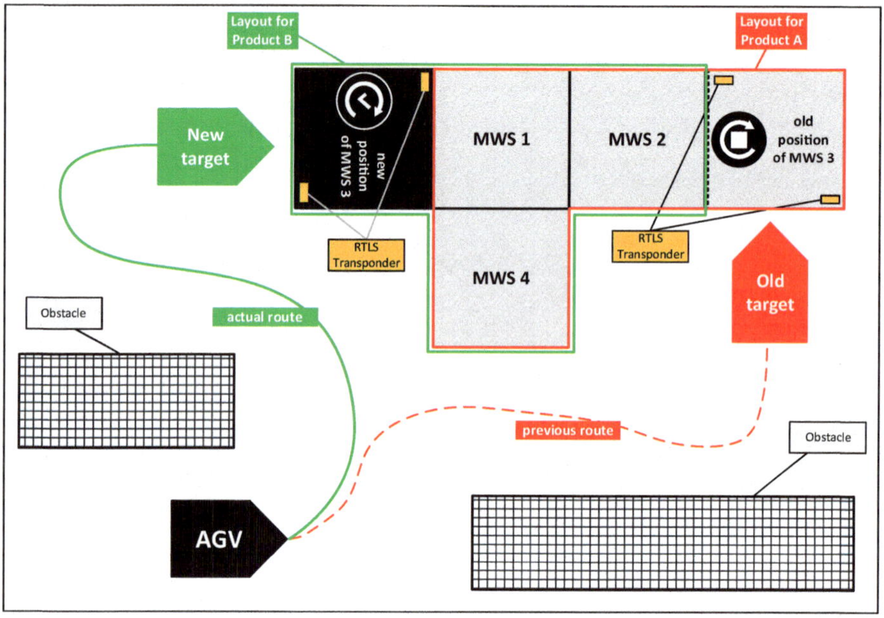

Fig. 3. Re-Routing to the actual position of the MWS 3

The test runs of the AGV started always at the same starting point, only the position of the particular target has been changed physically. The coordinates and orientation of the MWS 3, transmitted by the RTLS, were used for pre-positioning the AGV in front of the target with a defined offset and correct orientation to the target. The fine positioning was done by means of the AGV's laser scanners. So the changes in the software was done automatically by the entire system itself.

The results of the five final test runs showed that the inaccuracy of the RTLS (± 200 mm) affected the calculated target for the AGV and finally the test duration. Table 1 shows the deviations (orientation and coordinates) from the ideal target and the individual test durations.

Due to the deviations of the current target from the ideal target, the AGV needed in the worst case 4 s more time for the fine positioning with their laser scanners in front of the target MWS as in the best case. In all test runs the desired results could be achieved. So the successful interaction of the RTLS and the control system of the

Table 1. Results of the test runs

Test-Nr	Deviation of the correct orientation	Deviation of the goal coordinates x/y	Test duration
1	−4,72°	−4,5 mm / 5,0 mm	31 s
2	−3,37°	1,5 mm / 7,5 mm	30 s
3	8,25°	5,0 mm / 10,0 mm	33 s
4	1,76°	2,5 mm / 0,0 mm	29 s
5	−0,77°	-3,0 mm / 3,0 mm	29 s

AGV could be confirmed and another milestone for creating agile production systems has been accomplished.

6 Conclusion and Outlook

Within the framework of this research work, the approach to dynamically changing targets has been successfully implemented. This new milestone in the development of intralogistics will enable new applications with the effect of increased productivity and cost-effectiveness. The successful networking of the RTLS with the control software of the AGV revolutionizes the use of AGVs, in particular changes of targets in the map. From now on, target changes no longer have to be made by an expert of the control software, because the actual target coordinates are permanently provided by the RTLS and leads to a steadily updated map. This entire development is an important enabler for automatically reconfigurations of a modular and really agile production lines.

For future work, it is already planned to power up the wireless communication of the described setup. The actual WiFi communication will be replaced with the newest communication standard 5G. This will guarantee the necessary bandwidth and latency. The first step will be to substitute the currently used WiFi devices with 5G-capable equipment so that the existing 5G network of the smartfactory@tugraz can be used. Especially in terms of safety-features of AGVs, 5G can be compared to WiFi quite well considering the maximum latency of the system. Studies at Center Connected Industry have shown that the follow-up of an AGV receiving safety signals of an infrastructure ESPE is significantly shorter when using 5G instead of WiFi. Additionally, 5G as a communication technology paves the way for real time cloud or edge computing with its high data rates and low latency [4].

Acknowledgements. The project smartfactory@tugraz has been mainly funded by the Austrian Federal Ministry for Climate Action, Environment, Energy, Mobility, Innovation and Technology as well as 19 industrial partners. One of these partners is the company "incubed IT GmbH" which granted the AGV and relevant scientific support. The RTLS was provided with appropriate support by the partner company Siemens AG Austria.

References

1. Ramsauer, C., Kayser, D., Schmitz, C. (eds.): Erfolgsfaktor Agilität – Chancen für Unternehmen in einem volatilen Marktumfeld, 1st edn. Wiley-VCH, Weinheim (2017)
2. Pichler, R.: Forschungsraum smartfactory@tugraz – Die Sichtbarmachung einer agilen Fertigung. In: WINGbusiness, 3/17, pp. 28–31, Österreichischer Verband der Wirtschaftsingenieure (2017)
3. Lee, J., Bagheri, B., Kao, H.: A Cyber-Physical Systems architecture for Industry 4.0-based manufacturing systems. In: Manufacturing Letters 3, pp. 18–23. DOI: https://doi.org/10.1016/j.mfglet.2014.12.001 (2015)
4. Llanasas, R.: How 5G Will Accelerate Cloud Business Investment. In: Compare the Cloud, 06.11.2019. https://www.comparethecloud.net/articles/how-5g-will-accelerate-cloud-business-investment/. Accessed 22 June 2020
5. Möhring, R., Köhle, E., Gawrilow, E., Stenzel, B.: Conflict-free Real-time AGV Routing. In: Fleuren, H., den Hertog, D. & Kort, P. (eds.): Operations Research Proceedings 2004, vol. 2004. pp. 18–24. Springer, Heidelberg (2005)
6. Steinhardt, N., Leinen, S.: Data Fusion for Precise Localization. In: Winner, H. et al. (eds.): Handbook of Driver Assistance Systems. pp. 605–646. Springer International Publishing, Switzerland (2016)
7. Thallmaier, S.: Customer co-design. A study in the mass customization industry. Springer Fachmedien, Wiesbaden (2015)
8. Ullrich, G.: Fahrerlose Transportsysteme. Eine Fibel - mit Praxisanwendungen - zur Technik - für die Planung. 2nd edn. Springer Vieweg, Wiesbaden (2014)
9. Wagner, S., Silveira-Camargos, V.: Managing Risks in Just-In-Sequence Supply Networks: Exploratory Evidence From Automakers. In: IEEE TRANSACTIONS ON ENGINEERING MANAGEMENT, vol. 59, pp. 52–64. DOI: https://doi.org/10.1109/TEM.2010.2087762. IEEE, New York (2012)
10. Siemens: The answer to the thrilling question of the "where". Siemens, Karlsruhe (2019)
11. Siemens: Real-Time Locating Systems (RTLS) in Production. Siemens, Nürnberg (2019)
12. Siemens: Technical Data RTLS4030T, https://support.industry.siemens.com/cs/pd/-1266062?pdti=td&dl=de&pnid=25277&lc=deWW, last accessed 2020/09/18.
13. Potdara, P.K., Routroya, S., Behera, A.: Addressing the Agile Manufacturing Impediments using Interpretive Structural Modeling. In: Materials Today: pp. 1744–1751, Proceedings 4, ICMPC (2016)
14. Witczak, M., Majdzik, P., Stetter, R. et al.: Multiple AGV fault-tolerant within an agile manufacturing warehouse. pp. 1914–1919, IFAC PapersOnLine 52–13 (2019)
15. Fritzsche, A.: Implications of agile manufacturing in the automotive industry for order management in the factories – evidence from the practitioner's perspective. pp. 369–374, Procedia CIRP 72 (2018)

Safety and Operating Concept for Collaborative Material Flow Systems

Matthias Hofmann(✉)

Institute of Mechanical Handling and Logistics, University of Stuttgart, Stuttgart, Germany
matthias.hofmann@ift.uni-stuttgart.de

Abstract. The automotive industry, including their whole parts and component manufacturers, are facing challenges in a dimension not known yet.

The strong differentiation of the product portfolio in connection with specific production lines holds a high vulnerability to capacity fluctuations in certain lines due to volatile demand within a manufacturer's product spectrum. From the present point of view, the production of automobiles with fundamentally varying technical specifications requires a flexible and scalable production, which enables an efficient production of batch size one. The key elements of a new concept, developed at the IFT are collaborative intralogistics components. One such system called "Mobile Supermarket" specializes on the supply characteristics of frequently needed parts with high variety and works according to the goods-to-man principle, which requires a highly sophisticated safety and interaction concept.

1 Motivation and Statement of the Problem

Automobile producers and their ancillary industries, as well as their machine and plant building equipping companies are faced with unprecedented challenges. In addition to the clearly palpable structural transition in the forms of mobility, trade wars and epidemiological events in 2020 have additionally intensified the requirements for efficient, lot size independent manufacturing concepts. In particular, the rigid production line principle is generally lacking in production flexibility – and its accompanying output efficiency – for the big OEMs of the automotive industry, cf. [1]. In contrast, the small and medium-sized factories without existing flow production, where the manual processes of manufacturing and logistics, in conjunction with a short planning phase, which, on the one hand, create flexibility but, on the other hand, drive the production costs due to personnel commitment, record shrinking profit margins in international competition.

1.1 Cost Factor Material Provision

The processing steps of the material provision chain are particularly important when making the production logistics and assembly processes more efficient. For this reason, the order picking and supply of parts and materials to the installation location are particularly still strongly influenced by manual activities, and are thus significant cost factors in many sectors of machine and plant construction or for the entire metal and

P. Weißgraeber et al. (Eds.): *Advances in Automotive Production Technology – Theory and Application*, ARENA2036, pp. 441–450, 2021. https://doi.org/10.1007/978-3-662-62962-8_51

electrical industry due to personnel commitment. From the backdrop of the production of more individualized and variable standard products up to batch size one, this effect is further strengthened. This has caused there to be an urgent need for rationalization of the production-related logistics in all those companies where the deployment sequence for small production batches has to be created from a large range of parts as needed, cf. [1].

The automation of the internal material flow by means of static warehousing systems and conveyor lines can be ruled out, however, wherever layout flexibility is called for, cf. [2]. Fully automated solutions, on the other hand, have proven to be only partially, and/or not at all, useable when a high degree of flexibility and adaptability when dealing with parts is called for. A human, with his senses, cognitive abilities and muscular coordination, has always been superior to automation where a rapid situational detection and adaptability is needed. Approaches to efficiency enhancement can thus lie in the fragmentation of work flow with partially automated solutions, where the autonomously operating robot and humans actively work together, cf. [3]. The material supply concept, known as "Mobile Supermarket", which was developed by the IFT thus aims to particularly reduce the manual switching and picking processes at assembly workstations, and produce a high degree of structural flexibility through the mobility of small-scale components. The "Mobile Supermarket" is one mobile component of a new logistics concept designed by the IFT for a flexible and convertible automobile production, cf. [1], cf. [4, 5].

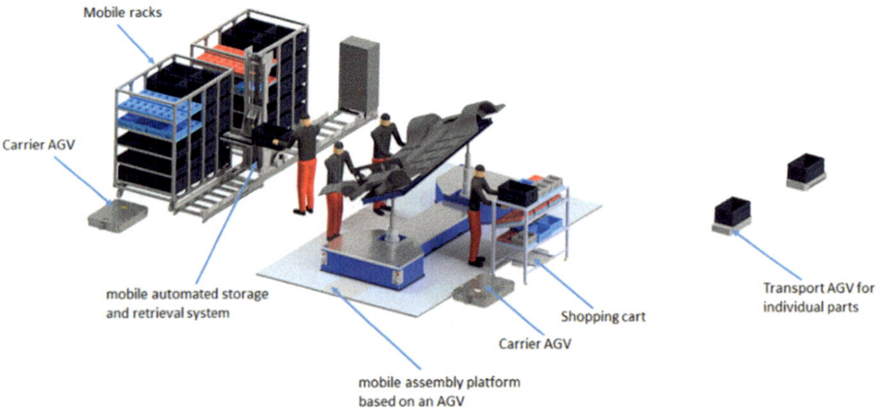

Fig. 1 Material flow systems based on mobile intralogistics components [1, 4].

1.2 Area of Tension Between More Efficient and Simultaneously Safer Man-Machine Collaboration

The operation of machines and plants is subject to country-specific regulations. The EU Machinery Directive 2006/42/EG [6] forms the legal framework for the national regulations of the member states of the European Union with regard to the safety of individuals and employees when operating machinery. In legal terms, there is an area of tension in the context of safety and prevention of bodily harm when operating mobile and fixed machinery. National and international norms, like the DIN EN ISO 12100 [7], are the direct result of the laws passed by the European Parliament and the European Council. The DIN EN ISO 12100 thus provides guiding principles and methods for the construction of safe machines and describes procedures for determining danger points as well as assessment of risks and potential dangers for the entire life cycle of a machine in all its operating phases. If it is not possible to assure the safety of the operating personnel and/or those persons in the vicinity of the machines while being operated through the inherently secure construction, appropriate safety measures and precautions must be planned. In this regard, compliance with safety margins, like those described in DIN EN 349 [8] and DIN EN ISO 13857 [9], are particularly relevant. For this purpose, many physically separating protective devices compliant with DIN EN ISO 14120 [10], or contact-free ones using optoelectronic protection systems compliant with DIN EN ISO 13855 [11], are used. As can be clearly recognized, a direct human-machine collaboration, which can contain a significant danger potential due to its operating principles, are subject to complex technical and normative requirements. This is especially true of collaborative work processes in which a machine's movements occur simultaneously with manual tasks, instead of sequential actions with appropriate breaks and/or downtimes. In this respect, there is, especially for such applications, a conflict of objectives for machine and process efficiency where people are collaborating with a machine and/or find themselves in close proximity to it.

2 "Mobile Supermarket" Concept

The IFT has developed a kind of small-scale portable Automatic Small Parts Storage System (ASPSS) with the specific aim of meeting the challenge on the supply of frequently needed parts with high variety making available without the necessity for time-consuming preliminary sequencing at the assembly point. Contrary to the conventional application scenario of an ASSPS in warehouse and picking areas – also called a logistics supermarket, cf. [12], in the automotive environment – it should be possible to employ the system in question directly at the installation point in the assembly process. In this respect, the parts supply system developed at the IFT represents a mobile supermarket, cf. [4, 5]. The "Mobile Supermarket" parts supply system concept consists of three individual components, which can also be utilized independently of each other, but in combination comprise a kind of mobile small parts warehouse. The Mobile Supermarket encompasses a compact automated guided vehicle (AGV) which transports a mobile shelf modules as well as a non-location-dependent picking unit to deal with the loading and unloading of small load carriers (VDA-SLC) from the mobile shelf modules, cf. Fig. 2.

The compact Carrier-AGV seen in Fig. 2 constitutes a proprietary concept and construction of the IFT. The AGV does not lift the shelves with a payload of 1200 kg, but drags them. For this reason it is, on the one hand, extremely compact, and, on the other hand, an unspecified design can be used. Unspecified in the sense that the AGV is not specifically designed for transporting the shelves, but can also be used in its basic form

Fig. 2 Prototype of the Mobile Supermarket [1, 4].

for transporting baskets of goods, special load carriers and small load carriers. Additionally, the compact AGV is suitable, because of its "carrier vehicle with accessory equipment" modularity, for additional application scenarios, whereby the type diversity of the AGV used can be reduced. As a result, the AGV has a wide range of application scenarios due to its design and can therefore be used cost-effectively.

The picking unit is a non-stationary small-scale storage and retrieval system (Mini-ASRS) which can be used for the loading and unloading of the shelf modules in one of the picking areas in front of the assembly area, as well as directly at the point of installation. In this application scenario the Mini-ASRS is intended in a human-machine collaboration, to provide the worker with the required materials for the pending work task according to the goods-to-man principle. In the final assembly lines of today's automotive production, the employees have to pick the required material themselves from shelves by using parts lists, while the range of parts in the shelves is matched to the planned throughput sequence of the workpieces in advance, cf. [13]. Short noticed changes in the production program, due to interruptions in the assembly process or postponement of the workpiece, effectuate a break in the pearl chain. Should there be changes at short notice in the production program due to interruptions in the assembly process or a postponement of the workpiece, the pearl chain will be broken. In a material flow system based on the principles of Just-In-Sequence or Just-In-Time it is not possible to handle such events in a cost-efficient way. Since the need in flexible and convertible production in regard to forms, time and location can vary, the assembly supply has to be oriented towards the real time needs – Just-In-Real-Time – of the individual workpieces. Real-time requirements, therefore, because the independent workpiece carriers, like those of the mobile assembly platform shown in Fig. 1, enable a pass through the assembly oriented on the technical production requirements, which can also be changed at short notice while the part is being produced. In consequence, the underlying material flow concept of the mobile supermarket focuses on the synchronization with the actual production program at the point where there is a physical need. For this reason, the mobile shelf modules are filled with a variant part portfolio instead of a pre-sequenced spectrum of the pearl chain.

3 Safety and Interaction Concept for the Mobile Supermarket

The mobile supermarket system not only stands out due to its reaction speed in the sense of a logistical context, but also in that the employee always has exactly the material necessary for the pending task due to the reduction in the use of manual picking and reloading processes. However, this requires forms of interaction between human and machine, which is therefore a form of man-robot collaboration (MRC). In order to meet the requirements of plant layout to be as open as possible, with a high handling capacity but, at the same time providing personal safety, line controls as well as the accompanying safety and operating concepts are necessary which exceed the current level of technology available. In order to provide a fluid and productive collaboration, there is, inevitably, a causal connection between safety and operation, making it essential to consider these elements together. With a payload of 60 kg, in connection with a maximum traverse axial speed (x axis) of 2 m/s, the potential danger from the Mini-ASRS is not to be classified in the same way as that of a collaboration robot, cf. [1] and cf. [3].

3.1 Sensor Concept

In order to meet the requirements resulting from a danger analysis with regard to personal safety, portable optoelectronic protective devices, in the form of two safety laser scanners, are mounted directly on the Mini-ASRS, in order to enable its unrestricted mobility. The laser scanner primarily ensures that no one is present during the travel motion in the horizontal and vertical axes by monitoring defined protective fields. The spatial extension of the protective fields are measured in compliance with DIN EN ISO 13855, primarily by means of the approach speed of people in the vicinity of the machine, as well as the follow-up time, which provides the time required to bring all the axes to a halt after an emergency cut-off command. This means that the measurements of the safety areas are particularly correlated with the axes speeds, the masses in motion, and the maximum allowable acceleration, as well as on the control side response behavior of the control components due to latency. Depending on the operating and driving status, a differentiated consideration is possible which is, finally, absolutely necessary to implement a collaboration for the material transfer. Differentiation between the machine movements, which serve the loading and unloading of the shelves and the material transfer process is necessary because the final handling of the SLC is only possible in the z axis (telescopic table) on the side facing the shelves. The resulting potential danger can be considered to be slight.

Only a warning field, rather than a protective field, is active on the laser scanner in this phase. Entering and leaving the warning field is an active component for the detection of completion of the material transfer. Leaving the warning field sets a trigger for the subsequent switching of the zone configuration to protective field. An automatic attempt of the Mini-ASRS for the subsequent transport order can be carried out. In order to avoid misinterpretation concerning the accidental entering and leaving of the warning field, an additional transfer position defined for a very limited acknowledgement zone is redundantly monitored by both laser scanners.

In order to economize on travel paths, the transfer positions should be as variable as possible, depending on the position where the employee needs the material. Measurement data from the laser scanners is gathered and evaluated for the necessary position detection, cf. Fig. 3. The person needing the material can intuitively influence the transfer point.

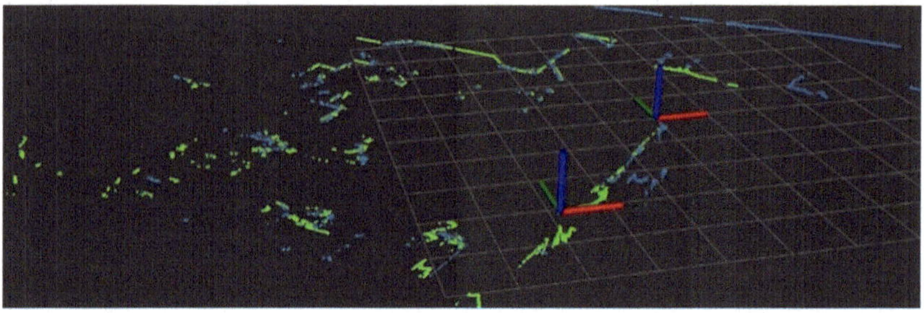

Fig. 3 Visualization of measurement data of the laser scanners. Outlines shown in blue (laser scanner 1) and green (laser scanner 2), cf. Fig. 5, are detected at a max. Distance of 25 m.

In order to achieve a sure differentiation between people and objects in line with this operating function, which is not technically possible using the laser scanner alone, additional 3-D cameras are used. In this respect, it is necessary to differentiate between safety relevant functions and those, which are standard for operation. The camera system is thus only an element to implement the operating functions and is not a suitable substitute for a laser scanner as certifiable person protection equipment, cf. [14]. Further operating elements, e.g. a tablet, smartwatch or picking wearable, may be needed, depending on the sort of material retrieval. This is primarily dependent on the desired materials flow architecture, and with it the question of whether an automated, centrally organized or decentrally-controlled allocation ensues. The degree of decentralization is inevitably correlated with the quantity and characteristics of the communication and interaction interfaces.

Pressure-sensitive safety mats, which are optional in the safety architecture of the mobile supermarket, are conceived to be spatially resolving, and are suitable for displaying safety functions while, at the same time, functioning as operating interfaces. If the safety mats are placed in the passageway for the material transfer, a more strongly differentiated gradation of the protective field switching on the laser scanners can take place. Stepping on the safety mats produces a position recognition, so that the laser scanners are only responsible for surveillance regarding the dynamic approach of persons in the direction of the danger zone. At the same time, the safety mats are also useful as tactile operating elements, in that the software functions of the operating procedure and/or the material flow can be actuated by foot.

3.2 Man-Machine-Interaction

While the possibility of intervention for the user is crucial for the operation of the procedures, the targeted information from the machine to the persons in the vicinity is assigned an equally great importance. This must, above all, be seen in the sense of efficient collaboration so that misunderstandings and dangerous situations can be avoided, not least because these can lead to an inadvertent emergency stop and, with it, downtime. An autonomously operating machine can cause people anxiety when the operating behavior is not predictable, cf [15]. Ultimately, this anxiety can pose a barrier to the safe and productive collaboration. In this respect, the concept for the safe man-machine interaction described here is intended to enable the machine to inform the users as well as those in the vicinity about its pending actions and operating state without frightening them. This is based on the human articulation, aimed, using optical and/or acoustic information, at people, and relies on the principle of dependability due to the unambiguity of the situation.

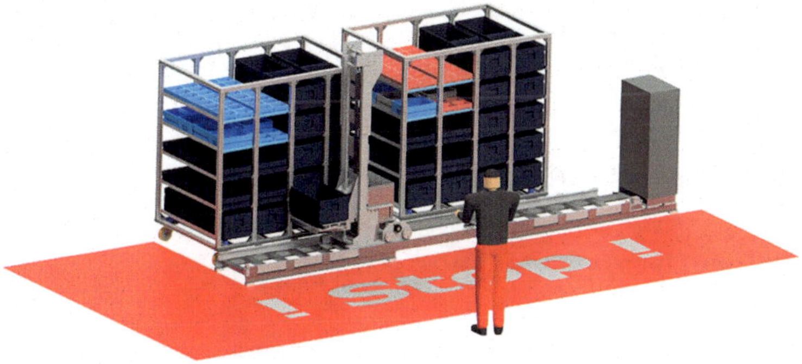

Fig. 4 Optical and/or acoustic information about pending actions, operating instructions and also safety relevant warnings for the persons in the vicinity.

The concept for interaction does not only affect the regular operational procedures, like signaling for the pending material transfer, but also problem-solving processes. These include situations where people or objects are in the safety field and blocking the machine's operation. The problem can be communicated to those people in the vicinity on a situational and demand-orientated basis, to get them to leave the safety zone or to remove the objects from it.

3.3 Machine-Machine-Interaction

A particularly sensitive process in the mobile supermarket's operating procedure, which requires a greater level of communication and interaction – in man-to-machine as well as machine-to-machine communication – is the delivery and removal of the mobile shelf modules by AGV. The safety field area created by the laser scanners of the Mini-ASRS have to be deactivated partially in order to move the shelves. A flexible access protection is needed which is implemented using an intelligent link to the safety functions of the transporting AGV and those of the Mini-ASRS in order to also unconditionally assure personal safety during these processes. It is a question of a form of "shared safety" in which the approach area monitoring of the AGV is simultaneously responsible, during the delivery and removal phase of one of the two shelf modules, for making sure that nobody is in the adjoining danger zone of the ASRS, cf. Fig. 5. If this safeguard is not in place, the operation of the Mini-ASRS has to be interrupted.

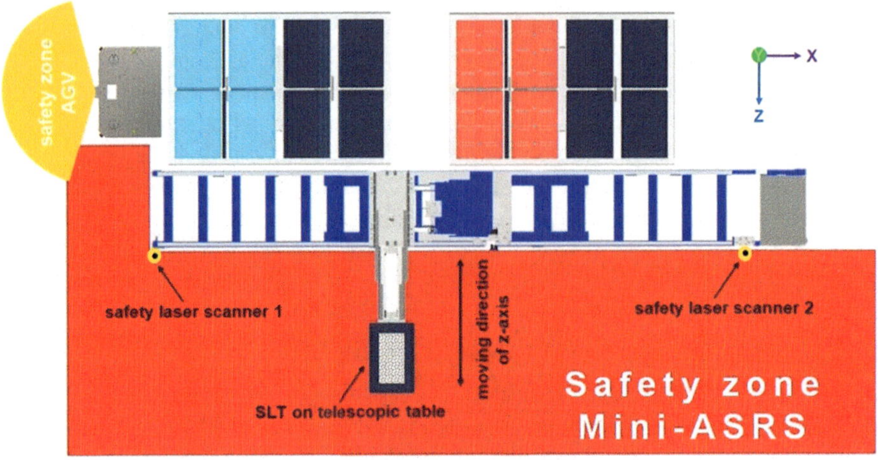

Fig. 5 Flexible access protection for the delivery and removal of the mobile shelf modules by AGV. Shared Safety between the AGV and the Mini-ASRS.

It isn't only a question of fulfilling the formal and normative requirements regarding personal safety, as the limits of the machine can be flexibly modified, but also to link the safety and machine controls of a number of pieces of equipment. This requires, among other things, creation of decentralized control structures and components, as well as efficient and redundantly safe wireless signal transmission.

4 Challenges on the Way to Flexible Production Plants with Collaborative Material Flow Systems

The establishment of a lot size independent manufacturing concept contains complex requirements. First and foremost, this involves the creation of flexible structures in the parts provision, which is based on a logistics concept that enables an object-specific parts provision for every workpiece, oriented on real-time requirements. The implementation requires disruptive approaches with regard to the material flow area as well as the intralogistics components. On the one hand, the required flexibility requires a departure from the well-known pearl chain principle and, on the other hand, forms of decentralization in the architecture of the material flow, in which the employees are not only the executive organ, but rather represent a decision-making authority for the triggering of a material provision. As each of these decisions has consequences for upstream and downstream steps in the process chain, this requires a complete networking and localization of all workpieces and resources, right down to the individual SLC. This also requires real-time-capable data transmission paths with high bandwidth in order to enable free navigation of AGVs and interaction from man to machine or machine to machine. Furthermore, data transmission must not be susceptible to interference due to structural facilities and a high number of participants. With regard to collaboration and shared-safety functions, redundantly safe wireless communication between the participants must also be guaranteed in order to implement the presented concept.

5 Conclusion

Since 2014, the IFT, helped by subsidies from the Baden-Württemberg government, has not only been able to develop concepts for convertible and flexible production for the efficient manufacture of goods in small quantities right up to batch size one, but rather ones for whose implementation necessary primary components can be transformed into prototypes. This includes, among other things, a kind of small-scale mobile ASPSS, which can be referred to as a mobile supermarket due to its mobility. This system stands out, not only in the sense of material supply, through the high degree of responsiveness and adaptability to the constantly changing production processes and procedures, in which the sequencing of components takes place at their destination, but also through the material transfer with man-machine interaction. In order to fulfill the necessary requirements and guidelines concerning personal safety in an industrial environment, an innovative safety and control concept has been developed which gives priority to demand-oriented linking of complete machines or subfunctionalities. Numerous research projects are currently being carried out at the IFT dedicated to the implementation of these goals. The results are being validated in prototypes in real-life test facilities at the research campus ARENA2036.

References

1. Bauernhansl, T., Fechter, M., Dietz, T., et al.: Entwicklung, Aufbau und Demonstration einer wandlungsfähigen (Fahrzeug-) Forschungsproduktion. Springer, Berlin (2020)
2. Popp, J. (2018): Neuartige Logistikkonzepte für eine flexible Automobilproduktion ohne Band. Dissertation Universität Stuttgart. https://doi.org/10.18419/opus-9987
3. Buxbaum H-J.; Kleutges M. (2020): Evolution oder Revolution? Die Mensch-Roboter-Kollaboration. In: Buxbaum H-J. (Hrsg.) Mensch-Roboter-Kollaboration. Springer Gabler, Wiesbaden. https://doi.org/10.1007/978-3-658-28307-0_2
4. Hofmann, M: Material flow systems and intralogistics components for a non-sequential, flexibly timed automobile production – First prototypes. XXIII International Conference on "Material Handling, Constructions and Logistics", Vienna, Austria 2019. ISBN 978-86-6060-020-4 (2019)
5. Hofmann, M.: Intralogistikkomponenten für die Automobilproduktion ohne Band und Takt – erste Prototypen. Logistics Journal : Proceedings, Vol. 2018. (urn:nbn:de:0009-14-47616) (2018)
6. European Parliament and European Council: Directive 2006/42/EC of the European Parliament and of the Council of 17 May 2006 on machinery (2006)
7. German Institute, D.I.N., for Standardization e. V.: DIN EN ISO 12100: 2011–03, Safety of machinery – General principles for design – Risk assessment and risk reduction (ISO 12100:2010). German version EN ISO **12100**, 2010 (2011)

8. DIN German Institute for Standardization e. V.: DIN EN 349: 2008-09, Safety of machinery – Minimum gaps to avoid crushing of parts of the human body; German version EN 349:1993+A1:2008 (2008)
9. German Institute, D.I.N., for Standardization e. V.: DIN EN ISO 13857: 2008–06, Safety of machinery – Safety distances to prevent hazard zones being reached by upper and lower limbs (ISO 13857:2008). German version EN ISO 13857:2008 (2008)
10. German Institute, D.I.N., for Standardization e. V. : DIN EN ISO 14120: 2016–05, Safety of machinery – Guards – General requirements for the design and construction of fixed and movable guards (ISO 14120:2015). German version EN ISO 14120:2015 (2016)
11. German Institute, D.I.N., for Standardization e. V.: DIN EN ISO 13855: 2010–10, Safety of machinery – Positioning of safeguards with respect to the approach speeds of parts of the human body (ISO 13855:2010). German version EN ISO 13855:2010 (2010)
12. Battini, D., Boysen, N., Emde, S.: Just-in-Time supermarkets for part supply in the automobile industry. J. Manage. Control **24**(2), 209–217 (2012)
13. Kratzsch, S.: Prozess- und Arbeitsorganisation in Fliessmontagesystemen. Schriftenreihe des IWF. Vulkan-Verlag GmbH, Essen (2000)
14. Korte, D., Neuschwander, B., Baumgarten, S., Lindermayr, J.: Konzept einer sicheren 3D-Umgebungssensorik. In: Book of abstracts – 1. Stuttgarter Automobiltagung zur Zukunft der Automobilproduktion (2019)
15. Oberle, M., Helfert, M., König, C., et al.: Experimentelle Untersuchung des menschlichen Einflusses auf die Unsicherheit in der Mensch-Technik-Interaktion. Z. Arb. Wiss. **72**, 44–45 (2017). https://doi.org/10.1007/s41449-017-0071-x

Combining Safe Collaborative and High-Accuracy Operations in Industrial Robots

Andreas Otto(✉), Shuxiao Hou, Antje Ahrens, Uwe Frieß, Marcel Todtermuschke, and Mohamad Bdiwi

Fraunhofer Institute for Machine Tools and Forming Technology (IWU), Reichenhainer Str. 88, 09126 Chemnitz, Germany
andreas.otto@iwu.fraunhofer.de

Abstract. We present a general approach for realizing safe collaborative behavior and high-accuracy operations in industrial robots. For example, collaborative behavior is required in human robot collaboration, whereas high-accuracy is needed in robotic machining. The operation sequence of a robot with corresponding often-contradictory demands on safety functions, control mode and accuracy is specified in a task-oriented path planner. The task planner generates optimized reference trajectories based on an environmental model, kinematic and task-specific constraints. The resulting reference path is fed into an online operation management, which monitors the workspace, ensures safety, and realizes different online control strategies especially for the dynamic tasks such as human robot collaboration or high-precision operations. For the high-accuracy tasks, we implemented a model predictive controller using a mechatronic model for predicting future configurations of the robot based on the measured present configuration. The model-predictive control is demonstrated for the compensation of deviations due to external forces that acts on the end effector of the robot.

1 Introduction

Using robots has great advantages concerning their workspace and the flexibility in executing different tasks with one machine. In the era of Industry 4.0 robots should serve contradictory goals. In some applications, such as robotic machining, industrial robots should perform complex tasks with high accuracy at the end-effector [1]. On the other hand, sometimes a collaboration of the same industrial robot with humans in a shared-workspace is desired [2], where dynamic path generation is required.

The objective in the first scenario is a fast and accurate motion of the industrial robot with the potential for high repeatability. However, the high flexibility of robots with multiple joints and the typically serial mechanical structure lead to compliant behavior limiting its accuracy. In particular, external forces, additional weights at the end effector or high accelerations may lead to deviations. There are two common strategies for increasing the accuracy. First, the deviations at the end effector of the robots can be

P. Weißgraeber et al. (Eds.): *Advances in Automotive Production Technology – Theory and Application,* ARENA2036, pp. 451–459, 2021. https://doi.org/10.1007/978-3-662-62962-8_52

compensated offline in a feedforward way. Typical examples are offline compensation of deviations due to cutting forces in milling [3, 4] or due to nonlinear effects such as friction, Coriolis force and hysteresis [5, 6]. On the other hand, an online compensation of the deviations by an additional or modified feedback controller is possible, which can be done by using a laser tracker for online measurements of the real position [7] or predicting the force on the end effector [8, 9]. In general, for the compensation strategy models for the dynamics or the inverse dynamics of the robot are necessary [8, 10, 11].

The objective in human-robot collaboration (HRC) is much different, because the focus lies on a cooperative, efficient and safe behavior of the robot. When the human and the robot work together in a shared workspace, the shared workspace is considered as a dynamic and unstructured environment. In such environments, the pre-programmed robot trajectory, which is generated by using traditional offline programming or online teach-in processes, could lead to design non-productive robot facilities and prohibit the robot to reach high level of interaction with human as described in [2]. In order to avoid collisions with human, the typical solution is to use safety sensors to monitor the shared workspace and to stop the robot if a human enters the defined danger area. Such approaches ensure the human safety but reduce the efficiency of the production lines. To solve this problem, an online collaborative path-planning framework has been proposed to adjust the robot's trajectory in real-time based on humans or any other dynamic obstacles according to the safety standards DIN ISO/TS 15066. The framework allows robots to efficiently perform their tasks while safety of humans is ensured.

The content of this paper is the combination of safe collaborative behavior and high-accuracy operations via using a model predictive controller. The paper is organized as follows. Sec. 2 describes the basic concept including a task-oriented path planner and an online operation management. The HRC and the model-predictive robot control for high-accuracy operations are presented in Sec. 3 and 4, respectively. Applications with some experimental results are given in Sec. 5, followed by conclusions in Sec. 6.

2 Basic Concept

The basic approach is presented in Fig. 1. A task-oriented offline path planner generates an optimized reference path of the robot for a desired task sequence. During online operation of the robot, the operation management system monitors the environment via external sensors and controls the robot motion. Two types of tasks are possible.

Static task: The robot directly executes the offline-optimized path (initial path) generated by the path planner. No error compensation or adaptive path generation is used. Static tasks are, for example, simple movements or pick and place operations.

Dynamic task: For collaborative tasks and high-accuracy operations, the initial robot trajectory generated by the offline path planner will be adjusted and optimized in real-time by an additional online feedback controller. In HRC tasks the robot path is adapted to human behavior, for example, to avoid collisions or realizing handing over [12] (see Sec. 0). In high-accuracy operations, the reference path from the offline path planner is fed into a model-predictive controller, which compensates dynamic or static errors due to process forces and large accelerations (see Sec. 4).

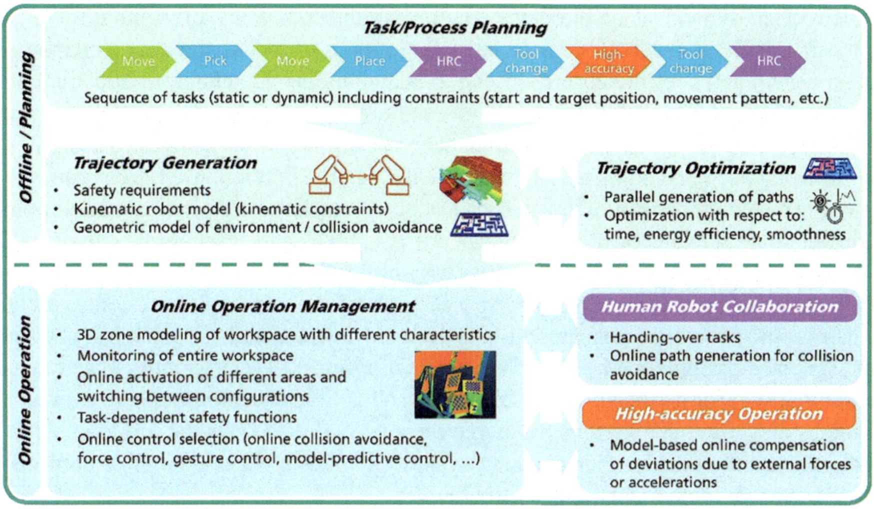

Fig. 1. Basic concept of the present approach

2.1 Task-Oriented Path Planner and Optimization

As illustrated in Fig. 1, firstly, the task-oriented path planner generates the required operation sequence of the robot based on pre-defined task-descriptions. It matches every task with the corresponding constraints. Typical task-related constraints are, for example, the start position, the target position, the speed of the tool center point, information about the orientation of the tool center point and the robot movement patterns (Linear, PTP). The global search-based path planner of the proposed approach will generate multiple collision-free paths, which fulfill the relevant task-related and kinematic constraints given by the kinematic robot model. Then, the proposed planner uses a multi-objective optimization algorithm to optimize each offline-generated path. It aims to ensure that each optimized path is locally optimal. The optimization in the offline phase could take into account different criteria, e.g. fastest-speed, lowest energy consumption, or smoothness (i.e. minimum jerk). The coupling between the global path planner and the local optimization algorithms ensures that the generated robot path is always globally optimal. These globally optimized robot trajectories will be fed into an online operation management module as initial motion trajectories for the robot in order to perform the required tasks. Furthermore, the task-oriented path planner classifies the proposed tasks either as static or dynamic one in the term of real time path adaptation.

2.2 Online Operation Management

The robot action requires task-specific functions and online control including data acquisition and analysis from external sensors. These functionalities are realized in the online operation management. The operation management is based on 3D zone modeling of the workspace, where different zones could have different characteristics and different control algorithms. Various parameters in every zone can easily be configured, e.g. a) dimensional/ dynamic characteristics, b) allowed objects/ events, c) required robot control using different feedbacks. When each operation is executed, the operation management system activates the corresponding area and applies the characteristics/functions of the zones to the executed operation:

Safety Functions: The operation management system uses sensors to monitor the entire workspace and continuously evaluate the safety situation in different zones. For example by recognizing the characteristics of objects in the activated zones, it determines whether some humans or objects appear in these zones and reacts to the situation. To obtain more accurate recognition results, various sensors have been used to detect and track the dynamic objects in the workspace, e.g. human and robot. For instance, a stereo camera could provide some features of the detected object. Another sensor, e.g. near-field vision system, could also be suitable for detecting more object features. Every sensor delivers this information in its coordinate system. Consequently, transformations are made to the coordinate systems to merge the acquired data from the different sensors. As a result, the operation management system provides a list of recognized objects in the workspace, including their detected features.

Control Selection Function: Since different operations need to be realized via different control strategies, the operation management system needs to switch between different control algorithms and ensure the stability of the entire control system. We integrate the control algorithms and the 3D zones to make the control strategy more flexible and stable. The control algorithms in our system include physical human-robot interaction based on force control, gesture control based on visual servoing, online path planning to avoid collisions between the robot and dynamic obstacles, online force compensation based on model-predictive controller, etc.

3 Human Robot Collaboration

In human-robot collaboration tasks, humans and robots share the same workspace: they can complete the same work or either each one completes their own tasks. In general, the position of the human can be only predicted for a short time horizon and the robot needs to adjust its trajectory in real time during its operation. Our online collision avoidance finds a collision-free trajectory of the robot in real-time according to human behavior, so that human and robot can safely complete their tasks without disturbing each other. In particular, the sensor system captures the current position, direction, and speed of the human. Then, the algorithm predicts the human movement in a short time horizon and generates a safe collision-free robot trajectory. The online collision avoidance continuously monitors human movements in the workspace based on updated sensor data and reacts with a path adaption if a collision with humans is predicted.

Based on the planned tasks in the offline task planner the control strategy can be changed by the zone-based operation management system to cooperate safely and efficiently with people. For instance, in a handing-over task, our system uses global sensors to find the human's position and plans a collision-free safe trajectory to approach the human. Then the robot uses a local sensor to track the position of the human's hand and hands the workpiece to the human worker [12]. Another example is gesture control, which also requires a vision-based control strategy (visual servoing).

4 High-Accurate Mode

In the high-accurate mode, we use some type of model-predictive control as an additional controller for the robots motion. It scans the current internal state of the robot (axes positions) and the external force on the end effector. A robot model uses this state together with the reference path to predict the future motion of the robot. The time-horizon for the prediction is one cycle of the robot controller (4ms). Based on the deviation between the predicted and the reference position, a compensated input is send to the robot controller for the next time step. In contrast to typical model predictive controllers, no numerical optimization is performed, because the compensated input values must be determined within a few milliseconds. Instead, we use an analytical formula to derive an optimized input at the next step from the predicted deviation.

The crucial part for such a control strategy is the model for the description of the dynamic behavior of the robot. For the simulation, a multi-body system is used, which is build in Matlab/Simulink/Simscape (see Fig. 2). It contains stiffness and damping values of the joints and a simplified model of the controller of each axis. For considering elasticity of the links, surrogate joints can be inserted in the middle of the long arm elements [8]. Tests with predefined loads on the end effector for several static poses of the robot are carried out while measuring the deviations of the robot with a Laser Tracker FARO ION. In addition, similar tests without loads but a predefined motion of separate robot axes are performed (cf. Fig. 3). The identification of the model parameters from the load experiments is very similar to [8]. A detailed description of the evaluation will be given elsewhere. The robot model from the Simscape environment is not suitable for an online implementation in the model-predictive controller. Therefore, we converted the model equations to C++. Currently, only a quasi-static model including stiffness values for each joint is implemented. A compensation of forces with the model is presented in Sec. 5.

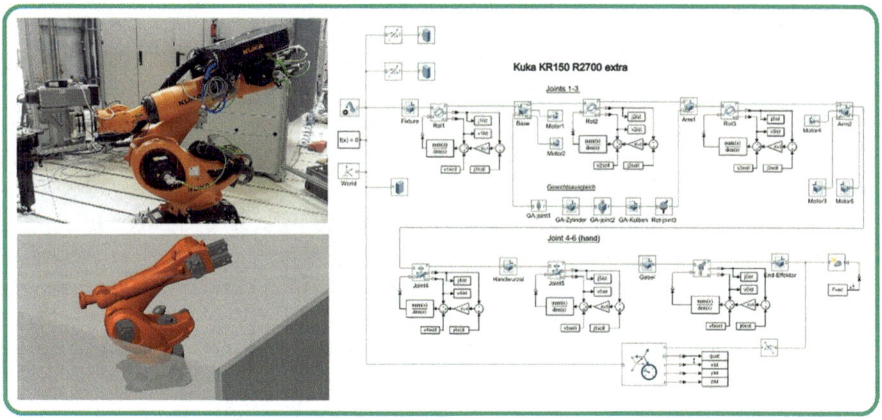

Fig. 2. Real robot and Simulink/Simscape model

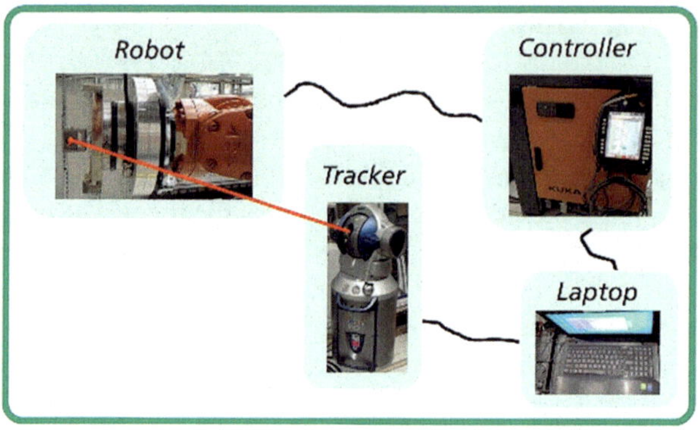

Fig. 3. Model identification with laser tracker

5 Applications

The task-oriented path planner can generate the optimal collision-free trajectory of robots for static tasks. An example is presented in Fig. 4a. The industrial robot picks up a large workpiece from a rack. Due to the large size of the workpiece, programming this picking process by manual teach-in usually takes tens of minutes, whereas our planner takes only a few seconds. Our planner optimizes the cycle time or energy consumption while avoiding collisions between the robot, the workpiece, and the rack.

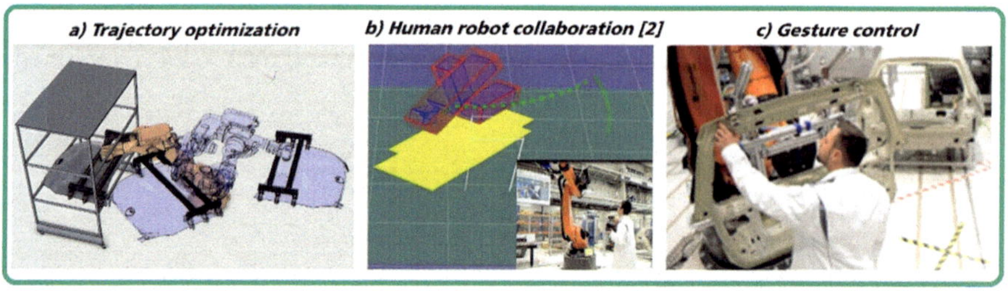

Fig. 4. a) Optimized picking operation, b) human robot collaboration, and c) gesture control.

A human robot collaboration task may follow the picking operation. In Fig. 4b, the robot brings the workpiece to the inspection zone where a human can control the workpiece quality. When the robot reaches the zone for human robot collaboration and the worker enters the inspection zone, the collaboration starts. The worker can guide the robot by using gesture or force control for adjusting the working height (see Fig. 4c). The online collision avoidance guarantees a safe trajectory for the complete process.

Regarding high-accuracy operations, we have done two experiments on a KUKA KR 150 R2700 extra with KSS Version 8.3.17, which demonstrate the need for the model-predictive control and show the first steps for compensating deviations. The

KUKA.RobotSensorInterface (RSI) was used for interaction of the additional model-predictive controller and the KUKA robot controller. The RSI allows receiving of position and force data from the robot sending compensated positions within one interpolation cycle (see Fig. 5). A force/torque sensor Schunk Omega160 SI-2500–400 was used for online measurement of forces during operation. In a first trial, various loads were manually fixed to the robot flange at a static pose (Fig. 6). A weight-driven deviation up to 2 mm was measured at the end effector. Then the model-predictive controller was turned on, and the complete deviation was compensated by the calculated correction values for each axis. In the second experiment, the robot moves along a linear reference path with a maximum speed of 250 mm/min. With the original KUKA controller and a weight of 50kg a significant deviation between the reference and the real path occurs. It is clear that an additional mass can be compensated separately by the controller or by a teaching process. However, this is no longer possible when process forces change along the path, for example, in a machining process. Therefore, the model-predictive controller was tested for compensating the displacements. In this case, the required correction values lie above the values that can be applied to each robot axis in one interpolation cycle and the control becomes unstable leading to oscillations. In a first workaround, a simple low-pass filter was applied and the gain of the correction was reduced. As a result, the deviation due to the external force cannot be fully compensated and some oscillations remain as can be seen in Fig. 7. In addition, Fig. 7 shows the general capability for compensating also deviations from high accelerations. The results were obtained from simulations of an exemplary trajectory via the model described in Sec. 4 and Fig. 2.

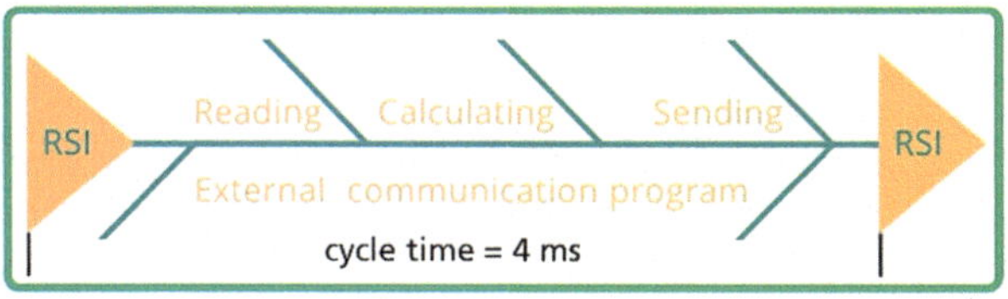

Fig. 5. Principle of KUKA RSI communication

6 Conclusion

We have presented a framework for realizing collaborative and high-accuracy operations with industrial robots. It combines well-known technologies, such as task-oriented path planning and offline optimization of robot paths, with recently emerging topics, such as collaboration of humans with industrial robots in a shared workspace and model-based online compensation of errors between the reference and the real path of the robot [8, 9]. Although the position error amounts in fact only to a few millimeters, it may still be a problem for high-accuracy tasks, as for example, robotic-based milling [1, 3, 4]. The objective here is the complete compensation of force and acceleration-induced deviations by using predictions of the robot behavior in a short time-horizon

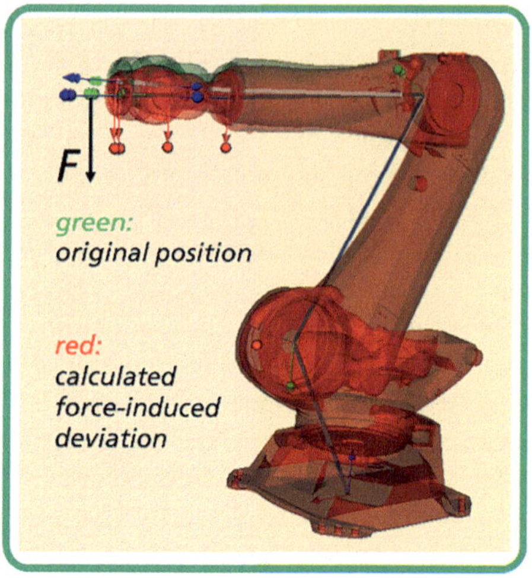

Fig. 6. Experiment with static loads

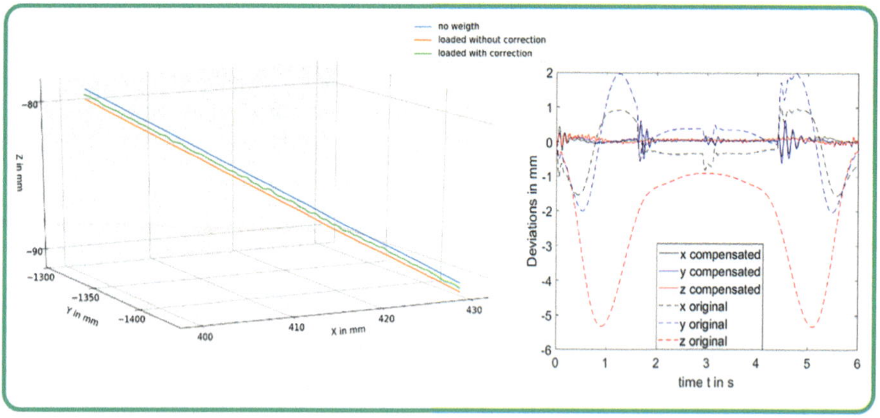

Fig. 7. Experimental results for compensation of external forces (left) and simulation results for compensation of inertia-induced deviations (right).

via accurate mechatronic models. The experiments have shown that there are several practical limitations regarding computational time for complex models, torque limits of the axes and stability of the control implementation. Therefore, further modifications of the model-predictive controller are planned, which take into account an efficient implementation of a dynamic robot model, constraints such as the maximum allowed state-dependent correction per each axis and interpolation cycle and the overall stability of the feedback control.

References

1. Ji, W., Wang, L.: Industrial robotic machining: A review. Int. J. Adv. Manuf. Technol. **103**, 1239–1255 (2019)
2. Bdiwi, M., Pfeifer, M., Sterzing, A.: A new strategy for ensuring human safety during various levels of interaction with industrial robots. CIRP Ann. **66**, 453–456 (2017)
3. Reinl, C., Friedmann, M., Bauer, J., Pischan, M., Abele, E., von Stryk, O.: Model-based off-line compensation of path deviation for industrial robots in milling applications. In: 2011 IEEE/ASME International Conference on Advanced Intelligent Mechatronics (AIM), Budapest, 2011, pp. 367–372 (2011).
4. Klimchik, A., Bondarenko, D., Pashkevich, A., Briot, S., Furet, B.: Compliance Error Compensation in Robotic-Based Milling. In: Ferrier, J.L., Bernard, A., Gusikhin, O., Madani, K. (eds.) Informatics in Control, Automation and Robotics, pp 197–216. Springer, Cham (2014).
5. Verdonck, W., Swevers, J.: Improving the dynamic accuracy of industrial robots by trajectory pre-compensation. IEEE International Conference on Robotics and Automation, Washington, DC **2002**, 3423–3428 (2002)
6. Cordes, M., Hintze, W.: Offline simulation of path deviation due to joint compliance and hysteresis for robot machining. Int. J. Adv. Manuf. Technol. **90**, 1075–1083 (2017)
7. Shi, X., Zhang, F., Qu, X., Liu, B.: An online real-time path compensation system for industrial robots based on laser tracker. Int. J. Adv. Robot. Syst. **13**, 1–4 (2016)
8. Zaeh, M.F., Roesch, O.: Improvement of the machining accuracy of milling robots. Prod. Eng. Res. Devel. **8**, 737–744 (2014)
9. SINUMERIK Run MyCC /ROCO. Siemens AG (2019).
10. Abele, E., Weigold, M., Rothenbücher, S.: Modeling and Identification of an Industrial Robot for Machining Applications. CIRP Ann. **56**, 387–390 (2007)
11. Klimchik, A., Pashkevich, A., Chablat, D.: Fundamentals of manipulator stiffness modeling using matrix structural analysis. Mech. Mach. Theory **133**, 365–394 (2019)
12. Bdiwi, M., Suchý, J., Winkler, A.: Handing-over model-free objects to human hand with the help of vision/force robot control. In: 10th International Multi-Conferences on Systems, Signals & Devices (SSD13), Hammamet, 2013, pp 1–6 (2013).

Industrial Indoor Localization: Improvement of Logistics Processes Using Location Based Services

Niklas Hesslein(✉), Mike Wesselhöft, Johannes Hinckeldeyn, and Jochen Kreutzfeldt

Hamburg University of Technology, Theodor-Yorck-Straße 8, 21079 Hamburg, Germany
itl@tuhh.de

Abstract. The function of logistics is to move objects from place A to place B Therefore it is an important task to determine the position of objects along the supply chain. Up to now, different technologies for indoor localization have been developed. During this technological evolution, two important aspects have been left behind, the communication between the systems and the central management of the data provided. Within the research project 'Industrial Indoor Localization', an open source software standard for environment modeling called 'Reference Architecture Indoor Localization' (RAIL) has been designed and tested. This protocol enables Location Based Services to allow querying location and object information quickly. Within the scope of the project, four services are being developed and exemplary described in this paper. At the beginning, a location-dependent order allocation algorithm for order picking is presented. This algorithm reduces the waiting time in narrow-aisle warehouses through forecasting. Secondly, an order orchestration service based on location data for picking control is evaluated. Furthermore, a functional area recognition to support picking and a service for finding points of interest is demonstrated, which includes navigation based on a routing algorithm. Using indoor localization, these services improve the intralogistical processes. These include increasing picking performance or reducing order throughput time by eliminating the need to scan the barcode. Finally, these services increase work safety, due to consideration of functional areas, and improve the transparency of the location of points of interest.

1 Introduction

The intensifying digitalization in production and logistics provides a wide range of possibilities for optimization of processes in manufacturing and transportation companies. Especially in the field of real time localization a great amount of data is collected and ready for further use. The combination of this data generated by different localization systems with other operational data is necessary to draw further conclusions and thus provides number of various interfaces. In addition to a high programming effort, these

P. Weißgraeber et al. (Eds.): *Advances in Automotive Production Technology – Theory and Application,* ARENA2036, pp. 460–467, 2021. https://doi.org/10.1007/978-3-662-62962-8_53

interfaces also provide a complex data structure, which results in a difficult way of extending the system with additional programs to gain further information from the gathered data. This problem can be tackled by an open standard which provides a single communication interface for different sensors as well as users. In the following paper an architecture tackling this problem will be introduced. It is shown that this architecture ensures an easy way of implementing Location Based Services (LBS), which uses different data sources to generate added value for different procedures with either production or logistics background.

2 State of the Art

Indoor localization is becoming an increasingly important topic in research, that promises considerable potential for improvement in various operating processes. Consequently, various localization systems are being developed. These can be camera-based, like motion capturing systems, or radio-based, like Bluetooth Low Energy (BLE) or Ultra-Wideband (UWB) [1]. However, localization systems are often proprietary solutions [2, 3]. As a result, only applications that have been developed by the manufacturer or certified partners can access the obtained localization data. There is thus an increasing risk of isolated solutions emerging in companies. This makes it more difficult to reuse these in a value-added way, since switching between different vendors is impossible.

Simultaneous use of two or more systems or the operation of the same technology from different manufactures therefore requires additional software that has interfaces to the respective systems. Since the realization has a high implementation effort, due to the large number of providers, research is currently being conducted on various localization standards. These include Omlox [4] and BearLoc [5] or RAIL [6]. The common aim of these standards is to develop a uniform communication protocol for storing and retrieving location positions in order to enable so-called LBS [2]. Accordingly, all three standards are designed to receive, store and merge location data from different technologies [4–6]. Each of these standards have different strength and weaknesses. BearLoc, for example, was developed with a low throughput rate of 1 Hz, which is not suitable for industrial applications [5]. In addition, detailed environment modelling is difficult to realize, as is the case with Omlox. This is due to the fact that both systems can only store the localization data [4, 5]. Additional data types or object information cannot be used. RAIL, on the other hand, was developed to react dynamically to a wide variety of data types, such as CAD, point clouds, or picking orders. In addition, this reference architecture offers the possibility of merging geometric object information with localization data, so that a real-time map can be easily generated. [6].

By providing localization data in combination with operating and control data, so-called LBS, can be developed to add value for the operational process in the industry [7]. The new services are considered one of the most interesting applications of Internet of Things [8]. LBS are functions and algorithms which are integrated in an application. They only provide information to the application, which handles interactions with other IT systems or the user. [9]. Different LBS can be combined in one application [9]. A well-known example with a combination of LBS is Google Maps. Navigation, tracking and traffic information are just a few functionalities for LBS. All in all, LBS can be divided into four categories for classification purposes: Tracking and navigation, security, information and transactions [10]. The range of functions varies from the transfer of location information to complex algorithms for merging location data with other information [10–12]. Looking at the indoor application of LBS, the number of LBS providers is considerably reduced, which also implies a much smaller number of available different services for the indoor environment. Therefore, the focus primarily lies on position determination and geofencing. Especially in the logistic context, the use of LBS is currently not commonly used, as the real time localization of logistic objects is associated with high investments. Since existing IT systems cannot fully cover the requirements of LBS, the procurement of new localization systems cannot usually be avoided. However, the use of geographic location data can offer considerable improvements in the process flow [13]. Especially the services route planning, tracking & tracing, localization, navigation and functional area recognition were predominantly evaluated as positive in a Delphi expert study [13]. As shown in the study [13], the application of LBS offers a high potential for the optimization of operational processes. Consequently, there is an increased need for research and development of LBS. To make this possible, it is also necessary to reduce the number of interfaces to localization systems. For this purpose, a communication standard must be developed that will become established in the industry.

3 Method

As part of the project IIL a prototype of the reference architecture RAIL has been developed to gather data from different sources. The data is further passed to applications which generate optimized results for different processes based on it. With that being said the following thesis was established: a localization standard is necessary to make best use out of localization data. To confirm this thesis three LBS have been implemented to show how one can simply integrate different sources into an application while maintaining just one interface for communication. Furthermore, it was analyzed how the designed LBS generate added value for companies. In order to analyze and prove the benefits of the LBS, these services have been validated and verified either by simulation or practical tests on a laboratory scale.

4 Location Based Services

4.1 Location-Dependent Allocation Algorithm

The location-dependent allocation algorithm (LDAA) addresses the problem of waiting time in narrow aisles. Based on current and future location data of the individual employees, the orders are to be transmitted to the order picker who can achieve the shortest order throughput time. The problem of existing forklift control systems is addressed that a narrow aisle is only considered to be passable as soon as the previous order picker has made a pick in the next aisle. Else there will be waiting time or detour. To solve this problem, the algorithm incorporates geographic location information from RAIL. At the beginning of the order assignment, a check is made to see whether an aisle overlap with other pickers is likely when the order is processed. This check takes place by determining all location positions of the employees at the calculated times of the picks. If there is no overlap, the order can be assigned to the next picker. Otherwise it is checked, if there will be a collision between the pickers. Restrictions therefore are the driving direction, estimated time of arrival and the storage location of the picks. If a simultaneous entry is possible according to the boundary conditions, the order is assigned. Otherwise the order is held back due to the waiting time and another one is allocated. After the order is allocated to the picker, the picking process begins. As already explained, the service controls the picking process as well. The difference to the first part is that only the next pick is considered and no longer the complete order. First, the system checks whether the next pick is feasible. If this is the case, the picking is continued. However, if it is not possible to take on the next pick, the system checks whether the pick sequence should be adjusted. If this leads to an improvement, picking continues according to the updated picking sequence.

Using a simulation, the LDAA was compared with an algorithm used in companies. The results show that the waiting time could be significantly reduced by using location information by up to 14%. This can be explained by the fact that the algorithm of the practice partner either makes the order picker wait or adjusts the picking order, but does not release the aisle earlier. The introduction of geographical location information, on the other hand, allows the aisle to be used a second time even though it is already occupied. The improved waiting time also increases the pick performance. As a result, the LDAA shows an increase in performance of up to 12% compared to a competing algorithm from a Warehouse Management System (see Fig. 1). It must be noted, however, that the algorithm of the practice partner does not need to have the complete picking list of an order at the beginning. This shows a higher flexibility in the process flow.

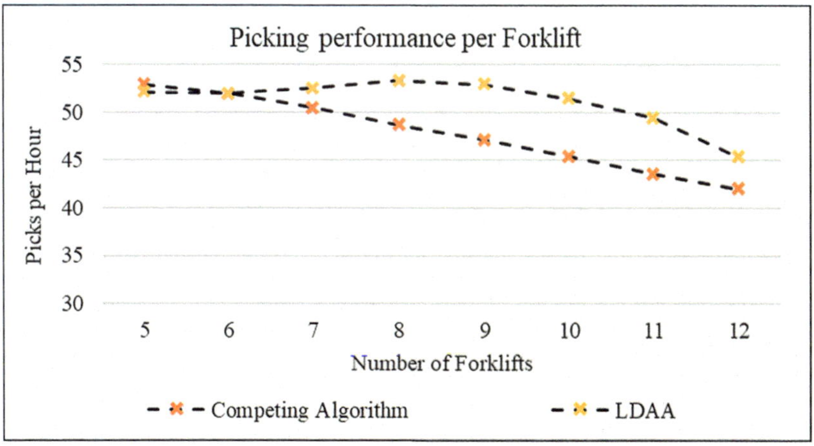

Fig. 1. Comparison of the picking performance between the competing algorithm and LDAA

According to the performance indicators, changes between the allocation algorithms only occur when a high number of pickers is active. This happens due to the fact that the LDAA is designed for a high probability of aisle overlap. The greatest improvement can be achieved with eight to eleven order pickers on 16 aisles. With more order pickers, the improvement drops, since there are fewer possibilities to avoid potential aisle overlaps and more frequent waiting time is required. An improvement of the algorithm could be achieved by using machine learning (ML). However, this was not the subject of this research. It is expected that ML can more accurately predict aisle overlaps through continuous training. In order to achieve this, the picking performance of individual order pickers can be considered to increase the accuracy of the prediction. By incorporating such information, waiting time could be further reduced. In future research the benefit of ML in this LBS should be investigated. All in all, it can be seen that the consideration of real time localization and operational control data both in order placement and during picking has a significant influence on picking performance in narrow-aisle warehouses. However, it is necessary to merge these two different types of data. In order not to develop a separate interface for each operating and localization system, it is necessary to introduce a reference architecture such as RAIL.

4.2 Order Orchestration

In addition to order allocation, the entire order picking process from order placement to completion can be supported by means of an LBS (Order Orchestration). Here, the localization information is received from RAIL and the operational data from the Warehouse Management System (WMS). The aim is to shorten the picking process by automatically transmitting all relevant information of the WMS. Relevant logistics processes are monitored by means of localization data, so that the current status is determined and then transmitted to the WMS. Thus, the use of a handheld is no longer necessary. The sequence of a picking process with a LBS is shown in Fig. 2.

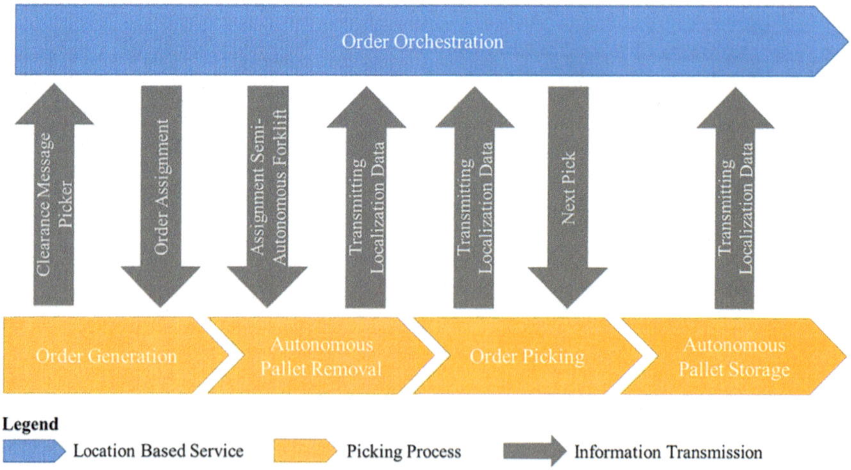

Fig. 2. Process of Order Orchestration

The developed LBS acts as an interface across all steps and continuously exchanges information with the systems used, such as the forklift. The process of the order orchestration is divided into 4 sections. After the picker has completed the last order, the LBS assigns to both a new order and a semi-autonomous forklift. At the beginning and end of the order, the forklift autonomously transports the pallet to the location designated by the WMS. The service updates the new position in the WMS, because of the real time location data out of RAIL. During the picking process the LBS transmits the respective picking list to the handheld. In addition, the service calls up the position of the order picker from RAIL in real time, as well as the location information of the storage compartment of the next pick. As soon as the employee is located directly in front of the storage compartment, the pick is transmitted to the WMS as completed and the order list is updated. This process is repeated until the order is completely processed.

As described, both localization data and information from the WMS are required, so that the LBS can automatically confirm all goods movements without scanning the barcode in the WMS. Since localization data is required directly from the forklift during pallet storage as well as from the order picker during order processing, the use of different localization systems such as lidar sensor technology and UWB is necessary. By integrating RAIL as middleware, both information can be merged and made available to the LBS in real time. Furthermore, RAIL also allows all data from the WMS, such as the picking order, to be made available centrally.

4.3 Feature Based Recognition of Points and Areas

Another LBS (Feature Based Recognition of Points and Areas) tackles the problem of locating different points of interest, such as a specific forklift, gates or sanitary facilities. In addition to this, functional areas can be integrated into the warehouse without adding any hardware. The integration runs via different types of IDs, which are unique for every object. Therefore, the data gathered in RAIL is used to generate a grid of the warehouse, which can be scaled to the necessary accuracy needed for a specific action (see Fig. 3).

The grid ID's are separated into three categories: negative, positive and floating-point numbers, which represent the following classes: impassable objects, points of interest and functional areas. The floating-point numbers are used to give a certain speed limit, for example if the number 0.5 is stored into a field the movement speed should be halved. With this ID-representation every point of interest in the warehouse can be easily retrieved and an agent can be guided towards it, while getting information about various locations in the warehouse.

To achieve this goal, it is necessary to collect data from different localization systems and combine them with operational data from the WMS. To solve this problem, normally

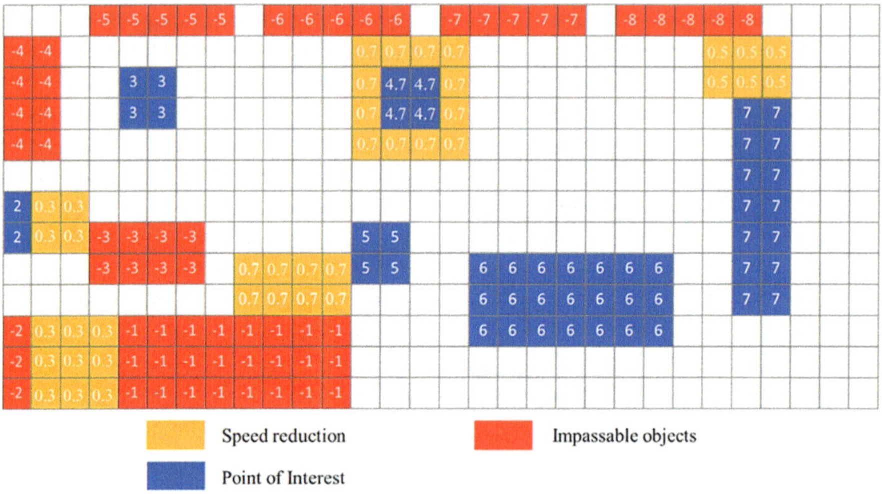

Fig. 3. Systematic Presentation of the Feature Based Recognition of Points and Areas

different interfaces have to be implemented for the communication between the LBS, Sensors and WMS. However, RAIL as a middleware offers the advantage that all data can be stored centrally and called up in a standardized manner. This significantly reduces the programming effort. Furthermore, the scalability of the grid means that different localization systems must be used for different applications. To allow this kind of flexibility it is important to communicate with these varying localization services by using RAIL. Additionally, e.g. data of heavily used routes can be transferred back towards RAIL to provide a basis for further LBS like heatmaps, order optimization algorithms etc.

5 Conclusion

Proprietary localization systems, such as those being developed today, can lead to isolated solutions within a company. This means that when LBS are introduced in an indoor environment, different interfaces to the information sources must be implemented. Especially the prototypical LBS LDAA and Order Orchestration require both localization and operational data. In the Feature Based Recognition of Points and Areas on the other hand, a dynamic use of different localization systems is assumed due to the scalability of the grid. The development of interfaces for each required data source involves significant additional expenditure of time and money, which makes the introduction of such systems unattractive. In order to solve this problem, a standardized middleware, such as the prototype RAIL, has to be developed, which is accepted by various companies. Without acceptance in the industry, no standard will be able to prevail, so that, for example, industry associations should be incorporate in the development. At the same time, the use cases of LBS should be further researched in order to demonstrate the added value of such systems to potential users.

Funding Information. The project is supported by the European Union and the State of Hamburg within the framework of the European Fund for Regional Development (EFRE). The project started in April 2019 and will be completed in September 2020.

References

1. Mautz, R.: Indoor positioning technologies. Habilitation, ETH Library (2012)
2. Zafari, F., Gkelias, A., Leung, K.K.: A Survey of Indoor Localization Systems and Technologies. IEEE Communications Surveys & Tutorial **21**(3), 2568–2599 (2019)
3. Greßmann, B., Klimek, H., Turau, V.: Towards ubiquitous indoor location based services and indoor navigation, 7th Workshop on Positioning, Navigation and Communication, pp. 107–112, Dresden (2010).
4. Schneider, D. T.: Wenn die Logistik den Unterschied macht. https://www.bvl.de/files/1951/1976/2115/Praesentation_TRUMPF_-_Wenn_die_Logistik_den_Unterschied_macht.pdf. Accessed 14. July 2020.
5. Chen, K, He, S., Chen, B., Kolb, J., Kath, R. H., Culler, D. E.: BearLoc: A Composable Distributed Framework for Indoor Localization Systems, Proceedings of the 2015 Workshop on IoT challenges in Mobile and Industrial Systems, pp. 7–12. Association for Computing Machinery, New York (2015)
6. Wendt, A., Brand, M., Schüppstuhl, T.: Semantically enriched spatial modelling of industrial indoor environments enabling location-based services. In: Schüppstuhl T., Tracht K., Franke J. (eds.) 3. Kongresses Montage Handhabung Industrieroboter, pp. 111–121. Springer, Berlin (2018)
7. Ryschka, S., Murawski, M., Bick, M.: Location-Based Services. Business & Information Systems Engineering **58**(3), 233–237 (2016)
8. Khelifi, F., Bradai, A., Benslimane, A., Rawat, P., Atri, M.: A Survey of Localization Systems in Internet of Things. Mob. Netw. Appl. **24**, 761–785 (2019)
9. Chow, C.-Y., Mokbel, M. F., Liu, X.: A peer-to-peer spatial cloaking algorithm for anonymous location-based service. In: Proceedings of the 14th annual ACM international symposium on Advances in geographic information systems, pp. 171–178, Association for Computing Machinery, New York (2006)
10. Barnes, S.J.: Location-based Services: The State of the Art. E-Service J. **2**(3), 59–70 (2003)
11. Chakravarty, T., Balamuralidhar P.: System and method for simultaneous wireless charging, tracking and monitoring of equipments. U.S. Patent No. 9,178,569. (2015)
12. Jeong, S.Y., Jo, H.G., Kang, S.J.: Fully distributed monitoring architecture supporting multiple trackees and trackers in indoor mobile asset management application. Sensors **14**(3), 5702–5724 (2014)
13. Dubucq, A., Hinckeldeyn, J., Kreutzfeldt, J.: Einsatz von Location Based Services in Produktion und Intralogistik. Zeitschrift für wirtschaftlichen Fabrikbetrieb **113**(5), 337–341 (2018)

Interface-Free Connection of Mobile Robot Cells to Machine Tools Using a Camera System

Johannes Abicht[1](✉), Torben Wiese[1], Arvid Hellmich[1], and Steffen Ihlenfeldt[1,2]

[1] Fraunhofer Institute for Machine Tools and Forming Technology IWU, 01187 Dresden, Germany
johannes.abicht@iwu.fraunhofer.de

[2] Technische Universität Dresden, Institute of Mechatronic Engineering, 01062 Dresden, Germany

Abstract. Part complexity increase the ratio of manual tasks on machine tools, i.e. feeding, to maintain high productivity. Mobile robot cells automate these tasks but require cost-intensive retrofittings for additional interfaces. This paper focuses on a new mobile robot cell without any mechanical or electrical connection to the machine tool. Therefore, the solution uses a camera system for wireless machine tool localization and signal transmission.

1 More Complex Parts Require Flexible Automation Solutions

More complex and individual parts as well as low batch sizes and varying boundary conditions require situationally optimal machines. As shown in the automation pyramid in Fig. 1, intelligent components offer additional skills for production systems to adapt to changes in the working processes.

Because conventional full automation is cost-intensive and only productive at full capacity under constant conditions, SMEs split the whole process chain into repeatable tasks on less flexible single-use machine tools and manual operations. With its cognitive abilities, humans still perform complex and variable tasks, like machine tending, part cleaning or deburring, where they are more flexible then automated systems [1].

The high level of manual tasks leads to productivity losses because of still standing machines in case of illness, vacation or job resignation of workers. Therefore, mobile robot cells are solutions to automate such manual tasks and thus maintain productivity in the case of worker absences [2]. However, necessary retrofittings of mechanical and control interfaces on the machine tools are great barriers for a large-scale usage of common concepts in the customer production facilities.

Therefore, this paper focuses on a new mobile robot cell that is applicable to machine tools without an electrical or mechanical coupling as *interface-free connection.*

For this purpose, a robust *camera-based localization* of the machine tool and *signal transmission* via existing HMI devices adds additional skills to the mobile robot cell.

P. Weißgraeber et al. (Eds.): *Advances in Automotive Production Technology – Theory and Application,* ARENA2036, pp. 468–477, 2021. https://doi.org/10.1007/978-3-662-62962-8_54

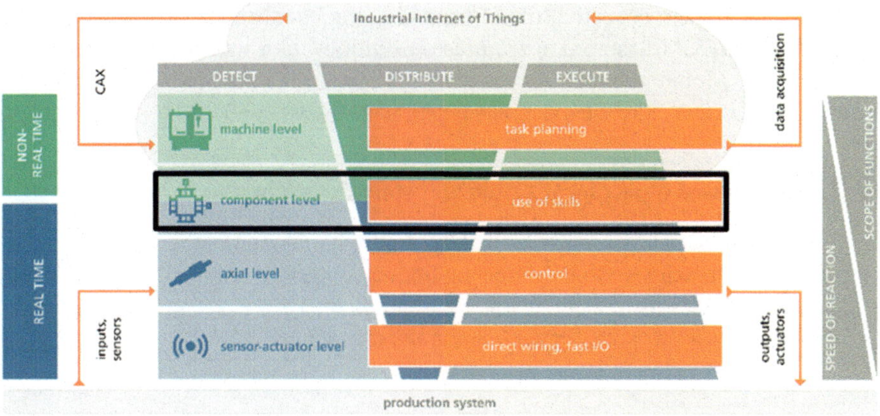

Fig. 1. Intelligent components offer skills for machine and process adaptability.

2 State of the art

2.1 Mobile robot cells for machine tool automation

Fig. 2 shows the components of mobile robot cells (MRC) for machine tool (MT) automation [3]. Most of the solutions are focused on machine tool tending applications [2–4]. Operators connect the mobile robot cell to a machine tool and configure a workflow consisting of pre-defined gripping and robot movements. As shown in Fig. 2, conventional solutions have specific docking stations (DS) to lock the relative position to the machine tool mechanically [5]. Furthermore, most solutions have wired control interfaces with standardized fieldbus connectors to send their current machine states for the workflow automation [5–7].

More flexible approaches are based on 2D or 3D vision for part and machine tool localization to adapt robot movements in case of position changes [8]. Technologies like ZigBee or Wi-Fi already enable wireless signal transmissions, but need additional gateways [9]. Some Cobot solutions use the machine control panel as signal input to automate the machine tending workflow while avoiding control interfaces [10].

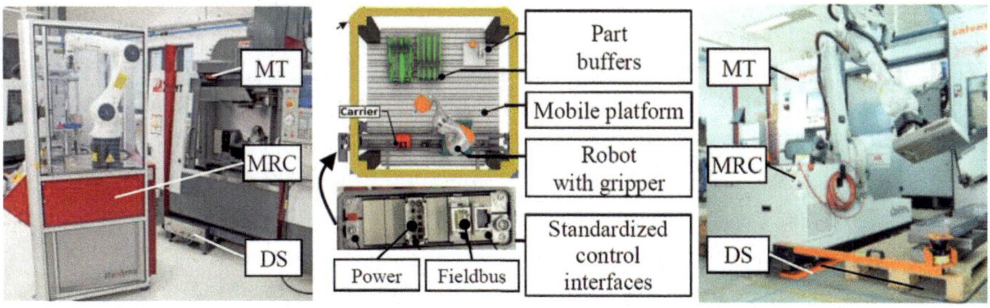

Fig. 2. Connection of mobile robot cells to machine tools [2–4, 7, 11].

Today, no mobile robot cell fulfils an “interface-free” connection to the machine tool. Especially the signal transmission requires retrofittings with wired or wireless control interfaces [11]. Mobile robot cells need an extension to a camera-based state detection through HMI devices, like the machine control panels [10].

2.2 Camera-based object localization and state detection

Camera-based object localization is widely used for pick and place applications and consists of two challenges. Object detection finds an object in the input image and object localization evaluates its translation and rotation relatively to a real world reference [8, 12]. Table 1 gives a short overview of common approaches.

Table 1 Camera-based object detection and localization approaches [8, 12].

Object detection		Object localization
2D	**3D**	**3D**
Convolutional image processing	3D shape detectors (Spin image, FPFH, SHOT)	Corresponding point clouds
Template matching (SIFT, SURF)		Template-based methods
Shape matching of contours		Voting-based methods
Marker-based (ArUco)		Regression-based methods
Deep learning based (DCNN, RCNN, YOLO)		

Commercially available solutions are usually limited to single image processing tasks that work well in static scenes. The development of OpenCV as an Open Source image-processing library enables the programming of application-specific, custom algorithms. Cheap Smart Cameras, such as Intel Realsense or ZED, encourage this trend and provide intelligent skills like auto-exposure against heavy lightning conditions [13]. 2D detection methods are suitable for fast applications with high accuracy demands. Here, convolutional processing is used for static image processing. Deep learning based methods work well in dynamic scenes, but need to be trained with large data sets. Great success has been achieved by shape matching algorithms using object contours [12]. If markers does not disturb the process, marker-based methods (i.e. ArUco markers) are robust and real-time capable alternatives through optimized algorithms.

For *camera-based state detection* of the machine tool, HMI devices feature optical information, like color and text signals (i.e. errors) on the control panel or the signal tower. [14] gives an overview of the current status in text and color signal detection. Deep learning based approaches become widely established in the field of text signal detection. They focus on an improved text segmentation with “segment and linking” methods. However, different fonts, distortions and character errors complicate a fast and reliable recognition in industrial environments. For the detection of color signals, there are traditional approaches from traffic light recognition, which use color thresholding, visual saliency, morphological filtering or edge/corner detectors [14]. To reduce the calculation time, they limit the ROI by giving position information (maps). After the detection of circular lights in the ROI, threshold values or key points evaluate the signal state [15, 16].

2.3 Goal of the work and research questions

The paper focuses on an "interface-free" connection of mobile robot cells to machine tools. In contrast to common solutions with dockings stations and control interfaces, a camera system performs a *machine tool localization* and state detection for *a signal transmission* through existing HMI devices. According to the state of the art in Section 2, there is no mobile robot cell using machine vision for both aspects. Therefore, the paper addresses the following research questions:

- How to robustly detect and localize machine tools through camera-based object localization in industrial environments?
- How to send control signals to machine tools and capture their current state through existing HMI devices by camera as signal transmission?

3 Robust 2D object detection as enabler for flexible robot cells

3.1 Additional camera skills for an interface-free machine tool connection

The mobile robot cell uses existing HMIs for signal exchange and combines them with established, camera-based object localization methods. For manual machine tending, the operator pushes buttons on the MCP to transmit start, stop or release signals to the machine tool. As shown in Fig. 3, the mobile robot cell adapts the operator movements by using additional tools to press the MCP buttons. A camera system on the robot arm evaluates optical signals on HMI devices (MCP, signal towers) as triggers for the robot cell workflow when a part is finished or errors occur. For repositioning, the integrated camera system locates the machine tool and HMI devices at static detection positions so that only a rough alignment is necessary. By adding these additional camera skills to the mobile robot cell, only existing interfaces to the machine tool are used.

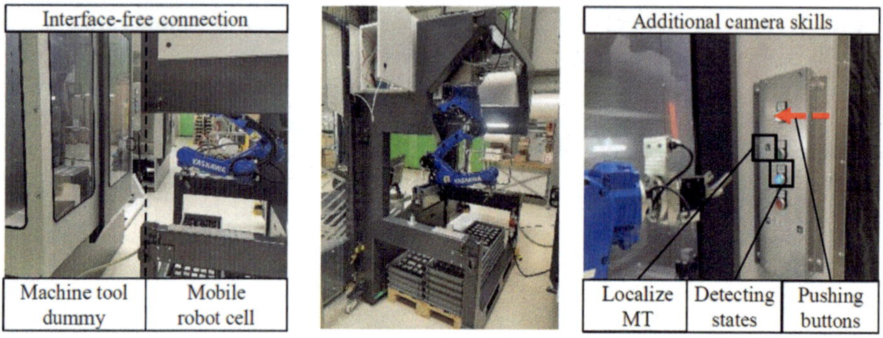

Fig. 3. Interface-free connection of mobile robot cells to machine tools.

3.2 Requirements for image processing and resulting conclusions

During operation, image processing has to handle reflective, metallic backgrounds (metal covers), changes in lighting conditions and soiled surfaces because of rough industrial environments.

Therefore, the mobile robot cell uses a Realsense D435i with RGB and depth camera module and a light exposure sensor. In addition to color images, this smart camera delivers corresponding point clouds. A software development kit and OpenCV libraries enable the programming in Python. For localization, the image processing performs a marker-based detection through ArUco-Markers with point clouds localization. The state detection approach is based on Tarkowski et al. [15].

3.3 Marker-based machine tool localization through point clouds

Concept. The marker-based machine tool localization allows a robot movement adaption in case of relative displacements. Therefore, the robot moves in specific robot user frames. Fig. 4 illustrates the concept for Pre- (1,3) and Fine-positioning (2) on the MCP. User frames are adjustable cartesian coordinate systems of the robot defined by three global points in the workspace. The camera system localizes these points by ArUco-Markers and the point correspondences at fixed global detection positions. The cell control transmits the coordinates to the robot. The robot then updates the user frame and transforms the taught movements to the new global user frame position.

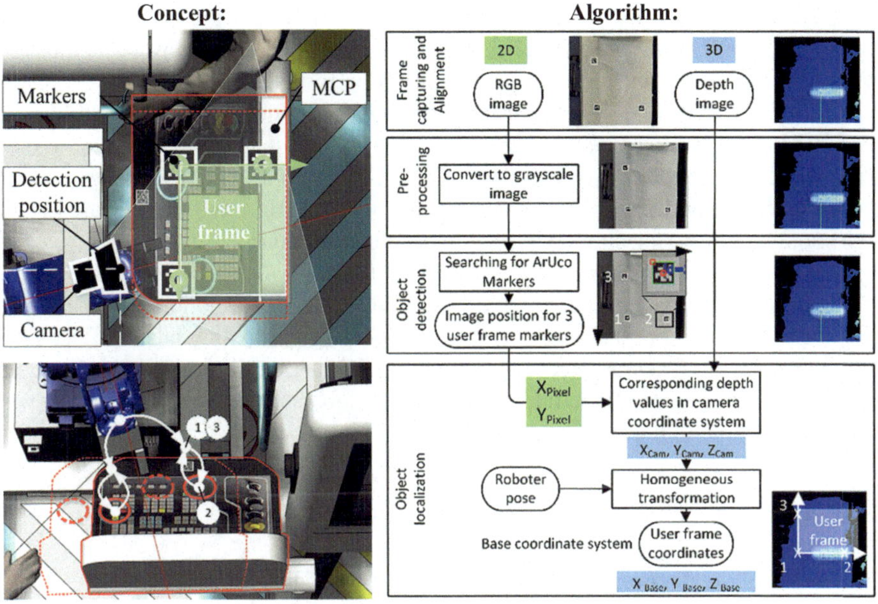

Fig. 4. Marker-based machine tool localization.

Algorithm. Fig. 4 shows the calculation of user frame points systematically. Firstly, the camera generates a frameset of two aligned RGB and depth frames. A conversion of the RGB image into a greyscale image follows as pre-processing. Step three consists of the ArUco marker detection according to [17]. Due to the frame alignment, the image processing searches point cloud correspondences $(X_{Cam}, Y_{Cam}, Z_{Cam})$ for (X_{Pixel}, Y_{Pixel}) in step 4. Finally, a homogeneous transformation translates the coordinates into the base coordinate system of the robot cell to get $(X_{Base}, Y_{Base}, Z_{Base})$.

For configuration of the machine tool localization, the operator attaches three ArUco markers for the user frame definition, teaches a global detection position and the necessary robot movements in the user frames.

3.4 Signal transmission through machine control panel

Concept. Fig. 5 describes the signal transmission between the mobile robot cell and machine tool. For signal output, the robot pushes buttons (i.e. Start NC program) with an operating tool through relative movements in user frames, based on Section 3.3. As shown in Fig. 5, color and text signals serve as signal input to define the machine tool state. These states are triggers for continuing the robot cell flow after a complete machining of a part. For the state detection, operators teach fixed detection positions for the HMI devices, like the MCP in Fig. 5, and region of interests (ROI) in the RGB-image. Basic color or text signal objects represent the signal inputs in Python. A machine configuration object combines them to a digital representation of the whole machine tool state. After serial evaluation of all signals, trigger states of the machine configuration object control the robot cell flow event-based.

Algorithm. According to Tarkowski et al., Fig. 5 describes the evaluation of basic color signals [15]. After capturing and pre-processing RGB frames over an adjustable evaluation time t_{detect} in step one, the algorithm performs a light intensity evaluation I_{Signal} and binarization for each frame in the ROI using a configured threshold value I_{Thres}. The binarization enables the assignment of a signal state for each frame stored in a signal list. In step three, the distribution of the color signals then determines the state On, Off or Blinking. Due to the signal sampling, detection time and frame rate has to be high enough to avoid aliasing effects and false interpretations.

The machine configuration object (MCO) in Python parameterizes the state detection algorithm and saves them into a text-based JSON file. A translation matrix assigns a machine tool state to a variable combination of optical signals. All signals have characteristic detection parameters as object attributes, e.g. ROIs or threshold values for the color signals. Then, the operator defines detection positions and teaches robot movements with the operating tool according to Section 3.3.

4 Experiments

Machine tool localization. For an initial evaluation of the concepts, a pre-test series on a machine tool dummy with the mobile robot cell of Fig. 3 was conducted. The test for the machine tool localization consists of nine set-up positions *(ΔX, ΔY)* within ± 100 mm. This tolerance is large enough for a rough positioning of the mobile robot cell by the operator. Three final measurements at position 1 analyze the influence of ambient light on the marker localization. For all tests, the measurement repeats 100 times.

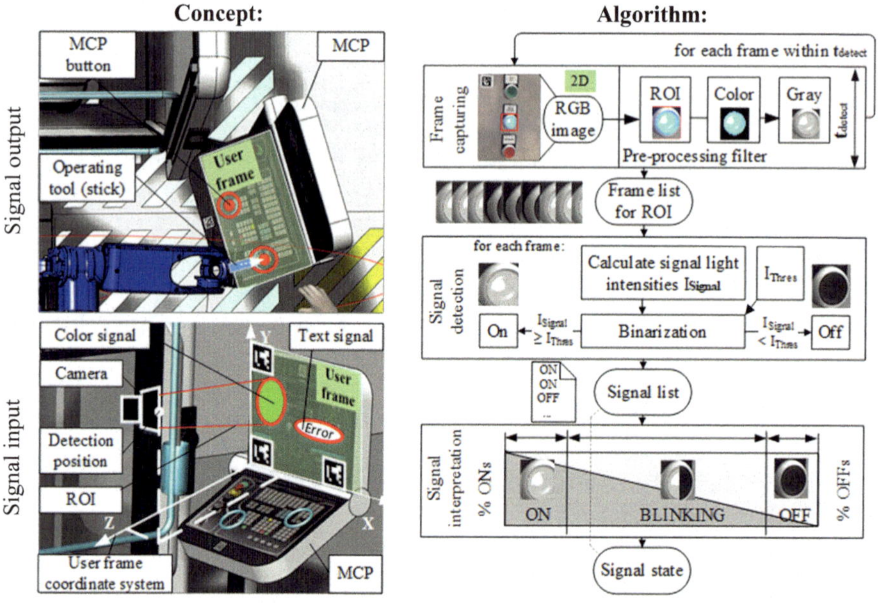

Fig. 5. Signal transmission through machine control panel.

The results for P1 in Fig. 6 show that the camera reliably detects the markers for each position with a standard deviation *s* of max. 0.25 mm. For all positions, the robot hits the green start button without errors. In comparison to the button dimensions (~20 mm), a sufficiently high measuring accuracy is given. The camera depth error leads to a higher deviation $s_{X.P1}$ in X direction. The artificial light does not influence the marker detection and localization for the tested light angles α.

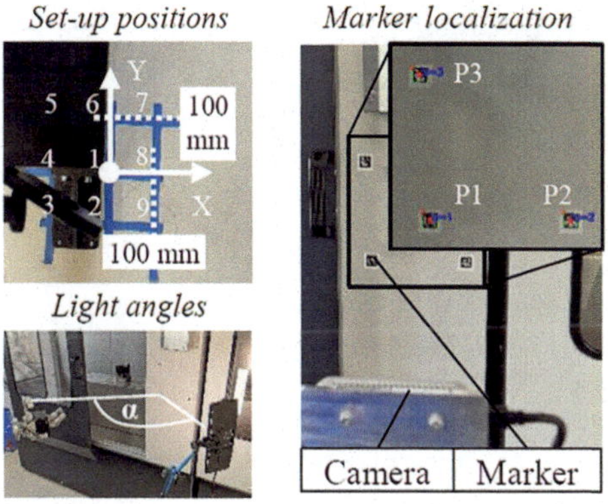

Pos.	α	$s_{X.P1}$	$s_{Y.P1}$	$s_{Z.P1}$
1	/ (Without artificial light)	0.25	0.02	0.04
2		0.19	0.04	0.03
3		0.22	0.06	0.03
4		0.15	0.01	0.02
5		0.21	0.01	0.03
6		0.19	0.07	0.03
7		0.03	0.09	0.01
8		0.10	0.01	0.02
9		0.13	0.03	0.02
1	0 °	0,13	0,07	0,02
1	45 °	0,12	0,06	0,02
1	90 °	0,12	0,06	0,02

Fig. 6. Experimental results of the pre-test series for machine tool localization.

Machine tool state detection. The pre-test for the state detection is carried out for four lighting conditions α at a taught detection point. Fig. 7 demonstrates the configuration of the blue color signal and translation into a machine state in the MCO exemplary. Pos_roi, width_roi and height_roi define the ROI for the blue color signal at the detection point for all camera frames. A initial light intensity measurement calculates the threshold value I_{Thres}. The test consists of 10 measurements for $t_{detect} = 4$ s.

According to the results, the algorithm repeatedly detects the "Finished" state within approx. $T_{state} = 4.3$ s. However, for $\alpha = 0°$ a false detection occurs in one case. One possible reason is the reflection of the artificial light on the MCP surface, which overexposes the image and distorts the light intensity measurement I_{Signal}. An adaptive adjustment of the exposure region of the Realsense camera or the detection angle to the MCP are possible strategies for stabilizing the algorithm.

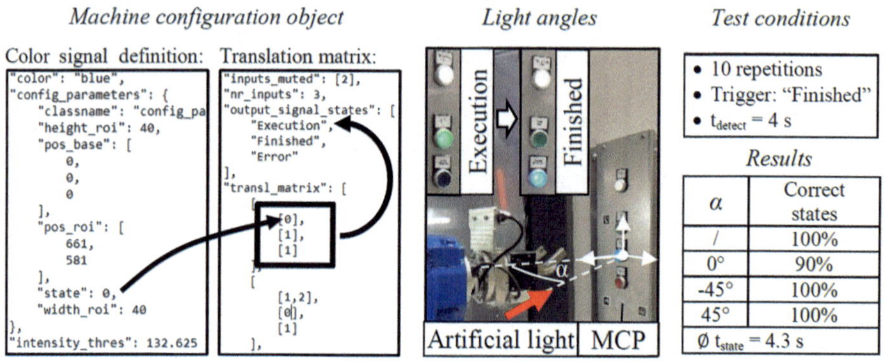

Fig. 7. Configuration and experimental results of the machine tool state detection.

5 Summary and outlook

This paper presents a new mobile robot cell without any additional mechanical and control interfaces to the machine tool. Therefore, a camera system adds additional skills for a machine tool localization and wireless signal transmission. The localization uses robot movements in adaptable cartesian coordinate systems. A robust 2D ArUco marker detection and localization in a corresponding point cloud allows the shifting of the user frame in case of relative displacements. For the signal transmission, the mobile robot cell pushes buttons on the MCP as signal output. A camera-based detection of optical signals on HMI devices serves as input for the robot cell workflow controlling. Successful pre-tests support the approach and motivate for further studies.

References

1. Seifermann, S.: Methode zur angepassten Erhöhung des Automatisierungsgrades hybrider, schlanker Fertigungszellen, p. 315. Shaker, Aachen (2018)
2. Bogue, R.: Europe continues to lead the way in the collaborative robot business. Ind. Robot. Int. J. **2016**(43), 6–11 (2016)
3. Nielsen, et al.: A methodology for implementation of mobile robot in adaptive manufacturing environments. J. Intell. Manuf. **28**(5), 1171–1188 (2017)
4. Manuel Alender. Working together as equals: Sensor solutions for robotics. https://www.sick.com/ag/en/sensor-solutions-for-robotics/w/robotics/. Accessed 10 July 2020.
5. Falkowski et al.: An Approach Towards High-Precision Docking of the Mobile Robots for Industrial Purposes. In: Automation 2020: Towards Industry of the Future, vol. 1140. Springer International Publishing, pp. 239–247 (2020)
6. Weyrich, et al.: Machine-to-Machine Communication. IEEE Softw. **31**(4), 19–23 (2014)
7. Wojtynek, et al.: Plug, Plan and Produce as Enabler for Easy Workcell Setup and Collaborative Robot Programming in Smart Factories. Künstl. Intell. **33**(2), 151–161 (2019)
8. Guoguang, D., et al.: Vision-based Robotic Grasping from Object Localization, p. 24. A Review, Pose Estimation, Grasp Detection to Motion Planning (2019)
9. Tan, et al.: M2M Communications in the Smart Grid: Applications, Standards, Enabling Technologies, and Research Challenges. Int. J. Digit. Multimedia Broadcast. **2011**(99), 1–8 (2011)

10. Cutting Tool Engineering: Cobot handles multiple-operation machining task for medical device. https://www.ctemag.com/news/industry-news/cobot-handles-multiple-operation-machining-task-medical-device (2019). Accessed 20 September 2020
11. Laura Schwarzbach: Flexible Roboterzelle optimiert CNC-Fräsprozess. https://www.maschinenmarkt.vogel.de/flexible-roboterzelle-optimiert-cnc-fraesprozess-gal-531683/?p=4#gallerydetail (2016). Accessed 2 July 2020
12. Astanin, et al.: Reflective workpiece detection and localization for flexible robotic cells. Robotics and Computer-Integrated Manufacturing **44**, 190–198 (2017)
13. Hofmann et al., 2019 - 2019. Object Detection, Classification and Localization by Infrastructural Stereo Cameras, in: Proceedings of the 14th International Joint Conference on Computer Vision, Imaging and Computer Graphics Theory and Applications. 14th International Conference on Computer Vision Theory and Applications, pp. 808–815.
14. Zou et al.: Object Detection in 20 Years: A Survey. IEEE (2019)
15. Tarkowski et al.: Efficient algorithm for blinking LED detection dedicated to embedded systems equipped with high performance cameras. In: IEEE EUROCON 2015, pp. 1–4 (2015–2015)
16. Yoneda et al.: Robust Traffic Light and Arrow Detection Using Digital Map with Spatial Prior Information for Automated Driving. Sensors (Basel, Switzerland) 20 (4) (2020)
17. Garrido-Jurado, et al.: Automatic generation and detection of highly reliable fiducial markers under occlusion. Pattern Recogn. **47**(6), 2280–2292 (2014)

The Fully Flexible Body Shop – A Holistic Approach for the Vehicle Production of Tomorrow

Marcel Todtermuschke[1](✉), Alexander Voigt[2], Rayk Fritzsche[1], Jens H. Lippmann[1], and Jörn Zastera[2]

[1] Fraunhofer Institut Für Werkzeugmaschinen Und Umformtechnik IWU, Reichenhainer Str. 88, 09126 Chemnitz, Germany
marcel.todtermuschke@iwu.fraunhofer.de
[2] Volkswagen AG, Berliner Ring 2, 38436 Wolfsburg, Germany

Abstract. Flexibility for all different models within one body shop production line could be achieved without changing jigs, grippers as well as joining tools. A comprehensive analysis of the process chain formed the basis for the necessary development steps. In addition, planning methods based on mathematical optimization algorithms were developed and applied to cope with the complexity of the task. All the individual aspects are obtained, and the newly developed components were combined into a complete plant concept and its economic efficiency has been proven by Volkswagen planning department.

1 Introduction

A significantly increasing number of variants of vehicle models as well as new markets and globalization are current challenges for all automobile manufacturers worldwide. To achieve this, it would have to be possible to implement completely flexible production from the press shop to final assembly. Flexibility is a very frequently used term, the definition of which – especially for automotive production – is not yet conclusively defined. In [1] a promising approach is comprehensively discussed and a method for determining a suitable degree of flexibility is presented. One of the bottlenecks for such an objective is the body shop, which has an automation level of more than 95%. Here, an average of 4 different models of the entire group are currently being produced on a single production line. This flexibility is achieved in today's car body construction on the one hand by model-specific grippers and welding guns, which are changed by the industrial robot by means of docking, and on the other hand by model-specific devices, which enable a model change in the so-called wild mix via changing systems, such as rotary drums. However, this leads to high investment costs, long commissioning times and large surface area requirements and ultimately to limited flexibility in the type diversity of a car body production line.

An industrially available solution of this kind is the Matrix Production concept from KUKA. This is based on a large number of cells of the same type, each of which is

P. Weißgraeber et al. (Eds.): *Advances in Automotive Production Technology – Theory and Application,* ARENA2036, pp. 478–485, 2021. https://doi.org/10.1007/978-3-662-62962-8_55

capable of adapting to different components by means of a fast and autonomous change of shape-bound tools. However, the tools of all components to be produced must be kept in one warehouse. [2].

However, there is still a desire for fully flexible operating equipment and processes, i.e. grippers, devices and joining processes as well as corresponding logistics concepts and planning tools to master the variety of variants and the shorter product runtimes. Daimler's cubeTEC concept pursues such a starting point. Here, the clamping technology based on commercially available numerically controlled axes already enables a geometrical adaptation to different components. [3].

As shown research and development activities focused primarily on the design, development and testing of flexible equipment and processes under large-scale production conditions. The use of industrial robots already allows a sufficiently high degree of flexibility for the movement of tools and components. Thus, new concepts for grippers, clamping devices and joining tools are primarily required to make the operating resources in car body construction more flexible. Therefore, it was necessary to develop software tools for an efficient plant planning and construction including visualization, which make the huge multi-criteria parameter variety controllable. The newly developed components should enable plug & play-capable control as well as predictive maintenance. These partial solutions were brought together in a concrete system concept using the example of a flexible door system, based on which an evaluation of functionality and – in particular – economic efficiency could be realized. This technical and ultimately scientific challenge led to the founding of the "Forschungsfabrik Karosseriebau" (FoFab-K) between the Volkswagen Group and the Fraunhofer Institute for Machine Tools and Forming Technology IWU in 2014 – which is a new type of research concept with direct reference to the application and the goal of a holistic, disruptive approach to fully flexible body construction. The objectives of FoFab-K are

- full type and derivative flexibility (in 'wild mix', i.e. each vehicle in the production sequence can be a different model)
- material flexibility (steel / aluminium)
- shortening of the plant construction and commissioning process to less than 2 months and this with
- consistently high quality and
- a rapid integration of new vehicle models into existing multi-model systems

2 Planning Processes

2.1 Definition of Requirements

For the development of a consistent flexibility concept for a car body construction plant, it was necessary to first analyse the current status of the process chain and above all to record it holistically and in the context of the existing boundary conditions. For this purpose, the interactions between the individual parts to be joined (including any geometric fluctuations arising from the forming manufacturing process), the joining technology and the clamping process that determines the geometry and thus quality are taken into account. From this, the forces occurring in the clamping process at the clamping points are determined, broken down according to the respective direction.

Fig. 1 shows an example of an examined clamping situation. With the knowledge of the clamping points that are particularly relevant regarding to quality influence from the experience of the designers and template makers of such devices, the number of measuring points could be limited to a minimum.

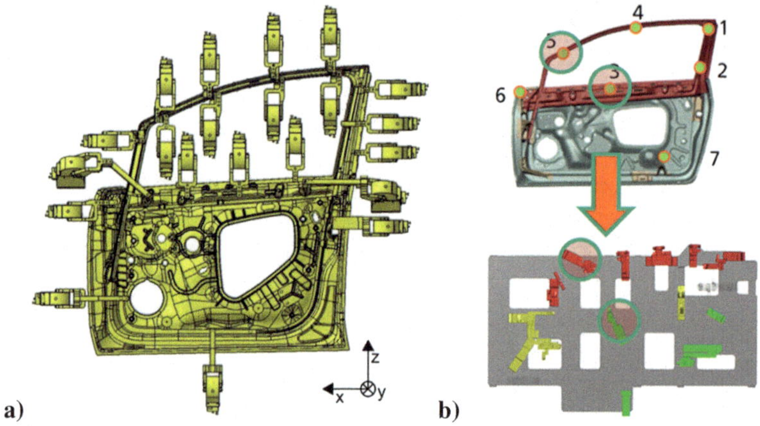

Fig. 1. Clamping situation of a selected geometry-determining device **a**) Total of all clamping points; **b**) analysed clamping points and determined forces (qualitative)

To investigate the relevance of individual clamping points, the clamping positions of these in the main direction of influence of the assembly geometry were also adjusted systematically and the effects on the resulting forces and finally on the component quality after joining were observed.

As a result of the investigations, it was found that a maximum load of 4.5 kN and a minimum stiffness of 4 mm/kN must be assumed for dimensioning the flexible clamping console to be developed. These assumptions are mainly used to realize the start-up process of new models in which usually larger geometric fluctuations of the individual parts occur than in the later series production. The actual clamping process is basically a predominantly static load, however, due to the number of clamping processes during the lifetime of such devices, cyclical loading must nevertheless be assumed.

2.2 Workspace Investigations for Flexible Clamping Technology

If *n* different, geometrically similar components are to be clamped and joined one after the other in a fixture, they can be aligned with each other in a variety of ways to determine the position of the clamping positions. This then results in different travels for the flexible clamping kinematics, see Fig. 2 and this ultimately determines not only the required working areas of these, but also the non-productive time required for changeover. Furthermore, certain positions, so-called reference points of the components are narrower in their tolerance than other clamping points, which in turn results in the shortest possible paths and stiff kinematics for this special clamping range. In addition, the smallest distances between adjacent clamping positions must also be taken into account, as these minimize the size of the clamping kinematics in order to avoid collisions and to accommodate all necessary clamping points in one fixture, while still allowing access to the

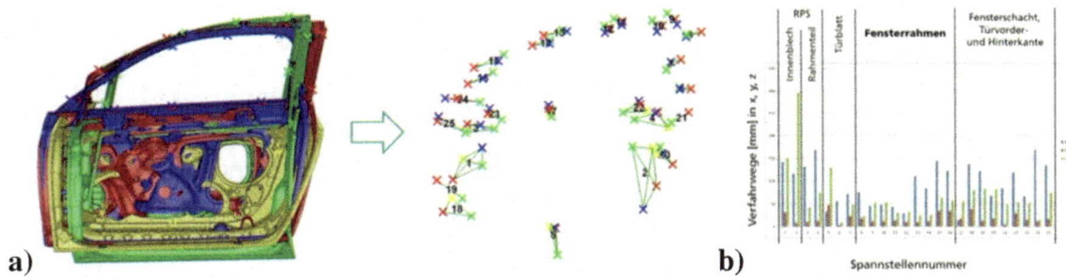

Fig. 2. Determination of the traverse paths of the flexible clamping kinematics **a**) analysed doors and their clamping points **b**) x-/y-/z-distances of all clamping points

joining tools to the joining points. The required degrees of freedom for the realization of all clamping points is the final boundary condition of this multi-criteria problem.

As a result, the minimum requirement for the working area of a clamping kinetics has been defined as 100 mm x 200 mm x 60 mm.

2.3 Software-supported Fixture Design

The complexity of the relative arrangement of the components and the related positioning and alignment of the clamping fixtures makes manual design of these impossible. There are simply too many interdependent requirements, so that it is only possible to find suitable design solutions through the targeted use of optimization algorithms. This requires a multi-stage procedure. First, a software calculates the position of the clamps under the mentioned boundary conditions of minimum travel distances and collision avoidance of the clamping kinematics. Then, based on the clamping kinematics, which are available as a modular system, an automated assembly of the fixture is carried out, which is finally implemented directly in the CAD system and then, with some manual adjustments, a suitable fixture is realized, see Fig. 3.

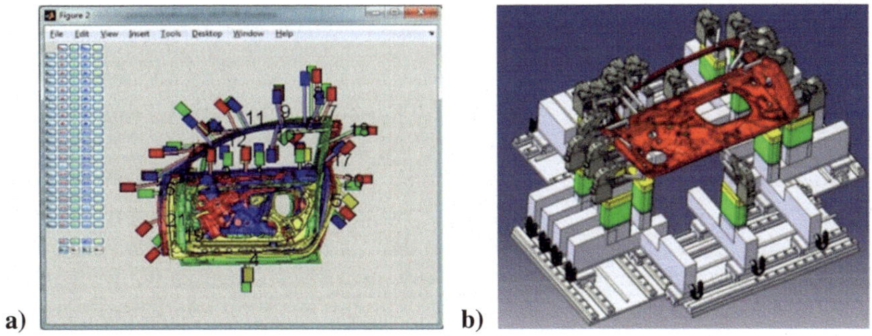

Fig. 3. Software-based fixture design **a**) multi-criteria optimization of the clamp position, **b**) modular model design in CATIA

3 Handling and Clamping Technology

Different concepts have been developed to make the clamping fixtures more flexible. One premise was that each individual clamping point should be linearly adjustable in all three spatial directions. This resulted in a basic concept in which each clamping element is on a separate three-axis kinematics. This enables automatic and model-independent adjustment of all clamping points and thus compliance with the high-quality requirements. Fluctuations in the individual part geometries from the manufacturing process can thus be compensated at any time during the clamping process across the individual joining steps. The preferred variant of the developed three-axis kinematics is shown in Fig. 4a.

Due to the narrow design, whereby the kinematics are not much wider than a standard bodywork clamp, accessibility of the joining tools is guaranteed, and several clamping points can be positioned close to each other. Process forces of up to ±4.5 kN in all three spatial directions can be absorbed at each clamping point. The axes can be controlled separately and moved simultaneously. A clamping element on the kinematics can be moved vertically by 60 mm and horizontally by 200 mm x 200 mm with the kinematics shown in Fig. 4 with an accuracy of 0.1 mm. The travel speeds are designed so that all positions within the working range are reached within 5 s.

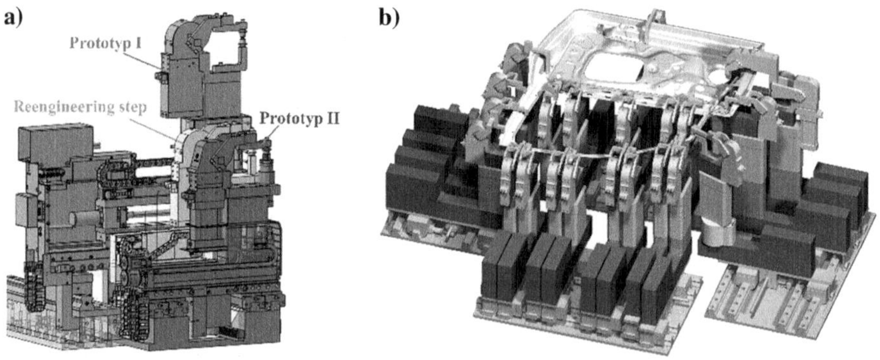

Fig. 4. **a**) Three axis kinematics; **b**) Model flexible clamping device for door production

The modular principle makes it quite easy to implement additional three-axis modules with longer travels. Fig. 4b shows an example of a fully flexible clamping device for the manufacturing of a car door. A relatively compact design is possible due to the cross-module use of the lower guide rails. To be able to absorb the high process loads, the lower guide slides are slightly wider than the rest of the module contour. To be able to move adjacent modules close together, they are arranged alternately on two parallel pairs of rails. [4] Such a device as shown in Fig. 4b, for example, consists of 25 modular three-axis modules. For the implementation of the control of these 75 axes, different concepts and the use of decentralized control components were examined.

With the help of the specially developed planning software, also the clamping points can be corrected so optimally that flat surfaces are always clamped. Angular deviations up to a certain degree can be compensated with so-called pendulum supports. If the angular deviations on a clamping contour between the different components are too large or the contour is uneven, special needle cushions have been developed which automatically adapt to the component contour during the clamping process and fix the component at a defined point. [5].

In addition to the solutions for the clamping fixtures, various flexibilization concepts and prototypes for handling different components with adaptable gripping systems were developed. For this purpose, two different principles are used fundamentally. On the one hand, end effectors such as clamps, suction cups or magnets are repositioned in a controlled manner on the gripping system with the aid of additional axes, and on the other hand, several suction cups or magnets are attached to the handling system in a non-specific and freely movable manner so that they can adapt to the respective component contour. Fig. 5 shows an example of a flexible gripper for different components.

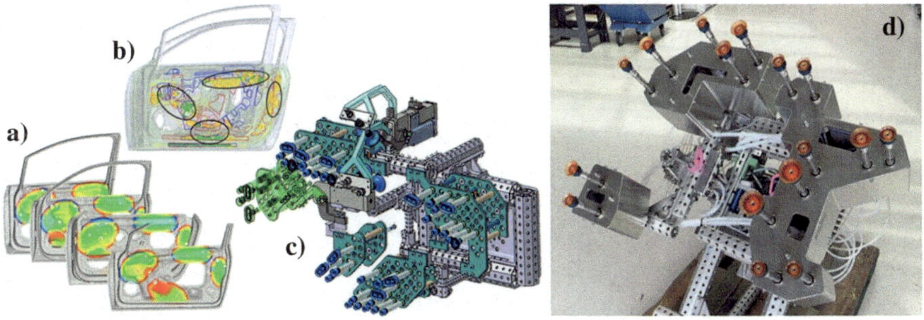

Fig. 5 Component-flexible gripper development: **a**) software-supported component analyses; **b**) software-supported cluster classification; **c**) prototype construction; **d**) suction cluster gripper ready for series production

The Fig. 5d shows a suction cluster gripper with spring-mounted suction elements. The arrangement of the suction cups was optimized using developed component analysis methods. When a component is picked up, the spring plungers initially adapt themselves freely to the component contour. The spring plungers are fixed at a defined position and the component is transported by the handling system in a secure position. Optical systems are preferably used for precise detection of the component position in the gripper. In this way, possible positional deviations can be compensated for by a robot path correction. As further variants, different handling grippers with additional linear axes for repositioning individual clamping points were implemented. These and other concepts, such as a magnetic cluster gripper and various sensor and position control systems, have been combined in a kind of modular system.

4 Flexible Joining

The boundary conditions of automobile production also require a comprehensive, holistic approach in the field of joining technology to make joining technologies more flexible and simpler. In addition, not only the demands for efficient production technologies are increasing for the car manufacturer, but also the social demands and the state requirements for resource conservation and climate protection. For this purpose, it is necessary to reduce the number of joining processes, e.g. by evaluating decisive criteria (benefits, costs, accessibility, etc.) for production. The metallic materials used in car body design are steel and aluminium in various alloy forms. With the boundary conditions of the automotive industry, laser beam and resistance welding have been developed as targeted joining technologies under the umbrella of the FoFab-K.

4.1 Laser Welding

In addition to resistance welding, laser beam welding offers the possibility of joining a wide variety of material combinations in car body construction with high quality. Laser beam welding offers an enormous application and performance potential for the joining of automobile-specific components. This is shown, among other things, by the contactless and tool-independent suitability of the process for welding the materials used, the low and locally limited energy input, the high welding speed or the large number of possibilities for beam shaping and positioning of the laser steel. The typical remote laser steel welding or the use of a so-called laser picker (laser welding tongs without lower clamping jaw) are predestined for such welding tasks, especially because of the one-sided accessibility. But also, the laser beam welding process with local clamping technology (laser welding guns from company Trumpf / Duering as well as IPG) could be developed for steel and aluminium joints in car body construction. This laser beam welding technology is particularly characterised by fewer clamping points on the welding device compared to conventional joining processes. The joining of foreign materials, e.g. steel with aluminium, is also possible with this technology.

4.2 Resistance Spot Welding

For resistance spot welding of steel-steel and aluminium-aluminium joints, a uniform electrode cap shape was successfully developed and tested. Intelligent welding process control has reduced both the contact of the electrode caps and the formation of spatter to the required level without subjecting the caps to excessive thermal stress.

5 Discussion

It could be shown that an economically efficient, flexible production system can be achieved by the holistic consideration of all boundary conditions and comprehensive software support for the planning, design and realization of the individual components.

The clamping modules, which have already been optimised, are currently being tested in series. Within a reengineering process, specific adaptations of individual functional and drive components were implemented with the help of cooperation with various component manufacturers. As a result, the installation space of the kinematics could be reduced by a total of 40% while maintaining the same performance.

Currently, the developed different gripping concepts are being verified close to series production with the help of prototypes. The suction cluster gripper already replaces a large number of different grippers in series production, which means that non-productive times for changing and additional space requirements for interchangeable grippers are no longer relevant.

The next step is the complete planning of a flexible system to prove the efficiency and robustness of these modular components as well as the construction of a modular system, which means a worldwide standard for all plants in the Volkswagen group.

6 Conclusion

The article describes a holistic approach of a fully flexible body shop line. This required the analysis of the existing process chain and a consistent development approach that included handling systems, fixtures, joining technology and finally also logistics.

It was shown that based on optimization algorithms new handling and clamping concepts can be realized, which allow geometrically similar assemblies to be produced in one manufacturing station without changing grippers or fixtures.

By using fewer but more flexible joining technologies, production start-up times are shortened, maintenance costs are reduced and the demands on the fixtures are significantly lower. With the reduction of the number of joining technologies and the further development of the processes, flexible and efficient joining methods are available to the body shop for future model series to meet the constantly changing boundary conditions.

Thanks to the comprehensive planning and integration of all the components developed into a complete system, the economic efficiency of this concept could be demonstrated at an early stage, which means that it can be expected to be implemented within the Volkswagen Group in the near future.

References

1. Kampker, A. et al: Identifying and evaluating flexible approaches for automobile body production areas. E3S Web of Conferences, Volume 95, International Conference on Power, Energy and Mechanical Engineering, ICPEME (2019)
2. Industrie 4.0-Matrix Production, KUKA Systems GmbH. www.kuka.com
3. Mercedes-Benz Cars: Der Rohbau der Zukunft ist flexibel., DAIMLER AG, media.daimler.com
4. Fritzsche R., Richter A., Putz M.: Product flexible car body fixtures with position-dependent load balancing based on finite element method in combination with methods of artificial intelligence. 11th CIRP Conference on Intelligent Computation in Manufacturing Engineering - CIRP ICME, (2017)
5. Fritzsche R., Voigt E., Schaffrath R., Todtermuschke M., Röber M.: Automated design of product-flexible car body fixtures with software-supported part alignment using particle swarm optimization. 13th CIRP Conference on Intelligent Computation in Manufacturing Engineering - CIRP ICME, (2019)

Development of an Integrated Data-Driven Process to Handle Uncertainties in Multi-Variant Production and Logistics: A Survey

Simon Dürr[2,3](✉), Raphael Lamprecht[1], Matthias Kauffmann[3], Jörg Winter[3], Heinz Alexy[3], and Marco Huber[1,2]

[1] Center for Cyber Cognitive Intelligence (CCI), Fraunhofer Institute for Manufacturing Engineering and Automation IPA, Nobelstraße 12, 70569 Stuttgart, Germany

[2] University of Stuttgart, Institute of Industrial Manufacturing and Management IFF, Allmandring 35, 70569 Stuttgart, Germany
simon.duerr@porsche.de

[3] Dr. Ing. h.c. F. Porsche AG, Porscheplatz 1, 70435 Stuttgart, Germany

Abstract. A key differentiator in customer satisfaction in the automotive industry is offering a choice of high-dimensional possibilities to customize an individual vehicle. In the luxury segment this equals more than one billion variants. Innovative data-driven processes are necessary for the planning and handling of vehicles in production and distribution in order to guarantee indispensable factors such as stability, flexibility and transparency across the entire supply chain and to deliver the right vehicle to the right place at the promised time. Highly complex business environments, multi-variant products, trends with effects on distribution networks and future mobility concepts confront manufacturers with new challenges. This paper provides a survey of currently used methods and technologies to handle the previously mentioned challenges in the area of the customer order management of an automotive manufacturer. A specific field of research is the concept of planned orders which are derived from early anticipated customer requirements and their utilization throughout the entire planning and order management process. In addition to the generic approaches for achieving integrated planning of sales and production programs as well as the resulting material requirements, artificial intelligence methods are investigated with regard to the concept of the planned orders. For this purpose, various existing approaches to anticipate planned orders and to effectively manage the supply chain are examined. In conclusion, this paper provides a survey of state of the art methods regarding artificial intelligence linked to current research on agile production systems. Despite existing uncertainty in the upcoming years the implementation of stable data-driven processes is

P. Weißgraeber et al. (Eds.): *Advances in Automotive Production Technology – Theory and Application*, ARENA2036, pp. 486–494, 2021. https://doi.org/10.1007/978-3-662-62962-8_56

crucial. In this context, the possibility to prepare an automotive manufacturer for the upcoming challenges in digitalization and globalization is evaluated.

1 Introduction

The automotive industry is currently undergoing a tremendous change that effects nearly all aspects of the business environment, ranging from new mobility solutions to innovative distribution channels. In order to stay competitive, automotive manufacturer have to operate highly complex production and supply chain networks which are subject to various risks and uncertainties. A recent example of these risks is the global COVID-19 pandemic that has brought supply and demand chains to a standstill early in 2020. This paper provides a survey on artificial intelligence and data-driven methods to handle uncertainties in multi-variant production and logistics. Furthermore, the paper provides an analysis of technological concepts in order to produce and to deliver the right vehicle to the right place at the promised time.

The paper is structured as follows: Within Sect. 2, the fundamentals of the concept of planned orders, supply chain management under uncertainty and artificial intelligence in multi-variant production systems are explained. In Sect. 3 a classification scheme is developed which is used in Sect. 4 to provide an evaluation of the reviewed literature. Sect. 5 concludes with a discussion and future research needs.

2 Scope of Review

2.1 Concept of Planned Orders

The aim of the order fulfillment [1] or Order-To-Delivery (OTD) [2] process in the automotive industry is to schedule, produce and deliver a customized vehicle to the right place at the promised time. In order to guarantee maximum flexibility, customers are allowed to change their configuration until shortly before the vehicles are scheduled for production [3]. This point is referred to as order freeze. A distinction is made between two strategies for order processing: Build-to-order (BTO) and build-to-stock (BTS). While in the BTO-strategy a direct assignment of a built vehicle to a customer before the order freeze exists [4], in the BTS-strategy the customer is unknown at the time the vehicle is released for production. This leads to the risk of lost sales due to vehicles which are built and do not meet customer needs [2]. To handle these challenges the virtual-build-to-order (VBtO) strategy [3] can be pursued. VBtO describes the possibility of finding the optimal allocation to a customer via a flexible decoupling point over the complete pipeline of already finished vehicles and those in production or planning stage. Based on individual decoupling points, desired delivery dates of customers can be better integrated into the calculation of the planned date in production than with ordinary OTD concepts. The aversion of customers to wait for their vehicle plays an essential role [2] as well as to increase the margin through customer proximity [3] and to reduce the stock of vehicles. A potential fundamental for the VBtO-system is the concept of planned orders to anticipate customer wishes by generating planned objects and assign real sales orders to planned orders in a multi-variant production environment [5]. This allows for integrating sales planning and production planning into the order fulfillment process and enables the derivation of material requirements.

2.2 Supply Chain Management Under Uncertainty

Dealing with uncertainty is a crucial part in effective supply chain (SC) management. The automotive industry is characterized by its complex, global supply chain networks following just-in-time and just-in-sequence logic. The product itself is highly customized resulting in multi-variant product configurations. In order to meet the resulting challenges, reducing the SC uncertainty plays an important role for operating successfully in the complex business environment. Drivers of uncertainty are manifold and have been intensively studied in the literature [6,7]. To classify the different sources of uncertainty the common approach of grouping them into supplier, manufacturing and demand uncertainty, initially proposed by [8] is followed. Manufacturing uncertainty arises from suboptimally configured manufacturing processes and the inherent stochasticity in the availability of resources and workers due to machine breakdown or illness. The stochasticity in material supplies is expressed by supplier uncertainty. The automotive customer order management is mainly driven by demand uncertainty. Uncertainty arises in demand forecasting and the resulting derivation of production programs. The distribution channel via intermediate dealer and regional sales organizations adds additional uncertainty. Reducing uncertainty in demand has enormous potential in cutting costs by reducing the inventory along the SC and allowing for more efficient production capacity utilization. In BTO fulfillment scenarios, forecasting can also help to better anticipate customer demand which therefore allows realizing higher contribution margins and customer satisfaction.

2.3 Artificial Intelligence in Multi-Variant Production

Leveraging artificial intelligence (AI) and data-driven methods to improve business processes can be identified in nearly all economic sectors. The focus of this paper is to review the potential of AI-based applications for the multi-variant automotive customer order process. The scope of technologies and methods that are summarized under the generic term of AI is very wide. Therefore, the survey is focused on AI-based and data-based methods that enable some form of autonomous reasoning, learning or planning and acting. Methods that aim for solving classical optimization problems in the area of mathematical programming or operations research are not subject to the review.

3 Taxonomy

3.1 Review Methodology

The literature considered in the survey was selected primarily by scientific database search focusing on automotive industry but also considering work in related areas. The databases Scopus and Web of Science were used based on their large scope and relevance in the field of production, logistics and applied AI. The literature was first collected by systematic keyword search. Relevant literature was selected based on the information in the abstract and filtered through a full text analysis. Furthermore, only peer-reviewed literature was considered as well as literature published later then 2000 due to the major breakthroughs in the recent years in the field of AI. After the selection process, a list of 14 references is considered in the survey.

3.2 Classification Scheme

In the following a scheme for the classification of approaches is developed. For the evaluation and differentiation of the considered work, the classification is divided into two components. First, the subject area of planning and ordering focusing on automotive industry is considered. Table 1 shows the classification scheme for planning and ordering. Based on the scope of review, the approaches are divided into three categories and refined into attributes to explore the main objectives. Firstly, order fulfillment in which the general approaches to the subject of matching of planned orders, VBtO, BTO, and BTS are separated and analyzed. Secondly, anticipating customer demand in which different approaches considering the generation of planned orders and sales planning are explored. Thirdly, challenges regarding the topics demand uncertainty, mass customization, multi-variant production as well as complex SC network are evaluated.

Following the evaluation of the approaches regarding planning and ordering processes, the AI-methods used in the reviewed papers are explored. Table 2 shows the classification scheme for the examined technologies. To classify the technologies considered in this review, the different AI-based challenges are grouped by reasoning, learning as well as planning and acting. Reasoning is differentiated into probabilistic graphical methods and rule-based fuzzy systems. Methods focusing on learning are classified by statistical learning methods and deep learning based methods. The term statistical learning methods is used to characterize machine learning methods excluding those, based on deep learning. Finally, in autonomous planning and acting, reinforcement learning concepts are considered.

Table 1. Classification Scheme: Planning and Ordering

Attribute	Description
Order fulfillment	
VBtO	Strategy to find the best product for the customer
BTO / BTS	Different production strategies
Matching planned orders	Matching confirmed customer order to planned order
Anticipating customer demand	
Generating planned orders	Anticipating customer demand using planned objects
Sales planning	Demand forecast and deriving parts by feature planning
Challenges	
Demand uncertainty	Uncertainty in customer demand
Mass customization	Mass production linked to individual specified products
Multi-variant production	Manufacturing products which are described by multiple characteristics
Complex SC network	Processes and organizational network to serve products to customers

Table 2. Classification Scheme: AI-Technologies

Attribute	Description
Reasoning	
Probabilistic graphs	Models representing complex probability distributions
Fuzzy logic	Models based on fuzzy logical statements
Learning	
Statistical leaning	Machine learning methods excluding deep learning
Deep learning	Methods based on deep artificial neural networks
Planning and Acting	
Reinforcement learning	Methods based on trial and error learning

4 Review of Approaches

In Table 3 the reviewed literature is classified following the previous introduced scheme. For evaluation purposes, the review can be loosely divided into two groups. The first four papers provide an overview of approaches with focus on the process-related description and challenges within the topics order fulfillment and anticipating customer needs. To further differentiate the literature, an evaluation metric indicates whether the reviewed work is focusing on the described attributes(●), touching the topics (◑) or not even considering them at all. The strategy BTO and BTS are introduced in [5,9,10], the benefits of VBtO-systems are listed in [9] and the use of the concept of planned orders from generation to the matching of planned orders to real orders is described in [11] and [5]. In addition, solutions to handle challenges the manufacturer faces today and in the future are proposed. The approaches from [10,11] are based on traditional mathematical optimizations as well as simulation and do not consider AI-driven methods, but are evaluated in this survey from a procedural view. The models for solving the problems in [5,9] are based on probabilistic graphs where [9] is using a Markov network to handle the challenges in VBtO-systems whereas [5] uses trees with conditional probabilities in order to generate planned orders. The remaining ten papers belong to the second group comprising approaches using different AI-methods to handle various challenges in the area of logistics and production. The works, which consider the described technology are marked (⊠). In [12–15]

Table 3. Literature Review

	[9]	[5]	[11]	[10]	[12]	[18]	[16]	[17]	[19]	[13]	[20]	[14]	[15]	[21]
Order fulfillment														
VBtO	●	◑												
BTO / BTS	◑	◑		◑			◑	◑						
Matching planned orders		●												
Anticipating customer demand														
Generation planned orders		●	●											
Sales planning							●	●						
Challenges														
Demand uncertainty	◑	◑	●	◑	●	◑	●	●		●	◑	●	●	◑
Mass customization	●						◑	◑					●	
Multi-variant production	◑	●		◑			●	●						
Complex SC network					●	●			◑		◑			◑
Reasoning														
Probabilistic graphs	⊠	⊠				⊠	⊠	⊠						
Fuzzy logic								⊠						⊠
Learning														
Statistical learning					⊠					⊠		⊠	⊠	
Deep learning					⊠								⊠	
Planning and Acting														
Reinforcement learning									⊠		⊠			

machine learning methods such as support vector machines and artificial neural networks are used for demand forecasting in SC networks by taking different factors of uncertainty into account. Whereas these authors focus only on low-variant products or single product demand, they do not address the challenges of multi-variant products and mass customization. Demand uncertainty for multi-variant products is subject in [16,17]. The authors are using Markov networks for learning and representing a probability distribution that stores historical sales data and technological restrictions to forecast future product configurations. Instead of modelling uncertainty from a demand and product configuration standpoint, the authors in [18] are using Bayesian networks to model uncertainty resulting from complex SC networks. Another conceptually different approach is taken by [19,20]. These papers describe simplified SC systems as a Markov decision process and use reinforcement learning to learn a policy for effectively operating the supply chain. In contrast to the dynamic system representations, [21] considers a decision support system for managing static SC systems based on fuzzy logic and regression trees.

5 Discussion and Conclusion

The importance of data and information has grown enormously in recent years. AI methods are omnipresent and essential for solving current dynamic challenges in multi-variant production systems. A review of the literature has shown that data-driven methods within complex SC networks are well described in research. Approaches in different areas of AI, such as reasoning, learning and planning and acting were investigated. In particular with the focus on the automotive industry and its linkage to process-related topics. However, none of the considered literature unites both fields in a satisfying manner. A hybrid model that describes the challenges from a procedural perspective such as the concept of planned orders as well as offering an AI-driven solution was not found. Image recognition, speech recognition and automation systems are already used in current agile production systems in the automotive industry which underlines the high potential of AI-driven and automated solutions. Besides the potentials, future challenges will arise to improve existing concepts from the field of AI in the automotive production, logistics and order management environment like providing proper data quality as well as choosing and adapting the correct analysis method, just to name two of them. Organizational challenges also often hinder the practical use of AI, despite the enormous chances that are offered. Especially within multi-variant and high-dimensional problems such as the concept of planned orders, AI methods can be used to anticipate customer wishes better, to generate planned objects, and to match them to real orders. Various approaches are conceivable to establish the link between recent AI technology and the requirements for an agile production system to develop data-driven processes that prepare the automotive manufacturers for the future in a targeted manner.

References

1. Staeblein, Thomas, Aoki, Katsuki: Planning and scheduling in the automotive industry: A comparison of industrial practice at german and japanese makers. Int. J. Prod. Econ. **162**, 258–272 (2015)
2. Holweg, Matthias, Miemczyk, Joe: Delivering the '3-day car'–the strategic implications for automotive logistics operations. J. Purchasing and Supply Manage. **9**(2), 63–71 (2003)
3. Brabazon, Philip G., MacCarthy, Bart: Virtual-build-to-order as a mass customization order fulfilment model. Concurrent Engineering **12**(2), 155–165 (2004)
4. Volling, Thomas., Matzke, Andreas., Grunewald, Martin, Spengler, Thomas S.: Planning of capacities and orders in build-to-order automobile production: A review. Eur. J. Oper. Res. **224**(2), 240–260 (2013)
5. Buergin, Jens., Beisecker, Julian., Fischer, Sebastian., Geier, Bettina., Tutsch, Hansjoerg., Mercamp, Stefan, Lanza, Gisela: Generation of planned orders and their matching with customer orders in multi-variant series production. Procedia CIRP **61**, 499–504 (2017)
6. Simangunsong, E., Hendry, L.C., Stevenson, M.: Supply-chain uncertainty: A review and theoretical foundation for future research. Int. J. Prod. Res. **50**(16), 4493–4523 (2012)
7. Peidro, David., Mula, Josefa., Poler, Raúl, Lario, Francisco-Cruz: Quantitative models for supply chain planning under uncertainty: a review. Int. J. Adv. Manuf. Technol. **43**(3–4), 400–420 (2009)
8. Davis, T.: Effective supply chain. Sloan Management Review/Summer, pp. 35–46 (1993)
9. Brabazon, P.G., MacCarthy, B.L.: On markovian approximations for virtual-build-to-order systems. J. Oper. Res. Soc. **61**(10), 1471–1484 (2010)
10. Volling, Thomas, Spengler, Thomas S.: Modeling and simulation of order-driven planning policies in build-to-order automobile production. Int. J. Prod. Econ. **131**(1), 183–193 (2011)
11. Feng, Keli., Rao, Uday S., Raturi, Amitabh: Setting planned orders in master production scheduling under demand uncertainty. Int. J. Prod. Res. **49**(12/13), 4007–4025 (2011)
12. Carbonneau, Real., Laframboise, Kevin, Vahidov, Rustam: Application of machine learning techniques for supply chain demand forecasting. Eur. J. Oper. Res. **184**(3), 1140–1154 (2008)
13. Wang, J., Zhu, W., Sun, D., Lu, H.: Application of svm combined with mackov chain for inventory prediction in supply chain. 2008 4th International Conference on Wireless Communications, Networking and Mobile Computing, pp. 1–4. IEEE (2008)
14. Sarhani, M., Afia, A.E.: Intelligent system based support vector regression for supply chain demand forecasting. 2014 Second World Conference on Complex Systems (WCCS), pp. 79–83. IEEE (2014)
15. Kim, M., Jeong, J., Bae, S.: Demand forecasting based on machine learning for mass customization in smart manufacturing. In Unknown, editor. Proceedings of the 2019 International Conference on Data Mining and Machine Learning – ICDMML 2019, pp. 6–11, New York, USA, ACM Press (2019)
16. Gebhardt, J., Klose, A., Detmer, H., Rügheimer, F., Kruse, R.: Graphical models for industrial planning on complex domains. In: Riccia, G.D., Dubois, D., Kruse, R., Lenz, H.-J. (eds) Decision Theory and Multi-Agent Planning, pp. 131–143. Springer, Vienna (2006)

17. Schmidt, Fabian., Gebhardt, Jörg, Kruse, Rudolf: Decomposable graphical models on learning, fusion and revision. In: Zadeh, Lotfi A., Yager, Ronald R., Shahbazova, Shahnaz N., Reformat, Marek Z., Kreinovich, Vladik (eds.) Recent Developments and the New Direction in Soft-Computing Foundations and Applications. Studies in Fuzziness and Soft Computing, vol. 361, pp. 439–452. Springer International Publishing, Cham (2018)
18. Garvey, Myles D., Carnovale, Steven, Yeniyurt, Sengun: An analytical framework for supply network risk propagation: A bayesian network approach. Eur. J. Oper. Res. **243**(2), 618–627 (2015)
19. Valluri, Annapurna., North, Michael J., Macal, Charles M.: Reinforcement learning in supply chains. Int. J. Neural Syst. **19**(5), 331–344 (2009)
20. Shin, Joohyun, Lee, Jay H.: Multi-timescale, multi-period decision-making model development by combining reinforcement learning and mathematical programming. Comput. Chem. Eng. **121**, 556–573 (2019)
21. Rodríguez, G.G., Gonzalez-Cava, J.M., Pérez, J.A.M.: An intelligent decision support system for production planning based on machine learning. J. Intell. Manuf. **31**(5):1257–1273 (2020)

Author Index

P. Weißgraeber et al. (Eds.): *Advances in Automotive Production Technology – Theory and Application,* ARENA2036, pp. 495–497, 2021. https://doi.org/10.1007/978-3-662-62962-8

Printed by Printforce, the Netherlands